Nanomedicine and the Cardiovascular System

Nanomedicine and the Cardiovascular System

Editors
Ross J. Hunter
Cardiology Research Fellow
St Bartholomew's Hospital
London
UK

Victor R. Preedy
Professor of Nutritional Biochemistry
School of Biomedical & Health Sciences
King's College London
and
Professor of Clinical Biochemistry
King's College Hospital
UK

CRC Press
Taylor & Francis Group
an **informa** business
www.crcpress.com

6000 Broken Sound Parkway, NW
Suite 300, Boca Raton, FL 33487
270 Madison Avenue
New York, NY 10016
2 Park Square, Milton Park
Abingdon, Oxon OX14 4RN, UK

Science Publishers
Jersey, British Isles
Enfield, New Hampshire

Published by Science Publishers, an imprint of Edenbridge Ltd.

- St. Helier, Jersey, British Channel Islands
- P.O. Box 699, Enfield, NH 03748, USA

E-mail: info@scipub.net Website: www.scipub.net

Marketed and distributed by:

CRC Press
Taylor & Francis Group
an informa business
www.crcpress.com

6000 Broken Sound Parkway, NW
Suite 300, Boca Raton, FL 33487
270 Madison Avenue
New York, NY 10016
2 Park Square, Milton Park
Abingdon, Oxon OX14 4RN, UK

ISBN 978-1-57808-726-6

Cover Illustrations: Reproduced by kind courtesy of the undermentioned authors:
Figure No. 1 from Chapter 7 by Guo-Jun Zhang
Figure No. 5 from Chapter 9 by Andrea Masotti
Figure No. 2B from Chapter 13 by Sarra de Valence et al.
Figure No. 8 from Chapter 14 by Toshiro Fukami et al.

Library of Congress Cataloging-in-Publication Data

Nanomedicine and the cardiovascular system / editors, Ross J. Hunter, Victor
R. Preedy.
p. ; cm. -- (Nanoscience applied to health and medicine)
Includes bibliographical references and index.
ISBN 978-1-57808-726-6 (hardcover)
I. Hunter, Ross, 1977- II. Preedy, Victor R. III. Series: Nanoscience
applied to health and medicine.
[DNLM: 1. Cardiovascular System. 2. Nanomedicine--methods. 3. Nanostructures--therapeutic use. WG 100]
LC classification not assigned
612.1'045--dc23

2011032738

Printed in the United States of America

Preface

The nanosciences are a rapidly expanding field of research with a wide applicability to all areas of health. They encompass a variety of technologies ranging from particles to networks and nanostructures. For example, nanoparticles have been proposed to be suitable carriers of therapeutic agents whilst nanostructures provide suitable platforms for sub-micro bioengineering. However, understanding the importance of nanoscience and technology is somewhat problematical as a great deal of text can be rather technical in nature with little consideration to the novice. In this series *Nanosciences Applied to Health and Biomedical Sciences* we aim to disseminate the information in a readable way by having unique sections for the novice and expert alike. This enables the reader to transfer their knowledge base from one discipline to another or from one academic level to another. Each chapter has an abstract, "key facts", applications to other areas of health and disease and a "mini-dictionary" of key terms and phrases within each chapter. Finally, each chapter has a series of summary points. In this book **Nanomedicine and the Cardiovascular System** we cover nanoparticle contrast agents, cell sheet nanotechnology, nanowire field-effect transistors, atomic force microscopy, transfusion medicine, nanoscale topography, nano wire sensor arrays, nanowire transistors, nanospheres, nanomonitor technology, nanospin probes, nanomatrices, vascular grafts, gene transfer using nanomicelles, nanobubbles, biodegradable nanofibers, cholesterol-lowering drugs, thromboses, nanorods, nitric oxide, iron oxide nano particles and stem cells.

Contributors to **Nanomedicine and the Cardiovascular System** are all either international or national experts, leading authorities or are carrying out ground breaking and innovative work on their subject. The book is essential reading for cardiologists, cardiovascular scientists, research scientists, medical doctors, health care professionals, pathologists, biologists, biochemists, chemists and physicists, general practitioners as well as those interested in heart disease and nano sciences in general. Nanomedicine and the Cardiovascular System is part of the *Nanoscience Applied to Health and Medicine* series.

The Editors

Contents

Section 1: General Methods and Applications

1

Nanoparticle Contrast Agents for Cardiovascular Medical Imaging

David P. Cormode,[1,a,*] *Ahmed Klink,*[1,b] *Zahi A. Fayad*[1,c] and *Willem J.M. Mulder*[1,d]

ABSTRACT

Over the last ten years, nanoparticles have become increasingly studied as contrast agents for medical imaging. This is due to their unique contrast-generating properties as well as their potentially long circulation half-lives, their high payload and the ease of integrating multiple properties. In particular, they are highly effective for molecular imaging, the non-invasive visualization of the levels of molecules or cell types. Cardiovascular diseases are particularly interesting to study with molecular imaging because of the multiple processes and stages involved and therefore the many cell types and molecules that are important in these diseases. Nanoparticle contrast agents that have been used in conjunction with cardiovascular magnetic resonance

[1]Translational and Molecular Imaging Institute, Mount Sinai School of Medicine, One Gustave L. Levy Place, Box 1234, New York, NY 10029.
[a]E-mail: davidcormode@gmail.com
[b]E-mail: ahmedklink@mac.com
[c]E-mail: Zahi.Fayad@mssm.edu
[d]E-mail: willem.mulder@mountsinai.org
*Corresponding author

List of abbreviations after the text.

imaging, computed tomography, nuclear imaging techniques or ultrasound include iron oxides, micelles, liposomes, emulsions, gold nanoparticles, quantum dots and lipoproteins. In addition to the above-mentioned technological aspects we highlight several examples of the use of nanoparticles in cardiovascular imaging as case studies at the end of this chapter.

INTRODUCTION

There has been a tremendous focus of research effort on nanotechnology in general and nanoparticles specifically in the past twenty years. Part of the reason for these efforts is that materials confined to nano-sizes (in the range of 1–100 nm in one or more dimensions) can exhibit unusual properties, be it optical, electronic, magnetic, fluorescent or catalytic. A pertinent example of this is quantum dots, semiconductor-based colloids that, when in the size range 1–10 nm, possess extraordinary fluorescent properties. Nanoparticles can be formed from a wide variety of materials, including inorganic materials such as gold, silver, platinum, iron oxide, bismuth and cadmium selenide. Such inorganic materials invariably possess some kind of organic coating, but nanoparticles can also be entirely composed of organic materials such as polymers, amphiphiles, sugars or proteins. Some of these coatings, such as polyethylene glycol, have been found to make nanoparticles highly biocompatible and non-toxic both *in vitro* and *in vivo*. Nanoparticles are now being explored in the biomedical field for drug delivery, gene therapy and medical imaging, among other applications.

Medical imaging is a highly important field, allowing for advanced diagnosis, monitoring of therapy and surgical planning. X-ray imaging is the oldest medical imaging technique, but it is still very widely used. Newer techniques include magnetic resonance imaging (MRI), positron emission tomography (PET) and X-ray computed tomography (CT). These techniques provide three-dimensional datasets of images of patients, as opposed to the two-dimensional images produced by X-ray imaging. Additionally, ultrasound is a widely used technique and fluorescence imaging systems are starting to become available. The details of these different systems are described in a subsequent section. A crucial aspect of all medical imaging systems is image contrast, i.e., visual differences between structures and tissues. There are many ways of generating image contrast, but one way is the use of contrast agents, substances used to enhance the contrast of structures or fluids within the body in medical imaging. This allows, for example, visualization of the vasculature or detection of liver tumors. Contrast agents are either small molecules or nanoparticles that incorporate some substance that produces contrast for the relevant imaging technique; for example, gadolinium is included in both small molecule contrast agents and nanoparticle formulations to produce contrast in MRI.

In the field of medical imaging, there are several reasons that nanoparticles are attractive for use as contrast agents, as opposed to small molecules. First, nanoparticles may provide unique contrast properties, such as the aforementioned fluorescence of quantum dots or the intense MRI contrast produced by iron oxide nanoparticles. Second, small molecules have a half-life of only a few minutes and the contrast they produce can therefore swiftly dissipate. Nanoparticles, on the other hand, can be designed to have extended half-lives, which can be valuable in cases of image-guided surgery, such as stent emplacement. The extended half-life of nanoparticles would avoid the need to repeatedly inject a small molecule contrast agent. Third, nanoparticles can deliver very high payloads, with as many as *millions* of contrast-generating atoms per nanoparticle as opposed to normally only one such atom for a small molecule. Last, it is relatively straightforward to include multiple components in nanoparticles, such as additional forms of contrast, targeting molecules or therapeutic materials. Inclusion of targeting molecules allows imaging of specific cell types, molecules (e.g., cell surface receptors) or biological processes. This field of targeted imaging is known as molecular imaging. Molecular imaging can allow advanced diagnoses, analysis of therapeutic efficacy and improved information on diseases.

Cardiovascular diseases are the primary cause of mortality in the Western world and consequently their study via imaging is of crucial importance for better understanding of fundamental disease processes, improved diagnoses and the evaluation of (new) therapies. The majority of the work that has been done in nanoparticle-based imaging of cardiovascular disease has focused on molecular imaging. Cardiovascular diseases are also often rather interesting to study, with many different cells and molecules involved and several stages of progression that have distinct levels of expression of cell types and molecular markers. For example, atherosclerosis is a progressive inflammatory disease in which there is a gradual buildup of fatty molecules in the arteries, recruitment of macrophages and other cell types to form tissue deposits known as atherosclerotic plaques (Fig. 1). Eventually, lipid-rich, necrotic or calcified cores develop in these plaques. These plaques sometimes rupture, whereupon a thrombus forms that may occlude the artery and block the blood flow. Ruptures of plaques in the coronary arteries cause 70% of heart attacks. The cell types, molecules and processes of interest include macrophages, endothelial cells, foam cells, smooth muscle cells, vascular cell adhesion molecule 1 (VCAM-1), intercellular adhesion molecule 1,

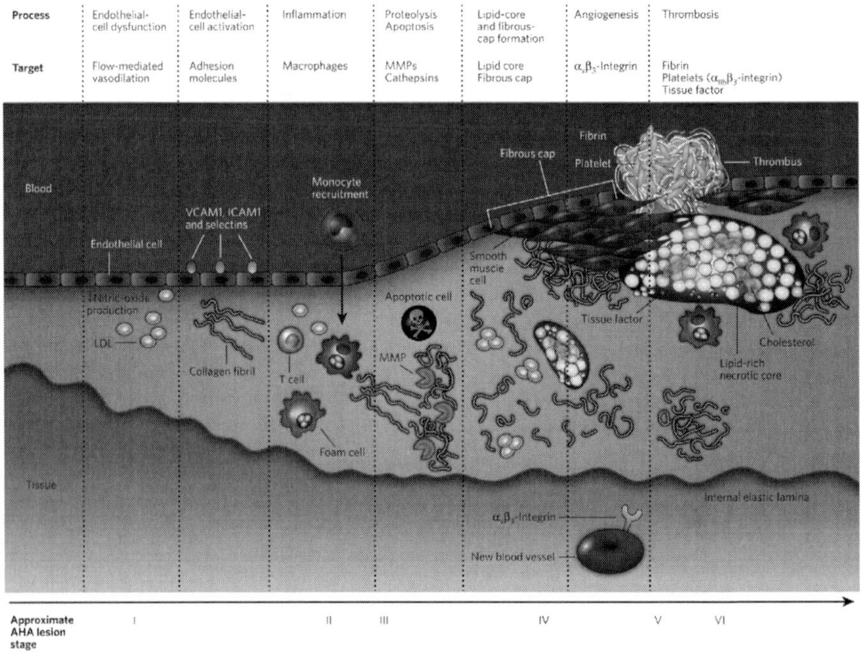

Fig. 1. A simplified depiction of the progression over time of an atherosclerotic lesion, indicating the different processes that occur and receptors expressed at different stages of lesion development. Reproduced with permission from Sanz and Fayad (2008).

Color image of this figure appears in the color plate section at the end of the book.

selectins, collagen, the $\alpha_V\beta_3$-integrin, matrix metalloproteinases, cathepsins, fibrin, angiogenesis, inflammation, apoptosis and thrombosis (Sanz et al. 2008). Imaging thromboses is discussed in depth in Chapter 17 on multifunctional nanoagents for the detection and treatment of thromboses. From a molecular imaging perspective, it is an interesting challenge to develop contrast agents for these widely differing species, which occur in different locations in the arteries. An additional challenge is the different locations of the vascular beds, i.e., the coronaries in the heart, the aorta and other arteries in the abdomen and the peripheral arteries. An imaging modality-nanoparticle combination that works well in the carotid arteries may not work so well for the coronary arteries, because of the motion of the heart and chest and differences in the surrounding tissue. For these reasons, cardiovascular diseases (e.g., atherosclerosis, myocardial infarction and abdominal aortic aneurysms) have drawn a great deal of interest in the molecular imaging field with many nanoparticles designed to investigate them.

In this chapter we discuss the different types of nanoparticles used as contrast agents in cardiovascular imaging. Subsequently, we discuss how the different imaging techniques work. Last, we give a series of examples of nanoparticles used in cardiovascular disease.

NANOPARTICLES

In this section we describe the different types of nanoparticles that have been used in cardiovascular imaging and highlight some of their most important features.

Nanoparticles can be synthesized from a wide variety of materials and have widely varying structures. Before covering nanoparticle types in depth, first we will discuss a generalized design of nanoparticles for molecular imaging (Fig. 2). These nanoparticles usually have a core and one or more layers of coating materials. The coatings are used to make the nanoparticle hydrophilic and biocompatible. Contrast-generating materials can be located within the core, in the coating or at the coating surface. In the case of gadolinium, for example, contrast is generated with these ions mainly via contact with water, so gadolinium is usually located at the nanoparticle surface, whereas fluorophores do not require contact with water and therefore can be located anywhere within the nanoparticle structure. The availability of different locations in which to load contrast-generating materials or other substances is often exploited to include two or even three types of contrast-generating material in the nanoparticle. In some cases, contrast-generating materials and therapeutics are combined

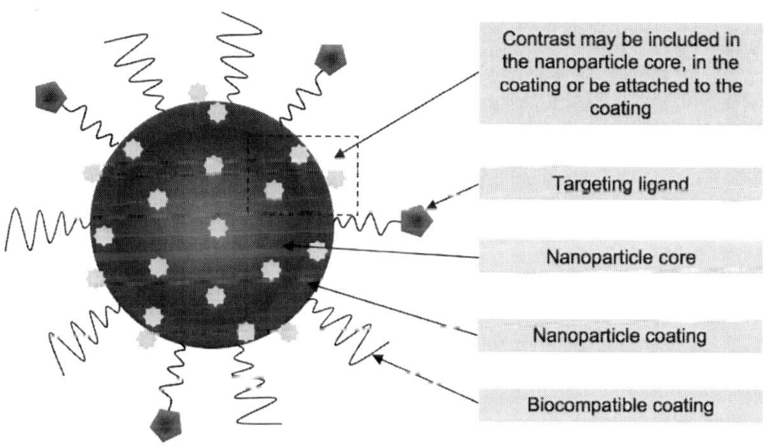

Fig. 2. Schematic depiction of a nanoparticle for use in cardiovascular molecular imaging, indicating the important components. Reproduced with permission from Cormode et al. (2009).

in one nanoparticle to create a "theranostic" formulation. Targeting ligands are normally located at the nanoparticle periphery, to allow the ligand to easily access its target molecule. Placement of the targeting ligand closer to the core could prevent accumulation of the nanoparticle at the desired target site.

The synthesis of nanoparticles can differ greatly, but generally a core or scaffold structure is first created and then subsequently modified. In many cases, the core structure has one or more layers of coating applied, or the coating is modified to make the nanoparticle water soluble and biocompatible. The coating is often designed to include chemical groups, such as amines, carboxylates or maleimides, which can be used to attach metal chelates, fluorophores or targeting ligands to complete the nanoparticle structure.

Iron Oxides

Dextran-coated iron oxide nanoparticles with the trade names Feridex or Sinerem are clinically approved for use as MRI contrast agents in humans. Iron oxide nanocrystals, when placed in a magnetic field, become magnetized and (usually) produce negative contrast in MR imaging because of disruption in the local field experienced by water as the nanoparticles tumble. Sinerem has been used to image macrophages in both animals and humans (see below) via passive targeting. Iron oxides can also be targeted with peptides or antibodies to specific receptors such as VCAM-1, as demonstrated by the group of Weissleder (McCarthy et al. 2007). Iron oxides are almost exclusively used as contrast agents for MRI. Iron oxides have also frequently been used to label cells for cell tracking via MRI, which is discussed in detail in Chapter 20 on iron oxide nanoparticles and cardiac stem cells. While the iron cores used are usually around 5–10 nm in diameter, the overall nanoparticle size can range from 7 to 3500 nm, as some iron oxide agents are embedded in large polymer matrices.

Micelles

Micelles are composed of aggregates of amphiphiles (molecules that have a hydrophobic tail and a hydrophilic headgroup) such as phospholipids (Fig. 3). In a micelle the amphiphiles are arranged such that their hydrophobic tails group together to form the nanoparticle core, while the hydrophilic headgroups face outward towards the aqueous environment. Micelles can be formed from mixtures of different amphiphiles. For example, amphiphiles can be synthesized to possess a metal chelate (containing gadolinium or a radioactive isotope such as copper-64) or a

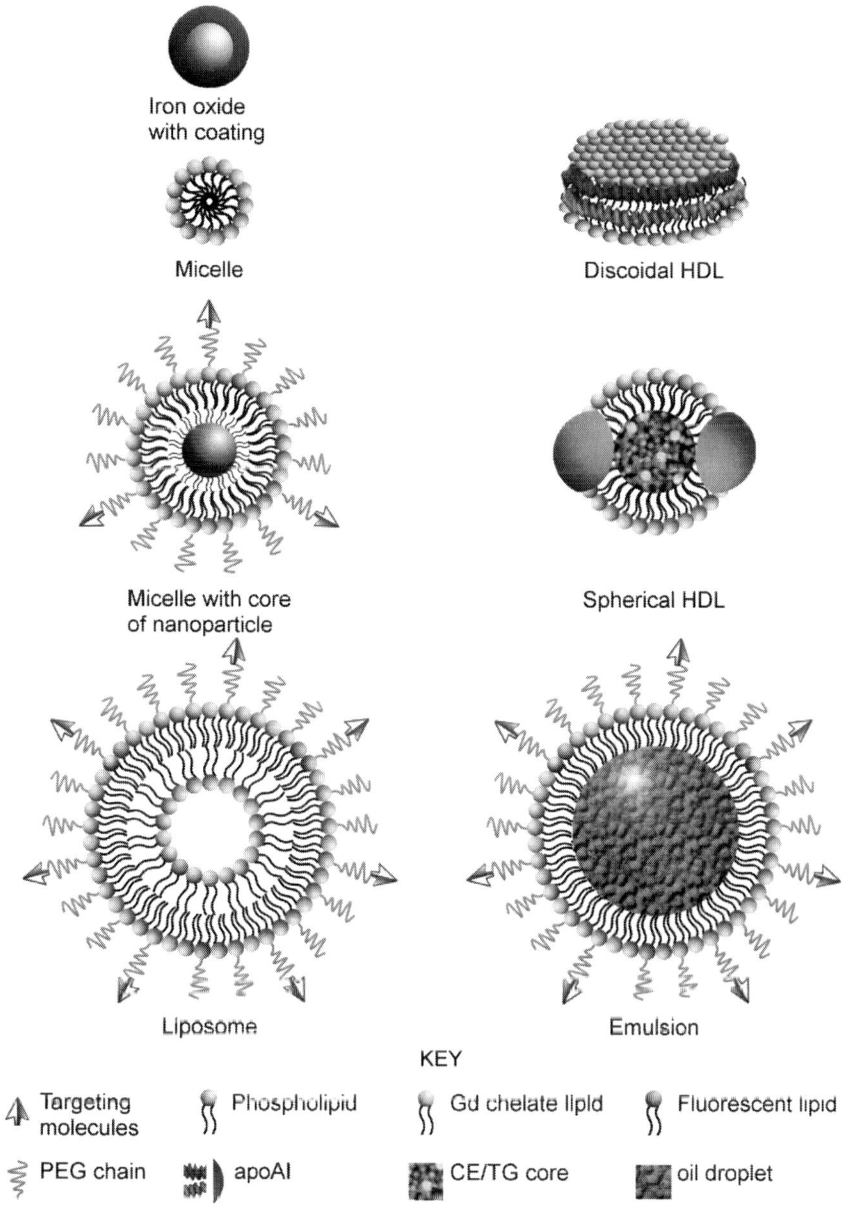

Fig. 3. Schematic depiction of many of the different nanoparticles discussed in this chapter for use in cardiovascular imaging. Reproduced with permission from Chen et al. (2011).

Color image of this figure appears in the color plate section at the end of the book.

fluorophore in their headgroup. These functionalized amphiphiles can be mixed with "carrier" amphiphiles to form the finished micelle. Carrier amphiphiles have no contrast-generating property, but contribute to the structural integrity of the micelle or possess biocompatible features such as polyethylene glycol chains. Additional amphiphiles may be included in the formulation that have functional groups in their headgroup to allow attachment or targeting ligands. Additionally, hydrophobic materials may be included in the micelle core. This may be used to incorporate another type of contrast-generating property in the micelle, because iron oxides, gold nanoparticles or quantum dots can be included in micelle-based contrast agents (Fig. 3). Micelles are generally around 20 nm in diameter, but may be larger if polymeric materials are used. Micelles have been reported as contrast agents for MRI, fluorescence and radioactivity-based imaging techniques.

Liposomes

Similarly to micelles, liposomes are aggregates of amphiphiles, but their structure is quite different, being composed of a lipid bilayer that encapsulates an aqueous core. Liposomes are sized in the 50–500 nm range. This structural change is due to the type of amphiphiles used. To form micelles, amphiphiles with a conical shape are used, with the headgroup larger than the tail. This favors the amphiphiles assembling into a small particle such as a micelle. Amphiphiles that are more cylindrical in shape, such that their headgroup and tail are similar in size, assemble into liposomes, which, being larger, have a lower surface curvature. Contrast-generating materials can be incorporated into liposomes using modified headgroup amphiphiles in the same way as for micelles (Mulder et al. 2009). While hydrophobic materials may be included in the lipid bilayer, this is not often done; it is more common for liposomes to carry hydrophilic contrast-generating materials, such as gadolinium chelates, in their core. Liposomes have been used as contrast agents for virtually every imaging modality.

Oil-in-water Emulsions

Oil-in-water emulsions are composed of oil suspended in water via amphiphiles. The amphiphiles solubilize the nano-sized oil droplets by surrounding the oil with their hydrophobic tails. Contrast-generating materials, therapeutics and targeting ligands can be incorporated in the same way as for micelles, with the additional feature that the oil used to form the emulsion can itself be active as a contrast-generating material. For example, the group of Lanza and Wickline have developed nanoparticle platforms based on perfluorocarbon oils that form emulsions (Lanza et al.

2006). These emulsions can be detected by ^{19}F MR imaging and have been used to investigate many processes in cardiovascular disease, notably angiogenesis and thrombus formation. Moreover, they have been also modified to be detected by ultrasound, CT, ^{1}H MRI methods, nuclear imaging and optical techniques. Interestingly, drugs have been included in these nanoparticles to make them both therapeutic and diagnostic, or "theranostic". By tuning the ratio of the oil to the amphiphiles and by altering the amphiphile type, a wide range of nanoparticle diameters can be achieved, i.e., 30 nm to 1 μm.

Gold Nanoparticles

Arguably, gold nanoparticles initiated the tremendous growth of the nanotechnology field seen over the past twenty years, owing to their unique optical, electronic and catalytic properties. They have been extensively studied, so their synthesis is very well developed. For example, their size can be tuned precisely in the 1–100 nm range. As contrast agents, they can be exploited in optical imaging–based techniques, but most reports on gold nanoparticle contrast agents used in cardiovascular applications have focused on CT. Gold nanoparticles are a relatively straightforward method to deliver the large quantities of radio-opaque material needed for CT imaging. There have been several reports on their application as long-circulating vascular phase agents, but more recent studies have been published in which gold nanoparticles have been used to perform targeted imaging of specific cell types. To achieve biocompatibility, their surface can be grafted with many different polymers or proteins. In addition, gadolinium chelates and fluorophores can be attached to the surface to yield additional contrast for MRI and fluorescence techniques.

Quantum Dots

Quantum dots are small (less than 10 nm) semiconductor crystals that have exceptional fluorescence properties (Medintz et al. 2005). They have broadband absorbance and narrow, tunable emission spectra. They do not photobleach like small molecule fluorophores. A significant disadvantage is that they are formulated with high toxicity metals such as cadmium, limiting their potential for use in patients. In addition to yielding fluorescence contrast, several groups have labeled quantum dots with gadolinium chelates for MRI contrast, thus forming bimodal contrast agents. They can be made biocompatible and targeted by the same strategies as the aforementioned agents. Although, as mentioned earlier, quantum dots themselves are small, they can be included in emulsions or other types of nanoparticles, resulting in nanoparticles with an overall diameter that is quite large.

Lipoproteins

There are several lipoproteins: chylomicrons, very low density lipoprotein (VLDL), intermediate density lipoprotein (IDL), low density lipoprotein (LDL) and high density lipoprotein (HDL). They range in size from 10 nm for HDL to over 250 nm for chylomicrons. Lipoproteins are nanoparticles responsible for cholesterol transport in the body, among other functions. They have cores of cholesterol esters and triglycerides, covered in a monolayer of phospholipids. HDL is depicted in Fig. 3. HDL is initially formed without a fatty core, so discoidal HDL (no core) is depicted as well as spherical HDL. Amphiphatic proteins are embedded in the phospholipid coating, which provides structural integrity and targeting.

LDL binds to the LDL receptor, which is overexpressed in cancer, and HDL binds to such receptors as the SR-B1 and ABCA1 receptors, which are overexpressed in macrophages, an important cell type in atherosclerosis. Adapting them as contrast agents allows these important natural nanoparticles to be tracked *in vivo* and provides a method to detect their target receptors (Skajaa et al. 2010). Lipoproteins can be adapted as contrast agents in the same ways as micelles. Lipoprotein contrast agents have been reported for MRI, CT and fluorescence imaging techniques. Despite possessing natural targeting, some formulations have been modified with peptides to be retargeted to other receptors or cell types.

MEDICAL IMAGING TECHNIQUES

Magnetic Resonance Imaging

Magnetic resonance imaging is a primarily non-invasive imaging technique that allows both the morphological and biochemical characterization of tissue. MRI exploits the resonance of atomic nuclei in a massive (0.5–10 T) magnetic field. The most important resonant nuclei for MRI are those of the hydrogen atoms (protons) in water, as water is the most abundant component of the human body. In addition, hydrogen is 100% resonant, while other elements such as carbon contain only a minority of resonant nuclei.

The spin of a proton can be aligned either with or against the magnetic field. In the presence of a very large magnetic field the energy gap between alignment and misalignment is high and a small majority of protons in a sample align their spin with the applied magnetic field. In MRI, first, the protons are excited from the ground state to the higher energy state with pulses of radiowaves at the resonance wavelength of protons. The protons subsequently relax to the ground state and emit radiowaves in the process. These emitted radiowaves are detected and converted into images. The rate of this relaxation is highly dependent on the environment of the proton and is described by the parameters T_1 and T_2 (longitudinal and transverse relaxation, respectively). Indeed, because of slightly different conditions that exist around the protons in different tissues, an appropriate MR sequence will distinguish tissues depending on their relaxation rate. For example, when using a T_1 weighted sequence, the shorter the T_1 value of a tissue, the faster the protons relax from the exited state to the ground state, the brighter the tissue looks. On the other hand, the shorter the T_2 value of a tissue, the darker the tissue appears on a T_2 weighted sequence.

MRI contrast agents act on the local environment of the protons by decreasing T_1 or T_2 and therefore alter the contrast in their vicinity. Metal ions that are paramagnetic, i.e., have a large number of unpaired electrons, cause protons to relax faster, mostly via coordination interactions. Paramagnetic metals are usually used as T_1 contrast agents. For example, the clinically approved MRI contrast agent Magnevist is composed of Gd^{3+} ions bound to diethylene triamine pentaacetic acid, a chelator. Most agents designed to reduce T_1 contain either Gd^{3+} or Mn^{2+}, which have seven and five unpaired electrons, respectively, and are therefore highly paramagnetic. For molecular imaging with T_1-based MRI, nanoparticles that include many (100-100,000) paramagnetic metal ions are used, as it is necessary to deliver a large amount of contrast agent to produce a signal change. To reduce T_2, iron oxide nanoparticles that are superparamagnetic (possessing a magnetic moment only when placed in a large magnetic field) are used, such as the commercially available agents Feridex or Sinerem. These agents produce fluctuations in the magnetic field around them, which causes the affected protons to relax faster. Only nanoparticles can create this effect. It is noteworthy that T_1 and T_2 are inversely proportional to the concentration of contrast agent. Therefore, as reduced T_1 values result in image brightening, T_1 contrast agents such as Magnevist produce brightness in images, whereas Feridex produces dark spots in images, as reduced T_2 values create image darkening.

Computed Tomography

CT was invented in the early 1970s and is the first medical imaging modality made possible by the computer. It is used to produce volumetric images that can be manipulated to reveal various structures in the body based on their density or their ability to attenuate X-rays. CT images are produced by passing X rays through the body, at a large number of angles via rotation of the X-ray tube around the subject on a single axis (Fig. 5). Multiple linear detector arrays opposite the X-ray source collect the transmitted X-rays and process the data into grayscale images. For

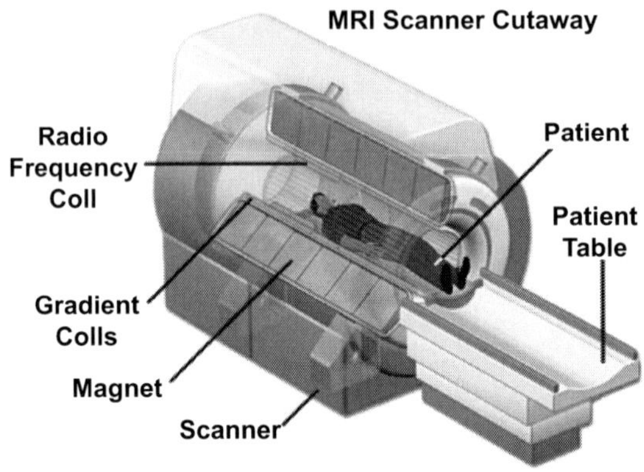

Fig. 4. Schematic of a magnetic resonance imaging system. Reproduced with permission from http://www.magnet.fsu.edu/.

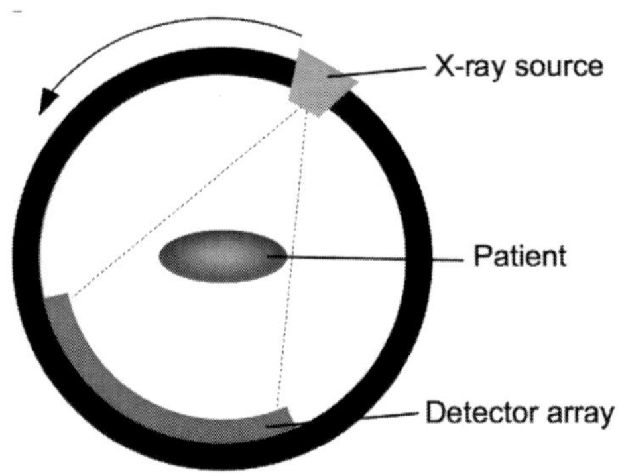

Fig. 5. Schematic diagram of a computed tomography imaging system (unpublished).

example, a dense structure, such as bone, will absorb most of the X-rays and allow only a small proportion of X-rays to pass. The result will be white pixels in the image. On the other hand, air will absorb very little of the X-ray beam, allowing a large number of X-rays to strike the detectors, the result being black pixels in the image.

Clinically approved contrast agents for CT are small molecules that include several iodine atoms and are applied to obtain images of the vasculature. Vascular CT imaging has particular value in cardiovascular disease for identifying atherosclerotic plaques that are occluding the coronary arteries. Iodine is highly X-ray attenuating and therefore results in image brightening in CT images. These agents are rapidly cleared from the circulation. Other elements, such as gold and bismuth, can also be used as a basis for contrast agents. Lately, several groups have reported nanoparticle-based contrast agents for CT (Cai et al. 2007). The advantages of nanoparticles for CT, compared to small molecules, are that they deliver a very large amount of contrast, they can have long circulation half-lives and they can be efficiently targeted.

Nuclear Imaging

Nuclear medicine describes the branch of radiology in which a chemical or compound containing a radioactive isotope is given to the patient. Rather than yielding information about the anatomy, nuclear imaging images provide information regarding the physiological condition of the patient. For example, cancerous cells have a rapid metabolism and consume a lot of glucose and also a radioactive analog of glucose, ^{18}F-fluorodeoxyglucose. Tumors with an increased glucose uptake will then appear as "hot spots" compared to the rest of the body. The main two nuclear imaging modalities are single photon emission computed tomography (SPECT) and positron emission tomography (PET).

SPECT

In SPECT, a nuclear camera records gamma-ray emissions originating from a radionuclide, from a series of different two-dimensional projections acquired at different angles. A tomographic reconstruction algorithm is then applied to the multiple projections, yielding a three-dimensional data set that can be manipulated to show thin slices along a chosen axis of the body, similarly to other tomographic techniques such as CT or MRI. Typically, a full 360-degree rotation is used to obtain an optimal reconstruction. The total scan time usually varies between 15 and 20 min, although multi-headed gamma cameras can provide accelerated acquisition. Radiotracers used for SPECT emit gamma radiation.

PET

PET is a kind of tomography made possible by the unique fate of positrons. When positrons undergo annihilation by combining with electrons, two 511-keV gamma rays are given off in opposite directions 180 degrees apart. In contrast to SPECT imaging that detects single events, in PET imaging, two detector elements on opposite sides of the subject are used to detect paired annihilation of positrons. PET scans are increasingly integrated alongside CT or MRI scanners (Townsend 2008), the combination giving both anatomic and metabolic information (Fig. 6). Because the two scans can be performed in immediate sequence during the same session, with the patient remaining in the same position, the set of images are precisely co-registered so the areas of PET activity can be correlated with the anatomy obtained on CT or MRI images.

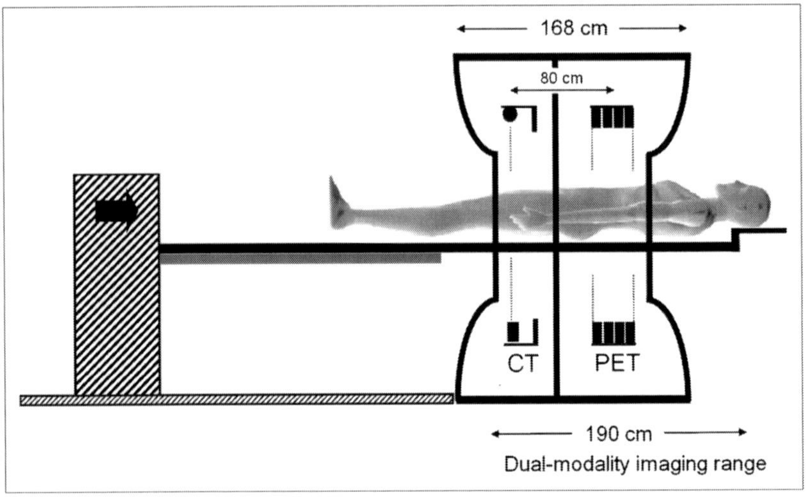

Fig. 6. PET/CT. Sample schematic of the configuration of a dual PET/CT imaging system. Image reproduced with permission from Townsend (2008).

Nanoparticles in Nuclear Imaging

Compared to MRI, CT or ultrasound, nuclear imaging techniques are highly sensitive to contrast agents, as background radiation is very low. Targeted imaging can be fairly easily performed by labeling an antibody or other ligand with a radioactive isotope. Therefore, some of the major motivations for using nanoparticles as contrast agents in MRI, CT or ultrasound, e.g., the delivery of large amounts of contrast or the creation of unique sources of contrast, do not apply in nuclear imaging. Nevertheless, there is some interest in using radiolabeled nanoparticles in nuclear imaging, owing to their favorable pharmacokinetics, the ease of integration of multiple properties in nanoparticles and the ease of validating their targeting with microscopy techniques. Additionally, it is very useful to label nanoparticles with radioactive isotopes in order to facilitate biodistribution studies.

Ultrasound

Ultrasound is a non-invasive diagnostic imaging tool that has improved the diagnosis and management of a number of diseases. Ultrasound has achieved an excellent patient acceptance because it is safe, fast, and inexpensive compared to other imaging modalities. Ultrasound technology produces sectional imaging or slices in multiple planes. An ultrasound machine consists of an ultrasound wave source, a transducer and a computer. The ultrasound transducer emits high-frequency sound waves, ranging from 1 to 10 MHz. Short bursts of these inaudible high-frequency sound waves are broadcast into the patient by a transducer, and some of the reflected sound waves from body tissues and structures are received by the transducer. The sound waves that are reflected directly back to the transducer are converted into images. Each organ and tissue has its own characteristic echo pattern. Solid organs have a homogeneous echo pattern, whereas fluid-filled organs and masses such as the urinary bladder, cysts and gallbladder have relatively fewer internal echoes. For cardiovascular imaging, ultrasound is frequently performed via use of a catheter inserted into the coronary arteries, a technique known as intravascular ultrasound.

Gas-filled microbubbles are currently the type of contrast agent most often used for ultrasound. These microbubbles oscillate, rapidly contracting and expanding, when they enter a high frequency ultrasound beam. This phenomenon generates sound waves that can easily be distinguished from the surrounding tissue. Inert gases such as nitrogen or perfluorocarbons are used to form the bubble. In order to protect the gas from diffusing into the bloodstream, the gas core is encapsulated in a shell. Materials such as albumin, lipids or synthetic polymers are used to form this coating. As the name suggests, microbubbles are larger than nanoparticles, but have many of the same design elements and can be specifically targeted.

EXAMPLES OF USE OF NANOPARTICLES IN MEDICAL IMAGING

There are many potential applications of nanotechnology in medicine in general and in medical imaging in particular. For example, iron oxide nanoparticle–based MR contrast agents such as Feridex and Resovist have multiple uses in the clinic. Feridex has a dextran coating, whereas Resovist is a carboxydextran-coated nanoparticle and both are large (60–180 nm) aggregates of coating material that contain multiple iron cores. These agents may be used for liver tumor diagnosis, for example. Sinerem is a smaller iron oxide agent (15–30 nm) and is often referred to as ultrasmall superparamagnetic iron oxides (USPIO). Because of its dextran coating, it has been shown to have an affinity for macrophages. This phenomenon was exploited to image macrophage expression in atherosclerotic plaques induced in rabbits using MRI (Ruehm et al. 2001). The success of such preclinical studies motivated investigators to carry out clinical trials, which proved the feasibility of imaging macrophage infiltration in the carotid atherosclerotic plaques in patients. In the ATHEROMA study, the investigators used the detection of macrophages with Sinerem-enhanced carotid MR imaging to evaluate the anti-inflammatory effect of high dose versus low dose of a lipid-lowering therapy, atorvastatin (Tang et al. 2009).

As shown in Fig. 7, the low dose of atorvastatin applied in patients for 12 wk did not eliminate the accumulation of USPIOs in carotid plaques (the dark spot in the post-USPIO images). However, the authors noticed that USPIO accumulation was eliminated in the plaque starting at 6 wk after the initiation of high-dose treatment, indicating less active macrophages and minimal inflammation in the plaque. At 12 wk, the effect was even more pronounced, as the injection of USPIOs did not provoke any visible MR signal loss in the carotid plaques (Fig. 7G and H). The study was carried out with 20 patients in each group and yielded significant differences in USPIO-defined inflammation, demonstrating the potential of USPIO-enhanced MRI to monitor anti-inflammatory therapies in patients with atherosclerotic lesions.

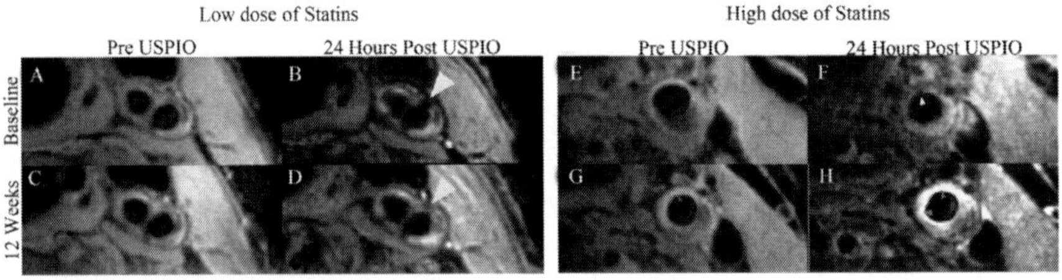

Fig. 7. Molecular imaging in patients. USPIO injection allowed monitoring of the effects of a low dose (left) versus a high dose of statins (right) in patients at 12 wk via MR imaging. A dark spot arising from USPIO uptake in the plaque is still visible after injection of USPIO at 12 wk in the low-dose group (C and D) compared to the baseline (A, B). The high-dose group shows a clear difference in USPIO uptake at 12 wk (G, H). The plaque contains no dark spots, showing the lack of USPIO uptake compared to baseline (E, F). Reproduced with permission from Tang et al. (2009).

Other investigators use iron oxide nanoparticles for cell tracking, as cells can be induced to internalize iron oxide nanoparticles by a variety of methods, which allows the cells to be tracked by MRI as they migrate through tissues. Owing to the high resolution of MRI and the biocompatibility of iron oxide particles, the MRI-iron oxide combination is a popular method for cell tracking, a technique that is of interest to insure proper cell administration and delivery in diseases for which cell therapy is needed, as described in Chapter 20 on iron oxide nanoparticles and cardiac stem cells.

Certain molecules involved in disease formation and progression represent attractive targets for nanoparticle-enhanced imaging, as their detection could lead to early recognition of diseases, better therapeutic management and enhanced understanding of disease progression. In order to develop these site-specific contrast agents, a targeting ligand toward the pathological biomarker of interest is often conjugated to the nanoparticle. For example, van Tilborg et al. formed paramagnetic and fluorescent micelles from gadolinium chelating lipids, fluorescent lipids, polyethylene glycol (PEG) carrier lipids and maleimide PEG lipids (Fig. 8) that were 22 nm in diameter. Annexin A5 (AnxA5) is a protein specific for phosphatidylserine, which is a phospholipid that is located on the inner side of the cell membrane and is exposed on the outer side of the cell membrane only during apoptosis. AnxA5 was conjugated to the bimodal micelles for the non-invasive assessment of apoptotic cells, which are considered to significantly contribute to atherosclerotic plaque instability. These AnxA5 micelles showed increased MR signal intensity in the aortic vessel wall of atherosclerotic, apolipoprotein knockout (apoE -/-) mice 24 h post injection compared to control micelles injected into apoE -/- mice or AnxA5-micelles injected into wild type (WT) mice, as shown in Fig. 8B-G. The presence of AnxA5-targeted micelles in the aorta was confirmed by *ex vivo* near-infrared fluorescence imaging while confocal fluorescence microscopy allowed the precise co-localization of AnxA5-micelles with apoptotic cells and macrophages.

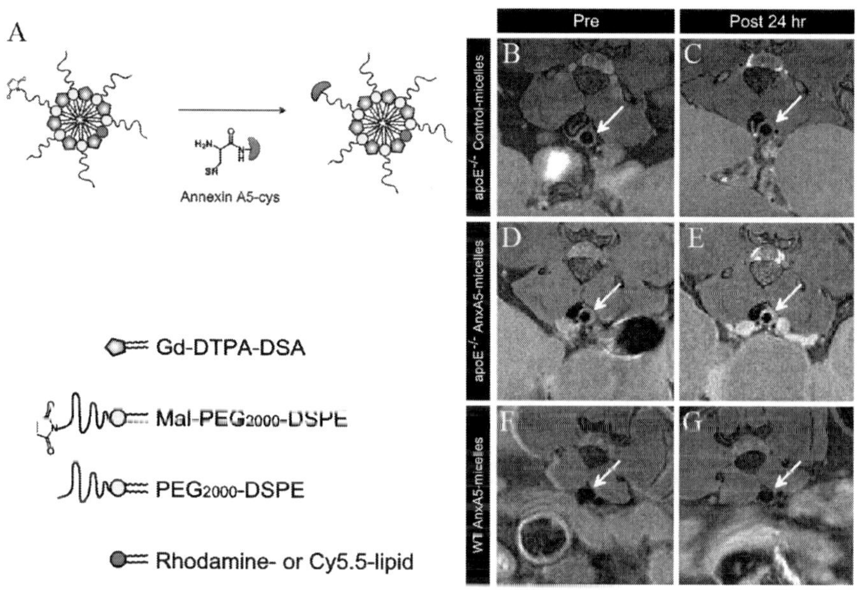

Fig. 8. Detection of apoptosis *in vivo*. (A) Schematic depiction of conjugation of AnxA5 to paramagnetic and fluorescent micelles. (B, C) Typical MR images obtained in atherosclerotic mice before and after injection of control micelles. No enhancement was noticed in that case, whereas injection of AnxA5-micelles produced a bright enhancement in the plaques of atherosclerotic mice (C, D). AnxA5-micelles did not produce any MR signal enhancement in healthy animals (E, F). Reproduced with permission from van Tilborg et al. (2010).

Color image of this figure appears in the color plate section at the end of the book.

Because of its fast imaging capabilities and high resolution, CT also represents an attractive modality for nanoparticle-enhanced imaging in cardiovascular disease. Nanoparticles developed for CT can be used as long-circulating blood pool agents in order to facilitate the visualization of the vasculature over extended periods. Cai et al. developed colloidal gold nanoparticles for imaging the vascular system using CT (Cai et al. 2007). These non-targeted small particles (38 nm) coated with PEG chains proved to be biocompatible and have a half-life of 30 h, which allowed the visualization of the vasculature system from immediately after injection up to 24 h after injection. The authors used these gold nanoparticles to visualize the major vascular structures of mice such as the heart and major arteries as well as smaller vessels using a small animal CT imaging system (Fig. 9A). They also demonstrated the ability of the particles to visualize the small and chaotic vasculature of tumors implanted in mice (Fig. 9B and C). Such nanoparticles could therefore be of great interest to monitor defects of the vascular system and for image-guided surgery.

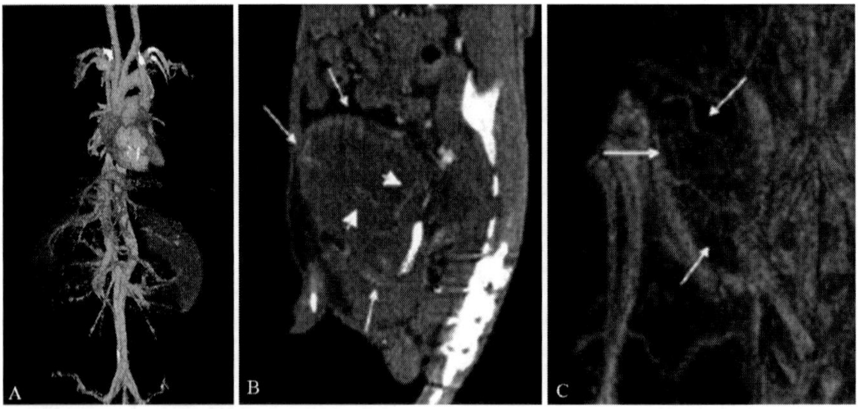

Fig. 9. Long-circulating CT contrast nanoparticles. Computed tomography angiography obtained with colloidal gold nanoparticles injected in mice (A). Gold nanoparticles allowed the clear visualization of the heart and major vessels as well as small branches. (B) Sagittal view of a tumor neovascularization (arrows) observed after injection of gold nanoparticles. (C) 3D reconstruction of the tumor vascularization allowing the visualization of small and chaotic vessels. Figure reproduced with permission from Cai et al. (2007).

Similarly to MRI, non-invasive visualization of macrophage expression in atherosclerotic rabbits can be performed using computed tomography. Hyafil et al. developed a radiodense nanoparticle contrast agent formed from iodinated cores dispersed in water using a surfactant. This agent was termed N1177 and was used for the detection of macrophages with CT in a rabbit model of atherosclerosis (Hyafil et al. 2007). These plaques typically contain a high level of macrophage infiltration. As can be seen in

Fig. 10, the authors showed that the enhancement of atherosclerotic plaques was significantly higher after the injection of N1177 compared to a small molecule CT contrast agent. The *in vivo* imaging results were corroborated by histological studies that showed a correlation between the density of the lesions assessed by CT after N1177 injection and the macrophage content in the corresponding histological sections. Importantly, the authors confirmed the presence of N1177 nanoparticles inside the macrophage lysosomes using transmission electron microscopy (Fig. 10E and F). Indeed, as shown in Fig. 10G, the electron-dense particles observed in the plaque macrophages exhibited the energy profile of iodine.

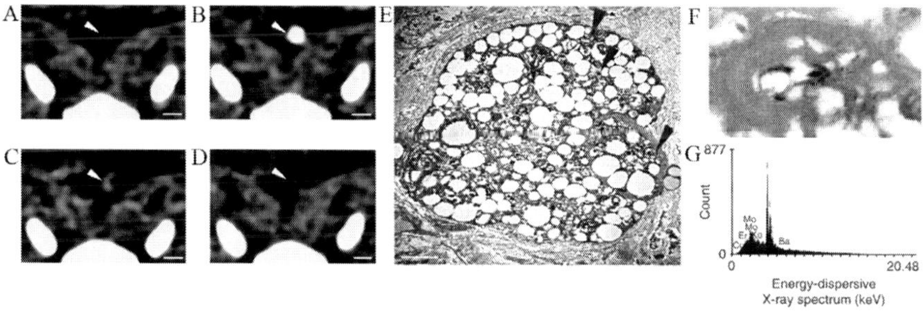

Fig. 10. Macrophage imaging with CT. (A, B, C) Typical images obtained before, during and 2 h after injection of N1177 respectively. Enhancement was observed in plaques 2 h after injection of N1177, whereas no enhancement was observed after injection of a conventional iodine contrast agent (D). (E, F) Transmission electron microscopy showing electron-dense features inside the lysosomes of macrophages 2 h after injection of N1177. (G) The energy profile of these features corresponded to iodine, confirming presence of N1177 in macrophages. Reproduced with permission from Hyafil et al. (2007).

APPLICATIONS TO AREAS OF HEALTH AND DISEASE

Early diagnosis and understanding of disease progression require improvements over conventional imaging techniques. Indeed, anatomical information is very limited and does not permit visualization of the biological processes occurring in the pathophysiology of diseases. Nanoparticles can be engineered to provide superior contrast and therefore facilitate the detection of key pathological information about disease progression. Nanoparticles can be targeted for molecular imaging, allowing the monitoring of disease processes and the effect of various drugs upon these processes, thus providing tremendously valuable information. In addition, nanoparticles can be used for several imaging techniques and can be applied to a wide variety of diseases such as cancer or atherosclerosis.

Key Facts

- Traditional imaging investigates the vasculature and provides anatomical information on the cardiovascular tissues.
- Nanoparticle-enhanced imaging can provide vascular imaging over an extended period or information on biological processes such as apoptosis that are occurring in the cardiovascular tissues. This latter technique is known as molecular imaging.
- Nanoparticles can be formulated to include contrast for MRI, CT, PET, ultrasound and fluorescence as well as therapeutics and targeting ligands, or combinations of these properties.
- Gadolinium or iron oxide is included in nanoparticles to create MRI contrast agents.
- CT nanoparticle contrast agents contain iodine, gold, bismuth or gadolinium.
- Contrast agents for ultrasound are composed of perfluorocarbon or nitrogen gas covered in a polymer, protein or phospholipid shell.
- Nanoparticles can be labeled with isotopes such as copper-64 or fluorine-18 so that they can be detected in PET imaging.
- Quantum dots are inherently fluorescent nanoparticles, but other nanoparticles can be made fluorescent by attaching small molecule fluorophores.

Definitions

Apoptosis: The process of programmed cell death, which occurs in various cardiovascular diseases, weakening tissue.

Computed tomography (CT): An anatomical imaging technique that exploits the varying X-ray absorbance of tissues to create volumetric images.

Macrophages: Inflammatory cells linked to the progression of atherosclerosis and adverse cardiovascular events.

Magnetic resonance imaging (MRI): An anatomical and, to some extent, functional imaging technique that requires a massive magnetic field. Images are formed by transmission to and reception of radiowaves from water protons.

Molecular imaging: "The *in vivo* characterization and measurement of biologic processes at the cellular and molecular level" (Weissleder et al. 2001).

Positron emission tomography (PET): A functional imaging technique that forms images from detection of pairs of gamma rays produced by the annihilation of a positron.

Single photon emission computed tomography (SPECT): A functional imaging technique that detects the emissions of gamma-rays from injected radiopharmaceuticals.

Ultrasound: An anatomical imaging technique that uses very high frequency sound waves.

Summary Points

- Nanoparticles can be formulated that produce image contrast for the majority of medical imaging techniques such as MRI, CT, PET, ultrasound and fluorescence techniques.
- The nanoparticles used in medical imaging can vary in size from 1 to 1000 nm, take a wide variety of morphologies and be composed of many different materials and components.
- Long-circulating nanoparticles can be used to highlight the vasculature for extended periods, as compared with traditional, small molecule contrast agents that are swiftly excreted.
- Nanoparticles can be targeted to specific cells or receptors. When the nanoparticles gather at the targets *in vivo*, imaging can provide an estimate of the level of expression. This is known as molecular imaging.
- Iron oxide nanoparticles are clinically approved and can be used for macrophage imaging in atherosclerosis.
- Preclinically, many processes can be followed in various cardiovascular diseases such as VCAM-1, angiogenesis, macrophages, and apoptosis.

Abbreviations

^{1}H	:	proton
^{19}F	:	fluorine-19
CT	:	computed tomography
HDL	:	high density lipoprotein
IDL	:	intermediate density lipoprotein
LDL	:	low density lipoprotein
MRI	:	magnetic resonance imaging
PET	:	positron emission tomography
SPECT	:	single photon emission computed tomography
VCAM-1	:	vascular cell adhesion molecule 1
VLDL	:	very low density lipoprotein

References

Cai, Q.Y., S.H. Kim, K.S. Choi et al. 2007. Colloidal gold nanoparticles as a blood-pool contrast agent for x-ray computed tomography in mice. Invest. Radiol. 42: 797–806.

Chen, W., D.P. Cormode, Z.A. Fayad and W.J.M. Mulder. 2011. Nanoparticles as magnetic resonance imaging contrast agents for vascular and cardiac diseases. WIREs Nanomed. Nanobiotechnol. 3, 2: 146–161.

Cormode, D.P., T. Skajaa, Z.A. Fayad and W.J.M. Mulder. 2009. Nanotechnology in medical imaging: probe design and applications. Arterioscler. Thromb. Vasc. Biol. 29: 992–1000.

Hyafil, F., J.C. Cornily, J.E. Feig et al. 2007. Noninvasive detection of macrophages using a nanoparticulate contrast agent for computed tomography. Nat. Med. 13: 636–641.

Lanza, G.M., P.M. Winter, S.D. Caruthers et al. 2006. Nanomedicine opportunities for cardiovascular disease with perfluorocarbon nanoparticles. Nanomedicine 1: 321–329.

Mccarthy, J.R., K.A. Kelly, E.Y. Sun and R. Weissleder. 2007. Targeted delivery of multifunctional magnetic nanoparticles. Nanomedicine 2: 153–167.

Medintz, I.L., H.T. Uyeda, E.R. Goldman and H. Mattoussi. 2005. Quantum dot bioconjugates for imaging, labelling and sensing. Nat. Mater. 4: 435–446.

Sanz, J., and Z.A. Fayad. 2008. Imaging of atherosclerotic cardiovascular disease. Nature 451: 953–957.

Skajaa, T., D.P. Cormode, E. Falk et al. 2010. High density lipoprotein-based contrast agents for multimodal imaging of atherosclerosis. Arterioscler. Thromb. Vasc. Biol. 30: 169–176.

Tang, T.Y., S.P.S. Howarth, S.R. Miller et al. 2009. The ATHEROMA (atorvastatin therapy: effects on reduction of macrophage activity) study: evaluation using ultrasmall superparamagnetic iron oxide-enhanced magnetic resonance imaging in carotid disease. J. Am. Coll. Cardiol. 53: 2039–2050.

Townsend, D.W. 2008. Dual-modality imaging: combining anatomy and function. J. Nucl. Med. 49: 938–955.

Van Tilborg, G.A.F., E. Vucic, G.J. Strijkers et al. 2010. Annexin A5-functionalized bimodal nanoparticles for MRI and fluorescence imaging of atherosclerotic plaques. Bioconjugate Chem. 21, 10: 1794–803.

Weissleder, R., and U. Mahmood. 2001. Molecular Imaging. Radiology 219: 316–333.

2

Cell Sheet Nanotechnology: Engineering and Applications to Cardiology

Yuji Haraguchi,[1,b] *Tatsuya Shimizu,*[1,c] *Masayuki Yamato,*[1,d] *Ross J. Hunter*[2] and *Teruo Okano*[1,a,*]

ABSTRACT

Cell-based regenerative medicine has emerged as one of the most promising therapies for patients suffering from severe heart failure and is already being performed in clinical practice. More recently, as an alternative to the direct injection of cell suspensions, advanced therapies have shifted towards the transplantation of tissue-engineered heart grafts. We have developed an original tissue engineering methodology termed "cell sheet engineering" that uses culture surfaces grafted with a temperature-responsive

[1]Institute of Advanced Biomedical Engineering and Science, TWIns, Tokyo Women's Medical University, 8-1 Kawada-cho, Shinjuku-ku, Tokyo, Japan.

[a]E-mail: tokano@abmes.twmu.ac.jp

[b]E-mail: yharaguchi@abmes.twmu.ac.jp

[c]E-mail: tshimizu@abmes.twmu.ac.jp

[d]E-mail: myamato@abmes.twmu.ac.jp

[2]Department of Cardiology, Barts & The London NHS Trust, London, UK.
E-mail: ross.hunter@bartsandthelondon.nhs.uk

*Corresponding author

List of abbreviations after the text.

polymer, poly(*N*-isoproplyacrylamide) (PIPAAm), that allows us to control the attachment and detachment of living cells via simple temperature changes. Hence, cultured cells can be harvested as intact cell sheets using the surfaces. Temperature-dependent cell attachment and detachment are controlled by alternating the nanoscale thickness of PIPAAm layers. This unique harvesting method can be performed without proteolytic treatment, preserving extracellular matrix (ECM) on the cell sheets. By layering cell sheets with preserved ECM, various three-dimensional tissues including pulsate myocardial tissues are easily constructed without using the scaffolds that are normally used for holding cells. By using such "intelligent" surfaces, confluent cardiomyocytes can be non-invasively harvested as intact and functional myocardial tissues, without disruption of nanoscale cellular structures, including ion channels and cell-to-cell junctions containing gap junctions. Therefore, electrical coupling within the cardiomyocyte sheet is completely conserved immediately after harvest. Interestingly, after layering, the cardiomyocyte sheets rapidly become electrically synchronized (after approximately 40 min). The preservation of nanoscale cell surface structures (namely gap junction precursors and ECM) are key in the rapid and complete electrical coupling of the layered cardiomyocyte sheets. In addition, electrical interactions between cardiomyocyte sheets can be prevented by inserting non-cardiomyocyte sheets. Conduction delay and conduction block between two cardiomyocyte sheets are observed when multi-layered fibroblast sheets are inserted between them, although this does not occur with a single fibroblast sheet. Polysurgery with cardiomyocyte sheets allows us to create viable and functional heart-like tissue, and transplantation of these multi-layer cardiomyocyte sheets onto the heart leads to functional coupling with native tissue. Transplantation of cell sheets has improved cardiac function in animal models of cardiac injury, with greater therapeutic effects compared to direct injection of cell suspensions. Cell sheet engineering using nanoscale surface chemistry produces cardiomyocyte sheets with preserved cellular surface structures, facilitating integration and electrical coupling with native tissue. There is enormous potential for this technique in treatment of cardiac patients, particularly those with conducting system disease and heart failure.

INTRODUCTION

Cell-based regenerative therapies are currently emerging as some of the most promising methods to treat damaged heart tissue. Regenerative therapy by the direct injection of suspended cells has been clinically performed and shown to produce some limited recovery from heart dysfunction (Menasche et al. 2001). However, with the injection of cell suspensions it is often difficult to control the shape, size, and location that the transplanted cells assume. In addition, significant cell loss due to physical stress, initial hypoxia, or cell wash-out has induced some problems (Zhang et al. 2001). To overcome these problems, myocardial tissue engineering has been viewed as the second generation in cell therapy (Akins 2002). Tissue engineering is currently based on concepts that three-dimensional (3D) scaffolds (for instance, polyglycolic acid, collagen gel, and gelatin) are used as an alternative for extracellular matrix (ECM), and cells are seeded into the scaffolds (Langer and Vacanti 1993). In contrast to these methods that use scaffolds for living cells, our laboratory has developed a unique approach using culture surfaces grafted with a temperature-responsive polymer, poly(*N*-isoproplyacrylamide) (PIPAAm), that can control the attachment and detachment of living cultured cells by simple temperature changes. By covalently immobilizing PIPAAm at nanoscale thickness onto conventional culture surfaces, changes in the surface properties can be controlled by simply varying temperature (Yamada et al. 1990; Okano et al. 1993). Above PIPAAm's lower critical solution temperature (LCST) of 32°C, living cells can adhere and proliferate on the surface. The multiplying cells then spontaneously detach themselves when the temperature is reduced below 32°C without any need for enzymatic digestion. Therefore, confluent cultured cells are non-invasively harvested as a contiguous cell sheet, keeping its intact cell-to-cell connections, simply by lowering the culture temperature (Matsuda et al. 2007; Masuda et al. 2008). Additionally, the harvested cell sheet can easily re-attach to other surfaces such as culture dishes, other cell sheets, and host tissues without any "adhesive" (Matsuda et al. 2007; Masuda et al. 2008).

Cardiomyocytes are electrically interconnected with gap junctions (GJs) to allow current to flow from cell to cell, ensuring simultaneous mechanical contraction. Cardiomyocytes cultured on these culture surfaces are also confluent and connect via GJs to synchronize contraction (Oyamada et al. 1994). Therefore, in myocardial tissues prepared by layering cardiomyocyte sheets, it is crucial that electrical and morphological communications are established between the layered cell sheets. In this chapter, we review cell sheet engineering based on nanoscale surface chemistry, the preservation of nanoscale cellular structures in cell sheets, and the characteristics of 3D myocardial tissues produced using cell sheet engineering. In addition, we describe the application of cell sheet engineering to myocardial tissue in the field of regenerative medicine.

TEMPERATURE-RESPONSIVE CULTURE SURFACE AND CELL SHEET ENGINEERING

Temperature-Responsive Culture Surface

A polymer, PIPAAm, exhibits temperature-responsive hydrophobicity changes in aqueous solutions (Heskins et al. 1968). We have succeeded in the fabrication of PIPAAm-grafted tissue culture polystyrene (TCPS) surfaces using techniques described previously (Yamada et al. 1990; Okano et al. 1993). *N*-isoproplyacrylamide (IPAAm) monomer in 2-propanol solution is spread onto TCPS surfaces. These surfaces are then subjected to electron beam irradiation, resulting in the polymerization and covalent grafting of IPAAm to the surface. The nanoscale PIPAAm-grafted layer onto TCPS also shows a temperature-responsive characteristic. At temperatures below 32°C, which is the LCST of PIPAAm molecules, because PIPAAm molecules are highly hydrated, PIPAAm-grafted surfaces become hydrophilic. And at temperatures above 32°C the surfaces abruptly change to hydrophobic because of the extensive dehydration of PIPAAm molecules. This alteration is completely reversible with temperature.

Preparation of Two Different PIPAAm-Graft Surfaces

Cells attach normally to hydrophobic surfaces better than hydrophilic surfaces. Therefore, we have applied the characteristics of PIPAAm-grafted surfaces to cell culture. In fact, cells can attach and proliferate to form a confluent cell monolayer on hydrophobized PIPAAm surfaces at 37°C, above the LCST of PIPAAm molecules, and the confluent cells are detached as a single cell sheet by lowering the culture temperature to 20°C below the LCST (Yamada et al. 1990; Okano et al. 1993; Matsuda et al. 2007; Masuda et al. 2008). The attachment and detachment of cells on PIPAAm-grafted surfaces can be controlled by the hydrophilic/hydrophobic changes of this polymer by simple temperature alterations. The amount of PIPAAm grafted onto TCPS surfaces significantly affects cell attachment (Sakai et al. 1996). A large amount of PIPAAm grafted onto TCPS surfaces inhibits cell adhesion. The detailed mechanism for this remains unclear. Therefore, we analyzed the detailed relationship between the amount of PIPAAm covalently grafted onto TCPS surfaces and the cell attachment/detachment behavior, using two different types of PIPAAm-grafted TCPS surfaces that have different graft amounts. The two different types of

PIPAAm-grafted surfaces were prepared by electron beam treatment after grafting two IPAAm solutions with different concentrations (Akiyama et al. 2004). The amount of polymer on the two PIPAAm-grafted surfaces was determined by attenuated total reflection Fourier transform infrared spectroscopy. The amount of polymer on each surface was 1.4 ± 0.1 μg/cm^2 (n = 4, mean ± SD) and 2.9 ± 0.1 μg/cm^2 (n = 4) (Akiyama et al. 2004). Therefore, those two surface types were designated as PIPAAm-1.4 and PIPAAm-2.9, respectively. Though both PIPAAm-grafted surfaces showed hydrophobic/hydrophilic property alterations in response to temperature, PIPAAm-1.4 surface was more hydrophobic than PIPAAm-2.9, both above and below PIPAAm's transition temperature. At 37°C the contact angle (θ) of PIPAAm-1.4 was 77.9 ± 0.60° (n = 3, mean ± SD) and that of PIPAAm-2.9 was 69.5 ± 1.20° (n = 3); at 20°C that the contact angle of PIPAAm-1.4 was 65.2 ± 1.20° (n = 3) and that of PIPAAm-2.9 was 60.0 ± 0.06° (n = 3) (Akiyama et al. 2004). Next, the thicknesses of the grafted PIPAAm layer of PIPAAm-1.4 and PIPAAm-2.9 were measured using a UV excimer laser and an atomic force microscope. First, each PIPAAm-grafted TCPS surface was cleaved by UV excimer laser to expose the TCPS surface. To detect the exposure of hydrophobic TCPS surfaces, the cleaved domains were stained with a hydrophobic fluorescent dye, DiIC18. Next, the thickness of the graft layer was measured by the atomic force microscope. The thickness of the grafted PIPAAm-1.4 or PIPAAm-2.9 was 15.5 ± 7.2 nm (n = 4, mean ± SD) and 29.5 ± 8.4 nm (n = 4) respectively (Akiyama et al. 2004). The thicknesses of two types of PIPAAm-graft surfaces and their hydrophobicities are summarized in Fig. 1.

Control of Cell Attachment and Detachment by Nanoscale Surface Chemistry

Next, cell attachment and detachment behavior on the two PIPAAm-grafted surfaces were examined. Cells attach to the surfaces of PIPAAm-1.4 and proliferate to form a confluent cell monolayer, which is harvested as a single cell sheet by temperature decrease from 37°C to 20°C (Akiyama et al. 2004). Conversely, cells hardly adhered to the surfaces of PIPAAm-2.9 (Akiyama et al. 2004). These results showed that alterations of nanoscale thicknesses of the PIPAAm-grafted layers play a crucial role in temperature-dependent hydrophilic/hydrophobic properties and cell attachment/detachment behavior. Since the cell adhesion protein fibronectin is a major component of ECM and also mediates cell adhesion on culture dish surfaces, the adhesive behavior of fibronectin on PIPAAm-1.4 and PIPAAm-2.9 was also examined. Fibronectin adhered to the surface of PIPAAm-1.4 at 37°C but not at 20°C (Akiyama et al. 2004). However, fibronectin was unable to adhere to the surface of PIPAAm-2.9 at 20°C or

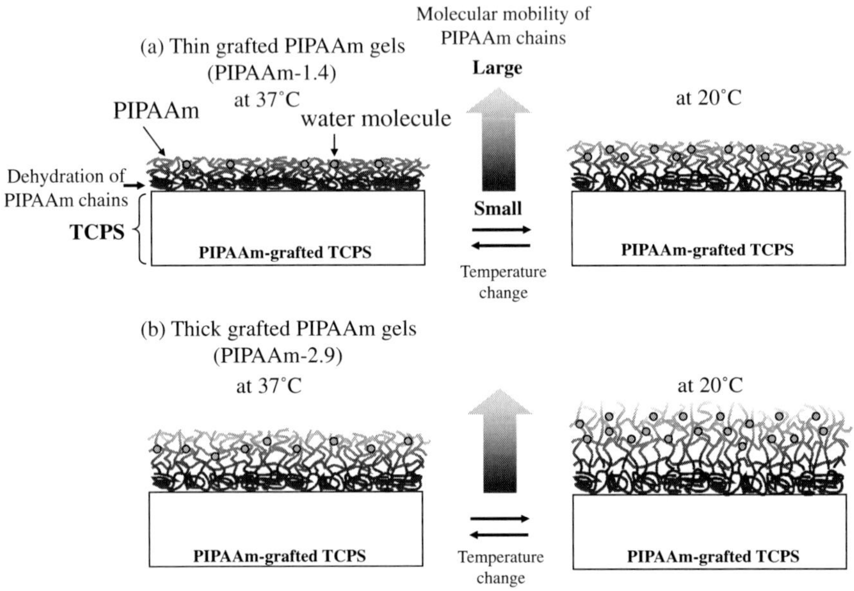

Fig. 1. Schematic diagrams of the influence of molecular mobility of grafted poly(*N*-isoproplyacrylamide) (PIPAAm) chains on the hydration of the polymer layers. The grafted PIPAAm layers are thin (a) and thick (b) at 20°C (right side) and 37°C (left side), respectively. The nearer the grafted PIPAAm layers to the interface of tissue culture polystyrene surface (TCPS), the stronger the hydrophobic and restriction of mobility obtained (black layers). Molecular mobility of the grafted polymer layers becomes larger according to the distance from TCPS interfaces (light brown layers). (Reprinted from Akiyama et al. 2004 with permission from American Chemical Society.)

37°C (Akiyama et al. 2004). These behaviors correlate with those of cell attachment, suggesting that cell attachment/detachment behaviors can be controlled by the adherence of cell adhesion proteins. The ability to mobilize the thin cellular layers is strongly influenced by the thicknesses of PIPAAm-grafted layers, since the dehydration of PIPAAm chains is enhanced by the underlying hydrophobic TCPS surface. Cell attachment and detachment are therefore influenced by these subtle differences in thicknesses of the polymer layers. Temperature-dependent cell attachment and detachment are only observed on PIPAAm-grafted surfaces that are approximately 20 nm thick, with no cell adhesion when the grafted layers are thicker than 30 nm. Cell sheets recovered using temperature-responsive "intelligent surfaces" have been applied to tissue engineering and regenerative medicine, and we describe the technology as "cell sheet engineering".

Preservation of Cell-to-cell Junctions and Extracellular Matrix within Cell Sheets

Cell sheets can be recovered by simple temperature alterations, which induce reversible nanoscale alterations in the thickness of PIPAAm grafted onto TCPS surfaces. Therefore, in the recovery of cell sheets, neither protease nor ethylenediaminetetraacetic acid (EDTA) is required, and consequently cell-to-cell junctions, which are susceptible to damage from protease treatment or the depletion of divalent cations, are preserved intact. Confluently cultured cells can be recovered as intact single cell sheets as shown in Fig. 2A. In addition to cell-to-cell junctions, the presence of fibronectin matrix (which is a major ECM component

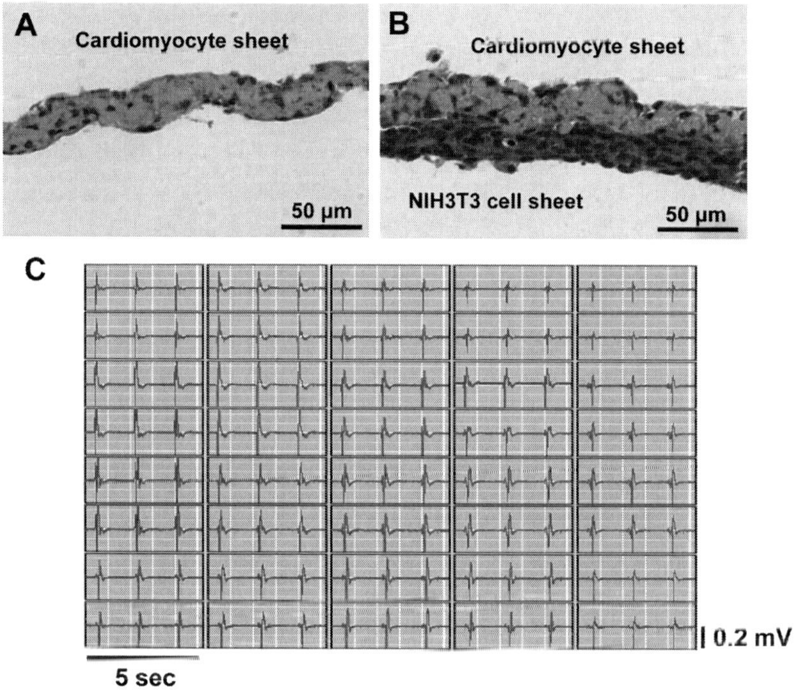

Fig. 2. Cross-sectional views of cell sheets and the electrograms of a cardiomyocyte sheet. A: The cross-sectional view of a single cardiomyocyte sheet. Hematoxylin and eosin staining of the cell sheet shows a cell dense structure. B: The cross-sectional view of a double-layered cell sheet, a cardiomyocyte sheet (the upper layer) and an NIH3T3 cell sheet (the lower layer). The tight binding of the two cell sheets is detected by hematoxylin and eosin staining. C: Electrical potentials of a single cardiomyocyte sheet just after detachment. Spontaneous electrical potentials are recorded by 40 individual electrodes beneath the cardiomyocyte sheet, and all electrical potentials are completely synchronized. (Unpublished material of the authors.)

Color image of this figure appears in the color plate section at the end of the book.

mediating cell adhesion onto dish surfaces) on cell sheets is detected by immunoblotting and immunohistochemistry (Kushida et al. 1999). The preservation of ECM on the basal surface of cell sheets is also shown by transmission electron microscopy (Kushida et al. 1999). Conversely, after protease treatment fibronectin is only faintly detectable (Kushida et al. 1999). This ECM that is produced during cultivation means cell sheets can be easily attached to other surfaces such as culture dishes, other cell sheets, and even host tissues without any "adhesive" (Matsuda et al. 2007; Masuda et al. 2008). The preserved ECM also means that 3D tissues can be easily created by layering cell sheets without scaffolds (Matsuda et al. 2007; Masuda et al. 2008). For example, in Fig. 2B, a double-layered cell sheet construct, which is fabricated by layering a cardiomyocyte sheet on the top of an NIH3T3 cell sheet, is tightly linked. The construct is very cell-dense since it has been fabricated without scaffolds. The preservation of nanoscale structures such as ECM on cell sheets is therefore important in fabricating 3D tissues without the need for scaffolds or "adhesive".

APPLICATION OF CELL SHEET ENGINEERING TO MYOCARDIAL TISSUE ENGINEERING

Electrophysiology of a Monolayer Cardiomyocyte Sheet

A sheet of neonatal rat cardiomyocytes detached from a temperature-responsive culture surface was examined histologically and electrophysiologically. The cardiomyocyte sheet is a very cell-dense tissue (Fig. 2A). As the cell sheet shrinks horizontally because of cytoskeletal tensile reorganization, the cardiomyocyte sheet develops several cell layers (Fig. 2A). Because the electrical properties are so important to the functioning of the cardiomyocytes, we examined how these properties are maintained during the recovery process from a temperature-responsive culture surface. The spontaneous action potentials of the cell sheet just after detachment were examined using a multiple-electrode extracellular recording system. The spontaneous action potentials of the cardiomyocytes comprising the sheet were found to be synchronized immediately after recovery (Fig. 2C). This indicates that electrical coupling within the monolayer cardiomyocyte sheet is completely conserved even just after detachment. Therefore, recovery of the cardiomyocyte sheet without the use of any proteases and EDTA allows the cell sheet to preserve its sarcolemmal ion channels and GJs between cardiomyocytes within the cell sheet. Conservation of nanoscale structures on the cell surface and in the extracellular space is important in maintaining tissue functionality. The gentle manner of this unique recovery processes is important as it enables the preservation of this aspect of cell sheets.

Rapid Electrical Coupling of Cardiomyocyte Sheets after Layering

The electrical interaction between two cardiomyocyte sheets after layering was also assessed using the multiple-electrode extracellular recording system (Haraguchi et al. 2006). To detect the action potentials of each cardiomyocyte sheet separately, there was partial overlap of layered cell sheets, with each sheet attached separately to electrodes (Fig. 3A, Haraguchi et al. 2006). Immediately after layering, the two cardiomyocyte sheets activate independently (Fig. 3B). However, after only 34 ± 2 min (n = 24, mean ± SEM), the two layered cardiomyocyte sheets synchronize (Fig. 3C). At this stage, though the excitation cycles of the two cell sheets are synchronized, slight conduction delay is evident. This conduction delay subsequently decreases in a time-dependent fashion (Haraguchi et al. 2006), with complete synchronization and no delay at 46 ± 3 min (n = 24) after layering (Fig. 3D). Electrical coupling of a triple-layered

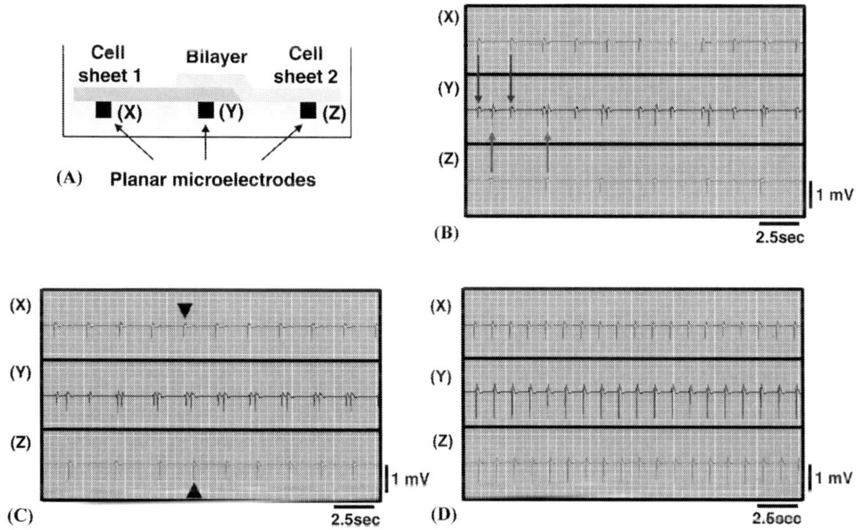

Fig. 3. Rapid synchronization of a layered cardiomyocyte sheet after layering. A: The schematic diagram of a double-layered cardiomyocyte sheet on multi-microelectrodes as viewed from the side. Two cardiomyocyte sheets were layered with partial overlap. Soon after layering, a double-layered cardiomyocyte is independently excited (B). Then, some microelectrodes beneath the monolayer portions detect the spontaneous electrical potentials of individual cell sheet (electrodes X and Z in this figure), while those beneath the overlaid portions detect the electrical potentials of both cell sheets simultaneously (electrode Y in this figure). Arrows indicate the corresponding electrical spikes of each cardiomyocyte sheet. At 34 ± 2 min, two cardiomyocyte sheets begin to couple electrically with slight delays (C). Arrowheads show the start of electrical coupling. At 46 ± 3 min, the excitation spikes of the two cardiomyocyte sheets are coupled completely (D). Data are presented as mean ± SEM. (Reprinted from Haraguchi et al. 2006 with permission from Elsevier.)

cardiomyocyte sheet is also rapidly established in much the same way at 31 ± 2 min after layering (n = 3, mean ± SD) (Haraguchi et al. 2010). These results show that the electrical coupling of layered cardiomyocyte sheets occurs rapidly and completely. This method of fabricating 3D myocardial tissues without any scaffolds allows rapid and complete electrical coupling of the constituent cells.

Gap Junction Formation in a Layered Cell Sheet

Because GJs are essential for the electrical coupling of cardiomyocytes, it was examined whether GJs form between two cardiomyocyte sheets after a short time period. GJ formation between two cardiomyocyte sheets was detected by fluorescent dye transfer assay (Haraguchi et al. 2006). A membrane-permeable dye, calcein AM, is cleaved within living cells to produce a low-molecular-weight green fluorescent dye, calcein, which is membrane impermeable, but small enough to pass through GJs (Lampe et al. 1994) of diameter approximately 1.5 nm. Calcein transfer from a calcein-loaded cardiomyocyte sheet to a calcein-free sheet was detected at 30 min after layering, and the fluorescence of the unloaded cell sheet increased and expanded time-dependently (Fig. 4A-D). The result suggests that GJ formation occurs within 30 min and increases time-dependently. Because no calcein transfer from the cardiomyocyte sheet to GJ-defective cells is observed (Fig. 4F), calcein should transfer between the layered cell sheets only through GJs, but not via non-specific diffusion of residual calcein AM. Rapid GJ formation is also detected by immunohistological analysis using GJ-related protein, connexin43 (Cx43) antibody (Haraguchi et al. 2006). In addition, Cx43 is detected on non-overlapping cell membrane as well as at cell-to-cell interfaces within a single cardiomyocyte sheet (Haraguchi et al. 2006). Cx43 on the non-overlapping cell membrane is thought to be GJ precursors. GJs are thought to be formed by the docking of two GJ precursors on the cell membranes on adjacent cells. This is dependent on the transcription of Cx43 RNAs, the translation and maturation of Cx43 proteins, vesicular transport, and other factors (Tadvalkar et al. 1983). Therefore, it is thought that electrical coupling between cardiomyocyte sheets is rapid, because cardiomyocyte sheets have GJ precursors already on the cell membranes and GJs at cell-to-cell interfaces. In addition, deposited ECM also promotes the intimate attachment between layered cell sheets and may accelerate the docking of GJ precursors. Preservation of nanoscale cell structures such as GJ precursors and ECM on the cardiomyocyte sheets is therefore important in the rapid and complete electrical coupling of the layered cell sheets. In the case of cardiomyocytes recovered using proteases, no such rapid electrical coupling is observed because both GJ precursors and ECM have been destroyed by these treatments.

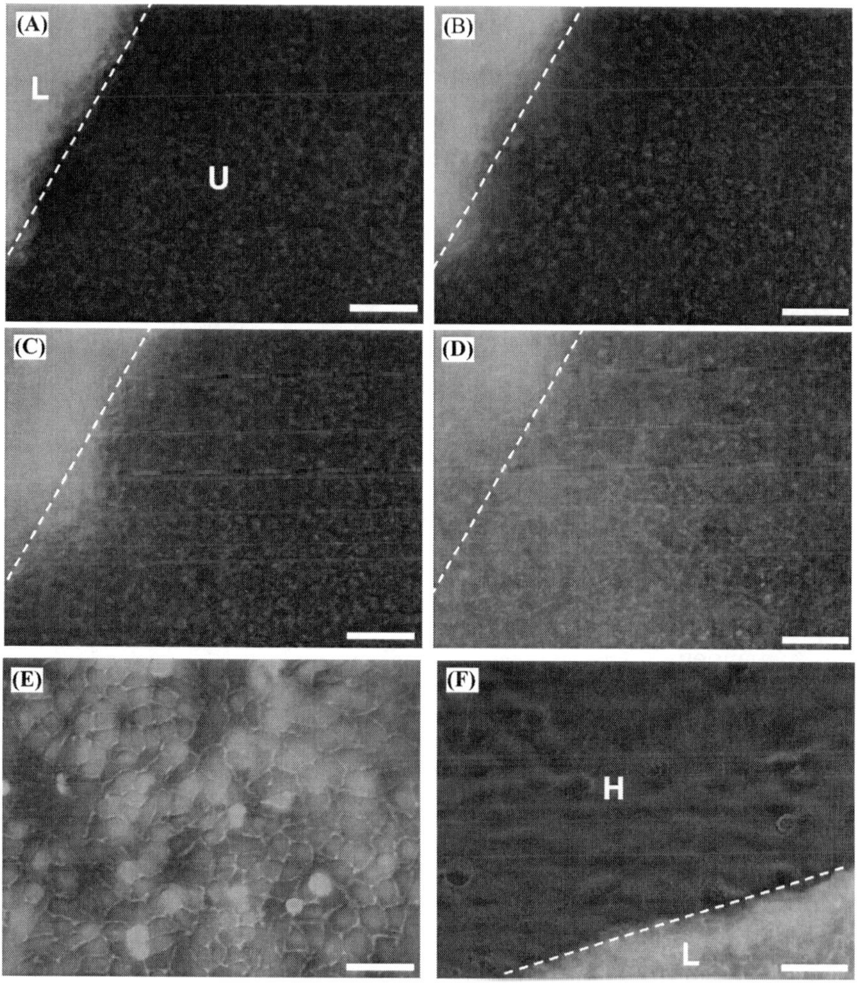

Fig. 4. Fluorescent dye transfer between a double-layered cardiomyocyte sheet. A calcein AM-loaded cardiomyocyte sheet was layered on top of an unloaded cardiomyocyte sheet. Fluorescent pictures at the border region of the layered cardiomyocyte sheet were taken at 0 min (A), 30 min (B), 60 min (C), and 120 min (D) after layering. Calcein transfer to the unloaded sheet is observed at 30 min and expanded time-dependently. Though HeLa cells, which are GJ-negative cells, have the ability to take in calcein AM directly (E), no dye transfer from a calcein-loaded cardiomyocyte sheet (L) to the confluent HeLa cells (H) is observed even after 120 min (F). Broken lines show the border regions between the calcein loaded cardiomyocyte sheet and unloaded cardiomyocytes or HeLa cells. L: calcein AM-loaded cardiomyocyte sheet; U: unloaded cardiomyocyte sheet; H: confluent HeLa cells. Scale bars show 50 µm. (Reprinted from Haraguchi et al. 2006 with permission from Elsevier.)

Color image of this figure appears in the color plate section at the end of the book.

Time Course of Electrical Coupling of a Layered Cardiomyocyte Sheet

The mechanism we have proposed for the electrical coupling of two cardiomyocyte sheets after layering is illustrated in Fig. 5. At first, the action potentials of two cardiomyocyte sheets are dissociated (Fig. 3B). This shows that no electrical coupling is yet established between two cardiomyocyte sheets, because there are too few GJs connecting them. The electrical excitations arising from the pacemaker cells of each cardiomyocyte sheet are conducted independently throughout the intact cell sheets. Therefore, two independent action potentials are detected at electrodes beneath the layered portion (Figs. 3B and 5A). The cardiomyocyte sheets synchronize electrically with slight delays at 34 ± 2 min after layering (Fig. 3C). After adequate GJs are formed between the two cardiomyocyte sheets, electrical excitation from the more rapidly depolarizing cell sheet, namely the pacemaker cell sheet (cell sheet 1 in Fig. 5B), is conducted to

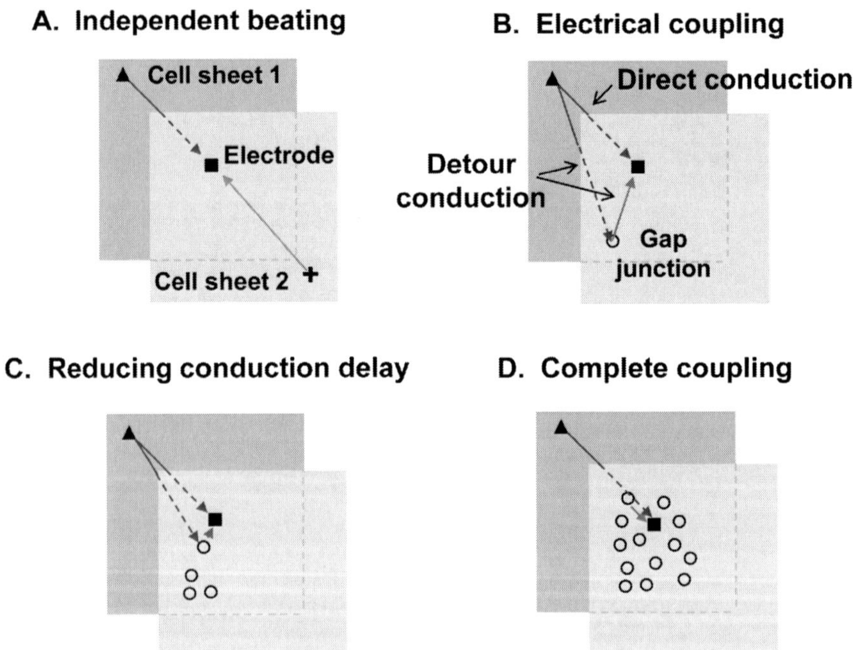

Fig. 5. Electrical conduction routes of coupled cardiomyocyte sheet layers. A: Before electrical coupling between two cardiomyocyte sheets. B: After electrical coupling with slight conduction delay. C: The conduction delays diminish in time-dependent manner due to random GJ formation. D: Complete coupling between two cardiomyocyte sheets. Arrows show electrical conduction in cardiomyocyte sheets. Squares and circles show electrodes and GJs, respectively. (This figure is our hypothesis and is unpublished.)

and propagates through the other cell sheet, namely the non-pacemaker cell sheet (cell sheet 2 in Fig. 5B). At this point, electrical signals reach electrodes through the two independent routes, both directly through the pacemaker cell sheet and indirectly through the non-pacemaker cell sheet having conducted through a small number of GJs (Fig. 5B). It is thought that these two conduction routes are the source of the observed delays. These delays decrease in a time-dependent fashion (Haraguchi et al. 2006), suggesting that (1) additional GJs are randomly formed at the overlapped part of the layered cell sheets and (2) the electrical impulse is conducted to the non-pacemaker cell sheet through the fastest possible route (Fig. 5C). At 46 ± 3 min after layering, the impulses propagate across both cell sheets to separate electrodes simultaneously, suggesting complete coupling has been established, by the formation of GJs throughout the interface between the layered cell sheets (Figs. 3D and 5D). During this process, GJs are formed by the docking of nanoscale GJ precursors on the surface of adjacent cell sheets. Greater understanding of basic tissue-to-tissue communications of myocardial tissues has allowed insight into this process.

Electrical Interaction of Separated Cardiomyocyte Sheets

In addition to electrical interactions between neighboring cardiomyocytes, interactions between separated cardiomyocytes are also important in the understanding of normal heart function and pathological states. Therefore, we analyzed the electrical interactions between separated cardiomyocytes prepared by cell sheet engineering (Haraguchi et al. 2010). Complete electrical coupling of two separated cardiomyocyte sheets occurs when an NIH3T3 mouse fibroblast sheet is inserted between them (Fig. 6A, Haraguchi et al. 2010). However, when a double-layered NIH3T3 cell sheet is inserted between two cardiomyocyte sheets, although they synchronize, the electrical coupling between them is incomplete and conduction delays are observed (Fig. 6B, Haraguchi et al. 2010). When a triple-layered NIH3T3 cell sheet is inserted, the electrical coupling of two cardiomyocyte sheets is completely blocked (Fig. 6C, Haraguchi et al. 2010). The physical distance between the two separated cardiomyocyte sheets prevents formation of GJs and results in failure of electrical coupling, and consequently conduction delay or block. Analyses of impulse propagation, conduction delay and block in cell sheets may provide an experimental model for the study of cardiac arrhythmia. The opportunity to examine the effect of antiarrhythmic drugs may help to elucidate their therapeutic mechanisms and aid further drug development.

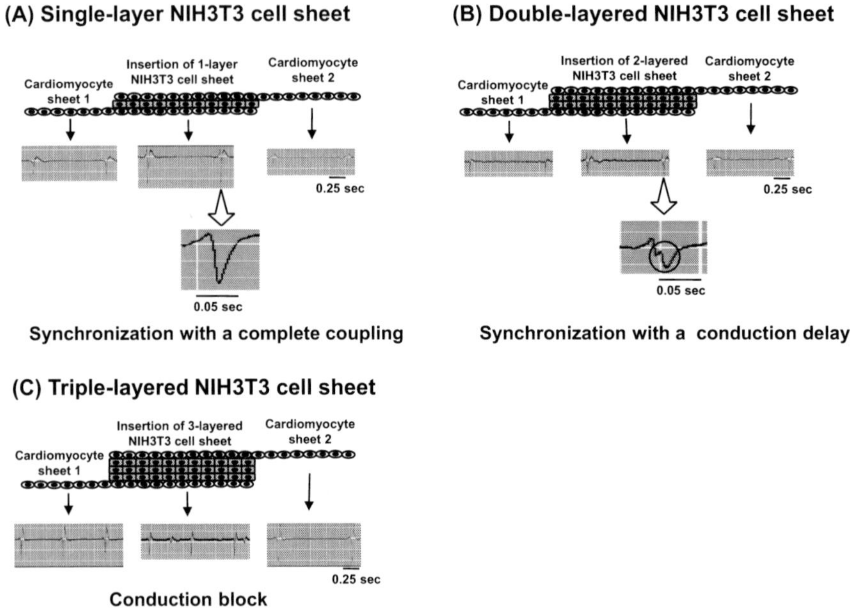

Fig. 6. Conduction delay and block between separated cardiomyocyte sheets caused by insertion of a layered NIH3T3 mouse fibroblast sheet. A single NIH3T3 cell sheet (A), a double-layered NIH3T3 cell sheet (B), or a triple-layered NIH3T3 cell sheet (C) was inserted between two cardiomyocyte sheets. Electrical potentials were monitored by a multiple-electrode extracellular recording system. In the case of single NIH3T3 cell sheet insertion, two cardiomyocyte sheets are completely coupled at 154 ± 41 min after layering (A) (n = 5, mean ± SD, Haraguchi et al. 2010). On the other hand, in the case of double-layered NIH3T3 cell sheets insertion, two cardiomyocyte sheets do not completely couple, though those cell sheets are synchronized (B) (Haraguchi et al. 2010). Two cardiomyocyte sheets are unable to synchronize when a triple-layered NIH3T3 cell sheet is inserted between them (C) (Haraguchi et al. 2010). A circle in (B) shows a conduction delay. (This figure is unpublished.)

3D MYOCARDIAL TISSUE RECONSTRUCTION *IN VIVO* AND THERAPEUTIC EFFECTS BY CELL SHEET TRANSPLANTATION

When layered cardiomyocyte sheets are transplanted into the dorsal subcutaneous tissues of nude rats (see Definitions), the beating of the transplanted grafts can be observed macroscopically and continue for up to 1 year (Shimizu et al. 2003). Interestingly, electrograms derived from transplanted grafts are detected independently from the host heart (Shimizu et al. 2003). In addition, repeated transplantations 24 h apart (or polysurgery) of triple-layer cardiomyocyte sheets produces a thick (approximately 1 mm) cell-dense myocardium with a well-organized microvascular network (Shimizu et al. 2006). The spontaneous and synchronized beating of the tissue graft can be macroscopically observed from the subcutaneous implant site, and this beating is observed to continue for several minutes even after surgical resection (Shimizu et al. 2006). This polysurgery, whereby cardiomyocyte sheets are repeatedly transplanted, allows us to create viable and functional myocardial tissue with normal tissue architecture. In addition, the transplantation of layered cardiomyocyte sheet onto rat heart leads to a functional coupling between transplanted graft and heart tissue via GJ formation (Sekine et al. 2006). Preservation of nanoscale cell structures on cell sheets is fundamental to this functional coupling.

Transplantation of cardiomyocyte sheets, myoblast sheets, or mesenchymal stem cell sheets can improve damaged heart function in rat, hamster, canine, and pig models (Memon et al. 2005; Miyagawa et al. 2005, 2010; Kondo et al. 2006; Hata et al. 2006; Miyahara et al. 2006; Hida et al. 2008; Sekiya et al. 2009). This therapeutic effect is greater than that brought about by injection of cell suspension, and this may be due to the lower rate of cell loss with cell sheet transplantation. Clinical trials using engineered cell sheets have already begun. Cell sheet engineering with preservation of nanoscale surface chemistry and cell structures may open a new paradigm in the fields of electrophysiology and regenerative treatment of cardiac failure, and it has an enormous potential impact on the lives of cardiac patients.

Key Facts

- Cell sheet engineering is a scaffold-free tissue engineering technique used to make sheets of cells suitable for grafting to other tissues. The cell sheets can be used in regenerative medicine of various tissues including heart tissue. Clinical trials have already begun in some tissues including heart tissue.
- A cell sheet is composed of confluent cells recovered from temperature-responsive culture surfaces. Cell-to-cell junctions and extracellular matrix (ECM) are completely preserved in the cell sheet. Cell sheets can be adhered to various tissues without suture because of the presence of ECM.
- Temperature-responsive polymer, poly(*N*-isoproplyacrylamide) (PIPAAm), is covalently bound to the cell culture surface. The surface is temperature-responsive in that the hydrophobicity/hydrophilicity of the surface and attachment/detachment of culture cells are controlled by simple temperature changes.
- The hydrophobic/hydrophilic properties of poly(*N*-isoproplyacrylamide) (PIPAAm) are changed by temperature alterations. At temperatures below the lower critical solution

temperature (LCST), the polymer becomes hydrophilic, and at temperature above LCST, the polymer changes to hydrophobic. These changes are completely reversible with temperature.
- A cardiomyocyte sheet is a cell sheet fabricated with cells derived from heart tissues. Cells in the cardiomyocyte sheets beat spontaneously and synchronously, and in layered cell sheets beating can be observed macroscopically.
- In animal models of heart disease, cell sheet transplantations can ameliorate disease states, including reduced fibrosis, enhancement of left ventricular function, inhibition of left ventricular dilatation, and prolongation of life expectancy.

Definitions

Beating (contraction) of cardiomyocytes: Cardiomyocytes beat (contract) by the interactions of two cardiac muscle proteins, actin and myosin. The interactions are closely related to Ca^{2+} concentration in cardiomyocytes.

Connexin43: A gap junction (GJ)-related protein. This protein is one of the connexin subtypes and makes up GJs in heart tissue.

Electrical coupling and synchronization: Mammalian cardiomyocytes are coupled electrically and beat synchronously. Ions pass through GJs producing an electrical current rapidly passing through the tissue to activate it.

Extracellular matrix (ECM): A complex network of polysaccharides and proteins secreted by cells. Some of the major constituents are fibronectin, collagen, laminin, vitronectin, proteoglycan, and glycosaminoglycan. ECM is closely related to various cellular functions, such as adhesion, growth, differentiation, and migration, in addition to its structural role.

Gap junction (GJ): Communicating cell-to-cell junction that allows ions and small molecules to pass between neighboring cells. Mammalian cardiomyocytes are coupled electrically by ion currents through GJ.

Gap junction (GJ) precursor: GJs are made from a connexin43 hexamer although other connexin subtypes may be involved. GJs are formed by the docking of two GJ precursors on adjacent cells.

Hematoxylin and eosin staining: Cell nuclei are stained to be blue-violet with hematoxylin, and the cytoplasms are stained red-pink with eosin.

Ion channel: Cell membrane proteins that allow ions to pass through them. They are often selective for one particular ion and may undergo conformational change to be activated only at certain membrane potentials. Concentrations of ions in cardiomyocytes, particularly Na^{+}, Ca^{2+}, and K^{+}, are closely related to the action potentials and contraction of cardiomyocytes.

Nude rat: A mutant animal that has a deteriorated or absent thymus, resulting in a poor immune system due to the greatly reduced numbers of T lymphocytes. It can receive different types of tissue grafts, because it shows a very weak rejection response.

Regenerative medicine: A field of medicine that aims to treat damaged tissues or organs using functional cells or growth factors.

Scaffold: Biodegradable polymers used for fabricating three-dimensional (3D) tissue-like grafts in scaffold-based tissue engineering. Collagen, gelatin, agarose, alginic acid, hyaluronic acid, polyglycolic acid, poly-L-lactic acid, and polycaprolactone are used as scaffolds in tissue engineering.

Spontaneous beating of cardiomyocyte sheets: Cardiomyocyte sheets beat spontaneously *in vitro* and *in vivo* without stimulation. The fastest-beating cell naturally becomes a pacemaker cardiomyocyte, which depolarizes spontaneously and time-dependently, in the cell sheet. Action potentials are propagated from the pacemaker cell to other cardiomyocytes in the cell sheet, and the spontaneous beatings of cell sheets are observed.

Tissue engineering: A technology invented to overcome some of the limitations of injecting dissociated cell suspensions. In scaffold-based tissue engineering, biodegradable 3D polymers are used as scaffolds for 3D tissue-like grafts grown *in vitro*. Cell sheet engineering is a scaffold-free tissue engineering methodology that can make 3D tissues. The 3D grafts can be transplanted into animals and humans. Transplanted cell survival rates are significantly increased by tissue engineering compared to the direct cell injection method.

Summary Points

- The attachment and detachment of cells on PIPAAm-grafted surfaces can be controlled by the reversible hydrophilic/hydrophobic changes of this polymer with simple temperature alterations. These reversible temperature-dependent hydrophilic/hydrophobic changes are controlled by nanoscale alternations of the thicknesses of PIPAAm layers.
- Cell sheets recovered from temperature-responsive culture surfaces preserve nanoscale cell surface structures and ECM. ECM is important in fabricating 3D constructs without any scaffolds or "adhesive". The nanoscale cell surface structures, in particular ion channels and cell-to-cell junctions containing GJs, allow functional integration of grafts into host tissues.

- Two cardiomyocyte sheets become electrically synchronized rapidly (approximately 40 min). The rapid coupling is thought to be related to the preservation of GJ precursors on cardiomyocyte sheets.
- Electrical coupling between separated cardiomyocytes can also be analyzed using cell sheet engineering. Long distances between two cardiomyocyte sheets induce conduction delay or block. Use of cell sheets may allow analysis of arrhythmia mechanisms.
- Polysurgery (repeated transplantation of layered cell sheets onto host tissues) with cardiomyocyte sheets allows us to create viable and functional myocardial tissue, like real heart tissue. Transplantation of cell sheet onto heart tissue leads to functional coupling between transplanted graft and heart.
- Transplantation of cell sheets can improve cardiac function in various animal models of heart disease. The therapeutic effects of transplanted cell sheets were greater than those achieved by injection of cell suspensions.
- Cell sheet engineering with preserved surface chemistry and nanoscale structures has enormous potential in electrophysiology and regenerative medicine and holds great promise to help patients suffering from heart failure and other forms of heart disease.

Acknowledgments

We appreciate the useful comments and technical criticism of Dr. Norio Ueno (Institute of Advanced Biomedical Engineering and Science, TWIns, Tokyo Women's Medical University). This work was supported by grants from the Center of Excellence (COE) Program for the 21st Century, the Global COE Program, Multidisciplinary Education and Research Center for Regenerative Medicine (MERCREM), and the High-Tech Research Center Program from the Ministry of Education, Culture, Sports Science, and Technology (MEXST), Japan.

Abbreviations

Cx43	:	connexin43
3D	:	three dimensional
ECM	:	extracellular matrix
EDTA	:	ethylenediaminetetraacetic acid
GJ	:	gap junction
IPAAm	:	*N*-isoproplyacrylamide
LCST	:	lower critical solution temperature
PIPAAm	:	poly(*N*-isoproplyacrylamide)
TCPS	:	tissue culture polystyrene

References

Akins, R.E. 2002. Can tissue engineering mend broken heart? Circ. Res. 90: 120–122.

Akiyama, Y., A. Kikuchi, M. Yamato and T. Okano. 2004. Ultrathin poly(*N*-isopropylacrylamide) grafted layer on polystyrene surfaces for cell adhesion/detachment control. Langmuir 20: 5506–5511.

Haraguchi, Y., T. Shimizu, M. Yamato, A. Kikuchi and T. Okano. 2006. Electrical coupling of cardiomyocyte sheets occurs rapidly via functional gap junction formation. Biomaterials 27: 4765–4774.

Haraguchi, Y., T. Shimizu, M. Yamato and T. Okano. 2010. Electrical interaction between cardiomyocyte sheets separated by non-cardiomyocyte sheets in heterogeneous tissues. J. Tissue Eng. Regen. Med. 4: 291–299.

Hata, H., G. Matsumiya, S. Miyagawa, H. Kondoh, N. Kawaguchi, N. Matsuura, T. Shimizu, T. Okano, H. Matsuda and Y. Sawa. 2006. Grafted skeletal myoblast sheets attenuate myocardial remodeling in pacing-induced canine heart failure model. J. Thorac. Cardiovasc. Surg. 132: 918–924.

Heskins, M., J.E. Guillet and E. James. 1968. Solution properties of poly(*N*-isopropylacrylamide). J. Macromol. Chem. A2: 1441–1445.

Hida, N., N. Nishiyama, S. Miyoshi, S. Kira, K. Segawa, T. Uyama, T. Mori, K. Miyado, Y. Ikegami, C. Cui, T. Kiyono, S. Kyo, T. Shimizu, T. Okano, M. Sakamoto, S. Ogawa and A. Umezawa. 2008. Novel cardiac precursor-like cells from human menstrual blood-derived mesenchymal cells. Stem Cells 26: 1695–1704.

Kondoh, H., Y. Sawa, S. Miyagawa, S. Sakakida-Kitagawa, I.A. Memon, N. Kawaguchi, N. Matsuura, T. Shimizu, T. Okano and H. Matsuda. 2006. Longer preservation of cardiac performance by sheet-shaped myoblast implantation in dilated cardiomyopathic hamsters. Cardiovasc. Res. 69: 466–475.

Kushida, A., M. Yamato, C. Konno, A. Kikuchi, Y. Sakurai and T. Okano. 1999. Decrease in culture temperature releases monolayer endothelial cell sheets together with deposited fibronectin matrix from temperature-responsive culture surfaces. J. Biomed. Mater. Res. 45: 355–362.

Lampe, P.D. 1994. Analyzing phorbol ester effects on gap junctional communication: A dramatic inhibition of assembly. J. Cell Biol. 127: 1895–1905.

Langer, R. and J.P. Vacanti. 1993. Tissue engineering. Science 260: 920–926.

Masuda, S., T. Shimizu, M. Yamato and T. Okano. 2008. Cell sheet engineering for heart tissue repair. Adv. Drug. Deliv. Rev. 60: 277–285.

Matsuda, N., T. Shimizu, M. Yamato and T. Okano. 2007. Tissue engineering based on cell sheet technology. Adv. Mater. 19: 3089–3099.

Memon, I.A., Y. Sawa, N. Fukushima, G. Matsumiya, S. Miyagawa, S. Taketani, S.K. Sakakida, H. Kondoh, A.N. Aleshin, T. Shimizu, T. Okano and H. Matsuda. 2005. Repair of impaired myocardium by means of implantation of engineered autologous myoblast sheets. J. Thorac. Cardiovasc. Surg. 130: 1333–1341.

Menasche, P., A.A. Hagege, M. Scorsin, B. Pouzet, M. Desnos, D. Duboc, K. Schwartz, J.T. Vilquin and J.P. Marolleau. 2001. Myoblast transplantation for heart failure. Lancet 357: 279–280

Miyagawa, S., A. Saito, T. Sakaguchi, Y. Yoshikawa, T. Yamauchi, Y. Imanishi, N. Kawaguchi, N. Teramoto, N. Matsuura, H. Iida, T. Shimizu, T. Okano and Y. Sawa. 2010. Impaired myocardium regeneration with skeletal cell sheets—A preclinical trial for tissue-engineered regeneration therapy. Transplantation 90. 364–372.

Miyagawa, S., Y. Sawa, S. Sakakida, S. Taketani, H. Kondoh, I.A. Memon, Y. Imanishi, T. Shimizu, T. Okano and H. Matsuda. 2005. Tissue cardiomyoplasty using bioengineered contractile cardiomyocyte sheets to repair damaged myocardium: their integration with recipient myocardium. Transplantation 80: 1586–1595.

Miyahara, Y., N. Nagaya, M. Kataoka, B. Yanagawa, K. Tanaka, H. Hao, K. Ishino, H. Ishida, T. Shimizu, K. Kangawa, S. Sano, T. Okano, S. Kitamura and H. Mori. 2006. Monolayered mesenchymal stem cells repair scarred myocardium after myocardial infarction. Nat. Med. 12: 459–465.

Okano, T., H. Yamada, H. Sakai and Y. Sakurai. 1993. A novel recovery system for cultured cells using plasma-treated polystyrene dishes grafted with poly (N-isopropylacrylamide). J. Biomed. Mater. Res. 27: 1243–1251.

Oyamada, M., H. Kimura, Y. Oyamada, A. Miyamoto, H. Ohshika and M. Mori. 1994. The expression, phosphorylation, and localization of connexin 43 and gap-junctional intercellular communication during the establishment of a synchronized contraction of cultured neonatal rat cardiac myocytes. Exp. Cell Res. 212: 351–358.

Sakai, H., Y. Doi, T. Okano, N. Yamada and Y. Sakurai. 1996. Thermo-responsive polymer surfaces for cell culture: analysis of the surfaces and control of the cell attachment/detachment. In: N. Ogata, S.W. Kim, J. Feijen and T. Okano (eds.), Advanced Biomaterials in Biomedical Engineering and Drug Delivery Systems. Springer, Tokyo, pp. 229–230.

Sekine, H., T. Shimizu, S. Kosaka, E. Kobayashi and T. Okano. 2006. Cardiomyocyte bridging between hearts and bioengineered myocardial tissues with mesenchymal transition of mesothelial cells. J. Heart Lung Transplant. 25: 324–332.

Sekiya, N., G. Matsumiya, S. Miyagawa, A. Saito, T. Shimizu, T. Okano, N. Kawaguchi, N. Matsuura and Y. Sawa. 2009. Layered implantation of myoblast sheets attenuates adverse cardiac remodeling of the infarcted heart. J. Thorac. Cardiovasc. Surg. 138: 985–993.

Shimizu, T., H. Sekine, J. Yang, Y. Isoi, M. Yamato, A. Kikuchi, E. Kobayashi and T. Okano. 2006. Polysurgery of cell sheet grafts overcomes diffusion limits to produce thick, vascularized myocardial tissues. FASEB J. 20: 708–710.

Shimizu, T., M. Yamato, A. Kikuchi and T. Okano. 2003. Cell sheet engineering for myocardial tissue reconstruction. Biomaterials 24: 2309–2316.

Tadvalkar, G. and P. Silva. 1983. *In vivo*, rapid assembly of gap junctions is induced by cytoskeleton disruptors. J. Cell Biol. 96: 1279–1287.

Yamada, N., T. Okano, H. Sakai, F. Karikusa, Y. Sawasaki and Y. Sakurai. 1990. Thermo-responsive polymeric surface: control of attachment and detachment of cultured cells. Makromol. Chem. Rapid Commun. 11: 571–576.

Zhang, M., D. Methot, V. Poppa, Y. Fujio, K. Walsh and C.E. Murry. 2001. Cardiomyocyte grafting for cardiac repair: graft cell death and anti-death strategies. J. Mol. Cell Cardiol. 33: 907–921.

3

Nanowire Field-effect Transistors and Their Applications to Cardiology

Chia-Chang Tsai,[1] *Colin R. Martin,*[2] *Yen-Bin Liu,*[3]
Chien-Yuan Pan[4,]* and *Yit-Tsong Chen*[5,]*

ABSTRACT

The heart beat is fundamental to maintaining the circulation system to support life and homeostasis. The rhythmic heart contraction is a self-generating process that involves the action potential propagation throughout the heart to facilitate cardiac cell contraction. This activity can be monitored to reveal the

[1]Institute of Atomic and Molecular Sciences, Academia Sinica, P.O. Box 23-166, Taipei 106, Taiwan; E-mail: ccjohntsai@gmail.com

[2]Chair in Mental Health, School of Health, Nursing and Midwifery, University of the West of Scotland, Ayr Campus, Beech Grove, Ayr, KA8 0SR; E-mail: colin.martin@uws.ac.uk

[3]Division of Cardiology, Department of Internal Medicine, National Taiwan University Hospital, Taipei, Taiwan; E-mail: yenbin@ntu.edu.tw

[4]Institute of Zoology & Department of Life Science, National Taiwan University, 1 Sec. 4, Roosevelt Road, Taipei 106, Taiwan; E-mail: cypan@ntu.edu.tw

[5]Department of Chemistry, National Taiwan University, And Institute of Atomic and Molecular Sciences, Academia Sinica, 1 Sec. 4, Roosevelt Road, Taipei 106, Taiwan; E-mail: ytcchem@ntu.edu.tw

*Corresponding authors

List of abbreviations after the text.

health status of the heart. Established monitoring techniques detect the overall response from the heart; however, detailed local information, especially at the single cell level, is not available. This review introduces the characteristics of nanoscale field-effect transistor (FET) and discusses the applications of this newly developed nanotechnology to monitor cardiac responses. The conductivity of FET is highly sensitive to the electrical potential and ionic strength of the measuring environment. When the dimensions of an FET are reduced to a nanoscale level, its sensitivity can be greatly enhanced to detect molecular interaction and extracellular potential. FET-based sensors have been used to determine the binding constant and interacting kinetics of proteins. The proteins secreted from a single cell can be detected by the specific antibody modified on the surface of the FET, thus allowing the FET sensor to probe the proteins released from damaged cardiac cells for an early diagnosis of cardiac dysfunction. The nanoscale FET can also detect the electric potential changes induced by a single cardiac cell. The device can be applied to monitor the cardiac contraction to construct a three-dimensional image of the heart action potential firing profile and to characterize the abnormality due to local damages or ion channel defects. In future, nanoscale FET will be an important diagnostic tool for the accurate identification of cardiac dysfunction.

INTRODUCTION

Membrane potential is a special property of biological cells. Maintaining this potential requires the unequal distribution of certain ions across the plasma membrane and is energy intensive. The relative concentration of Na^+ is higher in the extracellular milieu than inside the cell; in contrast, K^+ concentration is higher inside the cell. At rest, the plasma membrane is slightly permeable to K^+ and the efflux of K^+ makes the membrane potential negative. When cells are excited, Na^+ channels are opened and the influx of Na^+ depolarizes the membrane potential. More K^+ channels are then opened to let K^+ flux out of the cell and repolarize the potential. Rapid changes in membrane potential constitute the basic profile of an action potential and are essential to many systems for signal transmission or activity synchronization. Other than the Na^+ and K^+ channels, many other types of ion channels and transporters are involved in shaping the action potential profile (Rudy and Silva 2006).

The heart beat is a self-generation process that starts with the action potential firing of the pacemaker or the so-called sino-astrial node at the top of the right atrium, which then causes the contract of both atria. The action potential signal spreads through atria and then triggers the atrioventricular node to relay the signal to the apex of the heart. This signal then induces the ventricle contraction to pump blood out of the heart. The process needs the coordination of muscle cells at the atria and ventricles to contract accordingly. Since many cells contract in a short interval (1–2 sec), the summed changes in the membrane potentials can be detected at the surface of the chest as the electrocardiogram (ECG) (Chen et al. 2010).

The ECG pattern provides important information about how the heart contraction is coordinated and synchronized. Abnormal heart contraction can be diagnosed by the change in the ECG profile. Mutations in the Na^+ channel conductivity will change the duration of the action potential and QT interval (Amin et al. 2010). Most recording techniques monitor the overall ECG responses; a local defect due to the malfunction of a few cells and the action potential profile changed by ion channel dysfunction cannot be elucidated (Goldenberg et al. 2008). Therefore, new electric recording techniques with submicron resolution are required to evaluate the health condition of the heart at single cell level.

Tools have been developed to obtain detailed spatial information for biological studies; e.g., the techniques based on scanning probe microscopy (Cho et al. 2002; Schneider et al. 1997; Tsai et al. 2008) provide surface image at nanometer resolution. One-dimensional nanotubes (NTs) and nanowires (NWs) have large surface/volume ratios and can be used to configure FET devices with ultra-high detection sensitivity. For instance, carbon NT-FET (CNT-FET) and silicon NW-FET (SiNW-FET) are extremely sensitive to the changes of electric field on the NT/NW surface, when CNT-FET and SiNW-FET are subject to binding with even a small amount of target molecules. To date, CNT-FET and SiNW-FET have been widely used as biosensors for chemical analysis, biomedical diagnosis, and cellular research, because of their high sensitivity, selectivity, rapid response time, small analyte volume requirements, and biological compatibility (Cui et al. 2001; Patolsky et al. 2006a; Wang et al. 2007; Tsai et al. 2008; Lin et al. 2009, 2010). This chapter discusses the applications of CNT-FET and SiNW-FET in biological studies, with an emphasis on cardiology.

FIELD-EFFECT TRANSISTOR–BASED BIOSENSORS

An FET device is usually composed of three terminals: source, drain, and gate electrodes, in order to control the current of a semiconducting channel between the source and drain electrodes via the electric field applied through the gate electrode. The major charge carriers of an FET in the semiconducting channel can be electrons or holes, corresponding to an n-type or p-type FET device. For biosensory applications, biological receptors are chemically modified on the FET surface to recognize target analytes through their specific and strong interactions. As a consequence, a gating effect is generated because of the target-receptor interaction, which then varies the surface potential and channel conductance of the FET. These conductance changes inside the FET can finally be collected by a detection system.

FET-based biosensors have been employed for various biological applications and can be classified as enzyme-modified, cell-based, and immunologically functionalized FETs. Enzyme-modified FET comprises a redox active enzyme integrated with an electronic circuitry to give a real-time quantitative analysis of the enzyme substrate, e.g., sensing glucose from a catalytic reaction in the presence of glucose oxidase. Cell-based FETs were exploited to detect the released biochemical agents or real-time cellular responses from living cells, such as action potentials from neuron cells or electrical recordings from chicken hearts. The most frequently used biosensors are immunologically functionalized FETs, linking an antibody to the FET to detect the corresponding antigen.

Since their early applications as a sensory device (Tans et al. 1998), CNT-FETs have been widely used for the detections of gases (Collins et al. 2000; Kong et al. 2000), chemicals (Star et al. 2003a), proteins (Star et al. 2003b), DNA (Kim et al. 2009), and cells (Tsai et al. 2008; Wang et al. 2007). We have used CNT-FET to detect chromogranin A released from neurons (Wang et al. 2007) and single chromaffin cells (Tsai et al. 2008) with a detection sensitivity at sub-nanomolar level. Our results demonstrated that CNT-FET is a powerful sensor with high selectivity and sensitivity for biosensing detections. However, the sensing mechanism of a CNT-FET has been reported to be related to field effect, electron transfer (Gruner 2006), and Schottky barrier (Heinze et al. 2002). In contrast, the sensing mechanism of a SiNW-FET is simply dominated by the field effect caused by the interaction between target molecules and the receptor modified on the surface of SiNW-FET (Patolsky et al. 2006a). In this section, the discussion is focused on SiNW-FET.

Fabrication of SiNW-FETs

The SiNW-FET devices can be fabricated using a "bottom-up" (Patolsky et al. 2006a) or "top-down" (Li et al. 2004) technique. With the "bottom-up" method, SiNWs can be synthesized via the vapor-liquid-solid growing mechanism in a chemical vapor deposition reaction. The synthesis is usually assisted with metal nanoparticles that not only catalyze the SiNW formation, but also control the size of the as-synthesized SiNWs. Subsequently, the as-synthesized SiNWs are suspended in ethanol solution, dispersed onto a silicon substrate, and then electrically connected with electrodes in a standard lithographical procedure (Fig. 1a-b). Shown in Fig. 1c is the atomic force microscopic (AFM) topograph of an as-fabricated

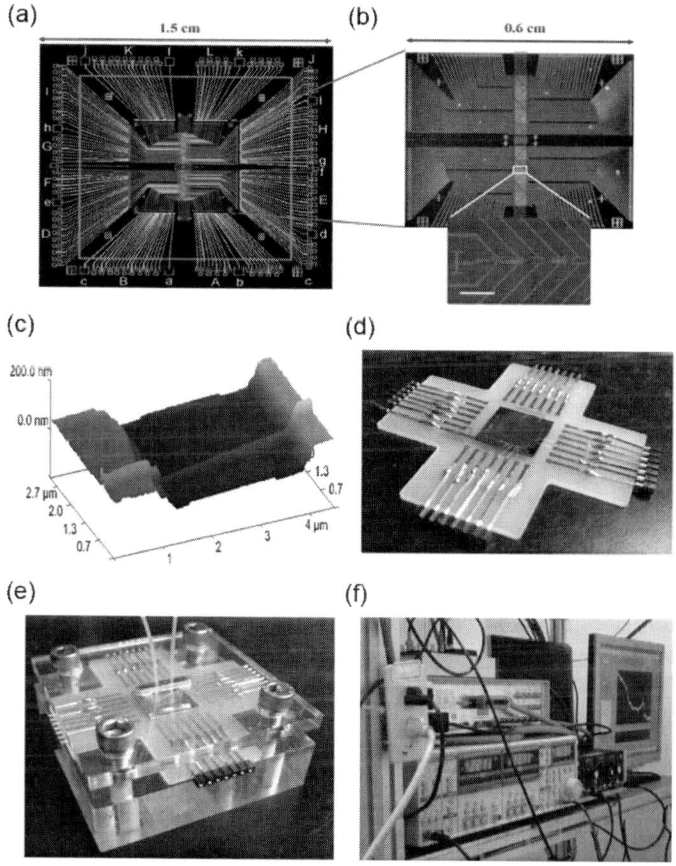

Fig. 1. Device fabrication and electrical measurement of SiNW-FETs in biosensory applications. (a) Mask design for the photolithographic fabrication of SiNW-FET device arrays. (b) Device arrays on a magnified scale. Top: image of the circuits in the area of the yellow square in (a); bottom: SEM image of a SiNW-FET array with a source-drain separation of 2 µm. The scale bar is 50 µm. In total, there are 180 pairs of source-drain electrodes on the silicon chip. (c) The topograph of a SiNW-FET scanned by AFM. A SiNW of ~50 nm in diameter is connected by two Ni/Al (70 nm/100 nm thick) electrodes of ~2 µm in separation. (d) The SiNW-FET device array on a silicon chip ($1.5 \times 1.5\ mm^2$) is connected to a plastic circuit board with aluminum wires (~30 µm in diameter). (e) A sample solution is delivered onto the SiNW-FET arrays through a polydimethylsiloxane microfluidic channel ($6.25 \times 0.5 \times 0.05\ mm^3$), which is designed to couple with the SiNW-FET arrays. (f) The variation of electrical signals is monitored by a detection system that combined a lock-in amplifier and a current preamplifier. (Wang et al. 2007. Reproduced with permission.)

Color image of this figure appears in the color plate section at the end of the book.

SiNW-FET, in which a SiNW of ~50 nm in diameter is connected by two Ni/Al electrodes of ~2 μm in separation. The SiNW surface can then be chemically modified with receptor molecules. The suitable SiNW-FET devices with decent modification of receptors are selected to be electrically connected to a plastic circuit board with ~30 μm diameter aluminum wires (Fig. 1d). A sample solution is delivered to the SiNW-FET array through a polydimethylsiloxane microfluidic channel (Fig. 1e) assisted with a syringe pump. Finally, the signal change due to the interaction between receptors and target molecules is recorded by a lock-in detection system (Fig. 1f).

Reusable SiNW-FET System

Application of SiNW-FETs to detect a particular target (e.g., an antigen) in biomedical diagnosis, the corresponding receptor (e.g., the antibody) is modified on the SiNW-FET surface prior to the detection. By virtue of the strong and specific binding affinity between antigen and antibody under normal physiological conditions, the receptor-modified SiNW-FET can serve as an extremely sensitive sensor with high selectivity. By the same token, because of this strong binding between the antigen and antibody, it is difficult to remove the antigen-antibody complex from the surface of SiNW-FET after detection, meaning that a SiNW-FET could be used only for a single measurement. With this limitation, consecutively quantitative analysis by a calibratable SiNW-FET is difficult to achieve.

To solve this problem, we developed a novel method to make the SiNW-FET devices reusable by taking advantage of the reversible interaction between glutathione (GSH) and glutathione S-transferase (GST), which has long been applied in protein purification. Taking our previous work as a representative example (Fig. 2) (Lin et al. 2009, 2010), 3-(aminopropyl) trimethoxysilane (APTMS) and 3-maleimidobenzoic acid N-hydroxysuccinimide ester (MBS) are two successive linkers that were chosen to immobilize GSH on a SiNW-FET. A self-assembled monolayer of APTMS is first modified to the silicon-oxide sheath on the SiNW; MBS then bonds APTMS with the formation of an amide bond. Subsequently, GSH is immobilized on the MBS-modified SiNW-FET via the reaction between the sulfhydryl group of GSH and the maleimide group of MBS, the product of which is referred to as GSH/SiNW-FET hereafter. This GSH/SiNW-FET is then ready to associate with a particular GST-fused protein (referred to as protein-GST) to screen interacting proteins. As shown in Fig. 2, in the end of each sensing measurement, the used protein-GST could be removed with ≥1 mM GSH washing solution, making the GSH/SiNW-FET return to its original state. As a result, this reversible association-dissociation between GSH and GST allows us to use a single SiNW-FET device for repetitive measurements.

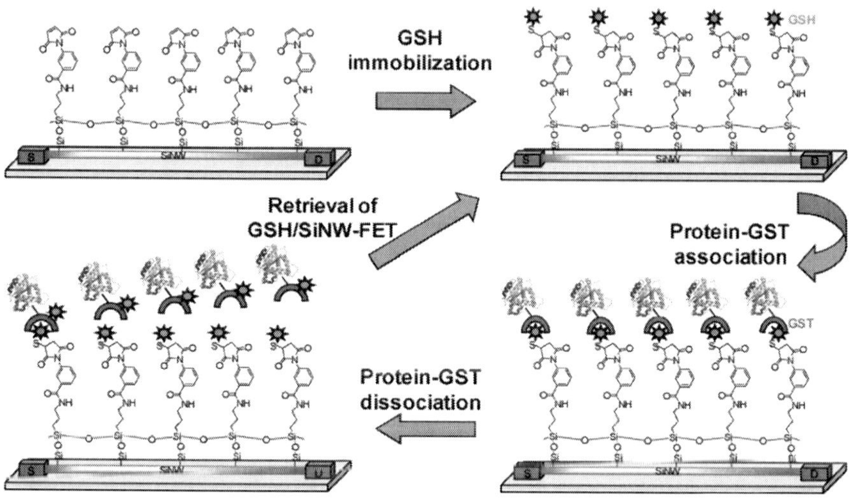

Fig. 2. Reversible high-throughput screening design of a reusable SiNW-FET biosensor. Upper left: The SiNW-FET is first modified with linkers prior to the immobilization of glutathione (GSH). Upper right and lower right: A protein fused with glutathione S-transferase (GST) is associated with the GSH/SiNW-FET to immobilize the protein on the SiNW-FET surface. Lower left: In the end of each measurement, the used protein-GSTs are removed with ≥1 mM GSH washing solution, making the GSH/SiNW-FET return to its original state. (Lin et al. 2009. Reproduced with permission.)

Color image of this figure appears in the color plate section at the end of the book.

Application of SiNW-FET in Biomolecular Detection

We have demonstrated that the calmodulin-modified SiNW-FET (referred to as CaM/SiNW-FET) exhibits selective electrical response to Ca^{2+} (>1 μM) and purified cardiac troponin I (~ 7 nM) (Lin et al. 2010). The conductivity change of CaM/SiNW-FET displays a linear dependence on the logarithm of the concentration of troponin I in a range from 10 nM to 1 μM. Furthermore, the CaM/SiNW-FET can also recognize the N-type voltage-gated Ca^{2+} channels (VGCC) that were expressed by cultured 293T cells as shown in Fig. 3.

N-type VGCCs located at the plasma membrane mediate the entry of Ca^{2+} into cells in response to membrane depolarization. It has been suggested that the Ca^{2+}-dependent inactivation (CDI) of several Ca^{2+} channel subtypes is mediated by the binding of CaM to the C-termini of these channels. To characterize the effect of CaM on N-type VGCC CDI, rat CaM is cloned and co-expressed with the N-type VGCC in 293T cells. Transfected cells are voltage-clamped in a whole-cell configuration at –70 mV and depolarized to +20 mV to evoke Ca^{2+} currents. Figure 3(a) shows that the representative current in cells expressing CaM decreases to ~15%

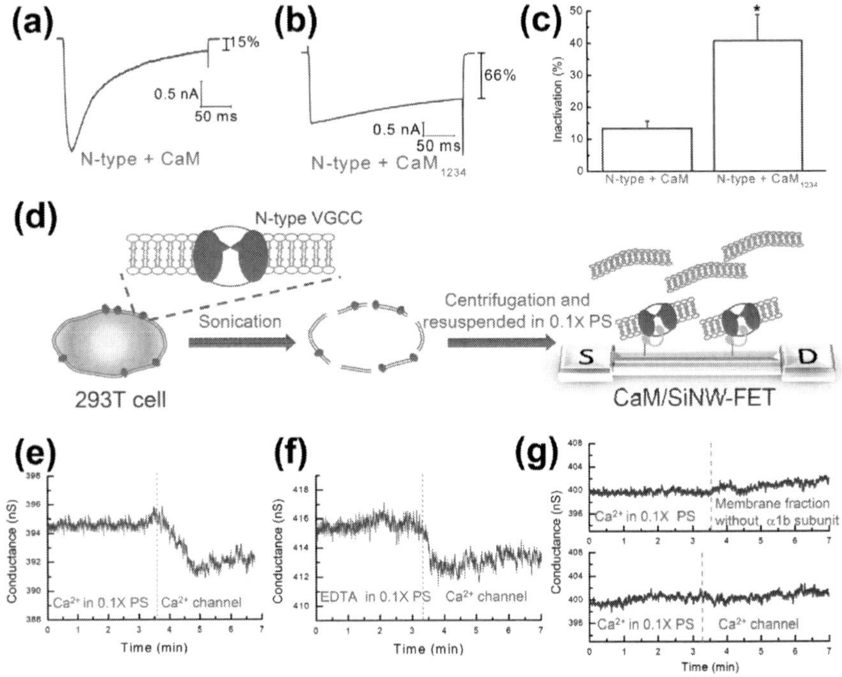

Fig. 3. Detection of N-type Ca^{2+} channels with a CaM/SiNW-FET. (a-c) Ca^{2+} currents in transfected 293T cells were recorded using a whole-cell patch configuration. The cells were voltage-clamped at –70 mV and depolarized to +20 mV for 250 ms to activate N-type VGCCs. Representative current traces from cells co-expressing N-type VGCCs with (a) CaM and (b) CaM_{1234} (a Ca^{2+} binding deficient mutant) are shown. Inactivation was obtained by normalizing the average currents between the 225 and 250 ms to the peak current during depolarization. (c) The bar graph depicts the average normalized inactivation values for cells expressing the different CaM constructs. The data represent the mean ± SEM. The sample numbers for N-type + CaM and N-type + CaM_{1234} are 7 and 5, respectively. *p value of < 0.05, as obtained by Student's t test. (d) Scheme for the detection of membrane fractions containing N-type VGCCs by CaM/SiNW-FET. (e-f) The electrical measurements represent the real-time detection of the binding of an N-type Ca^{2+} channel to the CaM/SiNW-FET in 0.1× PS (i.e., 0.1× phosphate solution, consisting of 0.76 mM Na_2HPO_4 and 0.24 mM NaH_2PO_4 at pH 7.4) supplemented with (e) 10^{-4} M Ca^{2+} or (f) 0.5 mM EDTA. (g) Two control experiments were conducted separately. Top graph: Sensing of the membrane fraction without the α1b subunit by the CaM/SiNW-FET in 0.1× PS supplemented with 10^{-4} M Ca^{2+}. Bottom graph: Sensing of a Ca^{2+} channel with a GST/SiNW-FET in 0.1× PS supplemented with 10^{-4} M Ca^{2+}. The protein concentration for the sensing experiments was 1.7 μg/μL (Lin et al. 2010. Reproduced with permission.)

Color image of this figure appears in the color plate section at the end of the book.

of the maximal current at the end of depolarization. On the contrary, this decline is ~66% in cells expressing CaM_{1234}, a mutant of CaM that lacks the ability to bind Ca^{2+} (Fig. 3b). As shown in Fig. 3c, the mean level of inactivation is 13.3 ± 2.3% (n = 7) in cells expressing wild-type CaM and 40.7 ± 8.1% (n = 5) in cells expressing CaM_{1234}. These results suggest that CaM is essential for the CDI of N-type VGCCs.

Figure 3d shows how N-type VGCCs can be detected with the CaM/SiNW-FET under physiological conditions. The N-type VGCCs prepared from 293T cells can be detected by CaM/SiNW-FET in the presence (Fig. 3e) or absence (Fig. 3f) of Ca^{2+}. These results are consistent with those reported earlier. Furthermore, the control experiments (Fig. 3g) convince us that the conductance of CaM/SiNW-FET is Ca^{2+} channel–specific and that CaM is essential for the detection of N-type VGCCs. These results demonstrated that SiNW-FET can be used as a biosensor for biomolecular detections under physiological conditions.

APPLICATION OF SINW-FET IN CARDIOLOGY

In addition to molecular recognition, SiNW-FET can be applied to detect the electric field generated by excitable cells such as neurons (Patolsky et al. 2006b) and cardiac cells (Offenhausser and Knoll 2001). Timko et al. have used SiNW-FET arrays as a powerful tool to study the beating of embryonic chicken hearts, which is an important achievement for the application of SiNW-FET in cardiology (Timko et al. 2009). In their experiment, they placed a freshly isolated heart on the top of a SiNW-FET device, where the electronic signals of the SiNW-FET corresponding to heart beat were recorded. From their demonstration, the SiNW-FET arrays fabricated with sub-10 μm spacing can be used in the electrical measurements with subcellular resolution.

To characterize the electrical activity over a heart, electromagnetic mapping systems (CARTO®, Biosense Webster Inc., U.S.A.) and three-dimensional navigation systems (EnSite®, St. Jude Medical, Inc., U.S.A.) have been developed. These techniques provide the electric information in a three-dimensional mode for physicians to evaluate the condition of the heart. Such monitoring can only be used in invasive electrophysiological study by point-to-point mapping and cannot be used as a long-term recording device. Additionally, such techniques have a spatial resolution at millimeter level and could not provide detailed action potential information at single cell level.

Offenhausser's group demonstrated that FET can be applied as a diagnostic tool to investigate the action potential of rat cardiac myocytes and the influence of drugs on heartbeat frequency (Offenhausser and Knoll 2001). Compared with short-term measurements, such as isolated whole organs (e.g., Langendorff preparation) and electrophysiological techniques (e.g., patch-clamp recordings), FET provides a long-term recording tool for drug test of living cells. Furthermore, NT/NW-FET also provides high temporal and spatial resolutions in characterizing the electric activity of heart at single cell level. Different channelopathies affect the action potential at various stages and have differential effects on the ECG profile (Szeliga et al. 2010; Campuzano et al. 2010). For specific points of interest, such as bundle of His, scar tissue, or arrhythmogenic foci, NT/NW-FETs can be used to elucidate the regional variation in signal transmission at single cell level. Also, NT/NW-FETs are able to offer new mechanistic insights into the electrical properties of the normal and diseased heart. In future, combined with the implantation technique and remote wireless sensor engineering, NT/NW-FETs may have their clinical implications in the early detection of pathological condition of cardiac tissue, such as arrhythmias and myocardial ischemia.

Summary Points

- The conductivity of FET is highly sensitive to the charges and potential on the sensor surface, which can be from environmental solution or approaching molecules.
- An FET sensor manufactured at nanometer scale has a very high sensitivity to detect surrounding molecules. However, the Debye-Huckel screening length caused by environmental electrolytic solution needs to be considered in the electrical measurements.
- NT/NW-FETs can monitor not only the interaction among biomolecules, but also the extracellular potential generated by a single excitable cell.
- Serum containing specific proteins released from damaged cardiac cells can be detected by NT/NW-FETs as an early diagnosis.
- By recording the change in the action potential firing profile with NT/NW-FETs at single cell level, the mutations in ion channels may be elucidated.
- NT/NW-FETs have high temporal and spatial resolutions and can be applied to examine local damage.

Definitions

Action potential: A quick change in membrane potential conducted by the changes in the membrane permeation to various ions.

Channelopathy: Diseases caused by mutations which interfere with the ion channel activities; these mutations can be in the ion channel main subunits or proteins that regulate ion channel functions.

Electric field potential: The fluxes of ions across the plasma membrane will not only change the membrane potential, but also induce a small change at the extracellular potential near the cell.

Ion channel: Membrane protein with specific pore-forming structure.

Nanowire: One-dimensional semiconductor material of <100 nm in diameter which is used as the conduction channel of a NW-FET.

Key Facts

- FET relies on an external electric field to control the conductivity of a channel of n- or p-type charge carrier in a semiconductor material.
- When the dimensions of an FET are reduced to a nanometer scale with comparable sizes to biomolecules, the sensitivity of the nanodevice, as with CNT-FET and SiNW-FET, is vastly enhanced because of the large surface/volume ratio in the sensing element, thus having great potential in ultrasensitive biological detection.
- When an FET is employed in biosensing measurements, the conductivity is affected by environmental parameters such as the buffer ionic strength, the molecular compositions, and surface potential of the FET.
- When an FET is employed in biosensing measurements, the screening effect on the FET detection due to environmental electrolytic buffer needs to be considered. The binding site on the FET surface, where target-receptor interaction occurs, is usually designated to be no longer than the Debye-Huckel length of the electrolytic buffer.

Abbreviations

APTMS	:	3-(aminopropyl) trimethoxysilane
CaM	:	calmodulin
CDI	:	Ca^{2+}-dependent inactivation
CNT-FET	:	carbon nanotube field effect transistor
ECG	:	electrocardiogram
FET	:	field-effect transistor
GSH	:	glutathione
GST	:	glutathione S-transferase
MBS	:	3-maleimidobenzoic acid N-hydroxysuccinimide ester
NT	:	nanotube
NW	:	nanowire
NT/NW -FET	:	nanotube/nanowire based field-effect transistors
SiNW-FET	:	silicon nanowire field-effect transistor
VGCC	:	voltage-gated Ca^{2+} channel

References

Amin, A.S., A. Asghari-Roodsari and H.L. Tan. 2010. Cardiac sodium channelopathies. Pflugers Arch. 460: 223–237.

Campuzano, O., P. Beltran-Alvarez, A. Iglesias, F. Scornik, G. Perez and R. Brugada. 2010. Genetics and cardiac channelopathies. Gene. Med. 12: 260–267.

Chen, P.S., B. Joung, T. Shinohara, M. Das, Z. Chen and S.F. Lin. 2010. The initiation of the heart beat. Circ. J. 74: 221–225.

Cho, S.J., K. Jeftinija, A. Glavaski, S. Jeftinija, B.P. Jena and L.L. Anderson. 2002. Structure and dynamics of the fusion pores in live GH-secreting cells revealed using atomic force microscopy. Endocrinology 143: 1144–1148.

Collins, P.G., K. Bradley, M. Ishigami and A. Zettl. 2000. Extreme oxygen sensitivity of electronic properties of carbon nanotubes. Science 287: 1801–1804.

Cui, Y., Q.Q. Wei, H.K. Park and C.M. Lieber. 2001. Nanowire nanosensors for highly sensitive and selective detection of biological and chemical species. Science 293: 1289–1292.

Goldenberg, I., W. Zareba and A.J. Moss. 2008. Long QT syndrome. Curr. Prob. Cardiol. 33: 629–694.

Gruner, G. 2006. Carbon nanotube transistors for biosensing applications. Anal. Bioanal. Chem. 384: 322–335.

Heinze, S., J. Tersoff, R. Martel, V. Derycke, J. Appenzeller and P. Avouris. 2002. Carbon nanotubes as Schottky barrier transistors. Phys. Rev. Lett. 89: 106801.

Kim, S., T.G. Kim, H.R. Byon, H.J. Shin, C. Ban and H.C. Choi. 2009. Recognition of single mismatched DNA using MutS-immobilized carbon nanotube field effect transistor devices. J. Phys. Chem. B 113: 12164–12168.

Kong, J., N.R. Franklin, C.W. Zhou, M.G. Chapline, S. Peng, K.J. Cho and H.J. Dai. 2000. Nanotube molecular wires as chemical sensors. Science 287: 622–625.

Li, Z., Y. Chen, X. Li, T.I. Kamins, K. Nauka and R.S. Williams. 2004. Sequence-specific label-free DNA sensors based on silicon nanowires. Nano Lett. 4: 245–247.

Lin, S.P., C.Y. Pan, K.C. Tseng, M.C. Lin, C.D. Chen, C.C. Tsai, S.H. Yu, Y.C. Sun, T.W. Lin and Y.T. Chen. 2009. A reversible surface functionalized nanowire transistor to study protein-protein interactions. Nano Today 4: 235–243.

Lin, T.W., P.J. Hsieh, C.L. Lin, Y.Y. Fang, J.X. Yang, C.C. Tsai, P.L. Chiang, C.Y. Pan and Y.T. Chen. 2010. Label-free detection of protein-protein interactions using a calmodulin-modified nanowire transistor. Proc. Nat. Acad. Sci. USA 107: 1047–1052.

Offenhausser, A., and W. Knoll. 2001. Cell-transistor hybrid systems and their potential applications. Trends Biotechnol. 19: 62–66.

Patolsky, F., G.F. Zheng and C.M. Lieber. 2006a. Fabrication of silicon nanowire devices for ultrasensitive, label-free, real-time detection of biological and chemical species. Nature Protocols 1: 1711–1724.

Patolsky, F., B.P. Timko, G.H. Yu, Y. Fang, A.B. Greytak, G.F. Zheng and C.M. Lieber. 2006b. Detection, stimulation, and inhibition of neuronal signals with high-density nanowire transistor arrays. Science 313: 1100–1104.

Rudy, Y., and J.R. Silva. 2006. Computational biology in the study of cardiac ion channels and cell electrophysiology. Q. Rev. Biophys. 39: 57–116.

Schneider, S.W., K.C. Sritharan, J.P. Geibel, H. Oberleithner and B.P. Jena. 1997. Surface dynamics in living acinar cells imaged by atomic force microscopy: Identification of plasma membrane structures involved in exocytosis. Proc. Nat. Acad. Sci. USA 94: 316–321.

Star, A., T.R. Han, J.C.P. Gabriel, K. Bradley and G. Gruner. 2003a. Interaction of aromatic compounds with carbon nanotubes: Correlation to the Hammett parameter of the substituent and measured carbon nanotube FET response. Nano Lett. 3: 1421–1423.

Star, A., J.C.P. Gabriel, K. Bradley and G. Gruner. 2003b. Electronic detection of specific protein binding using nanotube FET devices. Nano Lett. 3: 459–463.

Szeliga, M.A., P.L. Hedley, C.P. Green, D.V. Moller and M. Christiansen. 2010. Long QT syndrome—a genetic cardiac channelopathy. Kardiologia Polska 68: 575–583.
Tans, S.J., A.R.M. Verschueren and C. Dekker. 1998. Room-temperature transistor based on a single carbon nanotube. Nature 393: 49–52.
Timko, B.P., T. Cohen-Karni, G.H. Yu, Q. Qing, B.Z. Tian and C.M. Lieber. 2009. Electrical recording from hearts with flexible nanowire device arrays. Nano Letters 9: 914–918.
Tsai, C.C., C.C. Yang, P.Y. Shih, C.S. Wu, C.D. Chen, C.Y. Pan and Y.T. Chen. 2008. Exocytosis of a single bovine adrenal chromaffin cell: The electrical and morphological studies. J. Phys. Chem. B 112: 9165–9173.
Wang, C.W., C.Y. Pan, H.C. Wu, P.Y. Shih, C.C. Tsai, K.T. Liao, L.L. Lu, W.H. Hsieh, C.D. Chen and Y.T. Chen. 2007. *In situ* detection of chromogranin a released from living neurons with a single-walled carbon-nanotube field-effect transistor. Small 3: 1350–1355.

4

Atomic Force Microscopy and the Detection of Nanosized Blood Microparticles

Y. Yuana,[1,a] M.E. Kuil,[2] T.H. Oosterkamp,[3] R.M. Bertina[4] and S. Osanto[1,b,*]

ABSTRACT

Microparticles are small membrane vesicles, also named microvesicles, that are released from virtually all cells of the body upon activation or apoptosis. The detection of proteins specific

[1]Department of Clinical Oncology, Leiden University Medical Centre, Albinusdreef 2, 2333 ZA Leiden, The Netherlands.
[a]E-mail: y.yuana@lumc.nl
[b]E-mail: s.osanto@lumc.nl
[2]Leiden Institute of Chemistry, Leiden University, Gorlaeus Laboratoria, Einsteinweg 55, 2333 CC Leiden, The Netherlands.
E-mail: m.kuil@chem.leidenuniv.nl
[3]Leiden Institute of Physics, Leiden University, Huygens Laboratorium, Niels Bohrweg 2, 2333 CA Leiden, The Netherlands.
E-mail: oosterkamp@physics.leidenuniv.nl
[4]Department of Thrombosis and Haemostasis, Einthoven Laboratory for Experimental Vascular Medicine, Leiden University Medical Centre, Albinusdreef 2, 2333 ZA Leiden, The Netherlands.
E-mail: r.m.bertina@lumc.nl
*Corresponding author

List of abbreviations after the text.

for the cells from which they are derived has allowed their quantification and characterization. Microparticles are found in plasma of healthy donors and derived from platelets, leucocytes, erythrocytes, endothelial cells, and malignant cells. The number of blood microparticles increases under pathological conditions with elevated numbers reported in patients with cardiovascular disorders, systemic inflammatory disease, and cancer.

In recent years it became clear that microparticles may serve as potential mediators of intercellular communication. Because of their lipids, proteins and nucleic acids they are able to stimulate diverse physiological processes such as thrombosis and activate immune cells, endothelial cells and fibroblasts to produce cytokines, cell adhesion molecules, growth factors and tissue factor (TF), and modulate endothelial functions. Endothelial microparticles (EMPs) have emerged as a biomarker of vascular dysfunction and also seem to play a prominent role in inflammation, vascular injury, and angiogenesis.

Enumeration of microparticles by conventional flow cytometer is hampered by the limitation of the wavelength of the laser light employed, precluding accurate assessment of microparticles smaller than 488 nm. Atomic force microscopy (AFM) provides imaging of three-dimensional topography of a surface at nanometre resolution. This method has been used to enumerate microparticles with diameter (d_{sph}) ranging from 10 to 475 nm, allowing the detection of 1,000-fold the numbers of platelet microparticles than flow cytometry does.

INTRODUCTION

The presence of MPs in plasma of healthy donors is well documented and elevated numbers of specific subsets of MPs have been reported in cardioavascular diseases, systemic inflammatory disease, and cancer. The large majority of circulating MPs are mostly derived from platelets, but other blood cells such as leucocytes, erythrocytes, endothelial cells and even malignant cells also may shed MPs. MPs have been shown to contain DNA, RNA, and recently also miRNA (Hunter et al. 2008; Chen et al. 2010), and intracellular proteins and surface markers from their parental cells.

At present, standardization of pre-analytical and analytical methods for the measurement of MPs remains a challenge. Pre-analytical steps, including blood collection, anticoagulant, centrifugation speed, freezing and thawing, need standardization to optimally study (*in vivo*) circulating MPs (Yuana et al. 2011). Different analytical methods and combinations of methods have been used to measure the number and concentration of MPs. Two common methods are flow cytometry and capture-based assays (Aupeix et al. 1997; Nomura et al. 2009). An advantage of flow cytometry is that it can be used for the simultaneous detection of two or more antigens on MPs by using antibodies conjugated with different chromophores (Burnier et al. 2009; Jy et al. 2004). However, the laser light employed in conventional flow cytometers excites at a wavelength of 488 nm. As MPs are heterogeneous and can have sizes below 488 nm, flow cytometry measurement of MPs is inaccurate (Furie and Furie 2006; Van Der Pol et al. 2010; Yuana et al. 2010). For capture-based assays, multiwell plates are coated with a probe (antibody or annexin V) that specifically binds a subset of MPs from plasma. Captured MPs can then be quantified/characterized using a second probe (e.g., peroxidase conjugated antibody) (Nomura et al. 2009) or by measurement of the prothrombinase activity (Ay et al. 2009; Aupeix et al. 1997). Although these capture-based assays enable a high-throughput measurement of subsets of MPs, they will give no quantitative information on the size and total number of these MPs in plasma. Furthermore, there are some specific problems (e.g., soluble antigen, specificity/affinity of antibody, the amount of antibody, phosphatidylserine (PS)-negative MPs) that will influence the number of MPs captured in these assays.

AFM has been used for decades to image the surface topography of (bio)particles with a high-resolution down to subnanometre range (Parot et al. 2007). The lateral resolution of AFM can vary from the atomic and molecular scale up to sizes larger than 125 μm and its vertical range can be up to 8–10 μm (Ricci and Braga 2004). This makes AFM more suitable for size detection and quantification of MP numbers than flow cytometry and capture-based assays. Another advantage of employing AFM in the study of MP is that AFM can be operated in fluid-tapping mode (Siedlecki et al. 1999; Yuana et al. 2010). AFM imaging in aqueous fluids preserves biological samples in their physiological state. AFM was first used to image MPs by Siedlecki et al. (1999). They used MPs generated from surface and thrombin-activated platelets.

Recently, our group reported the use of AFM operated in fluid-tapping mode to detect platelet microparticles (PMPs) isolated from platelet-poor plasma. AFM detects 1000-fold the number of PMPs than flow cytometry does, and the size of most of these MPs is below the detection limit of conventional flow cytometer.

HYPOTHESIS OF MP FORMATION

In the plasma membrane, the distribution of phospholipids (inward or "flip" or outward or "flop" translocation) is regulated by specific transporters. Under resting conditions, flippase activity is dominant and phospholipids, mainly PS and phosphatidylethanolamine, are sequestered in the inner leaflet of the cell membrane. After stimulation/activation, flippase activity is reduced and floppase activity markedly increases, resulting in transport of phospholipids to the outer leaflet of the cell membrane. The integrity of the cytoskeleton and calcium fluxes are also involved in phospholipid transbilayer fluxes. Membrane budding ultimately leads to the release of MPs that may expose PS. Many authors restrict the definition of MPs to 0.1- to 1-μm membrane fragments exposing PS and consequently exclude other vesicles that do not expose PS.

BIOLOGICAL RELEVANCE OF MP

In recent years it became clearer that MP may serve as potential powerful mediators of intercellular communication. Because of their lipids, proteins and nucleic acids they are able to stimulate diverse physiological processes such as thrombosis and activate immune cells, endothelial cells and fibroblasts to produce cytokines, cell-adhesion molecules, growth factors and TF, and modulate endothelial functions (Ratajczak et al. 2006; Piccin et al. 2007). EMPs have emerged as biomarker of vascular dysfunction and also seem to play a prominent role in inflammation, vascular injury, and angiogenesis.

The membrane of most blood MPs is rich in PS, providing a surface that can support procoagulant reactions. MPs also may carry TF, the major trigger of blood coagulation, which is normally not present in an active form in the circulation. However, under pathological conditions MPs may bear active TF on their surface and thus provide a favourable environment to initiate coagulation and support thrombogenesis.

MP IN CARDIOVASCULAR DISEASES

Under various pathological conditions, numbers of circulating MPs are increased. For instance, several cardiovascular diseases associated with inflammation (in particular EMPs) or thrombosis are accompanied by elevated levels of circulating MPs, suggesting that MPs are linked to endothelial dysfunction. Elevated levels of circulating EMPs originating from EMPs, which represent only a small fraction of the total number of blood MPs, have been reported in the acute coronary syndromes, ischemic stroke and severe hypertension with end organ damage, and in thrombotic thrombocytopenic purpura, which are all conditions associated with endothelial injury and pro-thrombotic state (Lynch and Ludlam 2007; Boulanger 2010).

EMPs may not only represent a biomarker of arterial damage but also serve as an active trigger of vascular repair. EMPs are a major component of human atherosclerotic plaques and may be an important mechanistic factor in the process of atherosclerosis. EMPs are also able to decrease nitric-oxide-dependent vasodilatation, increase arterial stiffness, promote inflammation, and initiate thrombosis.

The increased numbers of procoagulant MPs within the occluded coronary artery of patients suggests a pathophysiological role in coronary artery thrombosis. Morel et al. (2009) studied a small cohort of 12 patients within occluded coronary artery of patients with ST-segment elevation myocardial infarction treated by primary angioplasty within 24 h of symptom onset. MP procoagulant activity and cellular origin were characterized within the occluded coronary artery before primary angioplasty and after restoration of the epicardial blood flow, and in blood collected from the femoral artery. Levels of leucocyte-derived CD11a-positive MPs, endothelial-derived CD105-positive MPs, and TF-bearing MPs were significantly higher within the occluded coronary artery than in peripheral blood samples. Restoration of the epicardial blood flow led to a significant reduction of procoagulant CD11a- and CD105-positive MPs by 30% and 42%, respectively ($p < 0.05$).

In another recent study (Nozaki et al. 2010), plasma EMP levels were identified as independently predictive risk factors for cardiovascular events. Plasma CD144-positive EMP levels were measured by flow cytometry in heart failure (HF) patients (70 ischemic, 99 non-ischemic HF) and in 31 healthy subjects. HF patients were followed for a mean of 30 months. Endpoints were a composite of cardiovascular events (myocardial infarction, stroke, re-hospitalization for HF, and cardiovascular death) and all-cause mortality. EMP levels increased significantly with degree of cardiac failure. Multivariate Cox regression analysis adjusted for clinical factors identified high EMP levels as an independent predictor of future cardiovascular events, but not for all-cause mortality in HF patients [hazard ratio (95% confidence interval): 2.423 (1.034–5.681), $P = 0.04$ for cardiovascular events; and 2.095 (0.825–5.323), $P = 0.12$ for all-cause mortality].

Huang et al. (2010) studied circulating CD31-/annexin V-positive apoptotic MPs and circulating endothelial progenitor cell (EPC) levels in hypertensive patients with microalbuminuria. They demonstrated that patients with microalbuminuria or macroalbuminuria had significantly increased vascular (CD31/annexin V) apoptotic MPs and decreased circulating EPC numbers. By multivariate analysis, CD31/annexin V apoptotic MP level was an independent predictor of urinary albumin excretion rate in hypertensive patients. Microparticles isolated from hypertensive patients with microalbuminuria or macroalbuminuria attenuated EPC proliferation, migration, and increased H_2O_2 production, cellular senescence and apoptosis in comparison with those from hypertensive patients with normoalbuminuria. Their findings suggest that hypertensive patients with microalbuminuria or macroalbuminuria have increased endothelial apoptotic MPs and decreased circulating EPC levels, which may contribute to atherosclerotic disease progression and enhanced cardiovascular risk in hypertensive patients with nephropathy.

APPLICATION TO OTHER AREAS OF HEALTH AND DISEASE

Scott Syndrome

Interestingly, the rare hereditary bleeding disorder Scott syndrome is characterized by a deficient PS exposure and plasma membrane vesiculation and shedding. These patients have a decreased number of MPs in their circulation (Sims et al. 1989; Satta et al. 1997; Zwaal et al. 2005; Piccin et al. 2007).

Diabetes

In patients with diabetes, MP abnormalities have been found. MP phenotype and procoagulant potential may vary according to the type of diabetes or level of diabetes control (Diamant et al. 2002) and high blood MP-associated procoagulant activity was detected in patients with poorly controlled HbA1c levels (Sabatier et al. 2002). Both EMPs and PMPs have been found to be elevated in diabetes type 1 patients, whereas the highest EMP levels were found in the patients with microvascular complications.

Cancer, Thrombosis, and Poor Prognosis

Cancer patients have a 4- to 7-fold increased risk of venous thromboembolism (VTE) (Blom et al. 2006; Heit et al. 2000) and they also have a poor survival compared to those who did not develop thrombosis (Sorensen et al. 2000).

Plasma MP carrying bio-active TF have been reported to be associated with patients with cancer-associated VTE and poor prognosis (Tesselaar et al. 2007, 2009). In the cohorts studied, cancer patients with elevated

MP-associated TF activity and poor prognosis also had a small fraction of MUC-1-positive MP in their pool of blood MP, suggesting that perhaps a subset of MPs may originate from tumour cells themselves.

In contrast, plasma MP-associated TF activity has not been associated with confirmed pulmonary thromboembolism in patients presenting with clinically suspected pulmonary thromboembolism (Garcia et al. 2010).

IMAGING METHODS OF AFM

The AFM was invented in 1986 by Binnig et al. (1986) and its usefulness in biological research was quickly realized: it allows high resolution three-dimensional imaging in a physiological environment as was shown conclusively by the first AFM images of living cells in 1991 (Häberle et al. 1991).

AFM uses a sharp tip with a radius of curvature smaller than 10 nm and it is brought in close proximity to the surface, such that its surface profile can be felt. The force experienced by the tip is measured by detecting the deflection of tiny leaf spring, the cantilever, on which the tip is situated.

To obtain information about the surface the probe is raster-scanned over a surface. As the tip is moved sideways over the surface, the tip height is constantly adjusted such that the interaction between tip and surface is kept unchanged. In contact mode, the tip actually touches the surface and the tip is adjusted in such a way that the repulsive force between the tip and the surface is kept constant. Forces as small as 1 pN can be detected. To avoid damage to the surface or the tip, forces applied should be smaller than 1 nN. Alternatively, in dynamic mode or tapping mode, the cantilever can be oscillated at its resonance frequency and its interaction with the surface is measured by monitoring the changes in amplitude and frequency of the resonator.

Contact mode has the benefit that the applied force is directly related to the deflection and can be easily calculated. It has the drawback that the deflection might drift over time with temperature, so that the applied force is not known anymore, and the probe can easily damage fragile structures on the surface as it scans them laterally. Dynamic mode is gentler as most of the time the probe is not in contact with the surface and it hardly exerts a lateral force. The instantaneous perpendicular force can be higher, however, and the applied force is not easily estimated.

Depending on the resolution required, the tip may be scanned faster or slower across the surface to be imaged. As explained above, the image must be scanned one line at a time. Because the maximum scan speed of commercial AFM is typically a few lines per second, an image consisting of 512 by 512 pixels takes several minutes to image. In the future, AFM that are up to 100 times as fast are expected to be brought to the market.

Obviously the tip is an important factor in determining the lateral resolution achievable with this technique. Many different types of probes can be bought commercially; most are manufactured using silicon etching techniques. The probe itself usually has a pyramidal shape with a tip with a radius from a few nm to tens of nm. The tip end may also be modified in a variety of ways to achieve sharper tips, higher aspect ratios, durability or a specific surface chemistry. The tip is mounted on a tiny leaf spring typically 50–200 μm long, the spring constant of which can range from 10 mN/m to >1 N/m. Softer probes are often used in contact mode to minimize the applied force; stiffer probes are often used in dynamic mode because it then is easier to excite them in liquid and their higher resonance frequency allows for higher scanning speeds

MP DETECTION AND QUANTIFICATION BY USING AFM

Because of the high resolution of AFM, the supporting surface for MP attachment has to be flat, preferably on a nanometre scale over a micrometre range, to distinguish single MPs from the roughness of the supporting surface. One of the suitable surfaces is muscovite mica. Muscovite mica, a non-conducting layered mineral, has an atomically flat surface (Muller et al. 1997). The mica layer can easily be cleaved to expose the hydroxyl groups on its surface (Maslova et al. 2004). This cleaved surface can be functionalized to bind proteins (Klein et al. 2003).

For capturing CD41-positive MPs the mica surface needs to be modified and coated with anti-human CD41 antibody. This is because cleaved mica itself is unable to immobilize MPs. Detailed information on the modification of mica surface and AFM measurement of PMPs can be found in Yuana et al. (2010). The different steps are illustrated in Fig. 1. As a first step, molecular sieve beads are added to a solution of ethanolamine in DMSO (70°C) to adsorb the water that will be formed later during the amine reaction on the mica surface. Freshly cleaved muscovite mica is then incubated overnight at 70°C in this ethanolamine solution. After the mica surface is rinsed several times in DMSO and in ethanol (RT), the surface is dried under N_2 flow. Subsequently, the amine-modified mica is incubated for 3 h in a chloroform solution containing EGTA and triethylamine. Afterwards, the modified mica is washed in chloroform and dried under N_2 flow. Finally, anti-human CD41 antibody is added to this mica surface and incubated for 30 min. Usually, we reach a density of about 2,000 molecules/100 μm^2 surface area. At this protein density the mica can capture up to ~530 CD41 positive-MPs/100 μm^2 (Yuana et al. 2010). Mica coated with an IgG1 isotype control was used to control for non-specific binding of MPs to the modified mica.

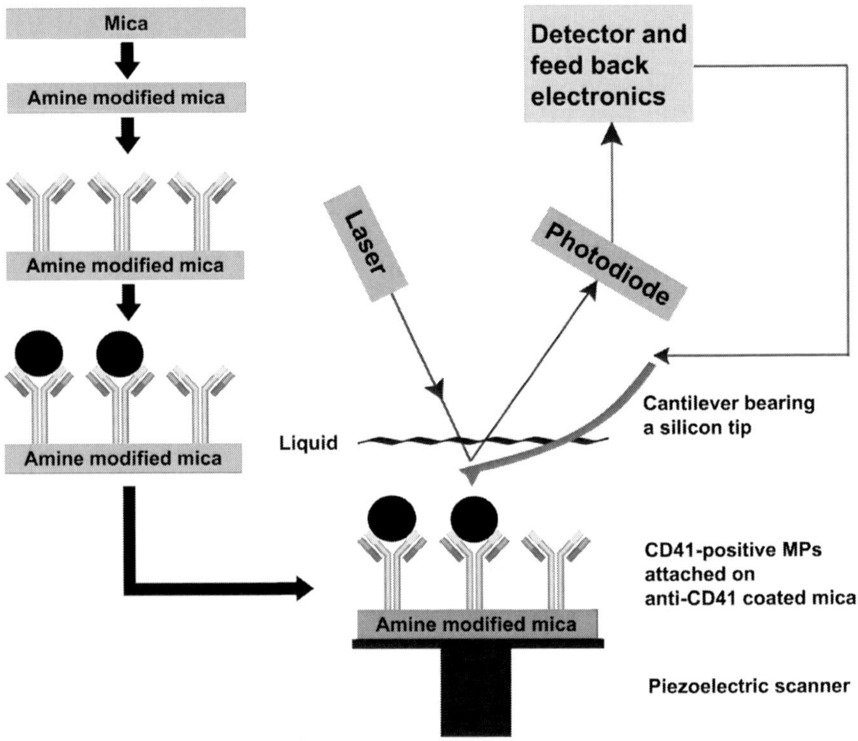

Fig. 1. Surface modification of a mica, antibody attachment and MP capture and measurement of CD41-positive MPs by AFM. After cleaved mica is amine modified, the anti-CD41 antibody/IgG1 isotype control is applied. MPs are incubated for 30 min on this coated surface and after unbound MPs are rinsed, this surface is imaged by AFM operated in fluid-tapping mode. Unpublished data.

Fresh platelet-poor plasma derived from citrated blood is used to prepare MPs. The procedures of blood collection, plasma, and MP isolation are described in detail by Yuana et al. (2010). The isolated MP suspension is applied to the CD41-coated mica surface and incubated for 30 min to allow the binding of MPs to anti-CD41. To remove the unbound MPs, the mica was washed twice with Hepes buffer (10 mM Hepes, 137 mM NaCl, 4 mM KCl, and 0.1 mM Pefabloc SC, pH 7.4).

The AFM glass fluid cell was used for imaging in fluid-tapping mode. Before being used, this cell should be cleaned by sonicating in sodium dodecylsulfate, rinsed with distilled water, and allowed to dry under N_2 flow. A microcantilever silicon tip with a spring constant of 2 N/m and 70 kHz resonance frequency is attached to the fluid cell. After mounting the sample on the piezoelectric scanner of AFM, Hepes buffer is added. This buffer preserves MPs in their physiological state. The fluid cell with the cantilever attached is then seated against the sample. With an imaging rate of 1.5 Hz, the surface is scanned at different positions to obtain four to 10 images with a scan size of 100 μm^2 per image (Fig. 2). When a single MP needs to be characterized, a smaller scan size is used (Fig. 3). The image scan contains the topography information (x, y, and z dimensions) of attached MPs/particles. The value of z dimension (height) is smaller than the value of x/y dimension (apparent diameter) of attached MPs indicating that bound MPs are flat. This phenomenon was also observed by other groups (Siedlecki et al. 1999; Richter and Brisson 2005). Assuming a spherical shape of unbound MPs in their physiological state, MPs should be analyzed in terms of their volume. Image scans are analyzed by using quantification software, SPMEDIATOR version 6.1 (courtesy of S.J.T. van Noort, Leiden University, Leiden, the Netherlands), which calculates automatically the volume and z value of a particle. From the MP volume the MP spherical diameter ($d_{sph} = \sqrt[3]{6V/\pi}$) can be calculated and a size distribution of MPs in the sample can be defined. From the z value the calculated MP diameter can be corrected from the tip-convolution effect of scanning. This software also gives the total number of MPs attached on 100 μm^2 surface area that is used to calculate the number of CD41-positive MPs/L plasma (for details see Yuana et al. 2010).

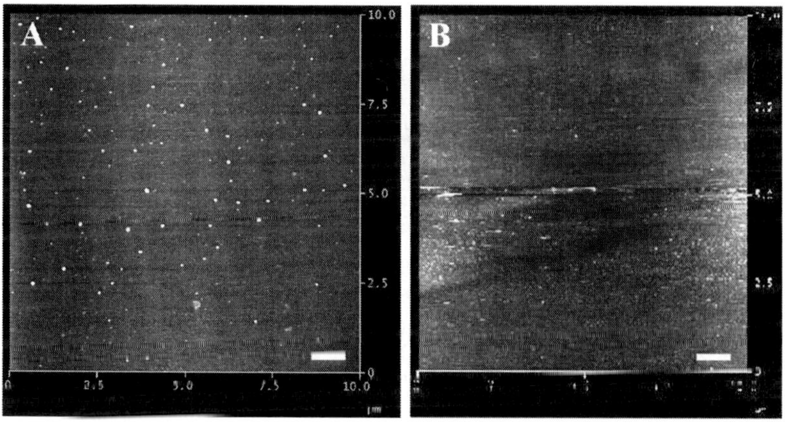

Fig. 2. Three-dimensional AFM topography of (A) CD41-positive MPs and (B) isotype control (IgG1)-coated mica. MPs were isolated from plasma as described in (Yuana et al. 2010). Twenty µl of MPs was added on to anti-CD41 or IgG1-coated mica. The scale bar in these images is 1 µm. Unpublished data.

Considering the number of analytical factors involved (mica modification, antibody density on mica surface, MP isolation and AFM detection), the observed number of CD41-positive MPs is highly reproducible (coefficient of variation = 16%). The number of MPs specifically bound to the mica was found to be linearly correlated to the

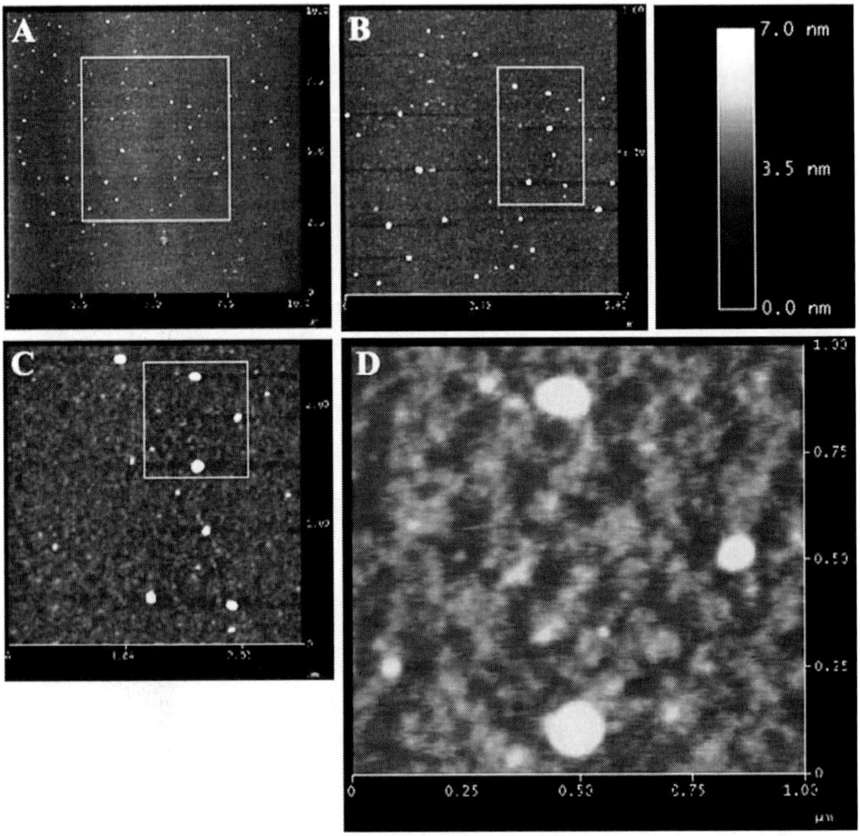

Fig. 3. Three-dimensional AFM topography of CD41-positive MPs. To characterize CD41-MPs the mica surface capturing CD41-MPs is scanned at 100 μm² (A), 25 μm² (B), 6.25 μm² (C) and 1 μm² (D). Colour bar indicates the height (z) of the object. Conditions as described in Fig. 2. Unpublished data.

MP concentration as long as the surface was not saturated with MPs. Importantly, AFM is more sensitive than conventional flow cytometry in the detection of CD41-positive MPs. AFM detects 1,000-fold the numbers of CD41-positive MPs that flow cytometry does (Yuana et al. 2010). AFM seems to be able to detect nanosized CD41-positive MPs with diameter sizes (d_{sph}) ranging from 10 to 475 nm (mean: 67.5 ± 26.5 nm) and from 5 to 204 nm (mean: 51.4 ± 14.9 nm) in the blood of donors and cancer patients, respectively. After passing the MP suspension through a 0.22 μm filter, AFM still detects CD41-positive MPs in the filtrate (d_{sph}: 10–120 nm), whereas flow cytometry does not.

In conclusion, AFM is a suitable method for the sensitive and reproducible detection of specific subsets of nanosized MPs. The present

AFM method can also be implemented to detect MPs derived from other cells than platelets by using other antibodies to coat the mica surface (e.g., anti-CD144 for endothelial cells, anti-CD14 for monocytes, anti-CD3 and/or CD4 for T lymphocytes, and anti-CD45 for leucocytes). This AFM method was, for example, implemented to capture TF-expressing MPs derived from cultured breast carcinoma cells, MDA-MB-231, stimulated with calcium ionophore (Fig. 4). However, as long as a proper external standard for the calibration of MP enumeration (e.g., MPs with a known concentration and size distribution) is not available, it is difficult to determine the absolute concentration of MPs in the sample. Also, the procedure from blood withdrawal until AFM measurement and analysis is labour-intensive (and time-intensive). Therefore, the throughput of this AFM method needs to be increased. Direct measurement of MPs in plasma is also preferred because it will prevent loss of MPs during the isolation procedure and preserve their morphology (Jy et al. 2004).

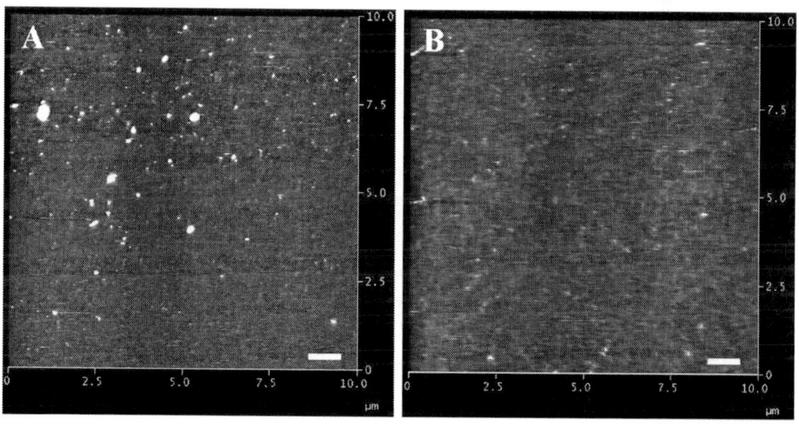

Fig. 4. Three-dimensional AFM topography of (A) TF-positive MPs and (B) isotype control (IgG1)-coated mica. MPs derived from cultured breast carcinoma cells, MDA-MB-231, stimulated with 10 μM calcium ionophore. The majority of TF-positive MPs is ~50 nm. The scale bar in these images is 1 μm. Unpublished data.

QUANTIFICATION OF THE NUMBER OF MPS USING OTHER METHODS THAN AFM

Non-optical Methods

Isolated MPs have been detected using different electron microscopy (EM) techniques (transmission, scanning, and cryo-electron microscopy). The size and size distribution of MPs can be obtained from EM images assuming that the size is not affected by the sample dehydration and vacuum processes needed in EM. In addition, information on the antigenic composition of MPs can be obtained by using immunogold-labelled antibodies (Aras et al. 2004; Heijnen et al. 1999). The magnification in EM makes it possible to quantify and observe both single MPs and a sufficiently large number of MPs in a single experiment. However, EM counts MPs semi-quantitatively and it is a laborious task to determine MPs size and size distributions of many samples.

Table 1. Summary of (potential) optical and non-optical methods for MP detection and quantification (unpublished data).

Methods	Size range (nm)	Size distribution	Concentration	Surface properties (e.g., using antibody)	Applied to MPs
AFM	1–500	Yes	Yes	Yes	Yes
SEM	1–500	Yes	Yes	Yes	Yes
TEM	1–500	Yes	Yes	Yes	Yes
CEM	1–500	Yes	Yes	Yes	Yes
CLS*	1–500	Yes	Yes	No	No
DLS	1–500	Yes	Yes	No	Yes
FCS	1–500	Yes	Yes	Yes	No
fluorescence-NTA	1–500	Yes	Yes	Yes	Yes

AFM, atomic force microscopy; CEM, cryo-electron microscopy; CLS, classical light scattering; DLS, dynamic light scattering; FCS, fluorescence correlation spectroscopy;
NTA, nanoparticle tracking analysis; SEM; scanning electron microscopy; TEM, transmission electron microscopy.

*in combination with DLS

Optical Methods—*(Fluorescence) Light Scattering of MPs in Solution*

In the interpretation of classical light scattering (CLS) results from MPs, the quantification of the data is complicated by the broad size distribution of MPs. Light scattering from a moderately broad population of otherwise homogeneous small vesicles has been described in detail by van Zanten and Monbouquette (1991), who obtained the size and bilayer thickness of vesicles by CLS using the Rayleigh-Gans-Debye theory for thin shells. However, in view of the broad size distribution of the MPs as determined in the AFM measurements of Yuana et al. (2010), accurate size distributions of submicron structures, while theoretically possible, cannot be easily obtained from light or other scattering methods (Coldren et al. 2003).

The fluctuations in the fluorescence or light scattering originating from a population of MPs can be used to derive the size distribution of the particles. At least four different approaches can be named: dynamic light scattering (DLS) or photon correlation spectroscopy, fluorescence correlation spectroscopy (FCS), and single molecule tracking/nanoparticle tracking analysis (NTA).

In all four techniques the observed fluctuations in the detected light intensity are explored to determine the diffusion of small particles. DLS and NTA have already been used to quantify the dynamics of MPs (Harrison et al. 2009; Lawrie et al. 2009).

For a detailed account of the possibilities in the characterization, assumptions have to be made on the contrast of MPs of a particular size. To calculate the scattering contrast at a fixed scattering angle, Mie theory can be used. The shell thickness and the refractive index of the MP interior and shell are needed for these calculations. Van der Pol et al. (2010) have calculated the overall scattering contrast and their calculations could be (easily) extended to convert DLS data to a size distribution. Similarly, the contrast mechanism of the fluorescence of labelled MPs has to be considered. Basically three approaches are possible: antibody labelling that will bring a fluorescent label just outside the lipid membrane, labelling with a membrane label, or labelling with a label specific for the interior of the MP. As an example, Gall et al. (2010) studied vesicles involved in endocytosis in plant cells by AFM and DLS and were able to determine the *in vivo* size (distribution) of endocytic vesicles of sub-microscopic size.

In all methods the presence of remaining platelets should be avoided as much as possible as the surface of the platelets contains the same surface antigens as the MPs; moreover, the size of the platelets results in an overwhelming light scattering contribution that prevents the accurate quantification of MPs.

Optical Methods—*Fluorescence Detection of Immobilized MPs on a Surface*

To be able to enumerate sub-diffraction-sized particles bound to a surface, we have to estimate the distance to nearest neighbours of individual particles and their distribution over the surface.

Clark and Evans (1954) devised a method to assess the nearest neighbour distance and the randomness of the planar distribution of objects, albeit in a completely different context. For a maximum surface density of 200–500 MPs per 100 μm^2, the average distance is computed as 0.2 μm or less, assuming a random distribution of MPs. In the linear working range of the MP capture, the particle surface density is about 1 MP per μm^2 and the individual MPs can be detected provided that a suitable non-bleaching probe can be used. Using QD-labelled antibodies with QDs in a size range from 5 to 20 nm in combination with the smallest MPs of about 25 nm will allow the counting of individual immobilized MPs.

Summary Points

- MPs may function as trigger and/or biomarker of the disease and its severity.
- MPs may function as surrogate endpoints in the future for assessing the efficacy of new drugs and therapies.
- AFM detects 1,000-fold the numbers of CD41-positive MPs that flow cytometry does.
- AFM detects nanosized CD41-positive MPs with diameters (d_{sph}) ranging from 10 to 475 nm (mean: 67.5 ± 26.5 nm) in blood of donors.
- The QD-labelled antibodies with QDs in a size range from 5 to 20 nm can be used to label MPs allowing the counting of individual MPs by using optical methods (e.g., NTA, FCS).

Definitions

Liquid/fluid-tapping mode: An AFM operation mode in which the cantilever bearing tip oscillates in liquid and is brought close enough to the sample so that it just barely hits or "taps" the sample surface at the lowest point of the oscillation to eliminate lateral forces (friction or drag).

Microparticles (MPs): Small membrane vesicles, also named MVs, released from virtually all cells of the body upon activation or apoptosis. In the past, they have been considered to be inert debris of cells ("cellular dust"). Depending on the definition used, MP may encompass MVs, apoptotic bodies and apoptotic MPs, and exosomes, and can be secreted from intracellular multivesicular bodies or released as blebbing membrane from the cell surface.

Quantum dots (QDs): Nanoparticles, or nanocrystals, of a semiconductor material ranging from 2 to 10 nm (10–50 atoms) in diameter. QDs are commonly made from zinc sulphide, lead sulphide, cadmium selenide and indium phosphide. QDs work in much the same way as energized electrons that move to a higher level and return to their lower and stable state by emitting energy as light corresponding to a particular frequency. QDs have been tested in most biotechnological applications that use fluorescence, including DNA array technology, immunofluorescence assays in cell and animal biology. For immunofluorescence, QDs are resistance to bleaching over long periods of time (minutes to hours) and only a small number of QDs is necessary to produce brighter signal than other dyes.

Tissue factor (TF): Also referred to as tissue thromboplastin or coagulation factor III, a 43kD to 45kD, single chain, transmembrane glycoprotein with 263 amino acid residues. It is composed of three distinct domains: the extracellular domain, the transmembrane domain, and the cytoplasmic carboxyterminal domain. TF is a constituent of both the subendothelial layer of the vascular wall and the extravascular tissue. TF has been known as the primary physiological initiator of blood coagulation.

Venous thromboembolism (VTE): A term used to describe a blood clot in a vein. VTE is a disease that includes both deep vein thrombosis and pulmonary embolism resulting from a combination of risk factors, including environmental and genetic sources.

Abbreviations

AFM	:	atomic force microscopy
CLS	:	classical light scattering
DMSO	:	dimethylsulfoxide
DNA	:	deoxyribonucleic acid
DLS	:	dynamic light scattering
EGTA	:	ethylene glycol tetraacetic acid
EM	:	electron microscopy
EMP	:	endothelial microparticles
FCS	:	fluorescence correlation spectroscopy
Hepes	:	4-(2-hydroxyethyl)-1-piperazineethanesulfonic acid
KCl	:	potassium chloride
miRNA	:	micro ribonucleic acid
MP	:	microparticle
MV	:	microvesicle
NaCl	:	sodium chloride
NP	:	nanoparticle
NTA	:	nanoparticle tracking analysis
PMP	:	platelet microparticles
PS	:	phosphatidylserine
QD	:	quantum dot
RNA	:	ribonucleic acid
RT	:	room temperature
SDS	:	sodium dodecyl sulphate
TF	:	tissue factor
VTE	:	venous thromboembolism

References

Aras, O., A. Shet, R.R. Bach, J.L. Hysjulien, A. Slungaard, R.P. Hebbel, C. Escolar, B. Jilma and N.S. Key. 2004. Induction of microparticle- and cell-associated intravascular tissue factor in human endotoxemia. Blood 103: 4545–4553.

Aupeix, K., B. Hugel, T. Martin, P. Bischoff, H. Lill, J.L. Pasquali and J.M. Freyssinet. 1997. The significance of shed membrane particles during programmed cell death *in vitro*, and *in vivo*, in HIV-1 infection. J. Clin. Invest. 99: 1546–1554.

Ay, C., J.M. Freyssinet, T. Sailer, R. Vormittag and I. Pabinger. 2009. Circulating procoagulant microparticles in patients with venous thromboembolism. Thromb. Res. 123: 724–726.

Binnig, G., C.F. Quate and C. Gerber. 1986. Atomic force microscope. Phys. Rev. Lett. 56: 930–933.

Blom, J.W., J.P. Vanderschoot, M.J. Oostindier, S. Osanto, F.J. van der Meer and F.R. Rosendaal. 2006. Incidence of venous thrombosis in a large cohort of 66,329 cancer patients: results of a record linkage study. J. Thromb. Haemost. 4: 529–535.

Boulanger, C.M. 2010. Microparticles, vascular function and hypertension. Curr. Opin. Nephrol. Hypertens. 19: 177–180.

Burnier, L., P. Fontana, B.R. Kwak and A. Angelillo-Scherrer. 2009. Cell-derived microparticles in haemostasis and vascular medicine. Thromb. Haemost. 101: 439–451.

Chen, T.S., R.C. Lai, M.M. Lee, A.B. Choo, C.N. Lee and S.K. Lim. 2010. Mesenchymal stem cell secretes microparticles enriched in pre-microRNAs. Nucleic Acids Res. 38: 215–224.

Clark, P.J. and F.C. Evans. 1954. Distance to nearest neighbor as a measure of spatial relationships in populations. Ecology 35: 445–453.

Coldren, B., R. van Zanten, M.J. Mackel, J.A. Zasadzinski and H.T. Jung. 2003. From vesicle size distributions to bilayer elasticity via cryo-transmission and freeze-fracture electron microscopy. Langmuir 19: 5632–5639.

Diamant, M., R. Nieuwland, R.F. Pablo, A. Sturk, J.W. Smit and J.K. Radder. 2002. Elevated numbers of tissue-factor exposing microparticles correlate with components of the metabolic syndrome in uncomplicated type 2 diabetes mellitus. Circulation 106: 2442–2447.

Furie, B., and B.C. Furie. 2006. Cancer-associated thrombosis. Blood Cells Mol. Dis. 36: 177–181.

Gall, L., R.C. Stan, A. Kress, B. Hertel, G. Thiel and T. Meckel. 2010. Fluorescent detection of fluid phase endocytosis allows for *in vivo* estimation of endocytic vesicle sizes in plant cells with sub-diffraction accuracy. Traffic 11: 548–559.

Garcia, R.P., H.C. Eikenboom, M.E. Tesselaar, M.V. Huisman, M. Nijkeuter, S. Osanto and R.M. Bertina. 2010. Plasma levels of microparticle-associated tissue factor activity in patients with clinically suspected pulmonary embolism. Thromb. Res.

Häberle, W., J.K.H. Hörber and G. Binnig. 1991. Force microscopy on living cells. J. Vac. Sci. Technol. B 9: 1210–1213.

Harrison, P., R. Dragovic, A. Albanyan, A.S. Lawrie, M. Murphy and I. Sargent. Application of dynamic light scattering to the measurement of microparticles. J. Thromb. Haemost. 7[Supplement 2]. 2009.

Heijnen, H.F., A.E. Schiel, R. Fijnheer, H.J. Geuze and J.J. Sixma. 1999. Activated platelets release two types of membrane vesicles: microvesicles by surface shedding and exosomes derived from exocytosis of multivesicular bodies and alpha-granules. Blood 94: 3791–3799.

Heit, J.A., D.N. Mohr, M.D. Silverstein, T.M. Petterson, W.M. O'Fallon and L.J. Melton, III. 2000. Predictors of recurrence after deep vein thrombosis and pulmonary embolism: a population-based cohort study. Arch. Intern. Med. 160: 761–768.

Huang, P.H., S.S. Huang, Y.H. Chen, C.P. Lin, K.H. Chiang, J.S. Chen, H.Y. Tsai, F.Y. Lin, J.W. Chen and S.J. Lin. 2010. Increased circulating CD31+/annexin V+ apoptotic microparticles and decreased circulating endothelial progenitor cell levels in hypertensive patients with microalbuminuria. J. Hypertens. 28: 1655–1665.

Hunter, M.P., N.Ismail, X. Zhang, B.D. Aguda, E.J. Lee, L. Yu, T. Xiao, J. Schafer, M.L. Lee, T.D. Schmittgen, S.P. Nana-Sinkam, D. Jarjoura and C.B. Marsh. 2008. Detection of microRNA expression in human peripheral blood microvesicles. PLoS. ONE. 3: e3694.

Jy, W., L.L. Horstman, J.J. Jimenez, Y.S. Ahn, E. Biro, R. Nieuwland, A. Sturk, F. Dignat-George, F. Sabatier, L. Camoin-Jau, J. Sampol, B. Hugel, F. Zobairi, J.M. Freyssinet,

S. Nomura, A.S. Shet, N.S. Key and R.P. Hebbel. 2004. Measuring circulating cell-derived microparticles. J. Thromb. Haemost. 2: 1842–1843.

Klein, D.C., C.M. Stroh, H. Jensenius, M. van Es, A.S. Kamruzzahan, A. Stamouli, H.J. Gruber, T.H. Oosterkamp and P. Hinterdorfer. 2003. Covalent immobilization of single proteins on mica for molecular recognition force microscopy. Chemphyschem. 4: 1367–1371.

Lawrie, A.S., A. Albanyan, R.A. Cardigan, I.J. Mackie and P. Harrison. 2009. Microparticle sizing by dynamic light scattering in fresh-frozen plasma. Vox Sang. 96: 206–212.

Lynch, S.F., and C.A. Ludlam. 2007. Plasma microparticles and vascular disorders. Br. J. Haematol. 137: 36–48.

Maslova, M.V., L.G. Gerasimova and W. Forsling. 2004. Surface properties of cleaved mica. Colloid J. 66: 322–328.

Morel, O., B. Pereira, G. Averous, A. Faure, L. Jesel, P. Germain, L. Grunebaum, P. Ohlmann, J. M. Freyssinet, P. Bareiss, and F. Toti. 2009. Increased levels of procoagulant tissue factor-bearing microparticles within the occluded coronary artery of patients with ST-segment elevation myocardial infarction: role of endothelial damage and leukocyte activation. Atherosclerosis 204: 636–641.

Muller, D.J., M. Amrein and A. Engel. 1997. Adsorption of biological molecules to a solid support for scanning probe microscopy. J. Struct. Biol. 119: 172–188.

Nomura, S., A. Shouzu, K. Taomoto, Y. Togane, S. Goto, Y. Ozaki, S. Uchiyama and Y. Ikeda. 2009. Assessment of an ELISA kit for platelet-derived microparticles by joint research at many institutes in Japan. J. Atheroscler. Thromb. 16: 878–887.

Nozaki, T., S. Sugiyama, K. Sugamura, K. Ohba, Y. Matsuzawa, M. Konishi, J. Matsubara, E. Akiyama, H. Sumida, K. Matsui, H. Jinnouchi and H. Ogawa. 2010. Prognostic value of endothelial microparticles in patients with heart failure. Eur. J. Heart Fail. 12: 1223–1228.

Parot, P., Y.F. Dufrene, P. Hinterdorfer, C. Le Grimellec, D. Navajas, J.L. Pellequer and S. Scheuring. 2007. Past, present and future of atomic force microscopy in life sciences and medicin. J. Mol. Recognit. 20: 418–431.

Piccin, A., W.G. Murphy and O.P. Smith. 2007. Circulating microparticles: pathophysiology and clinical implications. Blood Rev. 21: 157–171.

Ratajczak, J., M. Wysoczynski, F. Hayek, A. Janowska-Wieczorek and M.Z. Ratajczak. 2006. Membrane-derived microvesicles: important and underappreciated mediators of cell-to-cell communication. Leukemia 20: 1487–1495.

Ricci, D. and P.C. Braga. 2004. How the atomic force microscopy works, In: P.C. Braga and D. Ricci (eds.). Methods in Molecular Biology: Biomedical Methods and Applications. vol. 242. Humana Press Inc., pp. 3–12.

Richter, R.P. and A.R. Brisson. 2005. Following the formation of supported lipid bilayers on mica: a study combining AFM, QCM-D, and ellipsometry. Biophys. J. 88: 3422–3433.

Sabatier, F., P. Darmon, B. Hugel, V. Combes, M. Sanmarco, J.G. Velut, D. Arnoux, P. Charpiot, J.M. Freyssinet, C. Oliver, J. Sampol and F. gnat-George. 2002. Type 1 and type 2 diabetic patients display different patterns of cellular microparticles. Diabetes 51: 2840–2845.

Satta, N., F. Toti, E. Fressinaud, D. Meyer and J.M. Freyssinet. 1997. Scott syndrome: an inherited defect of the procoagulant activity of platelets. Platelets 8: 117–124

Siedlecki, C.A., I.W. Wang, J.M. Higashi, K. Kottke-Marchant and R.E. Marchant. 1999. Platelet-derived microparticles on synthetic surfaces observed by atomic force microscopy and fluorescence microscopy. Biomaterials 20: 1521–1529.

Sims, P.J., T. Wiedmer, C.T. Esmon, H.J. Weiss and S.J. Shattil. 1989. Assembly of the platelet prothrombinase complex is linked to vesiculation of the platelet plasma membrane. Studies in Scott syndrome: an isolated defect in platelet procoagulant activity. J. Biol. Chem. 264: 17049–17057.

Sorensen, H.T., L. Mellemkjaer, J.H. Olsen and J.A. Baron. 2000. Prognosis of cancers associated with venous thromboembolism. N. Engl. J. Med. 343: 1846–1850.

Tesselaar, M.E., F.P. Romijn, I.K. van der Linden, F.A. Prins, R.M. Bertina and S. Osanto. 2007. Microparticle-associated tissue factor activity: a link between cancer and thrombosis? J. Thromb. Haemost. 5: 520–527.
Tesselaar, M.E., F.P. Romijn, I.K. van der Linden, R.M. Bertina and S. Osanto. 2009. Microparticle-associated tissue factor activity in cancer patients with and without thrombosis. J. Thromb. Haemost. 7: 1421–1423.
Van Der Pol, E., A.G. Hoekstra, A. Sturk, C. Otto, T.G. Van Leeuwen and R. Nieuwland. 2010. Optical and non-optical methods for detection and characterisation of microparticles and exosomes. J. Thromb. Haemost. 8: 2596–2607.
Vanzanten, J.H. and H.G. Monbouquette. 1991. Characterization of Vesicles by Classical Light-Scattering. J. Colloid Interface Sci. 146: 330–336.
Yuana, Y., T.H. Oosterkamp, S. Bahatyrova, B. Ashcroft, R.P. Garcia, R.M. Bertina and S. Osanto. 2010. Atomic force microscopy: a novel approach to the detection of nanosized blood microparticles. J. Thromb. Haemost. 8: 315–323.
Yuana, Y., R.M. Bertina and S. Osanto. 2011. Pre-analytical and analytical issues in the analysis of blood microparticles. Thromb. Haemost. 105: 396–408.
Zwaal, R.F., P. Comfurius and E.M. Bevers. 2005. Surface exposure of phosphatidylserine in pathological cells. Cell Mol. Life Sci. 62: 971–988.

5

Nanobiotechnology-based Blood Substitutes and the Cardiovascular Systems in Transfusion Medicine

Thomas Ming Swi Chang

ABSTRACT

Red blood cell membrane contains blood group antigens, and typing and matching are needed before donor blood can be transfused into patients. This results in delays in emergency situations. Donor blood requires storage at 4°C using standard methods and is good for less than 42 d. Red blood cells cannot be sterilized to remove hepatitis viruses, HIV and other potential emerging infective agents. There is also a shortage of donor blood especially in emergencies, war or disasters. Nanobiotechnology has allowed the development of different types of red blood cell substitutes. The first-generation type is a simple modified hemoglobin oxygen carrier that has immediate potential for clinical uses. In particular, such substitutes do not have blood group antigens and can be

Director, Artificial Cells and Organs Research Centre, Professor Emeritus, Departments of Physiology, Medicine and Biomedical Engineering, Faculty of Medicine, McGill University, 3655 Drummond Street, 10th floor, Room 1004, Montreal, Quebec, Canada H3G1Y6; E-mail: Artcell.med@mcgill.ca

List of abbreviations after the text.

used immediately on the spot without matching or typing. Some can be stored for more than a year at room temperature. Infective agents such as HIV and other viruses and microorganisms can be removed. However, these first-generation red blood cell substitutes are only oxygen carriers and can have side effects in certain groups of patients and certain clinical conditions. New generations of red blood cell substitutes containing red blood cell antioxidant enzymes such as superoxide dismutase and catalase are being developed with promising results. The most advance red blood cell substitute is a nanodimension artificial red blood cell based on nanobiotechnology, but this is more complicated.

INTRODUCTION

Extracting the oxygen-carrying component, hemoglobin (Hb), from red blood cell (rbc) and purifying it to remove the membrane component would removed blood group antigens and also allow it to be sterilized. Unfortunately, Hb is a tetramer with two α subunits and two β subunits (α1β1α2β2) (Perutz 1989). When free Hb is infused into the body, the tetramer (α1β1α2β2) breaks down into toxic dimers (α1β1 and α2β2), which causes renal toxicity and other adverse effects. The challenge lies in making use of the excellent oxygen-carrying properties of Hb while preventing its toxic effects.

Nanobiotechnology has been defined as the assembling of biological molecules into structures of 1 to 100 nm (in diameter of complex, thickness of membranes or diameter of nanotubules). The first nanobiotechnology approach reported in the literature is the use of bifunctional agent to crosslink hemoglobin into nanodimension thickness polyhemoglobin (PolyHb) membrane for artificial cell membrane (Chang 1964, 1972). This can also be used to assemble and crosslink Hb with bifunctional agent, using glutaraldehye, into soluble PolyHb of nanodimension (Chang 1971) (Fig. 1). This principle has been developed into different types of blood substitutes. First-generation rbc substitutes in the form of glutaraldehye-crosslinked hemoglobins are oxygen carriers for the replacement of the oxygen-carrying function of rbc (Gould et al. 2002; Jahr et al. 2008; Moore et al. 2009). One of these is in routine clinical use in South Africa (Jahr et al. 2008). There are also other types of first-generation Hb-based oxygen carriers (Fig. 1). Further developments are being made towards new generations of more complete blood substitutes.

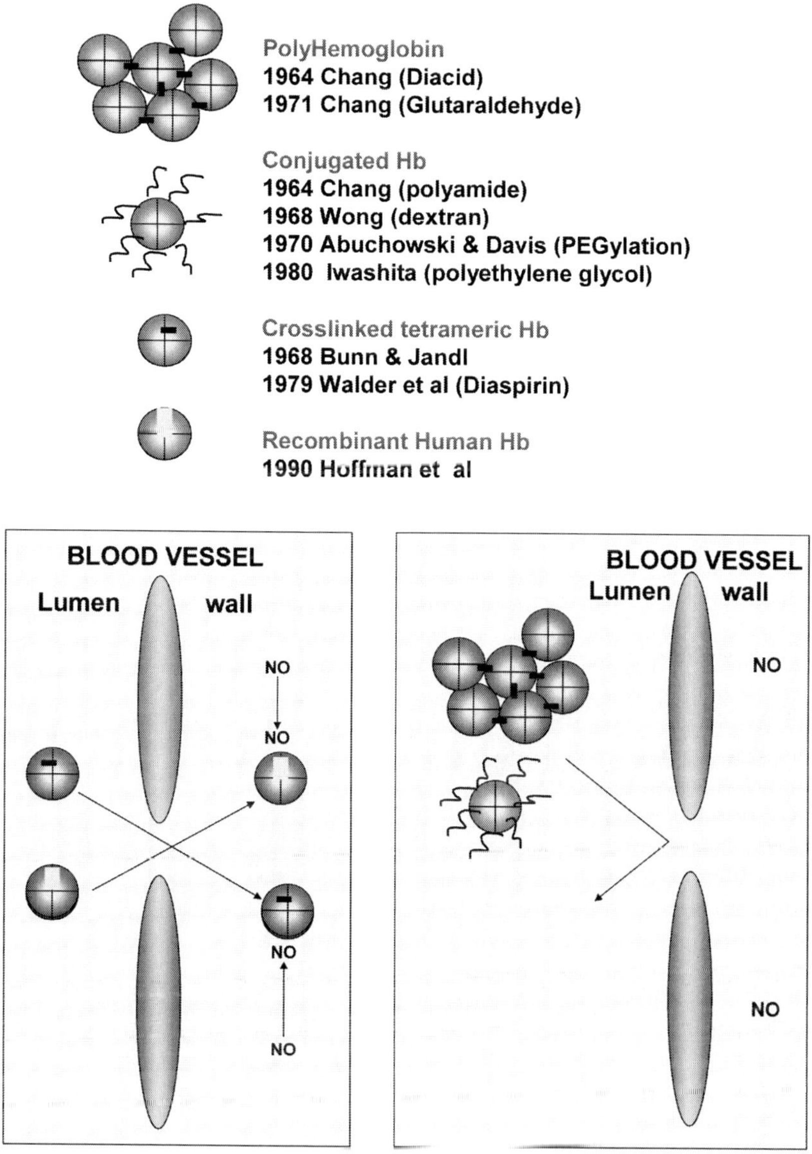

Fig. 1. Types of first-generation oxygen carriers as blood substitutes. Different types of first-generation blood carriers have been investigated and tested clinically. Single hemoglobin molecules resulted in much more vasopressor effects because they can cross the intercellular junction to bind and remove nitric oxide needed for normal vasoactivity (from Chang 2007, with permission).

FIRST-GENERATION OXYGEN CARRIERS

Figure 1 summarizes the four general methods of Hb modification. These are PolyHb based on nanobiotechnology, conjugated Hb, crosslinked tetrameric Hb and recombinant human Hb. As seen from Fig. 1, most of the basic principles for modified Hb were available by the 1960s. Unfortunately, there was no public demand or interest until after 1986, when human immunodeficiency virus (HIV) in donor blood became a major public concern. However, last minute developments met with many setbacks because of lack of prior research and basic knowledge. The present status of first-generation oxygen carriers is as follows (Table 1).

Polyhemoglobin is based on nanobiotechnology to assemble and crosslink the Hb molecules together into PolyHb. The first bifunctional agents used were sebacyl chloride (Chang 1964) and glutaraldehyde (Chang 1971). Later, two groups independently applied the glutaraldehyde-crosslinked PolyHb principle to develop oxygen carriers for clinical trials (Gould et al. 2002; Moore et al. 2009; Jahr et al. 2008). One of these has been approved for routine use in South Africa, where there is serious problem with HIV-contaminated donor blood (Jahr et al. 2008).

Conjugated Hb was first formed by the use of sebacyl chloride to crosslink diamine and Hb to form insoluble conjugated Hb (Chang 1964,1972) (Fig. 1). This has been extended to soluble conjugated Hb (Fig. 1) formed by linking one Hb molecule to soluble polymers. A maleimide-polyethylene glycol conjugated Hb was developed for clinical trial (Olofsson et al. 2006; Winslow 2006).

Intramolecularly crosslinked single tetrameric Hb was based on the use of a bifunctional cross-linker to crosslink the Hb molecules internally (intramolecular) (Fig. 1). This has been extensively developed and tested in clinical trials that showed vasopressor effect (Winslow 2006).

Recombinant human Hb is formed by the fusion of the two α subunits of each Hb molecule to prevent its breakdown into half molecule (dimers) (Fig. 1). A new recombinant human Hb has been prepared that does not bind nitric oxide thus obviating the problem with vasopressor effect (Winslow 2006). This is also a potential source of Hb for PolyHb and conjugated Hb and other future generation Hb based blood substitutes.

Vasopressor Effects

Intramolecularly crosslinked Hb and first-generation recombinant Hb blood substitutes contain 100% of molecular dimension modified Hb. Infusion causes vasopressor effects and also increased smooth muscle contractions. With another type of PolyHb that contains 36% molecular dimension modified Hb, significant vasoactivity and increased smooth

Table 1. Human donor blood vs nanobiotechnology-based blood substitutes.

Human donor blood		Nanobiotechnology-based blood substitutes
Infective agents	Rare in some regions using costly screening tests, but more frequent in other regions. If a new unknown infective agent appears, it might take years to develop a screening test for blood to be safe.	Infective agents can be sterilized and removed.
Source	Limited availability.	Unlimited because, in addition to human Hb (Hb), bovine Hb and recombinant human Hb can be used.
Blood group antigens	Blood group antigens need typing and cross-matching.	No blood group antigens.
Delay for use	Delay in use due to need for typing and cross-matching.	Can be used immediately because no typing or cross-matching is needed.
Storage stability	Can be stored for 42 da with standard refrigeration at 4°C.	Studies show that PolyHb can be stored for >1 yr at room temperature.
Circulation time	About 60 d depending on length of storage.	Circulate effectively with a half time of 24 h and are therefore good for short-term use such as surgery and in emergencies. There are ways to increase its length of function, including.
		repeated infusion, and combination with erythropoietin.
Function	Complete rbc functions. In addition to	First-generation ones function only as oxygen carriers and have several clinical uses.
	carrying oxygen, rbc also have antioxidant	Other uses require new generations of PolyHb (for example, PolyHb crosslinked to antioxidant.
	enzymes and enzymes for preventing MetHb formation.	enzymes). Future-generation nanodimension artificial rbc contain Hb and all of the enzyme systems of rbcs. Costs will increase with each new generation.
Vasopressor effects	No vasopressor effects.	To prevent vasopressor effects, small molecular-dimension Hb that can cross the intercellular junction has to be eliminated.

First-generation blood substitutes have a number of advantages over donor rbc. However, they are only oxygen carrier and do not have the complex enzyme system of rbc. New generations of blood substitutes are being developed to have the enzyme activities of rbc. Copyright permission (from Chang 2007 monograph).

muscle contractions were also observed when using larger volumes. On the other hand, systemic hypertension has not been a problem in clinical trials using PolyHb that contained < 1% molecular dimension modified Hb even when large volumes were infused (Gould et al. 2002). This has led to the proposal that the intercellular junctions of the endothelial lining of vascular wall allow molecular dimension Hb to enter into the interstitial space (Chang 1997; Gould et al. 2002) (Fig. 1). There, Hb acts as a sink in binding and removing nitric oxide needed for maintaining the normal tone of smooth muscles. This results in the constriction of blood vessels and other smooth muscles.

Design of Efficacy Animal Study for Hemorrhagic Shock

Various experimental designs were used in the testing of blood substitutes for the resuscitation of hemorrhagic shock. Those based on bleeding 30–40% of the blood volume are valid only for studying volume replacement, but this degree of blood loss does not require rbc replacement. Thus, this model is not sensitive enough for testing the efficacy for rbc substitutes in hemorrhagic shock. Testing the effectiveness of blood substitutes for both volume and rbc replacement requires a more severe model. For instance, using a modified Wigger's model with the removal of 2/3 of the total blood volume, we studied the effect of a single transfusion on the long-term survival of the animals (Chang 1997). Our results show that PolyHb blood substitutes are as effective as whole blood and more effective than 3 volume Ringers lactate, hypertonic-saline, dextran or albumin (Fig. 2).

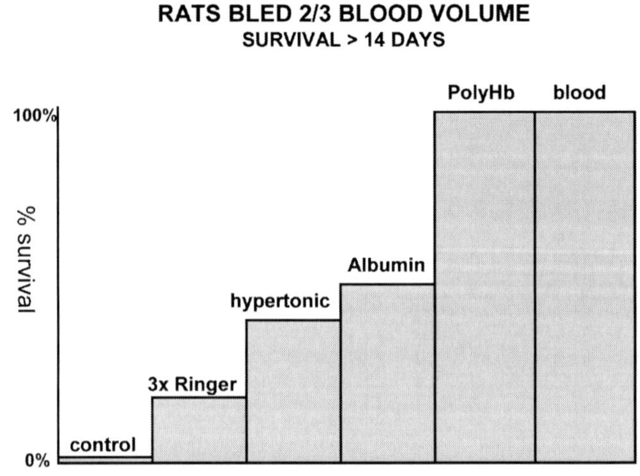

Fig. 2. Survival of hemorrhagic rat model. Long-term survival and recovery of 67% blood loss replaced with different types of solution. Only blood and PolyHb resulted in long-term survival (from Chang 2007, with permission).

Clinical Trials

Gould et al. (2002) reported their clinical trials on 171 patients using glutaraldehyde-crosslinked human PolyHb with < 1% unpolymerized Hb. They were able to replace extensive blood loss in trauma surgery and maintain the Hb level at the 8 to 10 g/dl needed for safe surgery with no reported side effects. Infusion of PolyHb in patients with Hb level as low as 2g/dl can raise the Hb level to within the 8 to 10 g/dl level with the patients recovering from surgery. Given that the supply of Hb from outdated donor blood is limited, a glutaraldehyde-crosslinked bovine PolyHb with < 3% unpolymerized molecular Hb has been developed and tested in clinical trials (Jahr et al. 2008) . For example, the team has carried out multicenter, multinational, randomized, single-blind, rbc-controlled clinical trials in patients undergoing elective orthopedic surgery. A total of 688 patients are randomized 1:1 to receive either the PolyHb or rbc at the time of the first perioperative rbc transfusion decision: 59.4% of the patients receiving PolyHb required no rbc transfusion all the way to follow-up; 96.3% avoided transfusion with rbc on the first postoperative day; and up to 70.3% avoided rbc transfusion up to day 7 after. This bovine PolyHb has been approved for routine clinical use in patients in South Africa, a region with higher incidence of HIV.

A multicenter clinical trial on the use of polyhemoglobin in pre-hospital emergencies (Moore et al. 2009) showed that no typing or matching is needed and PolyHb can be used on the spot. This clinical trial involved 714 pre-hospital emergency patients (Moore et al. 2009). These were trauma patients with blood pressure of less than 90 mmHg that were randomly divided into two groups: test group and control group. In this clinical trial, PolyHb was used on the spot in the ambulances, since, unlike donor blood, it required no typing or matching. The control group received standard management with saline infusion in the ambulance. The results showed that whereas the control group required rbc transfusion on admission into the hospital, the polyHb group required rbc transfusion only 14 h after admission into the hospital. However, there were side effects of myocardial infarction: 0.6% in the control group as compared to 3% in the PolyHb group. The authors of the study propose that PolyHb may be useful for patients at high risk of death when stored rbc are not available, even though there is some increase in frequency of adverse events (Moore et al. 2009).

This could be an important consideration in regions of the world with contaminated or insufficient donor blood. It could also be important in any emerging epidemic similar to HIV, or major civilian or armed conflicts. However, the FDA did not approve PolyHb for routine use in the United States, most likely because of the present availability of safe donor blood here. We also have to take regional and different situations into consideration. For example, in severe hemorrhagic shock with an 80% chance of death if there is not donor blood, what is the risk/benefit of PolyHb that would save lives (Fig. 2), but with a 3% chance of side effects of myocardial infarction, as compared to 0.6% when using donor blood? In any case, we have to take all points of view into consideration and to carry out further research to gain the much needed basic information on the complicated topic of rbc substitutes (2010b). One important area is vasoactivity of blood substitutes.

Vasoactivity

As mentioned earlier, some of the earlier blood substitutes that are small tetrameric Hb or those with a large proportion of tetrameric Hb can result in vasoconstriction and hypertension. A controversial meta-analysis combined the clinical results of different types of Hb, including intramolecularly crosslinked Hb that consists of 100% single molecules, and concluded that modified Hb causes vasoconstriction (Natanson et al. 2008). This meta-analysis did not take into consideration that modified Hb can be very different in composition and vasoactivity. On the other hand, we have to take all views into consideration and analyze all possibilities; further research is needed to clarify the issue of vasoactivity. Much research has been carried out on vasoactivity using different types of blood vessels, including hamster pouch microcirculation, aortic vessel ring perfusion, systemic blood pressure, and pulmonary artery perfusion. Since myocardial infarction is an important issue, we decided to test the effect of PolyHb on the coronary artery in rats. The heart rates in rats are very high and thus very sensitive to vasoconstriction, which results in a lack of oxygen supply. We prepared nanodimension PolyHb, each containing 0.4%, 16%, 38% and 78% of tetramers using the same glutaraldehyde crosslinking and characterized to ensure that they all had the same oxygen affinity (Yu et al. 2006; Chang 2007). The results showed that PolyHb with 0.4% of tetramer did not cause ECG changes. With increasing percentage of molecular dimension modified Hb, there is increasing elevation of the ST segment of the electrocardiogram. ST elevation is likely due to vasoconstriction of the coronary vessels resulting in decreased supply of oxygen to the heart. This result shows that vasoactivity varies extensively among different types of modified Hb. Small hemoglobin molecules cross the intercellular junction to remove nitric oxide resulting in vasoconstriction (Fig. 1). Further support of the role of nitric oxide in vasopressor effects has come from study showing that inhalation of nitric oxide can prevent vasopressor effects from infusion of tetrameric hemoglobin (Yu et al. 2008). However, this is not as simple as just the tetrameric Hb. Further studies (Yu et al.

2010) show that although PolyHb with < 1% tetrameric Hb does not cause vasoconstriction in normal mice, it causes vasoconstriction in diabetic mice or mice fed with high lipid diet. This supports my proposal at the 2009 International Symposium on Blood Substitutes for the need to have more narrow indications and the exclusion of patients and conditions with arteriosclerosis and other endothelial dysfunctions especially in situations of oxidative stress (Chang, 2010b). There is a substantial proportion of patients in the 2009 clinical trial who are between the age of 45 to 78 and thus more likely to have problems related to arteriosclerosis (Moore 2009).

SECOND-GENERATION RED BLOOD CELL SUBSTITUTES: OXYGEN CARRIERS WITH ANTIOXIDANT ACTIVITY

Ischemia-reperfusion

In order to use blood substitutes for some clinical conditions, such as arteriosclerosis severe sustained hemorrhagic shock, or stroke, we have to consider the issue of oxygen radicals. In prolonged severe hemorrhagic shock, stroke, and other ischemic conditions, there is insufficient supply of oxygen resulting in ischemia. Ischemia stimulates the production of hypoxanthine at a rate proportional to the duration of ischemia. When the tissue is again perfused with oxygen, reperfusion, hypoxanthine is converted into superoxide. By several mechanisms, superoxide results in the formation of oxygen radicals that can cause tissue injury (Chang 1997; D'Agnillo and Chang 1998; Alayash et al. 2007). The enzymes in rbc help to prevent this to a limited extent. Thus, superoxide dismutase (SOD) in rbc converts superoxide into hydrogen peroxide that is in turn converted by catalase (CAT) into water and oxygen. However, in severe and prolonged ischemia, the normal amounts of these enzymes in the rbc are not enough to prevent ischemia-reperfusion injury. We have therefore used nanobiotechnology to prepare a soluble PolyHb-CAT SOD complex with more CAT and SOD than those normally present in rbc (D'Agnillo and Chang 1998) (Fig. 3). This was done with a view to having an oxygen carrier with enhanced antioxidant enzymes to prevent ischemia-reperfusion injuries.

Oxygen Carrier with Antioxidant Activity

In the form of PolyHb-CAT-SOD, the enzymes stay in the circulation together with PolyHb that has a half life of 24 h in human. This is unlike the rapid removal of free SOD and CAT from the circulation (10 and 20 min, respectively). Furthermore, unless crosslinked to Hb, these enzymes

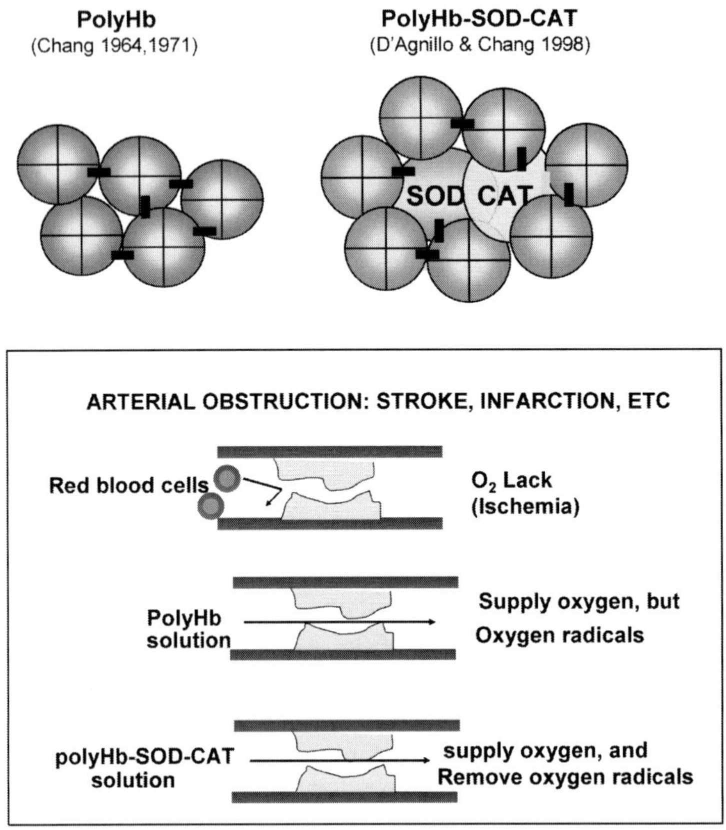

Fig. 3. Red blood cell and different types of blood substitutes in microcirculation. Arteriosclerosis causes narrowing and obstruction of blood vessels. Red blood cells, being 8 micron in diameter, have difficulty flowing through these partly obstructed vessels to supply the needed oxygen. PolyHb being a solution can perfuse partly obstructed vessels better than red blood cells to supply the needed oxygen. However, when ischemic tissue is perfused with an oxygen-rich solution, reactive oxygen species are generated, resulting in ischemia-reperfusion injury. Unlike PolyHb, PolyHb-SOD-CAT has the dual function of an oxygen carrier with the ability to remove oxygen radicals; like PolyHb, it is a solution and can perfuse partly obstructed vessels. (From Chang 2007, with permission.)

Color image of this figure appears in the color plate section at the end of the book.

are not located in close proximity to Hb and thus are less likely to give adequate protection to Hb-initiated oxygen radicals. *In vitro* studies reveal that crosslinked Hb-SOD-CAT is effective in removing oxygen radicals and hydrogen hydroxide, and in lessening damage to Hb and iron release.

Intestine and Ischemia-reperfusion

In severe sustained hemorrhagic shock, reperfusion even with rbc that contain the normal amounts of CAT and SOD may not be enough to prevent ischemia-reperfusion. Since PolyHb-CAT-SOD can contain more enzymes than rbc, we tested this preparation for reperfusion in an ischemia-reperfusion intestine model (Chang 2007). The use of PolyHb for reperfusion results in a substantial increase in oxygen radicals. On the other hand, the use of PolyHb-CAT-SOD markedly and significantly decreased the formation of oxygen radicals (Fig. 4).

The Brain and Ischemia-reperfusion

Arteriosclerosis is the most common cause of narrowing and obstruction of blood vessels, resulting in myocardial infarction or stroke. Red blood cells, being 8 micron in diameter, have difficulty flowing through these partly obstructed vessels to supply the needed oxygen to the brain (Fig. 3). PolyHb, being a solution, can perfuse partly obstructed vessels better than rbc to supply the needed oxygen. However, when an ischemic heart or brain is perfused with an oxygen-rich solution, reactive oxygen species are generated, resulting in ischemia-reperfusion injury. PolyHb-SOD-CAT is a soluble complex and can perfuse the narrowed arteries. However, unlike PolyHb, PolyHb-SOD-CAT has dual function of an oxygen carrier with the ability to remove oxygen radicals (Fig. 3). Cerebral ischemia is also present in some cases of severe trauma, where there is hemorrhagic shock plus brain damage.

We used a combined hemorrhagic shock and transient global ischemia-reperfusion rat brain model to compare the use of PolyHb-SOD-CAT solution with four other oxygen-carrying solutions: (1) PolyHb, (2) PolyHb with CAT and CAT added as solution without crosslinking to PolyHb, (3) purified Hb, and (4) oxygenated saline solution. Reperfusions were initiated 1 h after ischemia and the effects on blood-brain barrier and brain edema were followed for 6 h. Cerebral edema was measured as changes in cerebral water content. There was no significant difference between the PolyHb-SOD-CAT and sham control group when followed for 2, 4 and 6 h (Powanda and Chang 2002) (Fig. 5). On the other hand, significant increases in cerebral water content were observed after 4 h for PolyHb (234.8 ± 21.9 mg), free Hb plus SOD and CAT in free solution (152.3 ± 30.2 mg),oxygenated saline (171.3 ± 23.0 mg), and SF-Hb (279.8 ± 21.8 mg). Thus, except for PolyHb-CAT-CAT, all the solutions (oxygenated saline, SF Hb and PolyHb) resulted in cerebral edema. The result also shows that CAT and SOD have to be crosslinked into PolyHb-CAT-SOD to be effective in protecting against ischemia-reperfusion.

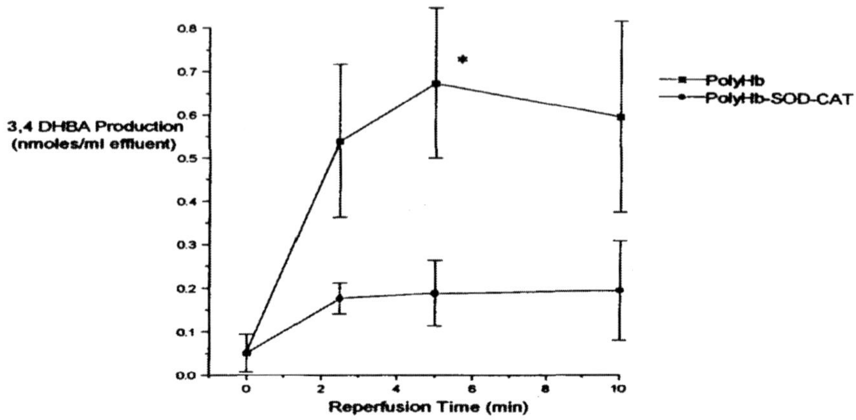

Fig. 4. Effects of perfusion of ischemic intestine on oxygen radicals. Oxygen radical measured as hydroxyl radical production assessed by the measurement of 3, 4 dihydroxybenzoate in ischemic intestine perfused with 5 g/dl of PolyHb () or 5 g/dl of PolyHb-SOD-CAT (). Intestinal effluent samples are collected during reperfusion at the indicated times. Data are presented as means ± SD. *Statistical significance, $p < 0.05$. (From Chang 2007, with permission.)

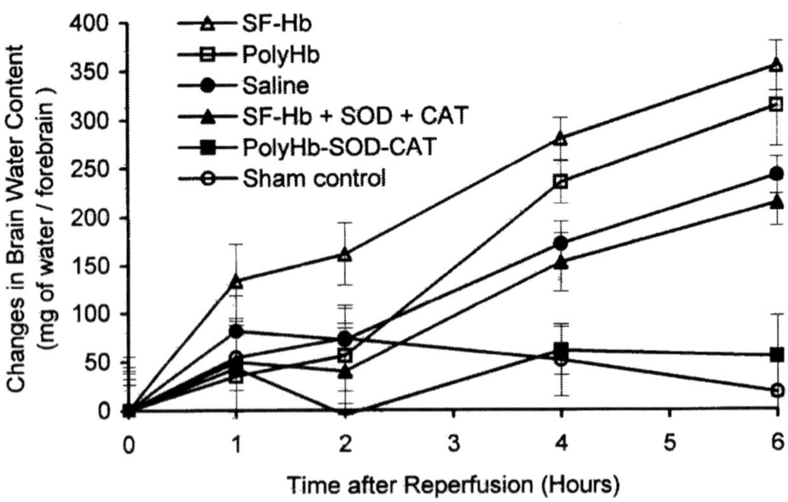

Fig. 5. Effects of reperfusion in ischemic brain on brain edema. This is followed as change in brain water content. The changes in brain water content of PolyHb-SOD-CAT treated animals are not significantly different from those of the sham control. The increase in water contents of saline, SF-Hb, SF-Hb + SOD + CAT, and PolyHb are significantly different from that of the sham control and PolyHb-SOD-CAT group by the 4th hour and increase thereon with time. Statistical significance is $P < 0.01$. (From Powanda and Chang 2003, with permission.)

Discussion

The above studies show that in ischemia-reperfusion the use of PolyHb-SOD-CAT resulted in significantly less production of oxygen radicals than PolyHb. This difference is much more obvious in the case of ischemia-reperfusion of intestine or brain, both organs being particularly sensitive to ischemia-reperfusion injury. On the other hand, the hind limb is much less sensitive to ischemia-reperfusion injury even when using PolyHb. Another area of potential application that has been studied is use of PolyHb-SOD-CAT in preventing ischemia-reperfusion damage to sensitive donor organs such as the kidney or liver (Chang et al. 2004). Another area investigated is cardiopulmonary bypass. Ischemia-reperfusion injury has been noted to occur after cardiopulmonary bypass in children undergoing surgery for congenital heart disease (Shum-Tim). An extension of our PolyHb-SOD-CAT approach is the development of synthetic mimics of SOD for complexing with modified hemoglobin in the form of polynitroxyl hemoglobin with covalently bound nitroxides (Buehler et al. 2004).

THIRD GENERATION: NANODIMENSION COMPLETE ARTIFICIAL RED BLOOD CELLS

The First Artificial Red Blood Cell

As discussed in the previous sections, the basic principle of nanobiotechnology-based blood substitutes (Chang 1964, 1971, 1972) has been developed into clinically useful PolyHb, conjugated Hb and PolyHb with antioxidant activities. We are now ready to continue with the earlier approach of a complete artificial rbc (Chang 1957, 1964, 1972) (Fig. 6). As discussed below, the low circulation time that was the single major obstacle to the practical realization of the original complete artificial rbc has now been partly solved.

Nanodimension Lipid Membrane Hb Vesicles

Increase in Circulation Time

Extensive research has been carried out by many groups using lipid membrane to encapsulate Hb into 200 nm lipid-encapsulated Hb (LEH) (Djordjevich and Miller 1980; Phillips et al. 1999; Tsuchida et al. 2006). The use of PEG-lipid membrane to form 200 nm diameter PEG-lipid membrane artificial rbc increased the circulation time to a clinically useful level (Phillips et al. 1999). This allowed the development of LEH for animal studies. Results showed that these were effective in replacing loss of blood

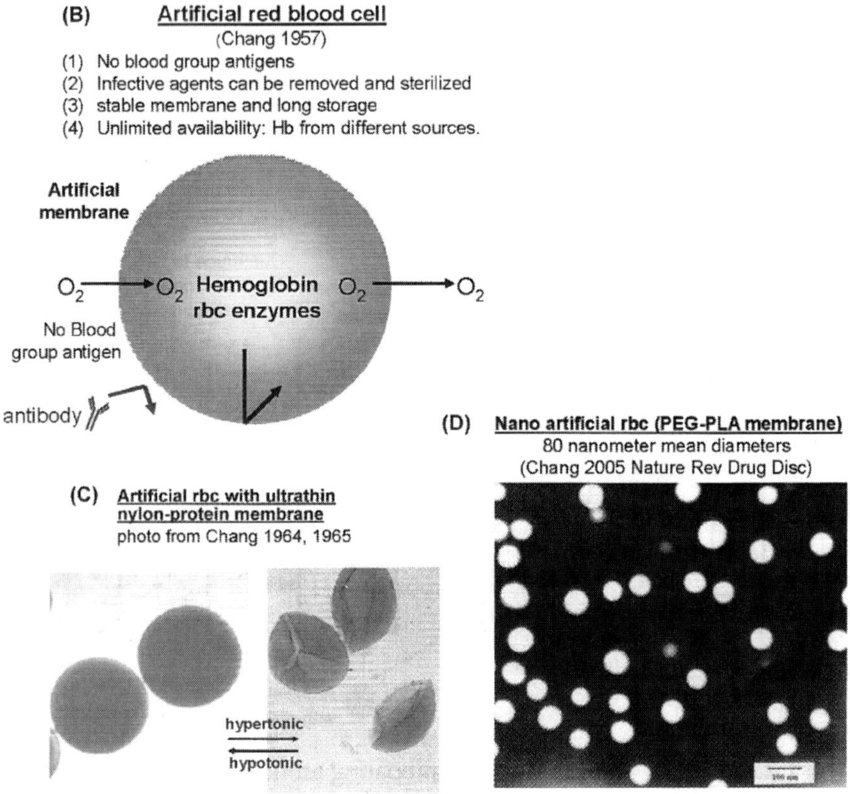

Fig. 6. Complete artificial red blood cells. (B) Schematic representation of artificial red blood cells. (C) Original micron dimension artificial red blood cell and (D) third generation nanodimension artificial red blood cell with 80 nm mean diameter. (From Chang 2007, with permission.)

Color image of this figure appears in the color plate section at the end of the book.

in hemorrhage shock and in exchange transfusion (Tsuchida et al. 2006). This is another important and major step in the development towards a more complete artificial rbc for clinical use.

Further Developments

LEH is useful for some clinical applications, but for other applications two further developments are needed: (1) The amount of lipid in the LEH may be too large. The smaller the particle, the larger will be the total surface area in the same volume of suspension. Thus, the total amount of lipid would be 10 times as high for the 1 ml suspension of 200 nm lipid membrane artificial rbc than for rbc. A large amount of lipid can decrease the phagocytic function of the reticuloendothelial system (RES). In traumatic hemorrhagic shock, the RES needs to be efficient in removing contaminating microorganisms. Furthermore, in ischemic reperfusion, lipid may induce lipid peroxidation. (2) The other problem with the use of LEH is the oxidation of Hb into MetHb that cannot carry oxygen (Tsuchida et al. 2006). Reducing agents present in the circulating blood cannot cross the lipid membrane to reduce the rate of MetHb formation. Glucose that is needed for the MetHb reductase system inside rbc also cannot cross the lipid membrane of LEH.

Nanodimension Biodegradable Polymeric Membrane Artificial rbc (Nano Artificial rbc)

The next step is to markedly reduce the amount of lipid needed for the membrane and replace it with biodegradable polymeric membrane. We used our earlier background in biodegradable membrane artificial cells (Chang 1972) to prepare nanodimension artificial rbc (nano rbc) of 80 nm mean diameter (Fig. 6) using PLA, PEG-PLA membrane and other biodegradable polymers (Chang 1997, 2005, 2007, 2009, 2010a, 2010c; Chang et al. 2003). The membrane thickness is 5–15 nm. Since polymer is stronger than lipid and is also porous, much less membrane material is required. Thus, we can replace most of the 6g/dl of lipid membrane in LEH with a smaller amount of biodegradable polymeric membrane material (Fig. 7). This marked decrease in the lipid component would lessen effects on the RES and lessen lipid peroxidation in ischemia-reperfusion. Furthermore, polylactic acid (PLA) is degraded in the body into lactic acid, then to carbon dioxide (Fig. 7). These are all normal body metabolites. For a 500 ml suspension of PLA nano artificial rbc, the total lactic acid produced is 83 mEq. The normal resting body lactic acid production is 1000–1400 mEq/day and the maximal body capacity to breakdown lactic acid is 7080 mEq/day. Thus, even in ischemic conditions, the amount of lactic acid produced can be well handled by the body.

Characteristics of Nano Artificial rbc Including MetHb Reduction

We can vary the Hb content in the PLA nano artificial rbc suspension from 3 g/dl to 15 g/dl (same as whole blood) (Fig. 7). This has normal P_{50}, Hill's coefficient and Bohr coefficients. Red blood cell enzymes such as carbonic anhydrase, CAT, SOD and the MetHb reductase system have been enclosed within these nano artificial rbc and retain their activities (Chang 1979, 2005, 2007; Chang et al. 2003). Unlike lipid membrane, biodegradable polymeric membrane is permeable to glucose. Thus, the inclusion of rbc MetHb reductase system prevents MetHb formation even at 37°C and we can also

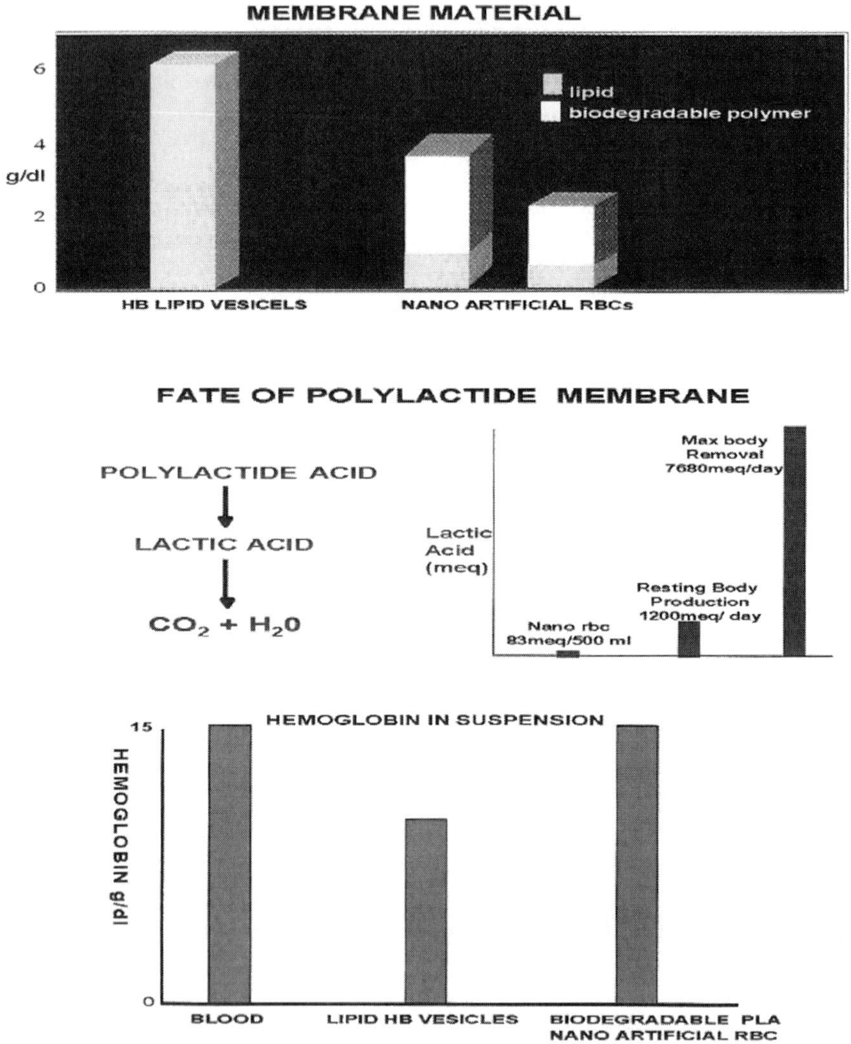

Fig. 7. Properties of nanodimension artificial red blood cells. Top: Amount of membrane material in hemoglobin lipid vesicles compared to polylactide membrane nano artificial red blood cells. Middle: Fate of polylactide membrane in PLA nano rbc compare to PLA metabolism in human. Lowe: Hb concentration reported. (From Chang 2007, with permission.)

convert MetHb to Hb at 37°C. Unlike lipid membrane, the nanocapsule membrane also allows plasma-reducing factors to enter the nanocapsules to prevent MetHb formation (Chang 2007).

Circulation Time and *In Vivo* Properties

In order to increase the circulation time, we synthesized a new PEG-PLA copolymer for the nano rbc membrane (Chang, 2005, 2007; Chang et al. 2003). After extensive research, we now have a circulation time in rats that is double that of glutaraldehye-crosslinked PolyHb. Since the RES in rat is much more efficient in removing particulate matter than in human, it is likely that the half life would be even longer in human. Nano artificial rbc did not have adverse effects on the histology and function of liver, spleen and kidney in rats when analyzed on days 1, 7 and 21 after infusions (Liu and Chang 2008).

Oxygen Carrier with Platelet-like Activity

In extensive blood loss, replacement of plasma volume and oxygen carrier is not enough. There is also the need to replace platelets and coagulation factors. We therefore use nanobiotechnology to develop a blood substitute that is an oxygen carrier with platelet-like properties.

Hb is crosslinked with fibrinogen to form a soluble nanobiotechnoligcal complex of PolyHb-fibrinogen (PolyHb-Fg) (Wong and Chang 2007). Exchange transfusion of 80% in rats of up to 98% using PolyHb-Fg did not result in clotting problems. The use of PolyHb alone resulted in some clotting problems after 80% exchange, worsening with further increase in blood replaced.

PRESENT STATUS AND FUTURE

Regional differences and the potentials of unknown infective agents must be included in any discussion of the prospect of blood substitutes and also in the degree of regulatory requirements (Chang 2010b). Blood substitute is urgently needed in regions of the world where there is severe shortage of donor blood because of cultural or religious beliefs that make people less willing to donate blood. It is also urgently needed in regions with higher incidence of infective agents such as HIV and thus higher potential for contaminated donor blood. It is less urgent in regions with lower incidence of HIV and where costly tests are being used to screen out infective agents in donated blood. On the other hand, it is also important to remember the past unexpected outbreak of HIV and hepatitis C and the resulting contamination of donated blood that persisted for years until proper screening tests were developed. If this should happen again with yet unknown agents (e.g., avian flu, bioterrorism) then it would again be disastrous if no blood substitutes, even first-generation blood substitutes, were immediately available. Past experience has shown that it takes many years to develop ideas on blood substitutes into products and that lack of basic information has resulted in much failure and delays. It is important to carry out basic research to gain important basic information needed for the simultaneous development of blood substitutes. Meanwhile, new generations of modified Hb are being developed that can modulate the effects of nitric oxide for those clinical applications that may have potential problems related to oxygen radicals. A further development is the use of PEG-lipids or PEG-biodegradable polymer membranes to prepare nanodimension artificial rbc containing Hb and complex enzyme systems.

APPLICATIONS TO AREAS OF HEALTH AND DISEASE

Blood substitutes have important implications in transfusion medicine. Blood substitutes do not have blood group antigens. They can be used immediately on the spot without the need for matching and typing. Red blood cell membrane contains blood group antigens and donor blood requires typing and matching before it can be transfused into patients. This results in delays in emergency situations. Some blood substitutes could be stored for more than a year at room temperature. Donor blood requires storage at 4°C using standard method and is good for less than 42 d. Infective agents such as HIV and other viruses and microorganisms can be removed from blood substitutes. Red blood cells cannot be sterilized to remove hepatitis viruses, HIV and other potential emerging infective agents. Blood substitutes can be prepared using hemoglobin from different sources, whereas donor blood depends on human donors and may be in short supply especially in emergencies, war or disasters.

Key Facts

- The author started research on artificial red blood cells in 1957.
- However, there was no initial interest in this area of research until the 1980s, when HIV resulted in contaminated donor blood.
- Many patients were infected with contaminated blood. Unfortunately, there was no blood substitute to replace the HIV-contaminated donor blood.
- The spread of HIV stimulated industries to produce blood substitutes, but this was not as simple as people thought. In the 30 years since that time, no blood substitute has yet been approved for routine patient use in North America.
- It is clear from that experience that blood substitutes are complex and cannot be developed into a product without more basic knowledge.

- The history of HIV contamination also suggests that we should intensify efforts to carry out both basic research and development in blood substitutes. We would not want to face the same problem of contaminated blood from another yet unknown epidemic. Suitable blood substitutes are also needed in some major surgeries and in cases of major disasters or war.

Summary Points

- Nanobiotechnology is the assembling of biological molecules into structures of 1 to 100 nm.
- First-generation blood substitutes have been developed. These are so-called polyhemoglobin (PolyHb) formed by the nanobiotechnological assembling of hemoglobin molecules into soluble complexes.
- Unlike red blood cells, PolyHb is only an oxygen carrier.
- Unlike red blood cells, it does not have blood group antigens and can be used immediately on the spot without the need for typing or cross-matching.
- PolyHb can be stored for more than one year at room temperature and can therefore be easily transported and stored. Donor blood has to be stored at 4°C and only for 42 d.
- Infective agents can be sterilized and removed from blood substitutes and therefore avoid problems related to contaminated donor blood.
- Unlike donor blood, there is an unlimited supply of blood substitutes because, in addition to human Hb (Hb), bovine Hb and recombinant human Hb can be used.
- Two types of PolyHb have been developed and tested in clinical trials in patients. Although one of them has been approved for routine clinical use in patients in South Africa, where there is higher incidence of HIV, in North America and other countries there is a need for further risk/ benefit ratio analysis and narrowing of use in more specific clinical conditions.
- PolyHb circulates effectively with a half time of 24 h and is therefore good for short-term use such as surgery and in emergencies. There are ways to increase its length of function, including repeated infusions and combination with erythropoietin.
- PolyHb is only an oxygen carrier and therefore cannot be used in clinical conditions with potential for ischemia-reperfusion injuries, as in sustained severe hemorrhagic shock, stroke or coronary ischemic diseases.
- Second-generation biotechnological blood substitutes are being developed to solve this problem. For example, hemoglobin can be assembled with antioxidant enzymes such as CAT and SOD to form a

soluble complex of PolyHb-SOD-CAT. Animal studies show that this can prevent ischemia-reperfusion injuries.

- In the meantime, complete artificial red blood cells of nanodimensions are being developed. One type is based on a bilayer lipid membrane artificial red blood cell of 200 nm. Another type is based on biodegradable membrane artificial red blood cell of 120-80 nm. These have much higher circulation time than PolyHb. They also contain all the enzyme systems normally present in red blood cells.

Definitions

Crosslinking: The use of chemical linkage to assemble hemoglobin molecules.

Ischemia-reperfusion injury: After severe lack of oxygen (ischemia), introduction of oxygen carriers (reperfusion) can result in the production of oxygen radicals and tissue damage.

Nanobiotechnology: The assembling of biological molecules into nanodimension structures of 1 to 100 nm in diameter, membrane thickness or diameter of nanotubules.

Polyhemoglobin: The assembling of hemoglobin molecules into nanodimension structures by crosslinking.

ABBREVIATIONS

CAT	:	catalase
Fg	:	fibrinogen
HIV	:	human immunodeficiency virus
Hb	:	hemoglobin
PolyHb	:	polyhemoglobin
rbc	:	red blood cells
SOD	:	superoxide dismutase

References

Alayash, A.I., F. D'Agnillo and P.W. Buehler. 2007. First-generation blood substitutes: what have we learned? Biochemical and physiological perspectives. Expert Opin. Biol. Ther. 7: 665–675.

Atoji, T., M. Aihara, H. Sakai, E. Tsuchida and S. Takeoka. 2006. Hemoglobin vesicles containing methemoglobin and tyrosine to suppress methemoglobin formation *in vitro* and *in vivo*. Bioconjugate Chem. 17: 1241–1245.

Buehler, P.W., C.R. Haney, A. Gulati, L. Ma and C.J. Hsia. 2004. Polynitroxyl hemoglobin: a pharmacokinetic study of covalently bound nitroxides to hemoglobin platforms. Free Radical Biol. Med. 37: 124–135.

Chang, E.J., T.H. Lee and K. Mun. 2004. Effects of polyhemoglobin-antioxidant enzyme complex on ischemia-reperfusion in kidney. Transplant Proc. 36: 1952–1954.

Chang, T.M.S. 1957. "Hemoglobin Corpuscles" Honours Physiology Research Report, 1957, Medical Library, McGill University.
Chang, T.M.S. 1964. Semipermeable microcapsules. Science 146: 524.
Chang, T.M.S. 1971. Stabilisation of enzymes by microencapsulation with aconcentrated protein solution or by microencapsulation followed by cross-linking with glutaraldehyde. Biochem. Biophys. Res. Common. 44: 1531–1536.
Chang, T.M.S. 1972. Artificial cells. Monograph. Charles C Thomas, Springfield, IL (full text available with free access at www.artcell.mcgill.ca).
Chang, T.M.S. 1976. Biodegradable semipermeable microcapsules containing enzymes, hormones,vaccines, and other biologicals. J Bioeng. 1: 25-32.
Chang, T.M.S. 1997. Monograph. Red blood cell substitutes: Principles, Methods, Products and Clinical Trials Vol I Karger/Landes Systems, Basel, Switzerland (free access on www.artcell.mcgill.ca).
Chang, T.M.S., D. Powanda and W.P. Yu. 2003. Ultrathin polyethylene-glycol-polylactide copolymer membrane nanocapsules containing polymerized Hb and enzymes as nano-dimension RBC substitutes. Artificial Cells, Blood Substitutes and Biotechnology 31: 231–248.
Chang, T.M.S. 2005. Therapeutic applications of polymeric artificial cells. Nature Review Drug Discovery 4: 221–235.
Chang, T.M.S. 2007. Monograph on "ARTIFICIAL CELLS: Biotechnology, nanotechnology, blood substitutes, regenerative medicine, bioencapsulation, cell/stem cell therapy" Singapore, World Science Publisher/Imperial College Press, pp. 1–452 (full text available with free access at www.artcell.mcgill.ca).
Chang, T.M.S. 2009. Nanobiotechnology for hemoglobin based blood substitutes. Critical Care Clinics 25: 373–382.
Chang, T.M.S. 2010c. Artificial Cells, Blood Substitutes and Nanomedicine. McGill University website www.artcell.mcgill.ca
Chang, T.M.S. 2010a. Blood replacement with engineered hemoglobin and hemoglobin nanocapsules. WIRE Nanomed. Nanobiotechnol. 2: 418–430.
Chang, T.M.S. 2010b. Editorial: Blood substitutes in 2010 Artificiall Cells, Blood Substitutes and Biotechnology 38: 1.
D'Agnillo, F., and T.M.S. Chang. 1998. PolyHb-superoxide dismutase. catalase as a blood substitute with antioxidant properties. Nature Biotechnol. 16: 667–671.
Djordjevich, L., and I.F. Miller. 1980. Synthetic erythrocytes from lipid encapsulated hemoglobin. Exp. Hematol. 8: 584.
Gould, S.A., et al. 2002. The life-sustaining capacity of human polymerized Hb when red cells might be unavailable. J. Am. Coll. Surg. 195: 445–452.
Jahr, J.S., C. Mackenzie, L.B. Pearce, A. Pitman and A.G. Greenburg. 2008. HBOC-201 as an alternative to blood transfusion: efficacy and safety evaluation in a multicenter phase III trial in elective orthopaedic surgery. J Trauma 64: 1484–1497.
Liu, Z.C., and T.M.S. Chang. 2008. Long term effects on the histology and function of livers and spleens in rats after 33% toploading of PEG-PLA-nano artificial red blood cells. Artificial Cells, Blood Substitutes and Biotechnology 36: 513–524.
Moore, F.F., F.A. Moore, T.C. Fabian, A.C. Bernard, G.J. Fulda, D.B. Hoyt, T.M. Duane, L.J. Weireter Jr., G.A. Gomez, M.D. Cipolle, G.H. Rodman Jr., M.A. Malangoni, G.A. Hides, L.A. Omert and S.A. Gould. 2009. Human polymerized hemoglobin for the treatment of hemorrhagic shock when blood is unavailable: The USA multicenter trial. J. Am. Coll. Surg. 208. 1–13.
Natanson, C., S.J. Kern, P. Lurie, S.M. Banks and S.M. Wolfe. 2008. Meta-analysis: risk of myocardial infarction and death: cell-free hemoglobin-based blood substitutes. JAMA 299: 2304–2312.
Olofsson, C., A. Torbjo, T. Johansson, S. Larsson, P. Nellga, S. Ponzer, B. Fagrell, R. Przybelski, P. Keipert, N. Winslow and R.M. Winslow. 2006. A multicenter clinical study of the safety

and activity of maleimide-polyethylene glycol–modified hemoglobin (Hemospan®) in patients undergoing major orthopedic surgery. Anesthesiology 105: 1153–1163.

Perutz, M.F. 1989. Myoglobin and Hb: Role of distal residues in reactions with haem ligands. Trends Biochem. Sci. 14: 42-44.

Phillips, W.T., R.W. Klipper, V.D. Awasthi, A.S. Rudolph, R. Cliff, V. Kwasiborski and B.A. Goins. 1999. Polyethylene glyco-modified liposome-encapsulated hemoglobin: a long circulating red cell substitute. J. Pharm. Exp. Ther. 288: 665–670.

Powanda, D. and T.M.S. Chang. 2002. Cross-linked polyHb-superoxide dismutase-catalase supplies oxygen without causing blood brain barrier disruption or brain edema in a rat model of transient global brain ischemia-reperfusion Artificial Cells, Blood Substitutes and Biotechnology 30: 25–42.

Tsuchida, E., H. Sakai, H. Horinouchi and K. Kobayashi. 2006. Hemoglobin-vesicles as a transfusion alternative. Artificial Cells, Blood Substitutes and Biotechnology 34: 581–588.

Winslow, R.M. (ed.). 2006. Blood Substitutes. Academic Press, Amsterdam 1–548.

Wong, N., and T.M.S. Chang. 2007. Polyhemoglobin-fibrinogen: a novel blood substitutes with platelet-like activity for extreme hemodilution. Artificial Cells, Blood Substitutes and Biotechnology 35: 481–489.

Yu, B.L., Z.C. Liu and T.M.S. Chang. 2006. PolyHb with different percentage of tetrameric Hb and effects on vasoactivity and electrocardiogram. Artificial Cells, Blood Substitutes and Biotechnology 34: 159–175.

Yu, B., M.J. Raher, G.P. Volpato, K.D. Bloch, F. Ichinose and W.M. Zapol. 2008. Inhaled nitric oxide enables artificial blood transfusion without hypertension Circulation 15(117): 1982–1990.

Yu, B., M. Shahid, E.M. Egorina, M.A. Sovershaev, M.J. Raher, C. Lei, M.X. Wu, K.D. Bloch and W.M. Zapol. 2010. Endothelial dysfunction enhances vasoconstriction due to scavenging of nitric oxide by a hemoglobin-based oxygen carrier. Anesthesiology 112: 586–594.

6

Collagen Scaffolds and Their Application to Cardiology—the Importance of Matrix Interactions and Nanoscale Topography

Lynn Donlon[1,a] and *Daniel Frankel*[1,b,*]

ABSTRACT

The ability of the heart to repair itself after injury is extremely limited, resulting in a high demand for corrective surgery and donor organs. Engineered cardiac tissue, in which cardiomyocyte cells are cultured on a scaffold *in vitro*, could provide a viable alternative. Tissue constructs may be natural or synthetic, but a large proportion are based on collagen, which is a key component of the extracellular matrix. Collagen has the advantage of specific biomarkers that promote cell adhesion and differentiation but can lack sufficient mechanical strength. The mechanical properties of synthetic scaffolds are easily tailored; however, they lack the necessary biomarkers.

[1]School of Chemical Engineering and Advanced Materials, Newcastle University, Merz Court, Newcastle, NE1 7RU.
[a]E mail: lynn.donlon@newcastle.ac.uk
[b]E-mail: d.j.frankel@newcastle.ac.uk
*Corresponding author

List of abbreviations after the text.

Collagen possesses a hierarchical structure ranging from the specific $(Gly\text{-}X\text{-}Y)_n$ amino acid sequence at the nanoscale, through individual tropocollagen molecules, to fibrils and fibres measuring tens of micrometres. The mechanical properties of collagen are related to the interplay between structures at two characteristic length scales. Alterations to the collagen structure lead to significant changes in mechanical properties. These structural changes can be deliberate (e.g., crosslinking) or a result of genetic modifications *in vivo*, such as Ehlers-Danlos syndrome.

A number of researchers have examined the use of collagen in myocardial tissue engineering, ranging from monolayer culture to 2D planar aligned scaffolds and 3D engineered constructs. The culture environment is extremely important in the differentiation of cardiomyocytes to the correct phenotype. The environment also plays a key role in cell-cell and cell-matrix interactions. Such interactions are vital in the development of synchronously beating myocardial tissue. A 3D environment results in myocardial tissue that provides a more accurate mimic for that found *in vivo*. Successful collagen-based constructs for *in vitro* myocardial tissue engineering range from gels consisting of a mixture of collagen and Matrigel™ to aligned tubular scaffolds. Mixtures of collagen and other extracellular matrix proteins such as elastin or synthetic polymers such as poly(ε-caprolactone) have also been used successfully. Alternative approaches involve *in vivo* implantation of acellular collagen grafts or decellularization of natural heart tissue followed by reseeding with cardiomyocytes. Forces within collagen gels can be measured quantitatively using microelectromechanical systems technology. The mechanical properties of collagen scaffolds can be improved by the introduction of nanoscale surface features such as nanochannels.

INTRODUCTION

Cardiovascular disease is a leading cause of death within the developed world. In the UK alone, 2.6 million people currently live with coronary heart disease, the most common cause of myocardial infarction (heart attacks) (www.bhf.org.uk). Once damaged, the heart has a limited ability to repair itself and, in the region surrounding infarction, healthy myocardium is replaced with fibrous non-contractile scar tissue. Scarring reduces contractile efficiency, forcing the heart to develop compensatory mechanisms. However, the additional strain placed upon the remaining healthy myocardial tissue can not be maintained indefinitely, resulting in eventual heart failure (Jawad et al. 2008).

Significant advances in cardiac surgery allow for the correction of minor defects in heart structure and function; however, in the most severe cases replacement of the whole organ remains the only option. Since the first procedure performed by Barnard in 1967, the success of organ transplantation has increased markedly. Progress can be attributed not only to advances in surgical techniques, but to improvements in post-operative care and more effective immunosuppressant drugs. However, demand for donor organs far outweighs supply and organ rejection still remains a significant problem. An alternative to the use of donor organs is clearly required.

CONCEPTS OF TISSUE ENGINEERING

Tissue engineering is a highly multidisciplinary field that combines engineering and the life sciences in the development of biological substitutes to replace or restore tissues and organs (Langer and Vacanti 1993). The concept of myocardial tissue engineering was developed in an attempt to overcome the shortage of donor organs. Usually, the process involves the integration of cells with biocompatible and biodegradable scaffolds *in vitro*, prior to transplantation into the body. Additional signalling molecules are often used to increase biocompatibility (Fig. 1). Tissue engineering constructs must fulfil a number of criteria (Jawad et al. 2008):

- The scaffold must be biocompatible and preferably biomimetic. The ideal construct will also promote cell adhesion and differentiation into cardiomyocytes.
- The scaffold material must be biodegradable, at the appropriate rate, to non-toxic byproducts.
- The construct must demonstrate the appropriate mechanical integrity and ideally contribute to continuous stretching/relaxing of myocardium tissue (heartbeat).

As well as providing replacement organs, engineered myocardial tissue could provide an ideal model for artificial organs in the screening of drugs that would traditionally have been tested in animal models. The effects of drugs in animals often do not accurately mimic the effect in the human body. Artificial tissue also provides a basis to study the effect of expression of specific genes, vital to a number of processes within the cardiovascular system.

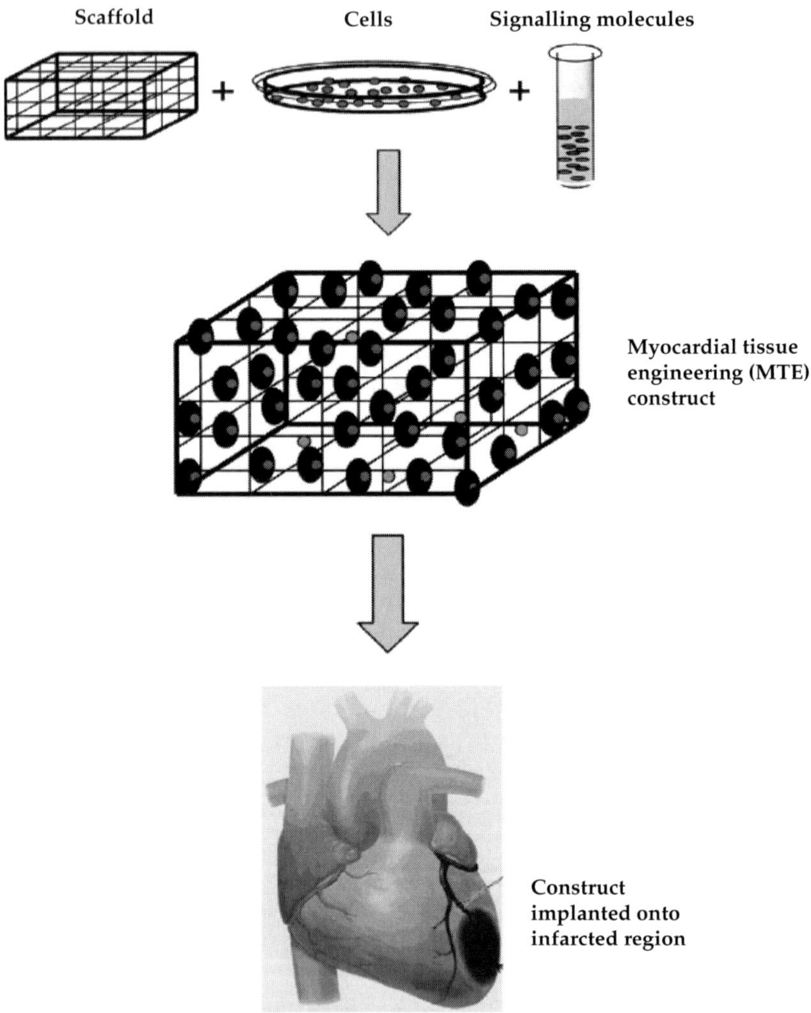

Fig. 1. Schematic representation of the principle of myocardial tissue engineering. Scaffold materials may be natural (e.g., collagen) or synthetic (e.g., poly(ε-caprolactone)). Signalling molecules are usually proteins found naturally within the extracellular matrix. Cells may originate from a variety of sources, but the most common are embryonic cardiomyocytes. Published with permission from Oxford University Press (Jawad et al. 2008).

SCAFFOLD MATERIALS FOR CARDIAC TISSUE ENGINEERING

A wide variety of materials have been examined as potential candidates for scaffolds in myocardial tissue engineering, ranging from natural constituents of the extracellular matrix (ECM) (such as collagen and elastin) to purely synthetic, biodegradable polymers such as poly(lactide) and poly(ε-caprolactone), PCL. Synthetic polymers have the advantage that their properties can be tailored to meet a given specification (e.g., morphology, mechanical properties, degradation kinetics); however, they do not possess the specific biomarkers of natural ECM. Natural materials such as collagen and gelatin promote cell adhesion and differentiation, but they often demonstrate weak and inconsistent mechanical properties when compared with synthetic analogues and can provoke an immune response *in vivo* (Li et al. 2000; Eschenhagen et al. 1997). Bio-inspired polymers that combine the advantages of both natural and synthetic components are a viable alternative and are the subject of much current research (Heydarkhan-Hagvall et al. 2008). Detailed discussion of synthetic, polymeric scaffolds is beyond the scope of this review, which will instead focus on the use of collagen in cardiovascular tissue engineering.

COLLAGEN: NATURE'S SUPPORT STRUCTURE IN THE EXTRACELLULAR MATRIX

The ECM plays a vital role in living tissues and organs, ranging from contribution to structural integrity to responsibility for cell adhesion, migration and differentiation (Koláčná et al. 2007). It consists of a complex mixture of proteins, which include collagens, fibronectin, elasin, laminin and entactin. It also contains proteoglycans such as perlecan. Collagens are the most abundant of these proteins and are essential in maintaining the structural integrity of organs and tissues. The biochemical and biophysical properties of the ECM have been studied extensively for a multitude of different soft tissues; however, the role of the ECM in cardiology remains poorly understood. Formovsky and co-workers present a review of the key ECM components, fibrillar collagen, elastin and proteoglycans, and their contribution to the cardiovascular system (Formovsky et al. 2010). ECM components are essential to signalling mechanisms and also make a key contribution to the mechanical properties of myocardium and pericardium as well as blood vessels and valves.

The Structure of Collagen

In order to develop effective biomimetic scaffolds for tissue engineering, it is vital to understand the structure and mechanics of collagen in detail. Like many biological materials, the structure of collagen is hierarchical, ranging from individual amino acids at the nanoscale through to fibres measuring tens of micrometres. Collagen exists in more than 20 different forms, classified according to their structure. All share the same characteristic amino acid sequence $(Gly\text{-}X\text{-}Y)_n$. Every third amino acid is a glycine, X is typically, though not always, a proline residue, while position Y is occupied by a variety of amino acids of which approximately every seventh residue is hydroxyproline. The secondary structure of collagen is α-helical, with three chains combining to form a triple helix structure. The chains may be identical (homotrimers such as type II collagen), a mixture containing two identical chains, the third being different (heterotrimers such as type I collagen) or even a mixture of three different chains (Koláčná et al. 2007). The α-triple helix chains are referred to as tropocollagen (TC) molecules and are typically ~280 nm in length and 1.5 nm in diameter. TC molecules arrange in a staggered fashion to form fibrils of the order of 1 μm, which further organize to form collagen fibres measuring tens of microns (Fig. 2) (Buehler 2008).

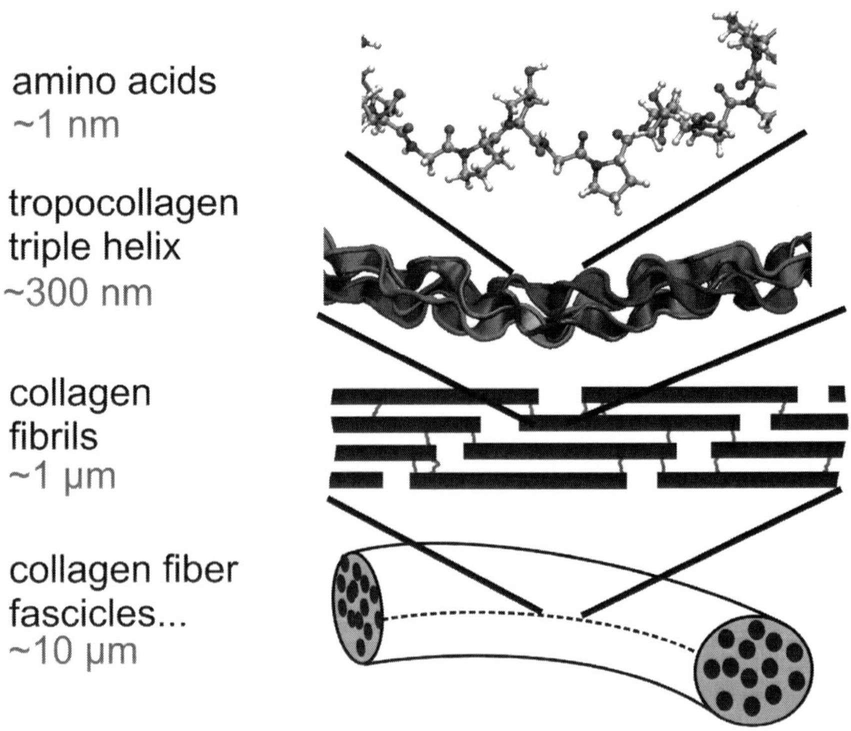

Fig. 2. Schematic representation of a selection of structural features of collagen. Collagen possesses a hierarchical structure with features that range from the characteristic amino acid sequence at the nanoscale, through tropocollagen molecules and fibrils at the sub-micrometre level to fibres measuring tens of micrometres. Published with permission from Elsevier Journals (Buehler 2008).

Color image of this figure appears in the color plate section at the end of the book.

The Mechanical Properties of Collagen at the Macro and Nanoscale

Detailed understanding of the mechanical properties of collagen requires consideration over a number of length scales and, in particular, the interplay between them (Buehler 2006). Most research to date has focused on the measurement of macroscopic mechanical properties of collagen fibres and scaffolds (Gentleman et al. 2003). Other studies have examined the mechanical properties of collagen fibrils using atomic force microscopy (Wang et al. 2005) and the localized mechanical properties of an aligned collagen scaffold using nanoindentation techniques (Chaudhry et al. 2009). However, these studies all consider a particular aspect of the collagen structure in isolation.

Buehler has used theoretical and molecular modelling to great effect to highlight the importance of the interplay between different levels of the hierarchical collagen structure. Up until this study, the natural design of collagen, in particular its underlying strength in relation to the length of individual TC molecules and staggered arrays of fibrils, remained poorly understood. Buehler demonstrates that the deformation mechanics of collagen fibrils can be classified in terms of three processes: intermolecular shear, slip-pulse propagation and fracture of individual TC molecules (Buehler 2006). A quantitative description of these mechanisms can be provided in terms of two characteristic length scales, χ_s and χ_R, with the governing deformation mechanism being a function of the ratio of the two, χ_s/χ_R.

Applications to Other Areas of Health and Disease

Changes to the underlying structure of collagen can have a severely detrimental impact upon mechanical properties. *In vivo* such changes usually have a genetic basis and are evidenced by diseases such as Ehlers-Danlos syndrome and osteogenesis imperfecta (brittle bone disease) (Myllyharju and Kivirikko 2001; Yalovaç and Ulusu 2007). To date, over 1000 mutations in a selection of genes have been identified as contributing to defections in the structure of collagen proteins. The consequences of such mutations cover a broad spectrum from the relatively minor to lethal disorders. The most common mutations involve substitution of glycine residues for larger amino acids. The positioning of glycine within the amino acid sequence is essential to the formation of the collagen triple helix and any substitution leads to disruption of the structure. The *in vivo* synthesis of collagen triple helix originates from the C terminus of the amino acid chain and, in general, mutations within this area lead to more severe disruption.

Fabrication of Fibrous Collagen Structures and Their Application to Cardiology

Several techniques exist for the fabrication of collagen and synthetic biocompatible polymers into fibrous structures, including electrospinning, free-form fabrication and extrusion. Each manufacturing process leads to variation in the underlying collagen structure, which manifests itself in the form of differing mechanical properties. Extrusion has been used by a number of researchers to form fibrillar structures for the specific purpose of determining collagen mechanical properties (Chaudhry et al. 2009; Wang et al. 2005).

Heydarkhan-Hagvall and co-workers examined a 3D electrospun scaffold that combined ECM proteins such as collagen with biodegradable synthetic polymers such as PCL (Heydarkhan-Hagvall et al. 2008). The combination of natural and synthetic materials eliminated the need for crosslinking of the fibrils, often necessary to ensure sufficient mechanical stability of purely natural proteins such as collagen. Overall fibre size was shown to increase as function of polymer/protein concentration accompanied by a corresponding decrease in pore size within the fibres. The effect of pore size on cell attachment and migration was investigated using stem cells. Gelatin/PCL hybrids showed improved cell migration when compared with collagen/elastin/PCL systems, despite the fact that initial cell attachment was similar in both cases. Gelatin/PCL hybrids also showed improved tensile strength when compared with the other systems studied.

Ishii and co-workers used a synthetic PCL, nanofibrous electrospun scaffold in the form of a mesh to mimic the topography of the ECM (Ishii et al. 2005). Small fibre diameters ensured maximum porosity, which proved beneficial to cell attachment and proliferation. The concept was extended to the formation of multilayered grafts, which provide additional thickness and rigidity. Layered constructs are a useful intermediate in the development of the fully vascularized grafts required for clinical applications, as they allow study of the effect of graft thickness on the permeation of oxygenated blood into the structure. In this study, five separate layers were combined to prepare grafts that showed synchronous, macroscopic contractions. Inter-layer fusion was also observed via formation of key intracellular connections (Fig. 3).

Modifying the Mechanical Properties of Collagen: The Effect of Crosslinking

Collagen can be crosslinked using a variety of techniques with the crosslinking process shown to have a profound effect on the mechanical properties and biodegradation rate of the material. Chemical methods of

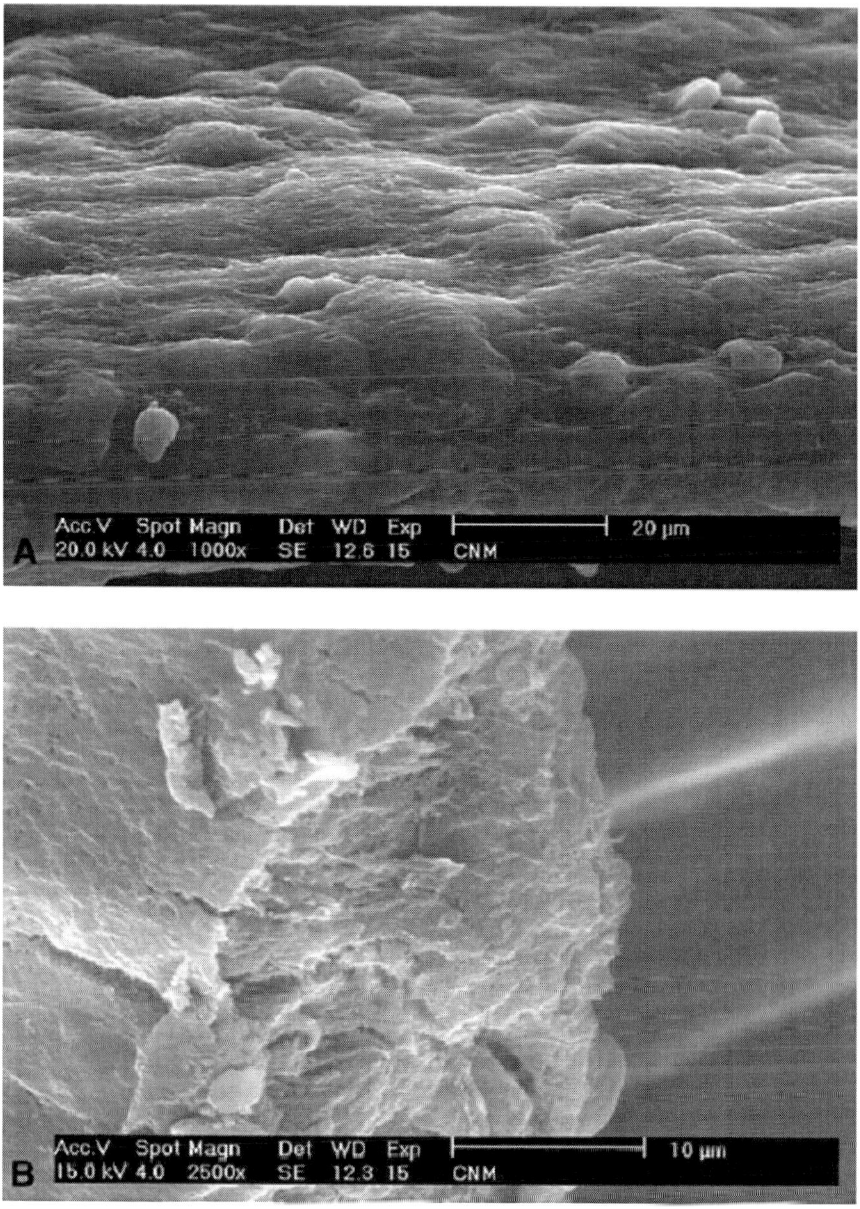

Fig. 3. Scanning electron microscopy (SEM) micrographs of multilayered poly(ε-caprolactone) grafts. (A) SEM micrograph of a five-layered graft, showing dense surface coverage of cells after 2 wk of culture. (B) SEM micrograph of a five-layered graft showing penetration of cells throughout the entire specimen. Non-degraded fibres of the scaffold are also observed. Published with permission from Elsevier Journals (Ishii et al. 2005).

crosslinking include reaction with formaldehydes (e.g., gluteraldehyde) or carbodiimides such as N-ethyl-N′-(3-dimethyl-aminopropyl) carbodiimide hydrochloride (EDAC). Alternatives include dehydrothermal treatment and irradiation with ultraviolet light. Chemical treatments are effective, but the process can leave behind toxic residues, and ultraviolet treatments provide a safer alternative.

A recent study by Madaghiele and co-workers examined the effect of the degree of crosslinking on collagen and gelatine-based films used as sealants for commercially available vascular prostheses (Madaghiele et al. 2009). A variety of crosslinking treatments were examined to determine the optimum degree of crosslinking and mechanical properties of the films. Dynamic mechanical analysis of the hydrated films showed elastic behaviour with shear moduli between 0.01 and 0.06 MPa, which is similar to that reported for crosslinked collagen gels (Binkman et al. 2003). Optimal values of crosslink density were determined to be ~1.2–1.3 x 10^{-5} mol/cm^3, which were achieved using both gaseous formaldehyde and dehydrothermal treatments.

It is interesting to note the difference between the bulk mechanical properties measured by Madaghiele using dynamic mechanical analysis and localized properties measured using nanoindentation reported by Chaudhry (Chaudhry et al. 2009). The nanoscale viscoelastic properties of an aligned collagen scaffold suggest elastic moduli of ~0.7 MPa, considerably higher than those in the bulk. Dynamic mechanical analysis provides useful information on the viscoelastic properties of a material, but it is appropriate only when the properties of a particular sample can be considered homogeneous. Nanoindentation lends itself to the study of materials that are inhomogeneous at the resolution scale of interest. An understanding of localized mechanical properties is vital to the study of cellular behaviour and is of key importance in tissue engineering.

Madhaven and colleagues considered crosslinked collagen gel constructs incorporating other components of the ECM such as chitosan and elastin for use in vascular grafts (Madhaven et al. 2010). Incorporation of these additional components led to significant improvements in the mechanical properties of the gel when compared with pure collagen constructs. Electron microscopy suggests that these improvements are linked to changes in the microstructure of the construct by either elastin fibres or chitosan sheets (Fig. 4). The study also examined the effect of crosslinking agents including formaldehyde, genipin and EDAC, which were shown to promote crosslinking via different mechanisms. Of the crosslinkers considered, genipin demonstrated improved biocompatibility and superior elongation, which can be attributed to the crosslinking mechanism and subsequent structure of the construct.

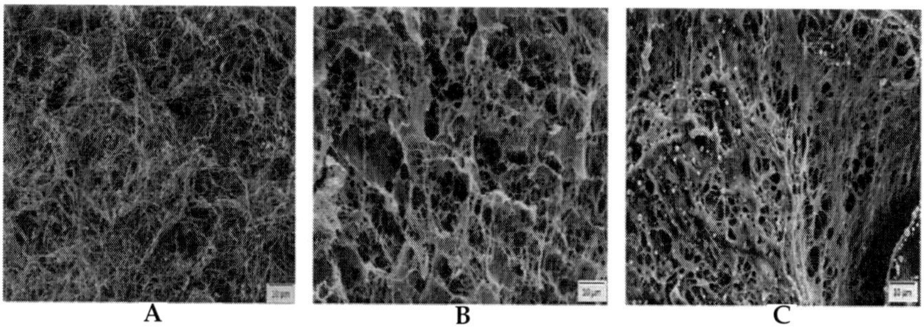

Fig. 4. Field emission scanning electron microscope images of collagen-based constructs. Constructs were prepared by crosslinking a selection of (A) collagen, (B) collagen-chitosan and (C) collagen, chitosan, elastin systems. Published with permission from Elsevier Journals (Madhaven et al. 2010).

PROPERTIES OF CARDIAC CELLS: UNDERSTANDING THE DIFFERENTIATION OF CARDIAC MYOCYTES

Cardiac myocytes undergo a number of phenotypical changes during development to their adult form and understanding these changes is vital to the development of functional artificial myocardial tissue (Evans et al. 2003). One key change is the transition from a proliferative (hyperplastic) to cell growth (hypertrophic) phenotype and subsequent withdrawal from the cell cycle. This transition is of particular importance when undifferentiated cells (e.g., stem cells) are used. Changes in cell phenotype may be characterized by a modification to cell shape. During the hypertrophic phase ventricular myocytes change from a rounded oval to rod shape, increase their overall size and form myofibrils; the cells also down-regulate the synthesis of DNA. The hypertrophic phase also plays host to the development of key cell-cell and cell-ECM interactions. In the heart, adhesions are crucial to the development, alignment and survival of cells and ensure coordinated and directional excitation and contraction.

IN VITRO CULTURE OF CARDIOMYOCYTES USING COLLAGEN SCAFFOLDS FOR APPLICATION TO CARDIOLOGY

The *in vitro* study of primary embryonic and neonatal cardiomyocytes is well established. In her classic study of 1955, Cavanaugh describes the behaviour and beating of embryonic chick heart cells, and the study was later extended by other researchers to encompass both chick and rat heart cells in monolayer culture (Cavanaugh 1955; Souren et al. 1992). These studies demonstrate the ability of myocardial cells to participate in spontaneously beating aggregates or organ cultures. However, accurate measurement of the beating rate in such aggregates is often difficult to perform and may be a function of diffusion in larger aggregate systems. Souren and co-workers describe the use of neonatal rat cardiomyocytes cultured on and within floating collagen matrix scaffolds (Souren et al. 1992). The ability of the gel to display rhythmic contraction is related to collagen concentration as well as cell localization and cell density. The authors rationalize the synchronous beating observed for myocytes seeded on to the surface of the gel in terms of key cell-cell and cell-matrix interactions as well as intercellular distance. When the cells are encapsulated within the collagen matrix, cell density is insufficient for the formation of the electrical contacts necessary to yield coherent contractile tissue.

A further disadvantage of planar, 2D tissue scaffolds for the *in vitro* culture of cardiomyocytes is the failure of the cells to mimic the differentiation and myocyte phenotype found *in vivo*. In a 2D culture environment, myocytes develop a flattened shape in preference to the rod-like *in vivo* phenotype. Simpson and co-workers tried to overcome this problem by using a planar aligned collagen scaffold, which resulted in a phenotype much more akin to that found *in vivo* (Simpson et al. 1994). However, a constrained 2D environment results in limited cell-ECM interactions and cell layering. The study also highlighted that a planar aligned scaffold is an unsuitable substrate for embryonic day 15 ventricular cultures as the scaffold does not provide an appropriate environment for differentiation and proliferation and results in poorly attached cells that do not initiate spontaneous contraction. The use of embryonic cell cultures is vital in the study of cardiomyocyte differentiation as they display enhanced DNA production in comparison with the corresponding adult cells.

A major breakthrough in myocardial tissue engineering was accompanied by the transition to culture in three dimensions. Cells grown in three dimensions have been shown to more closely resemble those *in vivo* in their phenotype, morphology and intracellular signalling pathways. The culture environment is especially important for cardiac myocytes (Eschenhagen et al. 1997; Evans et al. 2003). The use of 3D collagen-based tissue scaffolds was highlighted in 1997 by Eschenhagen and co-workers, who developed a coherently contracting cardiomyocyte culture scaffold based on type I collagen and Matrigel™, a complex mixture of proteins secreted by mouse tumour cells (Eschenhagen et al. 1997). The model heart tissue provides an ideal mimic to observe the effects of genetic and pharmacological manipulations on cardiac tissue. Traditionally, such studies have relied on primary cultures of cardiomyocytes in monolayer form, which avoided the interference of compensatory mechanisms found in animal models, but displayed the inherent problems and phenotypic differences associated with a planar environment. Primary cultures of cardiomyocytes have traditionally been used because of the lack of stable cardiomyocyte cell lines and the fact that they exhibit spontaneous contraction.

The cardiac myocyte populated matrices (CMPMs) developed by Eschenhagen were prepared via gelation of a collagen/cardiomyocyte mixture with the gels anchored between two Velcro-coated glass tubes to allow attachment to a force measuring system and subsequent measurement of the isometric contractile force. This system provides a number of advantages over those reported previously. First, the 3D collagen matrix allows the cardiomyocytes to remodel and develop into synchronously beating myocardial tissue, the first example of such a case. The CMPMs prepared display longitudinally orientated layers of cells towards the edges of the gel, which accurately mimic native embryonic myocardial tissue; the central section of the gel is more sparsely populated. A significant proportion of the cells (30–50%) show a fibrillar structure containing two nuclei, characteristic of differentiation in embryonic chick cardiomyocytes. The tissue also exhibits enhanced intracellular junctions typical of myocardium in the form of intercalated disks (longitudinal end-to-end connections), desmosomes (proteins that bind cells together in muscle tissue in a transverse fashion) and tight junctions that join together the cytoskeleton of adjacent cells. These features are either poorly defined or completely absent in cardiomyocytes prepared in monolayer culture.

Second, the constrained nature of the CMPMs ensures that the cardiomyocytes are subject to a stable isometric load or stress, either imposed by the cells themselves (c.f. the situation *in vivo*) or applied externally to study the effect of a specified mechanical stress on cardiomyocyte alignment and differentiation. The distribution of stress throughout the CMPM is likely to be asymmetric (due to a non-uniform distribution of cells), which accurately imitates the situation *in vivo* for which applied stress plays a key role in the regulation of growth of the developing heart (Eschenhagen et al. 1997). The CMPMs allow easy, direct and quantitative measurement of the forces of contraction within the tissue matrix. Force measurements are possible over hours or even days under appropriate conditions. The CMPMs show only a modest deterioration in contractile performance, in contrast to whole organ models for which the deterioration is more pronounced. The study also highlights that despite their simplicity, the CMPMs accurately reproduce the *in vivo* response of several inotropic agents, such as extracellular calcium. The response to other agents is less well defined, but is discussed in detail.

Finally, the CMPMs were used to study the effects of genetic manipulation of cardiomycoytes via gene transfer using the adenovirus.

A number of gene products have been identified as playing a role in the cardiovascular system but their effects remain largely unknown. Techniques in genetic engineering permit the over-expression or knockout of specific genes; hence, a model system in which their effects can be studied under controlled conditions is an important breakthrough.

The early work by Eschenhagen and co-workers provided an excellent basis for further studies on collagen-based scaffolds in cardiology. Within the Eschenhagen group, CMPMs were used to study the effect of mechanical stretch under controlled conditions on the behaviour of cardiac myocytes *in vitro* (Fink et al. 2000). Myocytes within engineered heart tissue subjected to stretch demonstrated phenotypic changes when compared with un-stretched analogues. Stretched cells displayed a rod-like morphology and demonstrated improved alignment into parallel arrays. Such cells were also characterized by increased cell dimensions (both length and width) and formation of longer myofilaments. Cells subject to stretch demonstrated superior contractile force of between two and four fold when stimulated with agents such as calcium. The Eschenhagen group also examined the use of engineered heart tissue as cardiac grafts in syngenic rates (Zimmerman et al. 2002). The engineered tissue differentiated to a cardiac phenotype became strongly vascularized and survived for at least 28 d following implantation into uninjured hearts.

Kutschka and co-workers used a collagen gel foam (GF)/Matrigel™ matrix as a scaffold and nutrient reservoir for rat cardiomyoblasts genetically modified with optical and fluorescent transporter genes (Kutschka et al. 2006). The tissue scaffold provided a robust mycocardial graft suitable for implantation into ischemic hearts (hearts with reduced blood supply), imageable using optical bioluminescence and small animal magnetic resonance imaging. This study was the first to use *in vivo* bioluminescence imaging as a high throughput and non-invasive technique to investigate myocardial graft survival as a function of time. The use of collagen GF/Matrigel™ tissue supports leads to enhanced survival of cardiomyocytes after transplantation, when compared with simple injection of cells or use of a plain collagen GF matrix. The GF/Matrigel™ grafts also improved left ventricular heart function when compared with controls, but the mechanism by which this improvement occurs has not been determined and provides an interesting challenge for further study.

Tissue constructs based on type I collagen and Matrigel™ have a number of advantages, but they have the drawback that the matrix proteins are poorly defined and not approved for clinical use. They are also extremely expensive. In an attempt to overcome these problems, Kofidis and co-workers developed a bioartificial myocardial tissue patch based on the clinically approved, collagen-based Tissue Fleece® matrix (Kofidis et al. 2002). Myocardial cells were shown to migrate into the fleece to form fibrils with the scaffold demonstrating comparable mechanical stability to native cardiac muscle. The engineered tissue also displayed continuous, rhythmic, macroscopically visible contractions over a period of 12 wk *in vitro*.

Li and co-workers used a biodegradable gelatin mesh for the culture of human and rat cardiac cells (Li et al. 2000). The extent of proliferation was found to vary according to cell type, ranging from simple cell survival in the case of human ventricular myocytes to the formation of spontaneous and rhythmically contacting grafts using rat ventricular cells. In all cases, cells preferentially migrated to the surface of the graft, forming a layer of between 300 and 500 μm.

Evans and co-workers introduced the concept of a tubular collagen scaffold, engineered from aligned collagen fibrils, to the detailed study of embryonic cardiac myocyte survival, attachment, DNA synthesis and differentiation *in vitro* (Evans et al. 2003). This study aimed to build on the early work of Simpson discussed previously, which used similar aligned collagen scaffolds but in a more constrained 2D environment (Simpson et al. 1994). A tubular scaffold geometry was selected to mimic the *in vivo* tubular environment of the embryonic heart. Myocytes grown on the scaffold demonstrated many of the characteristics of the hyperplastic to hypertropic transition observed in normal cardiac development, including development of a rod-like cellular morphology and essential cell-cell/cell-matrix interactions to permit synchronous contractions.

Gaballa and co-workers used a slightly different technique whereby they grafted a porous, acellular 3D collagen scaffold on to infarcted myocardium (Gaballa et al. 2006). The presence of the collagen scaffold as a support structure resulted in reduced cardiac remodelling and improved re-vascularization in the infarcted myocardium of rats. A novel approach developed by Ott and co-workers involved the decellularization of hearts using detergents by coronary perfusion (Ott et al. 2008). Although similar techniques had been developed previously for the decellularization of myocardial tissue, this study represented the first example using an entire organ. The decellularized hearts retained key ECM proteins such as type I collagen, laminin and fibronectin along with the fibrous structure of myocardial ECM. The hearts were then repopulated with cardiac cells, producing a contractile and drug-responsive biomimetic construct after 8 d in culture. With sufficient time for maturation, such bioengineered scaffolds could provide a source of donor myocardial tissue parts (such as valves or ventricles) or even whole organs for transplantation. The current study is limited to rat hearts but the authors reveal successful application of the decellularization techniques to porcine hearts that more closely resemble human size and complexity and also to other organs. The key challenge in the application of this technology lies in the cellular reseeding process, in particular the distribution of cells throughout the construct under the *in vitro* conditions necessary for maturation of the organ.

IMPORTANCE OF NANOSCALE TOPOGRAPHY AND INTERACTIONS IN CARDIAC TISSUE ENGINEERING

Zorlutuna and co-workers highlight the importance of cell organization and alignment at the nanoscale on the mechanical properties of vascular tissue (Zorlutuna et al. 2009). Collagen scaffolds were prepared containing nanochannels ranging from 330 to 650 nm in width, and seeded with smooth muscle cells isolated from the human saphenous vein. Cell alignment within the construct was confirmed using scanning electron and fluorescence microscopy (Fig. 5). The presence of nanopatterns as small as 332.5 nm resulted in improvements to both the ultimate tensile stress and Young's modulus of the scaffolds, generating values comparable to those for natural arteries and veins after 75 d in culture.

Intracellular and cell matrix interactions within a 3D environment have been shown to be vital in the development of tissues that mimic those found *in vivo*. Legant and co-workers studied the effect of cellular forces within collagen gels using microelectromechanical systems technology (Legant et al. 2009). They employed microcantilevers, both to constrain the remodelling of the collagen substrate and to report quantitatively on the forces generated during the remodelling process. At a cellular level, forces increased as a function of matrix rigidity, while the levels of ECM proteins expressed were closely related to mechanical stress. The study highlights the possibility of introducing gradients into engineered tissue in order to control protein expression and the organization of cells within a 3D environment.

Martuza and co-workers review the importance of the micro- and nanoscale environment in the application of stem cell technologies to cardiology (Martuza et al. 2009). The review highlights the problems of controlling precise collagen architecture both at the micro- and nanoscale as well as variance in biodegradation rate. The authors conclude that synthetic biomaterials that mimic the native ECM provide a better alternative to collagen for applications involving stem cells as precise control of their properties is straightforward and reproducible.

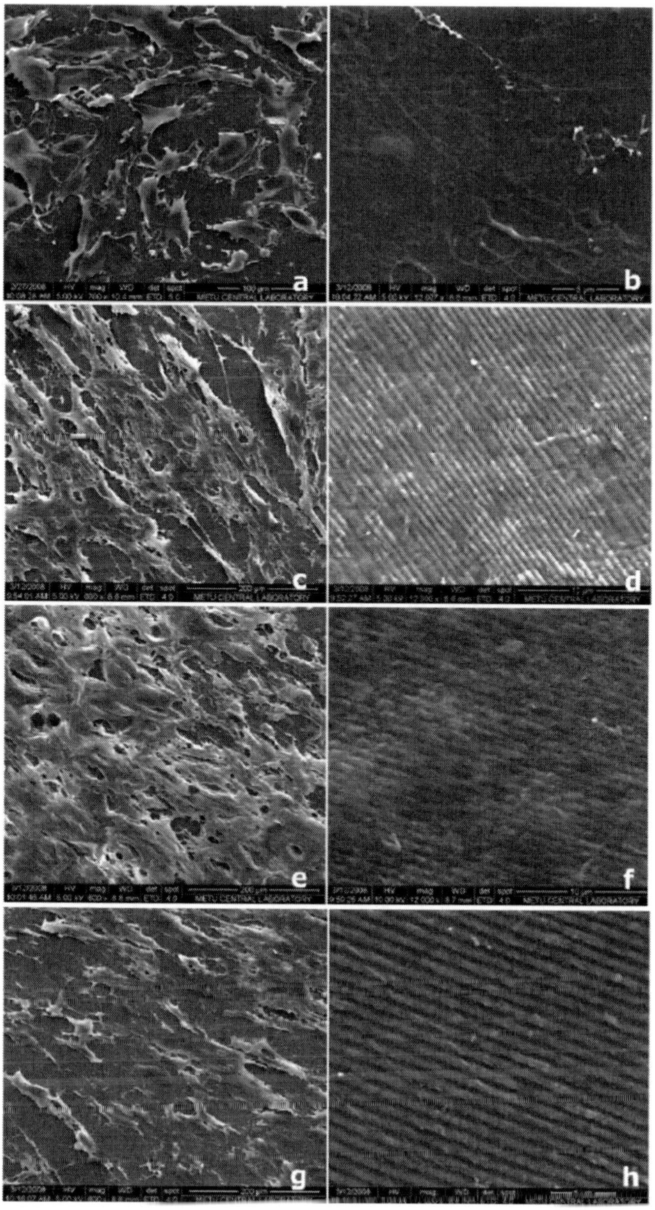

Fig. 5. Nanoscale features patterned on to a collagen scaffold for culture of cardiac cells. Scanning electron micrographs of day 14 vascular smooth muscle cells. Film surfaces covered by cells: (a) unpatterned film, (c) 332.5 nm patterned film, (e) 500 nm patterned film, (g) 650 nm patterned film. Region of sample unoccupied by cells: (b) unpatterned film, (d) 332.5 nm patterned film, (f) 500 nm patterned film, (h) 650 nm patterned film. Published with permission from the American Chemical Society (Zorlutuna et al. 2009).

Summary Points

- The ability of the heart to repair itself after injury is extremely limited, resulting in a high demand for corrective surgery and donor organs. Engineered cardiac tissue, in which cardiomyocyte cells are cultured on a scaffold *in vitro*, could provide a viable alternative. Tissue constructs may be natural or synthetic, but a large proportion are based on collagen, a key component of the extracellular matrix. Collagen has the advantage of specific biomarkers that promote cell adhesion and differentiation but can lack sufficient mechanical strength. The mechanical properties of synthetic scaffolds are easily tailored; however, they lack the necessary biomarkers.
- Collagen possesses a hierarchical structure that ranges from the specific $(Gly\text{-}X\text{-}Y)_n$ amino acid sequence at the nanoscale, through individual tropocollagen molecules to fibrils and fibres measuring tens of micrometres. The mechanical properties of collagen are related to the interplay between structures at two characteristic length scales. Alterations to the collagen structure lead to significant changes in mechanical properties. These structural changes can be deliberate (e.g., crosslinking) or a result of genetic modifications *in vivo*, such as Ehlers-Danlos syndrome. Over 1000 mutations in a selection of genes have been attributed to defections in collagen structure.
- Fabrication of collagen into fibrous structures is possible using a number of techniques. The most widespread of these are extrusion and electrospinning. Common techniques to determine the mechanical properties of collagen fibrils include atomic force microscopy and nanoindentation. Fibrils can also be prepared from mixtures of collagen and synthetic polymers such as PCL. The mechanical properties of such fibres often show an improvement to those containing collagen alone.
- Embryonic cardiomyocytes are the most widely studied cell type used in engineered heart tissue. Cardiomyocytes undergo a number of phenotypical changes during development. These changes are characterized by a modification to cell shape from rounded oval to rod-like. The cells also increase in size and form myofibrils.
- A number of researchers have examined the use of collagen in myocardial tissue engineering, ranging from monolayer culture to 2D planar aligned scaffolds and 3D engineered constructs. The culture environment is extremely important in the differentiation of cardiomyocytes to the correct phenotype. The environment also plays a key role in cell-cell and cell-matrix interactions. Such interactions are vital in the development of synchronously beating myocardial tissue. A 3D environment results in myocardial tissue that provides a more accurate mimic for that found *in vivo*.

- Successful collagen-based constructs for *in vitro* myocardial tissue engineering range from gels consisting of a mixture of collagen and Matrigel™ to aligned tubular scaffolds. Mixtures of collagen and other ECM proteins such as elastin or synthetic polymers such as PCL have also been used successfully. Alternative approaches involve *in vivo* implantation of acellular collagen grafts or decellularization of natural heart tissue followed by reseeding with cardiomyocytes. Forces within collagen gels can be measured quantitatively using microelectromechanical systems technology. The mechanical properties of collagen scaffolds can be improved by the introduction of nanoscale surface features such as nanochannels.

Key Facts

- Cardiac myocytes (also known as cardiomyocytes) are muscle cells found in the middle layer of the heart wall (myocardium).
- The *in vitro* study of cardiomyocytes is well established and dates back to early work by Cavanaugh in the 1950s.
- Cardiomyocyte cells have a high demand for supply of both oxygen and nutrients in order to survive and function adequately.
- *In vitro* culture of cardiomyocytes in a 3D environment results in cells that more accurately mimic the properties of those found *in vivo*.
- The ability of cardiomyocytes to display synchronous beating relies on cell-cell and cell-matrix interactions.

Definitions

Cell phenotype: The outward, physical manifestations of cell type, such as size or shape.

Cellular differentiation: The natural development process by which less specific cell types transform into mature cells to perform a specific function, e.g., muscle, fat, bone.

Crosslinks: Chemical bonds that link together polymer chains (natural or synthetic) to form a network structure.

Genipin: A plant extract and excellent natural crosslinker for proteins such as collagen.

Inotropic agents: Positive and negative inotropic agents increase or weaken the force of muscular contractions respectively.

Nanoscale: Structures or features on the scale of 1–100 nm.

Tissue scaffold: A support structure that can accommodate the growth of cells.

Abbreviations

2D	:	two dimensional
3D	:	three dimensional
CMPM	:	cardiac myocyte populated matrices
ECM	:	extracellular matrix
EDAC	:	N-ethyl-N′-(3-dimethyl-aminopropyl) carbodiimide hydrochloride
GF	:	gel foam
PCL	:	poly(ε-caprolactone)
SEM	:	scanning electron microscopy
TC	:	tropocollagen

References

Binkman, W.T., K. Nagapudi, B.S. Thomas and E.L. Chaikof. 2003. Photo-crosslinking of type I collagen gels in the presence of smooth muscle cells; mechanical properties, cell viability and function. Biomacromolecules 4: 890–895.

British Heart Foundation Statistics. 2010 (www.bhf.org.uk).

Buehler, M.J. 2008. Nanomechanics of collagen fibrils under varying cross-link densities: Atomistic and continuum studies. J. Mech. Behav. Biomed. Mat. 1: 59–67.

Buehler, M.J. 2006. Nature designs tough collagen: Explaining the nanostructure of collagen fibrils. Proc. Natl. Acad. Sci. USA 103: 12285–12290.

Cavanaugh, M.W. 1955. Pulsation, migration and division in dissociated chick embryo heart cells *in vitro*. J. Exp. Zool. 128: 573–587.

Chaudhry, B., H. Ashton, A. Muhamed, M. Yost, S. Bull and D.J. Frankel. 2009. Nanoscale viscoelastic properties of an aligned collagen scaffold. J. Mater. Sci.: Mater. Med. 20: 257–263.

Eschenhagen, T., C. Fink, U. Remmers, H. Scholz, J. Wattchow, J. Weil, W. Zimmermann, H.H. Dohmen, H. Schäfer, N. Bishopric, T. Wakatsuki and E.L. Elson. 1997. Three-dimensional reconstruction of embryonic cardiomyocytes in a collagen matrix: a new heart muscle model system. FASEB J. 11: 683–694.

Evans, H.J., J.K. Sweet, R.L. Price, M. Yost and R.L. Goodwin. 2003. Novel 3D culture system for study of cardiac myocyte development. Am. J. Physiol. Heart Circ. Physiol. 285: H570–H578.

Fink, C., S. Ergün, D. Kralisch, U. Remmers, J. Weil and T. Eschenhagen. 2000. Chronic stretch of engineered heart tissue induces hypertrophy and functional improvement. FASEB J. 14: 669–679.

Fomovsky, G.M., S. Thomopoulos and J.W. Holmes. 2010. Contribution of extracellular matrix to the mechanical properties of the heart. J. Mol. Cell. Cardiol. 48: 490–496.

Gaballa, M.A., J.N.E. Sunkomat, H. Thai, E. Morkin, G. Ewy and S. Goldman. 2006. Grafting an acellular 3-dimensional collagen scaffold onto a non-transmural infracted myocardium induces neo-angiogenesis and reduces cardiac remodelling. J. Heart Lung Transplant. 25: 946–954.

Gentleman, E., A.N. Lay, D.A. Dickerson, E.A. Nauman, G.A. Livesay and K.C. Dee. 2003. Mechanical characterization of collagen fibres and scaffolds for tissue engineering. Biomaterials 24: 3805–3813.

Heydarkhan-Hagvall, S., K. Schenke-Layland, A.P. Dhanasopon, F. Rofail, H. Smith, B.M. Wu, R. Shemin, R.E. Beygui and W.R. MacLellan. 2008. Three-dimensional electrospun ECM-based hybrid scaffolds for cardiovascular tissue engineering. Biomaterials 29: 2907–2914.

Ishii, O., M. Shin, T. Sueda and J.P. Vacanti. 2005. *In vitro* tissue engineering of a cardiac graft using a degradable scaffold with an extracellular matrix-like topography. J. Thorac. Cardiovasc. Surg. 130: 1358–1363.

Jawad, H., A.R. Lyon, S.E. Harding, N.N. Ali and A.R. Boccaccini. 2008. Myocardial tissue engineering. Br. Med. Bull. 87: 31–47.

Kofidis, T., P. Akhyari, B. Wachsmann, J. Boublik, K. Mueller-Stahl, R. Leyh, S. Fischer and A. Haverich. 2002. A novel bioartificial myocardial tissue and its prospective use in cardiac surgery. Eur. J. Cardio-Thorac. Surg. 22: 238–243.

Koláčná, L., J. Bakešová, F. Varga, E. Koštáková, L. Plánka, A. Nečas, D. Lucáš, E. Amler and V. Pelouch. 2007. Biochemical and biophysical aspects of collagen nanostructure in the extracellular matrix. Physiol. Res. 56 (Suppl. 1): S51–S60.

Kutschka, I., I.Y. Chen, T. Kofidis, T. Arai, G. Von Degenfeld, A.Y. Sheikh, S.I. Hendry, J. Pearl, G. Hoyt, R. Sista, P.C. Yang, H.M. Blau, S.S. Gambhir and R.C. Robbins. 2006. Collagen matrices enhance survival of transplanted cardiomyoblasts and contribute to functional improvement of ischemic rat hearts. Circulation 114: I167–173.

Langer, R. and J.P. Vacanti. 1993. Tissue engineering. Science 260: 920–926.

Li, R.K., T.M. Yau, R.D. Weisel, D.A.G. Mickle, T. Sakai, A. Choi and Z.Q. Jia. 2000. Construction of a bioengineered cardiac graft. J. Thorac. Cardiovasc. Surg. 119: 368–375.

Legant, W.R., A. Pathak, M.T. Yang, V.S. Deshpande, R.M. McMeeking and C.S. Chen. 2009. Microfabricated tissue gauges to measure and manipulate forces from 3D microtissues. Proc. Natl. Acad. Sci. USA 106: 10097–10102.

Madaghiele, M., A. Piccinno, M. Saponaro, A. Maffezzoli and A. Sannino. 2009. Collagen- and gelatine-based films sealing vascular prostheses: evaluation of the degree of crosslinking for optimal blood impermeability. J. Mater. Sci.: Mater. Med. 20: 1979–1989.

Madhavan, K., D. Belchenko, A. Motta and W. Tan. 2010. Evaluation of composition and crosslinking effects on collagen-based composite constructs. Acta Biomaterialia 6: 1413–1422.

Martuza, B., J.W. Nichol and A. Khademhosseini. 2009. Micro- and nanoscale control of the cardiac stem cell niche for tissue fabrication. Tissue Engineering: Part B. 15: 443–454.

Myllyharju, J. and K.I. Kivirikko. 2001. Collagen and collagen-related diseases. Ann. Med. 33: 7–21.

Ott, H.C., T.S. Matthiesen, S.K. Goh, L.D. Black, S.M. Kren, T.I. Netoff and D.A. Taylor. 2008. Perfusion-decellularized matrix: using nature's platform to engineer a bioartificial heart. Nat. Med. 14: 213–221.

Simpson, D.G., L. Terracio, M. Terracio and R.L. Price. 1994. Modulation of cardiac myocyte phenotype in vitro by the composition and orientation of the extracellular matrix. J. Cell. Physiol. 161: 89–105.

Souren, J.E.M., C. Schneijdenberg, A.J. Verkleij and R. Van Wijk. 1992. Factors controlling the rhythmic contraction of collagen gels by neonatal heart cells. In Vitro Cell. Dev. Biol. 28A: 199–204.

Wang, X., X. Li and M.J. Yost. 2005. Microtensile testing of collagen fibril for cardiovascular tissue engineering. J. Biomed. Res. Part A, 74A: 263–268.

Yalovaç, A. and N.N Ulusu. 2007. Collagen and collagen disorders. FABAD. J. Pharm. Sci. 32: 139–144.

Zimmermann, W.H., K. Schneiderbanger, P. Schubert, M. Didie, F. Munzel, J.F. Heubach, S. Kostin, W.L. Neuhuber and T. Eschenhagen. 2002. Tissue engineering of a differentiated cardiac muscle construct. Circ. Res. 8: 223–230.

Zorlutuna, P., A. Elsheikh and V. Hasirci. 2009. Nanopatterning of collagen scaffolds improve the mechanical properties of tissue engineered vascular grafts. Biomacromolecules 10: 814–821.

7

Cardiac Biomarker and Nanowire Sensor Arrays

Guo-Jun Zhang[1,2]

ABSTRACT

The detection of cardiac biomarkers is playing an important role in diagnosis of cardiovascular disease. In addition to traditional laboratory-based methods, the development of new devices that enable sensitive and rapid analysis of the biomarkers could help clinicians improve their diagnostic ability as well as provide an early and accurate indication of cardiac cellular necrosis. Silicon nanowire (SiNW)-based device is emerging as a highly sensitive, label-free, electrical biosensor for the direct detection of cardiac biomarkers. Here, we discuss fabrication of the SiNW biosensor using complementary metal-oxide semiconductor field-effect transistor–compatible technology, and its real-time detection of troponin T (cTnT) and multiplexed detection of three biomarkers, cTnT, creatine kinase MM (CK-MM) and creatine kinase MB (CK-MB), in serum simultaneously. The SiNW array format was produced through conventional optical lithography, etching and oxidation. Antibody was covalently immobilized on the

[1]Institute of Microelectronics, A*STAR (Agency for Science, Technology and Research), 11 Science Park Road, Singapore Science Park II, Singapore 117685;

[2]School of Clinical Laboratory Science, Hubei University of Chinese Medicine, Wuhan 430065, China; E-mail: gjzhangtj@hotmail.com

List of abbreviations after the text.

SiNW surface. The real-time detection of cTnT was successfully demonstrated in an assay buffer solution of concentration down to 1 fg/ml, as well as in an undiluted human serum environment of concentration as low as 30 fg/ml. For the multiplexed detection, three antibodies were spotted on the SiNW array surface. The multiplexed detection was accomplished by incubating the biomarkers to the various antibodies-spotted SiNW surface. The SiNW sensor arrays were demonstrated to have the required selectivity and sensitivity for multiplexed detection of 100 fg/ml cTnT, CK-MM, and CK-MB in untreated serum. The multiplexed detection method is independent of the ionic strength of the sample solution, allowing the SiNW sensor to directly analyze the biomarkers in serum simultaneously. The developed SiNW biosensor shows a promising potential of constructing a miniaturized device for point-of-care diagnostics.

INTRODUCTION

Cardiovascular disease (CVD) is a group of medical problems that affect the heart and surrounding blood vessels. It is becoming the number one cause of death globally, accounting for more than 50% of all deaths worldwide (Yach et al. 2005). More people die annually from CVD than from any other causes. An estimated 30 million people are troubled by CVD each year. According to the World Health Organization, by 2030, almost 23.6 million people will die from CVD.

Chest pain is often thought of as a classic symptom of CVD. Patients presenting in the emergency room with chest pain usually receive an electrocardiogram (ECG) test after the patient's history is interpreted and physical examination is accomplished. Although ECG is the single best test to identify patients with acute coronary syndromes (ACS) upon emergency room presentation, it cannot detect all myocardial injuries and lacks sensitivity (McQueen et al. 1983). Owing to the relatively poor sensitivity of the standard ECG, additional strategies are required. In 1954, the earliest documented use of biochemical markers in the study of myocardial infarction (MI) reported that glutamate oxaloacetic transaminase activity in serum increased above the reference range a few hours after acute myocardial infarction (AMI), reaching a peak after 2 to 3 d (Ladue et al. 1954). Cardiac biomarkers have since become the most commonly used test to identify patients with potential ACS.

Standard assays to detect cardiac biomarkers, such as enzyme-linked immunosorbent assay (ELISA), are sensitive (limit of detection or LOD > 10 pg/ml) (Muller-Bardorff et al. 1997), but suffer from important sample and reagent consumption in large-scale studies. Moreover, they are performed in central laboratories of clinics and hospitals and take a long time (> 6 h), which is highly incompatible with the quick decisions needed to save a heart attack patient. A simple, rapid and relatively inexpensive means defined as point-of-care (POC) device has emerged to reduce the therapeutic turnaround time and improve the medical and economic outcomes (McDonnell et al. 2009). The use of the POC device leads to a reduction in test ordering, sample transport to laboratories and data reporting. Several POC devices have been considered as alternatives to the lab analyzer, whereas some of them have been questioned regarding the accuracy of the technologies. Small bench-top analyzers and hand-held devices are the two main types of POC device formats available in the clinical setting. To date, the commercial cardiac biomarker POC devices, such as Nanogen's Cardiac STATus™ and CareStart™ Cardiac Markers, have helped medical professionals quickly establish cardiac marker levels from a finger prick of blood (< 15 min). However, these test kits lack the sensitivity (LOD >0.1 ng/ml) to take real advantage of their rapidity, and usually doctors have to test several times for different biomarkers to establish the diagnostic with enough certainty to engage in a much-needed treatment.

Silicon nanowires (SiNWs), which act as a field-effect transistor (FET) sensor, have attracted attention over the last several years. This electrical-based biosensor integrates biomolecules with complementary metal-oxide semiconductor (CMOS)–compatible technologies to allow the detection of change in charges when target analyte binds onto the immobilized probe. When a charged protein biomarker is captured by a specific antibody immobilized on their surface, the SiNWs can sense the resulting change in charge density around them, as it induces a change in the electric field at their surface. Owing to their large surface-to-volume ratio, tunable electrical properties and biocompatibility, the SiNWs have been proven to be a highly sensitive, label-free and specific detection for a wide spectrum of species, which includes metal ions (Cui et al. 2001; Zhang et al. 2007), proteins (Chua et al. 2009; Cui et al. 2001; Stern et al. 2007; Zhang et al. 2009b; Zheng et al. 2005), nucleic acids (Bunimovich et al. 2006; Cattani Scholz et al. 2008, Gao et al. 2007; Hahm and Lieber 2004; Li et al. 2004; Zhang et al. 2008a, b, 2009a, 2010a, b), and viruses (Patolsky et al. 2004), coupled with the versatility of single or multiplexed format. Moreover, they are fabricated through a completely CMOS-compatible process that is highly amenable to mass production and cost savings. Their fast response time and low detection limit (up to femtomolar range), in addition to the elimination of labeling, make this mode of detection intriguing. In this chapter, we focus on the use of the SiNW biosensor as a promising tool to detect cardiac biomarkers involving troponin T (cTnT), creatine kinase MM (CK-MM) and creatine kinase MB (CK-MB) with high sensitivity and specificity, aiming at development of the POC device for CVD diagnostics.

APPLICATIONS TO OTHER AREAS OF HEALTH AND DISEASE

This chapter is particularly focused on detection of cardiac biomarkers using SiNW biosensors. Nevertheless, the SiNW biosensors have also been applied to other areas involving various applications. Owing to the unique properties of the SiNW materials, the SiNW biosensor has attracted attention with respect to disease diagnostics. As surface chemistry on silicon surface has been intensively studied, various receptor molecules can be functionalized on the SiNW surface. As a result, the use of SiNW devices for the detection of multiple prostate cancer protein biomarkers simultaneously in a single, versatile detection platform has been explored (Zheng et al. 2005). Likewise, the SiNW biosensor modified with specific antibodies has yielded selective detection of different viruses such as influenza A and adenovirus in parallel (Patolsky et al. 2004). In addition, the SiNW covalently modified with peptide nucleic acid (PNA) have led to direct detection of microRNA in total RNA extracted from Hela cells related to cervical cancer (Zhang et al. 2009a). Furthermore, the PNA-functionalized SiNW biosensor is capable of detecting dengue virus quickly with high sensitivity, so it can be miniaturized into a portable device by integration with other components (Zhang et al. 2010a).

SILICON NANOWIRE BIOSENSOR FOR DETECTION OF CARDIAC BIOMARKERS

Nanostructures, such as nanowires (NWs), carbon nanotubes (NTs) (Kong et al. 2000), and nanoparticles (Park et al. 2002), have been widely investigated because of their diameters comparable to the sizes of chemical and biological species to be sensed, and thus their potential for biosensing applications. Nanoparticles, as alternative labeling instead of fluorescent dyes, have intensively been used for the detection of biosamples with high sensitivity. However, this method still suffers from the demand of target-labeling, which is time-consuming and probably causes conformational alteration or steric hindrance induced by the additional label. One-dimensional structures such as the NWs and NTs offer opportunities for real-time and label-free sensing applications, and more attractively their suitability for large-scale and high-density integration. The SiNW-based biosensors, in particular, have attracted much attention for their capabilities in sensitive detection of biological and chemical species, and their suitability for mass production benefiting from mature fabrication technologies.

Working Principle of Silicon Nanowire Biosensor

The SiNW sensors are typical FET-based devices, containing source, drain and gate electrodes. The SiNW connected between source and drain in the semiconductor channel serves as sensing component of the device. To detect a specific target, a receptor molecule that recognizes the target molecule is immobilized on the SiNW surface. The typical structure of the SiNW biosensor for detection of cardiac biomarker is illustrated in Fig. 1A. For immunological-modified SiNW biosensor, the semiconductor channel has a uniform conductance determined by the electron density in the n-type SiNW sensor, as shown in Fig. 1B. When the n-type nanowire sensor with surface receptors is exposed to the target molecule carrying negative charges, specific binding will lead to a decrease in conductance, and in turn an increase in resistance. In short, the sensing mechanism by the SiNW can be understood in terms of the change in charge density, which induces a change in electric field at the SiNW surface.

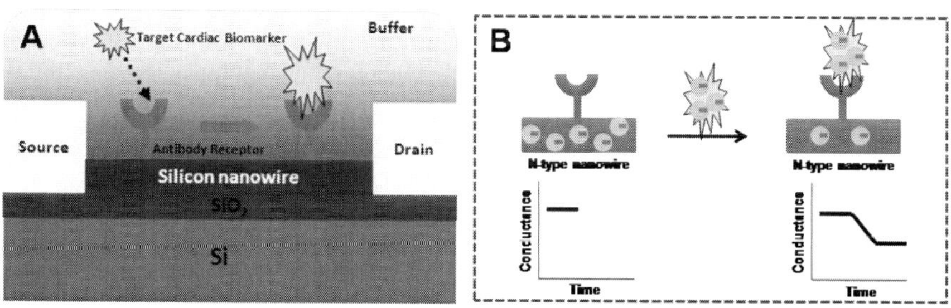

Fig. 1. Principle of silicon nanowire (SiNW) biosensor. (A) SiNW serving as the semiconductor channel is connected to the source and drain electrodes. Antibody immobilized on the SiNW surface is capable of specifically recognizing and capturing the cardiac biomarker. (B) When a negatively charged biomarker binds to the antibody, it repels the negative charges in the region of nanowire, resulting in a decrease in conductance. (Unpublished materials.)

Fabrication of Silicon Nanowire Biosensor

Two main fabrication approaches, top-down and bottom-up, have been employed to produce semiconducting NWs for the use of the SiNW biosensor. The performance of the SiNW biosensor is influenced by many factors, such as dimensions, carrier densities and mobilities, and choice of surface chemistry. It was found that smaller dimensions of the NWs resulted in a greater sensitivity (Stern et al. 2007a). This finding indicates the importance of NW fabrication of small dimension in order to achieve high levels of sensitivity.

Bottom-up techniques such as vapor-liquid-solid to grow SiNWs have been widely reported along with their applications as bio-sensors (Lu and Lieber 2006). 20 nm n-type or p-type SiNWs in diameter have been produced by the vapor-liquid-solid method using Au nanoclusters as a catalyst and diborane or phosphine as dopant precursors. In principle, the technique allows the growth of NWs on a wide range of substrates, i.e., the most common substrate is a Si wafer. High quality of NWs could be produced, but the NWs grown on the substrate are in random orientation and vary in dimension, leading to poor device uniformity and low fabrication yield. So it is a challenging task by itself to bring such techniques to manufacturing maturity. To reap the potential of SiNW in sensor applications on a large scale, it is important to develop technologies that can reliably produce them.

Top-down technologies use nanofabrication tools like e-beam lithography, or photolithography combined with size-reducing strategies, such as self-limiting oxidation, or superlattice NW pattern transfer (SNAP) method to produce NW structures. The resulting NWs are uniform and well aligned. Moreover, the technologies produce NWs in high yields and in a well-ordered orientation on the Si substrate, simplifying them into various functional devices. Various NWs have been produced by the top-down methods. The NWs fabricated by e-beam lithography have width of 50 nm and lengths ranging from 20 μm to 1 mm (Stern et al. 2007a, b; Li et al. 2004). The SNAP method created highly aligned, 20-nm-wide NWs on the n- or p-doped silicon-on-insulator wafer (Bunimovich et al. 2006).

Although there is a great amount of data on SiNW sensors, CMOS compatible SiNW devices are definitely challenging in terms of their manufacturability and commercialization. We have developed a top-down approach to the precise fabrication of individually addressable SiNWs in a perfectly aligned array format using CMOS-compatible technology. The SiNWs are formed through conventional optical lithography, etching and oxidation and thus are fully compatible with CMOS technology. A wafer consisting of 72 individual SiNW sensor chips is depicted in Fig. 2A. The chip for bio-sensing is designed with 40 clusters of five nanowires each. These 40 clusters are divided into two chambers, where the main test chamber has 36 clusters of NWs while the control chamber has 4 clusters. Each cluster contains five individually addressable NWs. Scanning electron microscopy micrographs (Fig. 2B) show that the SiNW arrays with the contact lines are well shaped, highly uniform and well aligned, averaging 90 μm in length, with 2 μm spacing in between two wires. The clusters are pitched at 200 μm. Aluminum metal bond pads are 300 μm × 300 μm and pitched at 500 μm. An atomic force microscopic image further

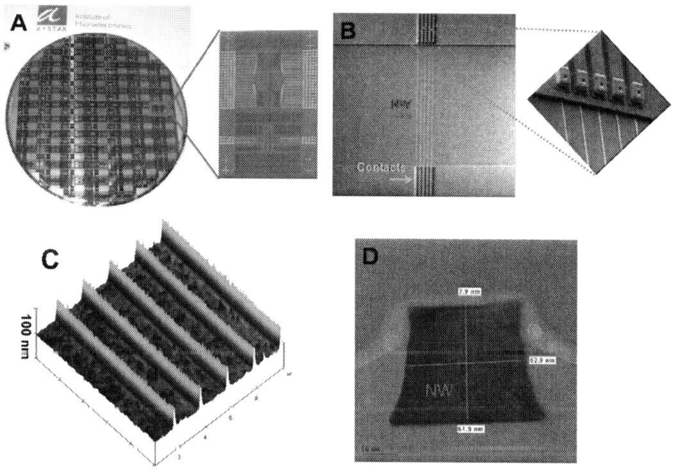

Fig. 2. Silicon nanowire (SiNW) biosensor array chip. (A) Optical images of an 8-inch wafer having 72 SiNW sensor chips (left) and zoom of a single SiNW chip (right) used in the experiment. (B) Scanning electron microscope image of a nanowire cluster composed of five individual nanowires, which are individually addressable. (C) Atomic force microscope image of nanowire arrays. (D) Cross-sectional transmission electron microscope image of a nanowire sensor. (Fig. A, B, D are unpublished materials, Fig. C is reprinted with permission from Gao et al. 2007. Copyright 2009 American Chemical Society.)

Color image of this figure appears in the color plate section at the end of the book.

visualizes the individual nanowires and the nanowire array (Fig. 2C). The width of each SiNW is appropriately 60 nm visualized by cross-sectional transmission electron microscopy image (Fig. 2D).

Real-time Detection of Troponin T by Silicon Nanowire Biosensor

Functionalization of Silicon Nanowire Surface

The surface of the SiNW has to be modified with a probe molecule so that the biosensor is capable of recognizing the specific target molecule. In order to attach the probe molecule on the surface, two approaches, electrostatic adsorption and covalent binding, have been adopted. Electrostatic adsorption is an attractive force responsible for adsorbing ionic solute on an oppositely charged adsorbent. Heath and co-workers demonstrated DNA detection in electrolyte solution using a SiNW sensor, in which a primary DNA strand was electrostatically adsorbed onto an amine-terminated organic monolayer atop the NW surface (Bunimovich et al. 2006).

The covalent binding method is based on the binding of the probe molecule on the SiNW surface by covalent bonds. The native oxide on the SiNW surface can be etched away by submerging the NWs into dilute hydrofluoric acid. The stable Si-C bonds can subsequently be formed on the generated hydrogen-terminated SiNW via photochemical hydrosilylation, resulting in SiNW coated with amino groups. Afterwards, the probe molecules can be immobilized on the SiNW surface to detect protein and DNA (Stern et al. 2007a; Zhang et al. 2008a, b).

Because the oxide grows on the SiNW surface naturally, a number of methods depend on functionalization of the oxide layer. In general, the most well-known linker molecules are alkoxysilane, of which 3-aminopropyltriethoxysilane (APTES) is the popular one. Amino groups are yielded on the surface by the reagent, which can be used to immobilize PNA, DNA and antibodies for various sensing (Chua et al. 2009; Zhang et al. 2009, 2010a, b). Likewise, reaction of the SiNW surface with 3-mercaptopropyltrimethoxysilane (MPTMS) gives rise to a thiol-termination. This -SH terminated SiNW biosensor can be used to detect DNA by immobilizing DNA probes modified with acrylic phosphoramidite at the 5′ end (Li et al. 2004). In addition, another common reagent terminated with aldehyde groups, 3-(trimethoxysilyl)propyl aldehyde, can be functionalized on the SiNW surface. The aldehyde-terminated SiNW surface can directly be immobilized with PNA, DNA and antibodies (Cui et al. 2001; Hahm and Lieber 2004; Zheng et al. 2005). Besides the alkoxysilane, phosphonate derivatives are the alternative molecules to functionalize the SiNW surface with hydroxyl groups, enabling PNA to be attached (Cattani-Scholz et al. 2008).

Figure 3 diagrams the functionalization method of antibody cTnT on the SiNW surface. APTES was used to bind to the hydroxyl-terminated SiO_2 surfaces of the NWs, resulting in an amino-functionalized surface. A bifunctional linker, glutaraldehyde, was then covalently attached onto the amino-terminated surface. Antibody attachment on the SiNW surface was carried out by applying antibody to the aldehyde-modified SiNW surface. In order to prevent non-specific binding of proteins in the detection step, the unreacted aldehydic groups on NW surface were passivated by ethanolamine.

Fluidic Exchange for Analyte Solution Delivery

To rapidly deliver the analyte to the SiNW surface, an efficient fluid exchange system is highly desirable. Two main methods for fluid exchange, close microfluidic channel and open chamber, have been developed. A polydimethylsiloxane (PDMS)-made microfluidic channel is commonly used and placed on the top of the SiNW sensor. It directs the solution over the NW and allows a very small volume of sample to be analyzed (Bunimovich et al. 2006; Cui et al. 2001; Hahm and Lieber 2004; Zheng et al.

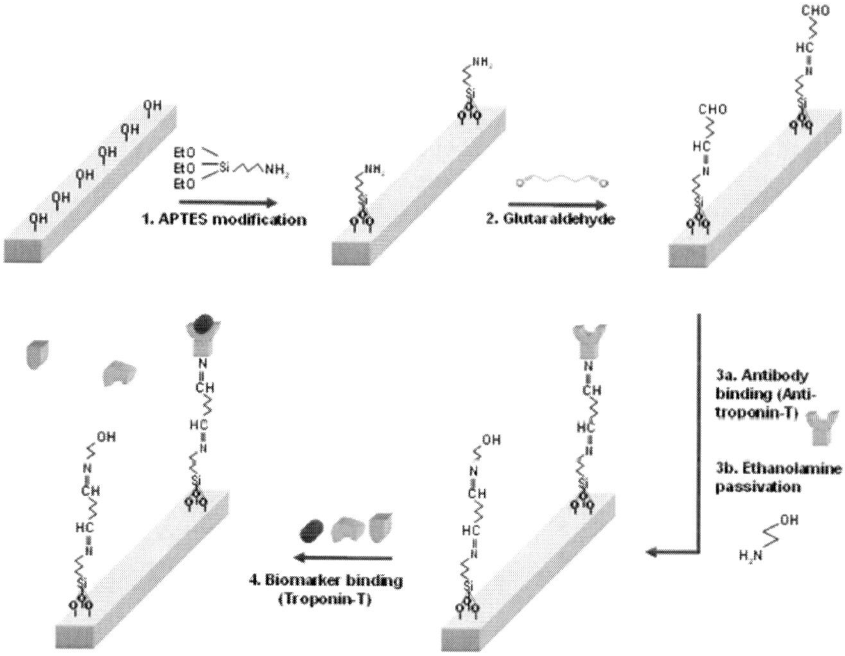

Fig. 3. Schematic diagram of chemical process for surface functionalization of one silicon nanowire (SiNW) sensor. The hydroxyl-terminated silicon dioxide surface of the nanowire binds to the ethoxy groups of 3-aminopropyltriethoxysilane. Glutaraldehyde converts the amino-terminated surface to an aldehyde-terminated one, which is able to bind with the N-terminus of the anti-troponin-T antibody. Antigen-antibody interactions cause troponin-T to bind specifically to the anti-troponin-T probes on the surface, causing changes in SiNW conductance. (Reprinted with permission from Chua et al. 2009. Copyright 2009 American Chemical Society.)

2005). However, laminar microfluidic flow partly restricts the ability of the analyte molecule to reach the SNW surface. Also, the target biomolecule may adsorb on the PDMS sidewalls, which prevents the SiNW sensor from achieving a very high sensitivity.

A rectangular macro-scale solution chamber (inner size: ~$0.8 \times 0.4\ cm^2$), made of acrylic, instead of a microchannel, was used for the real-time detection. Although it is not continuous flow, the solution is delivered from the top tube to the bottom NW sensor, which avoids diffusion issue, but increases detection sensitivity. In addition, a larger amount of solution can be applied. In one word, the simple design involves a reservoir that contains a volume of buffer solution sufficient to maximize binding efficiency while avoiding the diffusion limitations that microchannel systems inherently suffer from.

Choice of Buffer for Detection

The most commonly used buffer solution for measurement is phosphate buffered saline (PBS), which has a pH value of 7.4 and 0.15 M electrolyte. The cTnT contains negative charges at a pH 7.4 PBS buffer solution as the pI of the cTnT is ~ 5.0. The charge on the protein molecule may potentially be screened by the counterions that exist within the buffer solution. Various studies have clearly demonstrated the importance of the Debye screening effect on field effect sensors (Chua et al. 2009; Stern et al. 2007b).

The Debye length is an important parameter affecting the device performance. It is estimated using the formula $\lambda_D=0.32(I)^{-0.5}$, where I is the ionic strength of the buffer solution used for sensing. The Debye length is the distance over which significant charge separation can take place. On one hand, Debye length is expected to be long enough to ensure less charge screened by using dilute buffer solution with low electrolyte concentrations. On the other hand, excessive dilutions resulting in a low salt concentration may not retain biological activity of proteins. To investigate the influence of different buffer ionic strengths on detection sensitivity, a p-type SiNW biosensor was used to study biotin-streptavidin binding using different buffer ionic strengths (Stern et al. 2007b). The results demonstrate that the electrolyte concentration of buffers is critical for the sensitivity of the FET nanosensor, following 0.01×PBS (λ_D~7.3 nm) > 0.1×PBS (λ_D~2.3 nm) > 1×PBS (λ_D~0.7 nm).

Real-time Detection

Real-time detection was performed using an Alessi REL-6100 probe station (Cascade Microtech, Beaverton, OR) in conjunction with an HP parameter analyzer (HP-4155A). A two-probe system was set up in which a controlled voltage could be applied to the source electrode of the SiNW device with the drain electrode referenced to ground. A short pulse voltage of 0.1V was applied to the SiNW sensor and the current through the SiNW was measured. One data point was taken every 1 sec to reduce heating effects due to power dissipation through the SiNW sensor. A buffer solution was injected into the SiNW device region through the inflow tube. The current through the SiNW device while immersed in buffer solution was monitored for 100–200 sec after buffer was added in order to establish the baseline current relative to which changes in conductance would be measured. Upon establishment of the baseline current, the biomolecule analyte solution was added into the SiNW device region through the inflow tube, resulting in a change in conductivity due to specific binding of the biomolecule to the functionalized surface. The current through the sensor was monitored for 100–200 sec after analyte solution was added in order to obtain a stable current representing the conductance of the SiNW after the binding event.

To develop a SiNW array chip able to detect ultralow concentrations of cTnT in biological samples, detection and differentiation between various concentrations of cTnT in buffer solution were carried out. Solutions of cTnT of various concentrations were prepared by dissolving and serially diluting pure cTnT in 0.01 × PBS. SiNW sensors were functionalized with anti-cTnT antibodies and individually used in experiments to detect cTnT in these analyte solutions. Figure 4A demonstrates the real-time response of conductance upon injection of various concentrations of cTnT, where cTnT concentrations are decreased from 1 ng/ml to 1 fg/ml. The data was normalized by computing $|G/G_{t=0}|$ and plotted on the same axes for an effective comparison of the relative change in conductance for the SiNW sensors used to detect the different concentrations of cTnT. Known concentrations of cTnT were added after a stable reading with 0.01 × PBS buffer was achieved. Addition of 1 ng/ml, 1 pg/ml, 100 fg/ml, and 1 fg/ml solutions of cTnT resulted in conductance decreases of ~27%, ~20%,

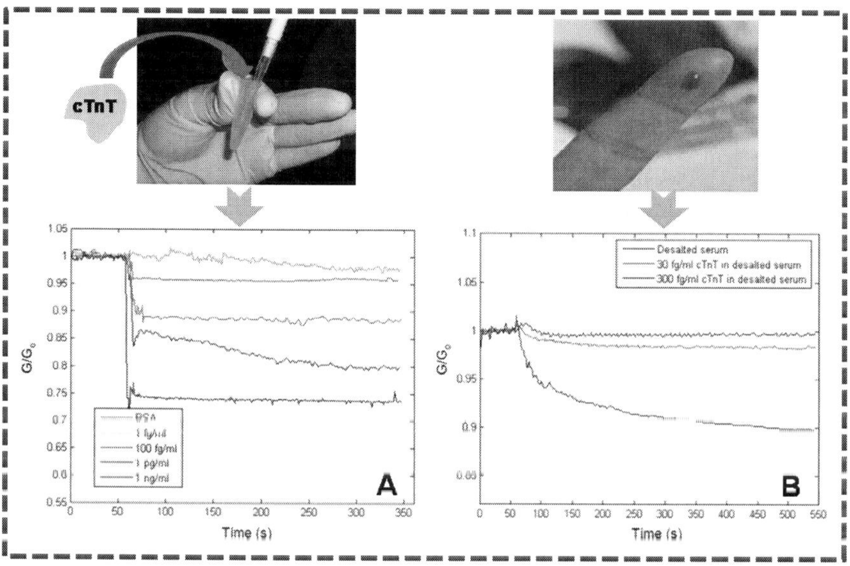

Fig. 4. Real-time detection of troponin T (cTnT) using silicon nanowire (SiNW) biosensor. (A) Real-time conductance response from a SiNW sensor functionalized with anti-cTnT. The response reduces with decreasing cTnT concentration. Addition of 1 μg/ml serum albumin solution leads to negligible signal, pointing to little non-specific binding. (B) Conductance of SiNW sensors functionalized with anti-cTnT and used in experiments to detect cTnT in desalted human serum. Serum solution without cTnT registered negligible change. (Reprinted with permission from Chua et al. 2009. Copyright 2009 American Chemical Society.)

Color image of this figure appears in the color plate section at the end of the book.

~11%, and ~5%, respectively. To evaluate the noise signal, blank buffer (0.01×PBS) without cTnT was injected into the liquid chamber after the baseline was stabilized in the same buffer. A blank sensor response of ~1% was obtained. The sensor response of ~5% caused by the 1 fg/ml cTnT is well above three times the blank response. We thus defined the detection limit for cTnT as 1 fg/ml based on the signal that exceeds the background by three times.

It is especially important to the development of biosensing sensors for medical POC applications that the sensor is able to accurately detect low concentrations of biomarkers in human blood serum. Experiments designed to test the SiNW sensor detection capabilities of cTnT in a human blood serum environment were therefore carried out. A sample of raw human serum was obtained from a healthy individual with no prior history of heart disease for use in experiments to demonstrate cTnT detection in serum. The raw serum was desalted using a microcentrifuge filter as reported (Zheng et al. 2005) and diluted back to the original protein concentration with 0.01×PBS. Various quantities of cTnT were spiked into aliquots of the desalted serum to obtain analyte solutions of required concentrations. A graph of SiNW conductance against time is given in Fig. 4B. An initial baseline for the conductance of the SiNW device was established by injecting desalted serum solution not containing cTnT into the acrylic reservoir. Upon subsequent addition of the cTnT-containing serum solution, the conductance of the sensor immediately decreases and eventually stabilizes at a lower value. We obtained a significant change for the 300 fg/ml cTnT concentration levels and a smaller change as expected for the 30 fg/ml cTnT concentration solution. To verify that the conductance change observed in the SiNW sensors is indeed due to cTnT binding other than interference of other proteins in serum, we applied a blank serum solution in lieu of the cTnT-spiked solutions used for detection experiments and obtained a virtually negligible change in conductance before and after injection. It should be noted the conductance change incurred by the injection of 30 fg/ml cTnT is distinguishable from the control signals. We have thus demonstrated the label-free and real-time detection of cTnT in an undiluted serum environment down to 30 fg/ml, which is 3 orders of magnitude lower than ELISA-related detection methods.

Multiplexed Detection of Cardiac Biomarkers with SiNW Sensor Array

The reliance on a single biomarker for diagnostic purpose has in many cases proven unsatisfactory. It is noted that simultaneous assessment of multiple biomarkers would provide complementary information and enable doctors to stratify risk more effectively and rapidly among patients with ACS. In patients with ACS, different cardiac events are predicted by cTnT, CK-MM and CK-MB. Although each of these markers appears to reflect a unique axis in the pathobiology of ACS, little is known about their utility in combination. By using a POC device, doctors can take the multiple-marker-detection approach to rapidly stratify patients for adverse cardiac outcomes.

Principle of Multiplexed Detection

The SiNW biosensor comprising NWs in an array configuration was produced as aforementioned. As shown in Fig. 5, three different cardiac biomarker-linked antibodies involving monoclonal antibody CK-MM

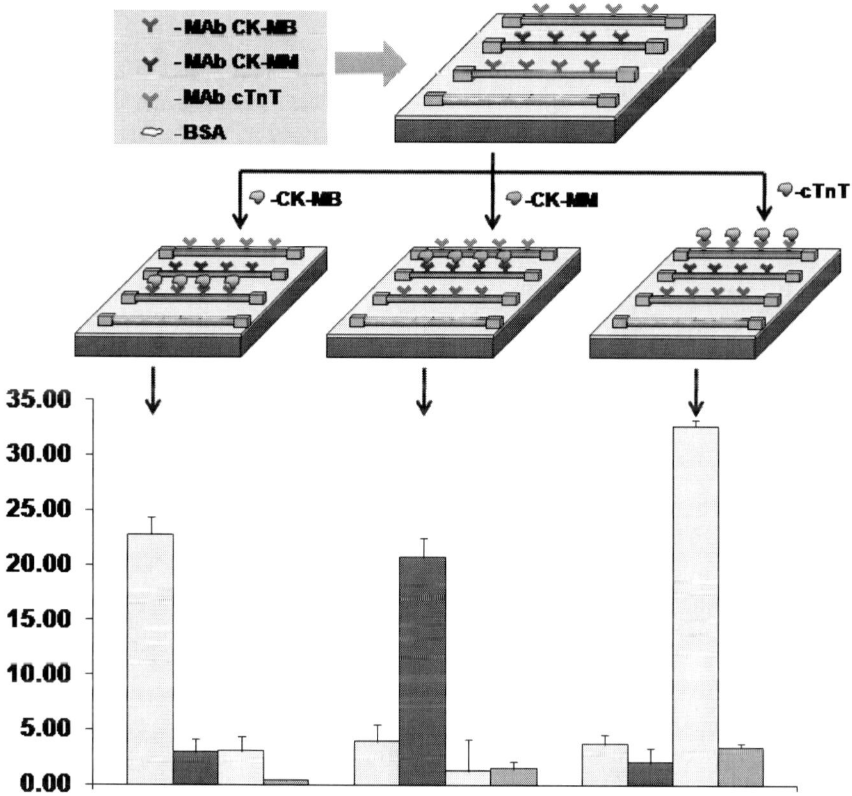

Fig. 5. Specificity of the silicon nanowire (SiNW) biosensor in a multiplexing format. 1 ng/ml of individual biomarker (troponin-T, creatine kinase-MM and creatine kinase-MB) is separately incubated with the SiNW sensor spotted with the three antibodies and serum albumin solution. (Reprinted with permission from Zhang et al. 2009b. IEEE Copyright 2009.)

Color image of this figure appears in the color plate section at the end of the book.

(mAb CK-MM), monoclonal antibody CK-MB (mAb CK-MB), and monoclonal antibody cTnT (mAb cTnT), were separately spotted on the SiNW chip array to allow selective multiplexed detection. As a control, BSA was also spotted beside the three antibodies.

Surface functionalization of the SiNW array chips with different antibodies was carried out using a conventional silane chemistry that has been described above. In order to reduce non-specific interactions of other proteins in the serum, a polyethylene glycol layer was used to block the unreacted aldehyde groups after antibody immobilization because it is well established that the polyethylene glycol layer at the surface is a useful route for obtaining low protein adsorption and low cell adhesion characteristics.

As discussed earlier, a low ionic strength buffer, possessing a long Debye length to prevent screening of the electrical signals in the SiNW FET biosensor, is essential. To analyze the cTnT in serum in real time, the serum has to be desalted for purpose of replacing the high ionic strength buffer in serum with the low ionic strength buffer. Owing to its sophistication, real-time detection is not applicable for direct analysis of multiple biomarkers in serum. To eliminate the desalting process for serum and achieve the direct detection of biomarkers in real sample, we developed an ultrasensitive and selective multiplexed detection of cardiac biomarkers involving cTnT, CK-MM and CK-MB in untreated blood serum using the SiNW array biosensor. The new detection method is independent of the ionic strength of the sample solution, allowing the SiNW sensor to directly analyze the biomarkers in raw samples.

Electrical Detection for Multiplex Platform

A steady-state measurement is adopted for multiplexed detection instead of real-time detection. To enable direct detection of multiple proteins in serum simultaneously, the serum is directly incubated with the SiNW biosensor functionalized with various antibodies. For antibody-antigen binding, the reaction could be conducted in an untreated serum that has high ionic strength. However, the measurement takes place in the presence of 0.01×PBS before and after binding. The sensor response is referred to as resistance change before and after cardiac biomarkers bound to the antibodies spotted on the SiNW surface.

Electrical measurements with the probe station were conducted on the SiNWs by connecting a fixed voltage source (100 mV) across the source and drain electrodes of the SiNW using the same probe station as described earlier. The measurements were carried out in an aqueous environment of 0.01×PBS buffer solution, which was added into an acrylic reservoir attached onto the chip. The interactions of antigen-antibody were verified by recording the resistance change before and after binding of proteins to each antibody in the above-mentioned measuring buffer. A total of 15 wires on one chip for each antibody-antigen interaction were measured for data analysis. The sensor response in resistance was analyzed by using the formula $[(R-Ro)/Ro]\times100$.

The cardiac biomarkers are negatively charged in a neutral PBS buffer (pH 7.4) with pIs of ~5, ~5.8 and ~6.5 for cTnT, CK-MB and CK-MM, respectively. Binding of cTnT, CK-MM or CK-MB to the SiNW surface increases the resistance of the SiNW sensor.

Specificity of Multiplexed Detection

The total protein in serum is abundant; its concentration is 40 to 65 mg/ml. The presence of other proteins may interfere with the binding of the three proteins to the antibody-functionalized SiNW surface. To investigate the specificity of the antibody-functionalized SiNW biosensor, three different cardiac biomarker-linked antibodies (mAb CK-MM, mAb CK-MB and mAb cTnT) and BSA were spotted onto the SiNW clusters. Selective binding of three cardiac biomarkers (cTnT, CK-MM and CK-MB) to the antibodies was realized by measuring the resistance change before and after the binding event. The binding of cardiac biomarkers to the surface-immobilized antibodies takes place only on the SiNW clusters with matching antibodies and vice versa. As seen from Fig. 5, while 1 ng/ml of cTnT, CK-MM and CK-MB in serum were separately incubated with the antibodies-functionalized SiNW, obvious resistance change was obtained to each corresponding specific antibody, whereas negligible change was seen in case of binding of the individual protein to any of the non-specific antibodies. As a control, negligible response change was visible from the BSA-spotted SiNW clusters in case any of the proteins was introduced to them. Specific binding recorded change of ~23% for CK-MB, ~22% for CK-MM, and ~33% for cTnT, respectively, while non-specific binding caused change of less than 5%. These results indicate that the impurities in serum do not affect the specific binding of each protein to its corresponding antibody.

Sensitivity of Multiplexed Detection

To investigate the sensitivity of the multiplexed detection platform, three antibodies were spotted on the SiNW array surface (Fig. 6A). Serum spiked with the three proteins at various concentrations from 1 ng/ml to 10 fg/ml was incubated with the three antibodies-spotted SiNW sensors. The sensing response versus concentrations of the proteins was plotted in Fig. 6B. Remarkable resistance change (~21% for CK-MB, ~19% for CK-MM, and ~27% for cTnT) obtained was found when serum spiked with 1 ng/ml of cTnT, CK-MM and CK-MB was applied. Resistance change reduced

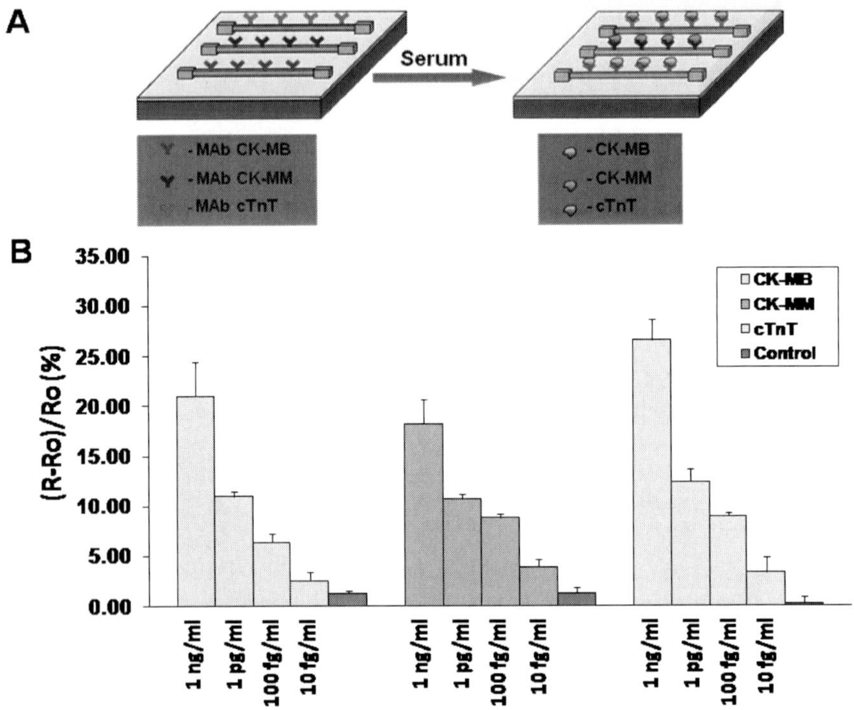

Fig. 6. Sensitivity of the silicon nanowire sensor in a multiplexing format. Response in resistance change versus various concentrations of the three biomarkers (troponin-T, creatine kinase-MM and creatine kinase-MB) spiked in untreated serum. No biomarker is spiked in serum with control. (Reprinted with permission from Zhang et al. 2009b. IEEE Copyright 2009.)

Color image of this figure appears in the color plate section at the end of the book.

but was still detectable (~7% for CK-MB, ~9% for CK-MM, and ~8.5% for cTnT) for serum spiked with 100 fg/ml of cTnT, CK-MM and CK-MB. A control serum without spiking any proteins exhibited resistance change of less than 2% to the three antibodies. The above results demonstrate that the antibodies-functionalized SiNW sensor is capable of multiplexed detection of proteins in undiluted and untreated blood serum with high sensitivity and selectivity down to the fg/ml concentration.

Summary Points

- Arrays of highly ordered SiNWs have been fabricated using CMOS-compatible technology, allowing the ultrasensitive, label-free, electrical detection of cardiac biomarkers.

- An integrated SiNW array biosensor capable of real-time detecting 1 fg/ml human cTnT in an assay buffer solution, as well as 30 fg/ml cTnT in a desalted serum environment, has been developed.
- The SiNWs spotted with anti-cTnT, anti-CK-MM, anti-CK-MB, and BSA show specific response to the corresponding proteins in the multiplex format.
- Highly sensitive and multiplexed detection of cardiac biomarkers by the SiNW biosensor have been achieved: 100 fg/ml for cTnT, CK-MM, and CK-MB in untreated serum, which is 2 orders of magnitude higher than the well-established ELISA method.
- Not only can the developed SiNW biosensor be applied to diagnose heart disease to assist doctors with rapid disposition and treatment for patients suffering from heart attack, but also it can be miniaturized towards the development of a POC device.

Key Facts

- CTnT is a part of the troponin complex. It binds to tropomyosin, interlocking to form a troponin-tropomyosin complex.
- CTnT is a key protein biomarker present in elevated concentrations in the bloodstream of a patient suffering from AMI.
- CTnT is more sensitive and specific than other cardiac biomarkers for myocardial injury.
- CTnT is released from the cell within 3 to 6 h following symptom onset.
- CTnT remains elevated for 7–10 d after injury.

Definitions

Biosensor: An analytical device that uses specific biochemical reactions mediated by isolated enzymes, DNA, immunosystems, tissues, organelles or whole cells to detect chemical compounds usually by electrical, thermal or optical signals.

Cardiac biomarker: Biomarkers are substances found in body fluids such as blood, saliva, or urine. They are often used for diagnostic purpose by reflecting the presence of diseases or organ damage, guiding the doctor in therapeutic decision. Cardiac biomarkers are protein components of cell structures that are released into circulation when myocardial injury occurs and the heart muscle is damaged. They play a pivotal role in the diagnosis, risk stratification, and treatment of patients with chest pain and suspected ACS.

Creatine kinase: A dimeric molecule composed of two subunits (M or B). CK was shown to exist in three molecular forms MM, MB, and BB. CK-BB is expressed in all tissues at low levels and has little clinical relevance. When the heart muscle dies during MI, it releases many molecules into the bloodstream, one of the more abundant being CK. Serial measurement of CK and CK-MB over the first 24 h after symptom onset has been suggested to have a high sensitivity and specificity for detecting myocardial injury.

Creatine kinase MB: CK-MB levels, along with total CK, are tested in persons who have chest pain to diagnose whether they have had a heart attack. Since a high total CK could indicate damage to either the heart or other muscles, CK-MB helps to distinguish between these two sources. Increased CK-MB can usually be detected in heart attack patients about 3-4 h after onset of chest pain. In the setting of AMI, CK-MB levels rise to twice normal within 6 h and peak within 12-24 h.

Silicon nanowire: One of the semiconducting NWs, can be prepared as single-crystal structures with several nanometers. Compared to planar silicon devices produced by microelectronic technology, they can be prepared as p- or n-type materials and configured as field effect transistors that exhibit higher sensitivity due to very small dimension of NWs constrained to tens of nanometers or less, and thus resulting in large surface-to-volume ratio.

Acknowledgments

The author would like to thank Mr. Huiyi Chua Jay, Ms. Ru-Ern Chee, Mr. Zhan Hong Luo Henry, and Ms. Min Huang Joon for biotesting; Dr. Ajay Agarwal, Ms. She Mein Wong, and Mr. Guang Kai Tay Ignatius for chip fabrication. The author also acknowledges financial support from the Singapore Economic Development Board (COY-15-POC/1457-3).

Abbreviations

ACS	:	acute coronary syndromes
AMI	:	acute myocardial infarction
APTES	:	3-aminopropyltriethoxysilane
BSA	:	serum albumin solution
CK	:	creatine kinase
CK-MB	:	creatine kinase MB
CK-MM	:	creatine kinase MM
CMOS	:	complementary metal-oxide semiconductor
cTnT	:	troponin T
CVD	:	cardiovascular disease
ECG	:	electrocardiogram
ELISA	:	enzyme-linked immunosorbent assay

FET	:	field-effect transistor
LOD	:	limit of detection
MI	:	myocardial infarction
MPTMS	:	3-mercaptopropyltrimethoxysilane
NT	:	carbon nanotube
NW	:	nanowire
PBS	:	phosphate buffered saline
PDMS	:	polydimethylsiloxane
PNA	:	peptide nucleic acid
POC	:	point-of-care
Si	:	silicon
SiNW	:	silicon nanowire
SNAP	:	superlattice NW pattern transfer

References

Bunimovich, Y.L., Y.S. Shin, W.S. Yeo, M. Amori, G. Kwong and J.R. Heath. 2006. Quantitative real-time measurements of DNA hybridization with alkylated nonoxidized silicon nanowires in electrolyte solution, J. Am. Chem. Soc. 128: 16323–16331.

Cattani-Scholz, A., D. Pedone, M. Dubey, S. Neppl, B. Nickel, P. Feulner, J. Schwartz, G. Abstreiter and M. Tornow. 2008. Organophosphonate-based PNA-functionalization of silicon nanowires for label-free DNA detection. ACS Nano 2: 1653–1660.

Chua, J., R.E. Chee, A. Agarwal, S.M. Wong and G.J. Zhang. 2009. Label-free electrical detection of cardiac biomarker with CMOS-compatible silicon nanowire sensor arrays. Anal. Chem. 81: 6266–6271.

Cui, Y., Q.Q. Wei, H.K. Park and C.M. Lieber. 2001. Nanowire nanosensors for highly sensitive and selective detection of biological and chemical species. Science 293: 1289–1292.

DeWinter, R.J., J.G. Lijmer, R.W. Koster, F.J. Hoek and G.T. Sanders. 2000. Diagnostic accuracy of myoglobin concentration for the early diagnosis of acute myocardial infarction. Ann. Emerg. Med. 35: 113–120.

Gao, Z.-Q., A. Agarwal, A.D. Trigg, N. Singh, C. Fang, C.H. Tung, Y. Fan, K.D. Buddharaju and J.M. Kong. 2007. Silicon nanowire arrays for ultrasensitive label-free detection of DNA. Anal. Chem. 79: 3291–3297.

Hahm, J.-I. and C.M. Lieber. 2004. Direct ultrasensitive electrical detection of DNA and DNA sequence variations using nanowire nanosensors. Nano Lett. 4: 51–54.

Kong, J., N.R. Franklin, C. Zhou, M.G. Chapline, S. Peng, K. Cho and H. Dai. 2000. Nanotube molecular wires as chemical sensors. Science 287: 622–625.

Ladue, J.S., F. Wroblewski and A. Karmen. 1954. Serum glutamic oxaloacetic transaminase activity in human acute transmural myocardial infarction. Science 120: 497–499.

Li, Z., Y. Chen, X. Li, T.I. Kamins, K. Nauka and R.S. Williams. 2004. Sequence-specific label-free DNA sensors based on silicon nanowires, Nano Lett. 4: 245–247.

Lu, W. and C.M. Lieber. 2006. Semiconductor nanowires. J. Phys. D: Appl. Phys. 39: R387–R406.

McDonnell, B., S. Hearty, P. Leonard and R. O'Kennedy. 2009. Cardiac biomarkers and the case for point-of-care testing. Clin. Biochem. 42: 549–561.

McQueen, M., D. Holder and N. El-Maraghi. 1983. Assessment of the accuracy of serial electrocardiography in the diagnosis of acute myocardial infarction. Am. Heart J. 105: 258–261.

Muller-Bardorff, M., K. Hallermayer, A. Schroder, C. Ebert, A. Borgya, W. Gerhardt, A. Remppis, J. Zehelein and H.A. Katus. 1997. Improved troponin T ELISA specific for cardiac troponin T isoform: assay development and analytical and clinical validation. Clin. Chem. 43: 458–466.

Park, S.-J., T.A. Taton and C.A. Mirkin. 2002. Array-based electrical detection of DNA with nanoparticle probes. Science 295: 1503–1506.

Patolsky, F., G. Zheng, O. Hayden, M. Lakadamyali, X. Zhuang and C.M. Lieber. 2004. Electrical detection of single viruses. Proc. Natl. Acad. Sci. USA 101: 14017–14022.

Stern, E., J.F. Klemic, D.A. Routenberg, P.N. Wyremebak, D.B. Turner-Evans, A.D. Hamilton, D.A. LaVan, T.M. Fahmy and M.A. Reed. 2007a. Label-free immunodetection with CMOS-compatible semiconducting nanowires. Nature 445: 519–522.

Stern, E., R. Wagner, F.J. Sigworth, R. Breaker, T.M. Fahmy and M.A. Reed. 2007b. Importance of the Debye screening length on nanowire field effect transistor sensors. Nano Lett. 7: 3405–3409.

Yach, D., S.R. Leeder, J. Bell and B. Kistnasamy. 2005. Global chronic diseases (Editorial). Science 307: 317.

Zhang, G.-J., A. Agarwal, K.D. Buddharaju, N. Singh and Z.Q. Gao. 2007. Highly sensitive sensors for alkali metal ions based on complementary-metal-oxide-semiconductor-compatible silicon nanowires. Appl. Phys. Lett. 90: 233903–233905.

Zhang, G.-J., J. Chua, R.E. Chee, A. Agarwal, S.M. Wong, K.D. Buddharaju and N. Balasubramanian. 2008a. Highly sensitive measurements of PNA-DNA hybridization using oxide-etched silicon nanowire biosensors. Biosens. Bioelectron. 23: 1701–1707.

Zhang, G.-J., G. Zhang, J. Chua, R.E. Chee, E.H. Wong, A. Agarwal, K.D. Buddharaju, N. Singh, Z.Q. Gao and N. Balasubramanian. 2008b. DNA sensing by silicon nanowire: charge layer distance dependence. Nano Lett. 8: 1066–1070.

Zhang, G.-J., J. Chua, R.E. Chee, A. Agarwal and S.M. Wong. 2009a. Label-free direct detection of MiRNAs with silicon nanowire biosensors. Biosens. Bioelectron. 24: 2504–2508.

Zhang, G.-J., Z.H. Luo, M. Huang, G.K. Tay, E.J. Lim and Y. Chen. 2009b. Highly sensitive and selective label-free detection of cardiac biomarkers in blood serum with silicon nanowire biosensors. IEDM Tech. Dig. P. 607–609.

Zhang, G.-J., L. Zhang, M.J. Huang, Z.H.H. Luo, G.K.I. Tay, E.-J.A. Lim, T.G. Kang and Y. Chen. 2010a. Silicon nanowire biosensor for highly sensitive and rapid detection of dengue virus. Sens. Actuators B: Chem. 146: 138–144.

Zhang, G.-J., Z.H.H. Luo, M.J. Huang, G.K.I. Tay and E.-J.A. Lim. 2010b. Morpholino-functionalized silicon nanowire biosensor for sequence-specific label-free detection of DNA. Biosens. Bioelectron. 25: 2447–12453.

Zheng, G., F. Patolsky, Y. Cui, W.U. Wang, and C.M. Lieber. 2005. Multiplexed electrical detection of cancer markers with nanowire sensor arrays. Nat. Biotechnol. 23: 1294–1301.

8

Electrical Recording from Cardiac Cells and Tissue Using Nanowire Transistors

Tzahi Cohen-Karni,[1] *Bozhi Tian*[2] and *Charles M. Lieber**[,3]

ABSTRACT

Nanowire field-effect transistors (NWFETs) represent diverse and powerful nanostructures for building nanoscale bioelectronic interfaces with cells and tissue. NWFETs exhibit exquisite sensitivity in chemical and biological detection and can form strongly coupled interfaces with cell membranes. We present a general scheme to assemble NWFETs with rationally designed composition and device geometry on inorganic and biocompatible flexible plastic surfaces. These devices can be used to measure extracellular and intracellular signals from heart tissue and cardiomyocytes. The measured extracellular signals are in millivolt-range, which are equal to or substantially greater than those

[1]School of Engineering and Applied Sciences, Harvard University, Cambridge, MA 02138 USA; E-mail: tzahi@cml.harvard.edu

[2]Koch Institute for Integrative Cancer Research, MIT, and Department of Anesthesiology, Children's Hospital Boston, Harvard Medical School, Cambridge, MA 02138 USA; E-mail: tianbz@cml.harvard.edu

[3]Department of Chemistry and Chemical Biology and School of Engineering and Applied Sciences, Harvard University, Cambridge, MA 02138 USA; E-mail: cml@cml.harvard.edu

*Corresponding author

List of abbreviations after the text.

recorded with either planar FETs or multielectrode arrays (MEAs), while the NWFET detection area is several orders of magnitude smaller. Basic studies showing the effect of device sensitivity and cell/substrate junction quality on signal magnitude are presented. Moreover, our ability to design and implement high-density arrays of NWFETs enables us to map signals at the subcellular level, a functionality not enabled by conventional microfabricated devices. Finally, we describe a synthetic breakthrough and novel fabrication to realize the first new electronic measurement tool for intracellular measurements since patch-clamp of the 1970s, a truly three-dimensional nanoscale transistor. We present novel hairpin-like nanostructures that separate the active nanoscale transistor and the connections to the outside world, and show how this can be readily configured as a three-dimensional, nanoscale transistor capable of non-invasive intracellular measurements. The exceptional synthetic control and flexible assembly of nanowires provides powerful tools for fundamental studies and applications in the life sciences broadly defined, and opens up the potential of merging active transistors with cells such that the distinction between non-living and living systems is blurred.

INTRODUCTION

Recording electrical signals from cells and tissue is central to areas ranging from the fundamental biophysical studies of function in, for example, the heart and brain, through medical monitoring and intervention. Over the past several decades, studies of electroactive cells and tissue have been carried out using a variety of recording techniques, including glass micropipette intracellular and patch-clamp electrodes (Sakmann and Neher 1984), voltage-sensitive dyes (Efimov et al. 2004; Scanziani and Hausser 2009); multielectrode arrays (MEAs), (Halbach et al. 2003; Reppel et al. 2004), and planar field-effect transistors (FETs) (Ingebrandt et al. 2001; Yeung et al. 2001). The latter two use well-developed microfabrication methods to allow for multiplexed detection on a scale not possible with micropipette technology, although the MEAs exhibit limited signal-to-noise (S/N) and relatively large detection areas that make cellular and subcellular recording challenging (Banks et al. 2002).

A new class of molecular-scale electronic interfaces can be formed with cells and tissue using chemically synthesized semiconductor nanowires (NWs) as functional elements. These NWs have received intense interest in recent years, leading to the development of structures with rationally controlled geometry, composition, and electronic properties (Morales and Lieber 1998; Hu et al. 1999). These characteristics have enabled semiconductor NWs to emerge as powerful building blocks for the bottom-up assembly of functional devices with application areas from nanoelectronics (Cui and Lieber 2001; Lu and Lieber 2007) to the biosciences (Cui et al. 2001; Patolsky et al. 2004; Zheng et al. 2005, Ragmir et al. 2010).

Nanowires as Biological Sensors

Underlying biological detection using semiconductor NWs (Cui et al. 2001; Patolsky et al. 2004; Zheng et al. 2005) is their configuration as FETs, which exhibit a conductance change in response to variations in the charge or potential at the surface of the channel region (Fig. 1A) (Sze and Ng 2007). This property makes FETs natural candidates for electrically based sensing since binding of a charged or polar biological or chemical species to the gate dielectric is analogous to the conventional case of applying a voltage gate using a metallic gate electrode (Fig. 1B). Significantly, NWFETs are more sensitive sensors than their planar counterparts because of their one-dimensional nanoscale morphology. An analyte binding to the surface of a NW leads to depletion or accumulation of carriers in the "bulk" of the one-dimensional nanometer-diameter structure, versus only a shallow region near the surface in the case of a planar device. This unique feature of semiconductor NWs enables exquisite charge sensitivity that opens up new opportunities for interfaces with chemical and biological systems.

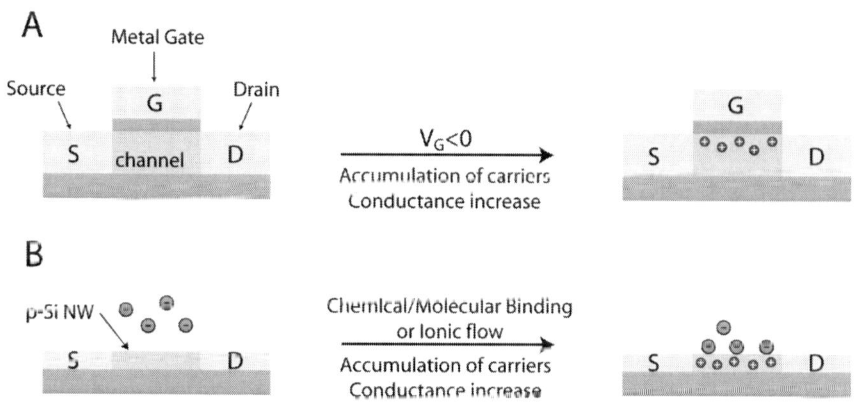

Fig. 1. Planar and NW transistors. (A) Schematic of a p-type planar FET device, where S, D and G correspond to source, drain and gate electrodes, respectively. (B) Schematic of electrically based sensing using a p-type NWFET, where binding of a charged biological or chemical species to the chemically modified gate dielectric induces carriers in the channel.

Applications to Areas of Health and Disease

NWs elaborated as well-defined n- or p-channel FETs have been used as ultrasensitive sensors of chemical and biological species. In 2004 we achieved the limit of biological detection—single particle sensitivity—by detecting, in real time, the reversible and selective binding of virus particles to antibody-modified NWFETs (Patolsky et al. 2004). The exquisite sensitivity of NWFET was further demonstrated by multiplexed sensing of cancer marker proteins down to approximately 2 fM (Zheng et al. 2005). This represents a sensitivity limit 10^4–10^9 times below that afforded by state-of-the-art planar FETs (Schoning and Poshossian 2002), and was verified using top-down fabricated NWs (Stern et al. 2007).

Advantages of Nanoscale Morphology for Extracellular Interfaces

A major advantage of NWFETs relates to the coupling between the NW and cells. The formation of a tight junction between a cell or cellular projections and the semiconductor surface is important in determining sensitivity, when measuring local field changes given the high ionic strength of the extracellular medium. On a flat surface the junction gap is on the order of tens of nanometers, a result of the finite size of adhesion/transmembrane proteins. NWs, however, are freestanding structures that *protrude* from the surface of the substrate, and thus naturally form tighter junctions with the cell membrane than is possible with a planar device (Fig. 2A).

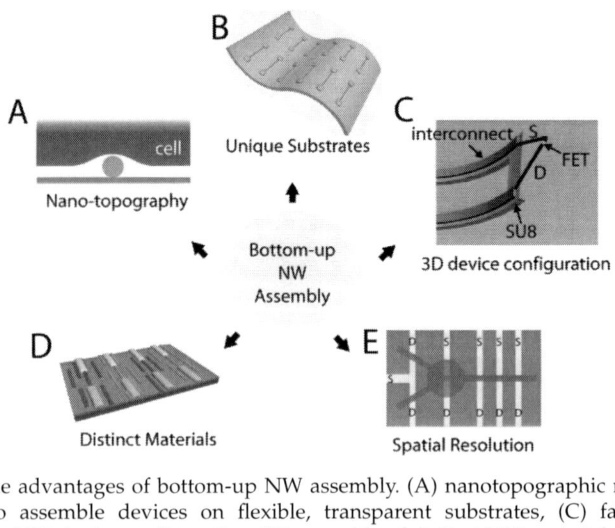

Fig. 2. Unique advantages of bottom-up NW assembly. (A) nanotopographic morphology, (B) ability to assemble devices on flexible, transparent substrates, (C) fabrication of unconventional 3D device configuration, (D) assembly of distinct NW materials on the same chip, and (E) high spatial resolution of NW devices.

Color image of this figure appears in the color plate section at the end of the book.

Cellular adhesion and guidance may be further enhanced by unique interactions between nanostructures and cell membranes. For example, it has been demonstrated that carbon nanotube networks and etched silica nanostructured surfaces promote cellular adhesion, spreading, and guidance, even in the absence of conventional adhesion factors (Sniadecki et al. 2006). Moreover, sparse NWs arrays were shown to direct axonal growth over large areas by promoting the formation of focal adhesions, which are critical for tight cell-substrate junctions (Kotov et al. 2009). Taken together, these findings indicate that NWFETs can serve as both sensitive electronic devices and nanoscale interfaces that promote unique and favorable substrate-cell interactions.

One final aspect of nanostructured surfaces that warrants emphasis is their effect on ion-channel function. For example, results suggest that neurons cultured on nanotubes exhibit enhanced activity (vs. planar control surfaces), and thus that conductive, nanostructured surfaces enhance membrane excitability (Kotov et al. 2009). While these studies warrant further mechanistic investigations, they provide additional evidence that nanostructured surfaces form unique interactions with cells not realized previously with planar electronic devices.

EXTRACELLULAR INTERFACES OF NWFET AND TISSUE/CELLS

NWFET Device Fabrication

Our bottom-up approach involves growth of NWs (Morales and Lieber 1998; Hu et al. 1999), and then assembly and interconnection of the NWs on a substrate (Patolsky et al. 2006). A true advantage of our approach is that NWs can be assembled on nearly any type of surface, including those that are typically not compatible with standard semiconductor processing, such as flexible plastic substrates (Fig. 2B) (McAlpine et al. 2003; Javey et al. 2007; Timko et al. 2009). A second advantage is that sequential patterning and assembly steps enable fabrication of distinct NWFET devices on a substrate, as indicated by the blue and green NWs in Fig. 2D (Patolsky et al. 2006). A third advantage is the ability to form unique three-dimensional (3D) device geometries that cannot be achieved otherwise (Fig. 2C).

Device arrays are fabricated by aligning NWs on the surface of a substrate using either fluid-based or contact printing techniques (Javey et al. 2007). Both yield highly aligned NWs with rationally varied pitch from tens of microns to as small as tens of nanometers. Interconnects are subsequently defined to form contacts with the NWs (Patolsky et al. 2006). Using this overall methodology, 10s to 100s of devices can be obtained on a single chip, enabling many experiments in parallel. Notably, high-density arrays enable electronic interfaces to individual subcellular features such as axons and dendrites, a spatial resolution difficult to achieve by other methods (Fig. 2E).

Device Characteristics

Conductance (G) versus water-gate (V_g) measurements for NW devices submerged in electrophysiology medium (Timko et al. 2009) show that p-type NWFETs all turn off close to or positive of $V_g = 0$. In the on-state, the device sensitivities, dG/dV_g, typically have values of 10s of μS/V.

NW/Heart Interfaces

In whole tissue, for example, the heart, NW devices can provide unique electrophysiological information. Activation sequences across the surface of the heart have been measured using macroscale metallic electrodes (Taccardi et al. 2008), optical microscopy of dyed tissue (Efimov et al. 2004), and MEAs (Reppel et al. 2004), but none of these techniques has been miniaturized to achieve single-cell resolution. Given the complexities of activation sequences, cellular or subcellular-level interfaces with cardiac tissue could be crucial for better understanding cardiac dysfunction, for example, arrhythmia (Cohen et al. 1989).

Characterization of Heart/NW Interface

We used live hearts from embryonic chickens (E10-E15 stage) as a model system for studying tissue/NW interfaces. In a typical experiment with a planar NWFET chip configuration (Fig. 3 A and B), a freshly isolated heart was placed on top of the active device region of a heated sample chamber (Timko et al. 2009). The hearts beat spontaneously at a typical frequency of 1–3 Hz. Representative data of simultaneous recording from a NWFET and a conventional electrode inserted into the heart show close temporal correlation between initial sharp peaks, although the pipette peak occurs ca. 100 ms before the NWFET peak (Fig. 3C). The consistent time difference is expected since the pipette was inserted into a spatially remote region with respect to the NWFET devices. Examination of individual NW signals reveals an initial fast phase (full-width at half maximum, FWHM = 6.8 ± 0.7 ms) followed by a slower phase (FWHM = 31 ± 9 ms), where these two phases can be attributed to transient ion-channel current and mechanical motion, respectively (Timko et al. 2009).

The peaks recorded with our NWFETs (Fig. 3C) exhibit excellent S/N, and the conductance changes associated with these peaks depends on the device sensitivity. NWFET results from a beating heart with device sensitivity varied by the water-gate (–0.4 to 0.4V) showed a decrease in the magnitude of the fast transient conductance change from ca. 55 to

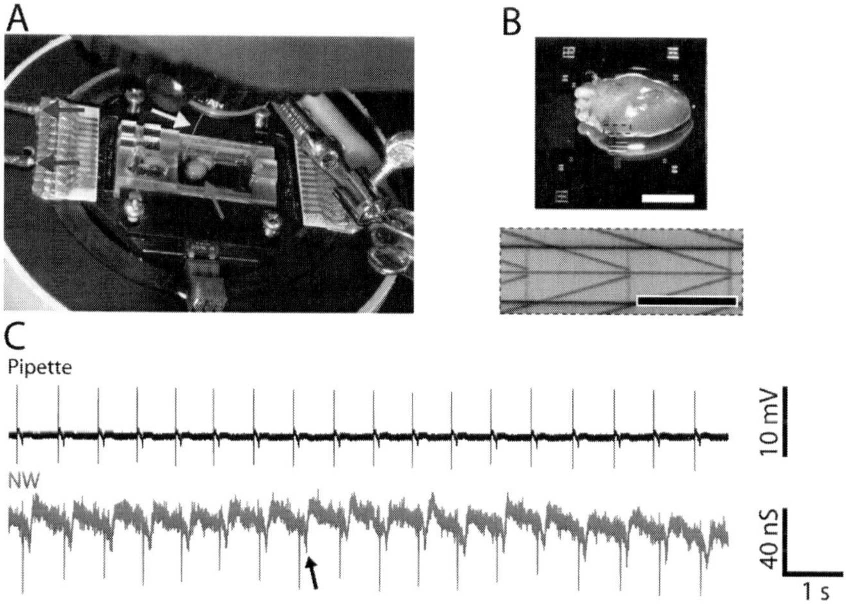

Fig. 3. NW/Heart interfaces. (A) Photograph of experimental setup showing heart on NWFET chip in temperature regulated cell. Arrows show position of heart (red), reference electrode (yellow) and interconnect wires (blue). (B) (top) Image of heart on surface of a planar chip; scale bar is 4 mm. (bottom) Zoom of dotted region in upper image showing three pairs of NWFETs; scale bar is 150 µm. (C) Simultaneous recordings from a glass pipette (black trace) and a NW device (red trace). Adapted from Timko et al. (2009) (American Chemical Society, copyright 2009).

Color image of this figure appears in the color plate section at the end of the book.

11 nS, which is correlated with the decrease in device transconductance over this same range of water-gate potentials. Notably, the voltage-calibrated signal determined using the device transconductance was constant at 5.1 ± 0.4 mV. These results confirm the stability of the interface between the NWFETs and beating heart (Timko et al. 2009).

Transparent and Flexible Substrates

NWs and carbon nanotube FETs can be fabricated on flexible plastic substrates (Javey et al. 2007; McAlpine et al. 2003) and thus open up the possibility of making chips that can be readily deformed to tissue and organs or used for *in vivo* studies. We have explored this concept by assembling active NWFETs on 50 µm-thick flexible and transparent plastic substrates. These flexible and transparent NWFET chips enable simultaneous optical imaging and electronic recording in configurations that are not readily accessible with traditional planar device chips, yet advantageous for producing diverse, functional tissue-device interfaces. A bent device chip with concave surface facing a beating heart immersed in medium (Fig. 4A) illustrates this capability. We note that the chip is readily integrated into an upright microscope and allows for both visual inspection and higher-resolution imaging through the transparent substrate while recording

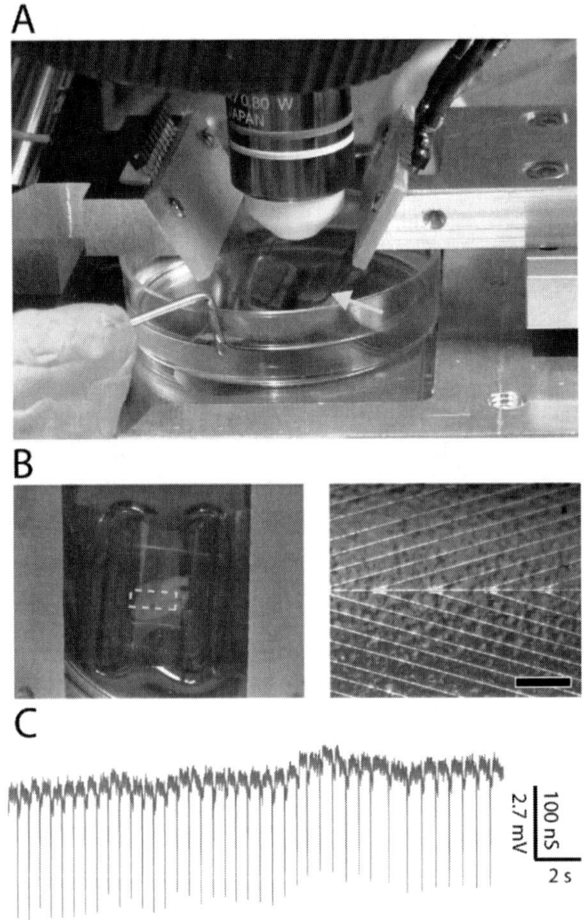

Fig. 4. NWFET array recording in a bent configuration. (A) Photograph of heart (yellow arrow) underneath the concave face of the NFET substrate. (B) (left) Photograph of same system showing registration between heart and markers on the substrate. (right) Optical image taken showing the heart surface versus position of individual NW devices located along the central horizontal axis. Scale bar is 150 μm. (C) Recorded conductance data from a NWFET in this concave configuration. Adapted from Timko et al. (2009) (American Chemical Society, copyright 2009).

Color image of this figure appears in the color plate section at the end of the book.

from NWFET devices (Fig. 4B). Notably, recording from a representative NWFET device in this inverted configuration (Fig. 4C) demonstrated excellent S/N peaks correlated with the spontaneously beating heart. In addition, similar recording was achieved on beating hearts in which bent chips were oriented with convex NWFET surface wrapped partially around the heart. Taken together, these results demonstrate that our flexible and transparent NW chips can be used to record electronic signals from organs in configurations not achievable by conventional electronics (Timko et al. 2009).

Our studies of neurons (Patolsky et al. 2006) and heart tissue (Timko et al. 2009) demonstrate that NW devices can be used as a general platform to investigate electrogenic biological systems. A separate group also demonstrated that lithographically patterned NWFETs can be used to record signals from rat cardiomyocytes and monolayers of rat aortic smooth muscle cells with high S/N and mV amplitudes (Pui et al. 2009). In our own laboratory, we investigated cultured cardiomyocyte monolayers in a new manner and probed the relationship between interfaces and signal magnitude (Cohen-Karni et al. 2009).

NW/Cardiomyocytes Interfaces

Characterization of NW/Cardiomyocyte Interfaces

We developed a flexible scheme for interfacing cardiomyocytes and cells in general with NWFETs. Chips were fabricated using the same process used for neuron and heart studies (Patolsky et al. 2006; Timko et al. 2009) (Fig. 5A). Separately, cardiomyocytes were cultured on thin, optically transparent and flexible pieces of polydimethylsiloxane (PDMS) (Fig. 5B) to form cell monolayers, and then a PDMS/cardiomyocyte substrate was aligned over the NWFETs on the chip (Fig. 5C), to enable measurements from spontaneously beating cardiomyocytes at the single cell level (Fig. 5D) (Cohen-Karni et al. 2009).

The ability to manipulate the PDMS/cell substrate independent of the NWFET chip enables us to identify specific cardiomyocyte regions using an optical microscope, place the desired cell or cell network over NWFET devices, and then record from the desired region. Importantly, our ability to identify and register specific cellular regions over NWFET elements opens up the possibility of a number of interesting studies, including multiplexed recording from well defined multicellular configurations as well as multiplexed measurements at the single-cell level for subcellular resolution investigations (Cohen-Karni et al. 2009).

These basic results for NWFET recording from PDMS/cardiomyocyte samples can be compared to previous studies of cardiomyocytes. For example, studies of cardiomyocyte monolayers cultured on conventional

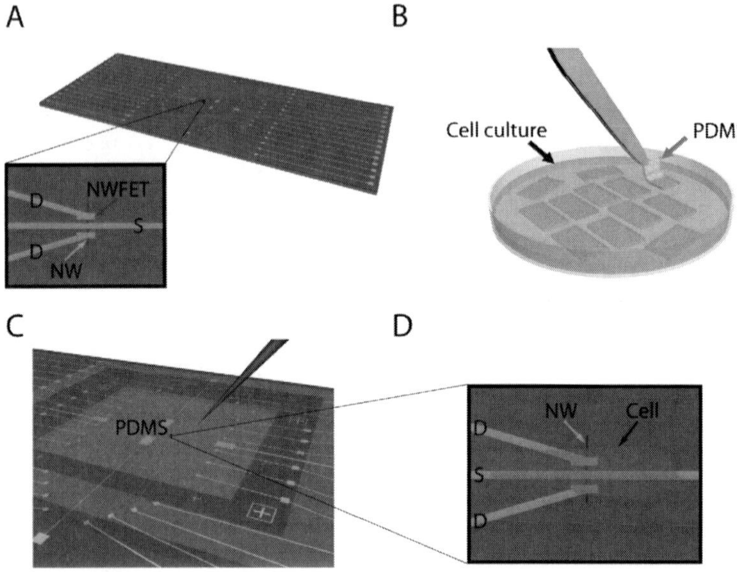

Fig. 5. Controlling nanowire/cardiomycytes interfaces. (A) NWFET chip, where NW devices are located at the central region of chip. The linear gold features correspond to contact and interconnect metal. Zoom shows source (S) and two drain (D) electrodes connected to a NW (blue arrow) defining two NWFETs. (B) Cardiomyocytes cultured on thin flexible PDMS. (C) PDMS substrate with cultured cells oriented over the NWFET chip. The green structure is the probe used to manipulate the PDMS. (D) Schematic of a cardiomyocyte (black arrow) oriented over a NWFET (green arrow). Adapted from Cohen-Karni et al. (2009) (National Academy of Sciences, copyright 2009).

Color image of this figure appears in the color plate section at the end of the book.

planar FET devices have yielded peaks with S/N of 2–6 and amplitudes from 0.2 to 2.5 mV (Ingebrandt et al. 2001; Yeung et al. 2001). In general, our NWFETs yield better S/N with values > 4 and a maximum of 25 observed. In addition, the typical calibrated voltages recorded in our measurements are similar to or greater than the largest values reported previously. The improvement in peak amplitude is consistent with the fact that the nanodevices protrude from the plane of substrate and hence can increase NW/cell interfacial coupling (Sniadecki et al. 2006; Kotov et al. 2009).

Cell Substrate Displacement Experiments

The ability to manipulate the PDMS/cell substrate independent of the NWFET chip also opens up new opportunities compared with cells cultured directly on device arrays. For example, we investigated the relationship between recorded signal magnitudes when a micropipette displaced the

PDMS a fixed distance (Fig. 6A). A direct comparison of single peaks recorded for increasing displacement values (Fig. 6B) shows a consistent monotonic increase in peak amplitudes. In addition, this high-resolution peak comparison demonstrates that there is no observable change in peak shape or peak width over this > 2x change in amplitude, and that the peak width is consistent with time-scales for ion fluxes associated with ion-channels opening/closing (Ingebrandt et al. 2001). A summary of data obtained for increasing/decreasing displacements (Fig. 6C) further

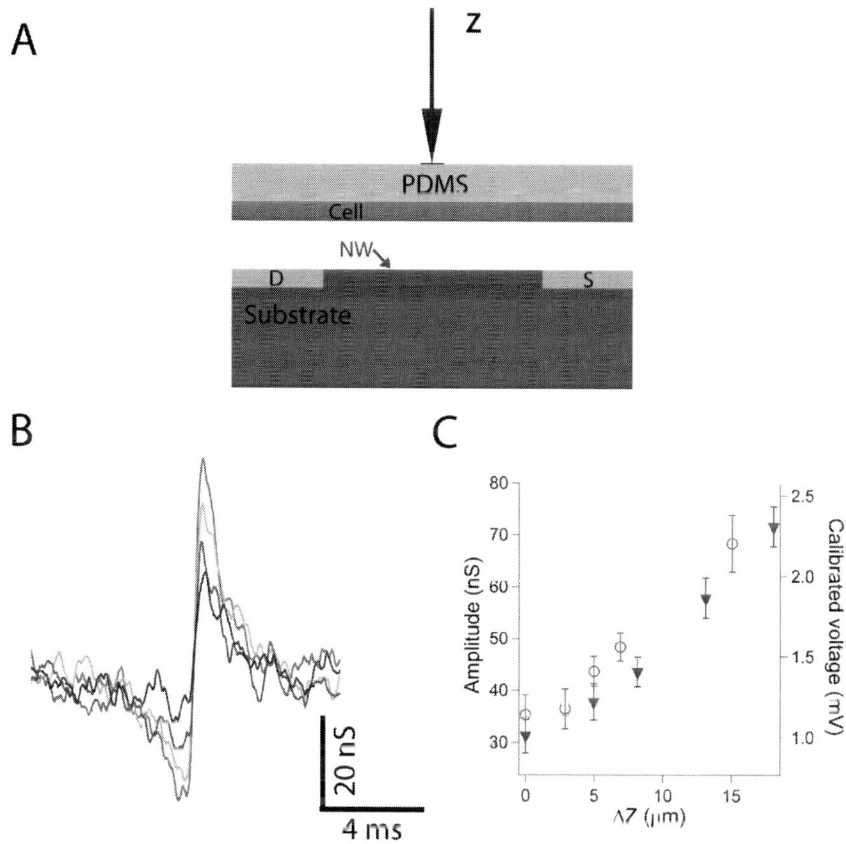

Fig. 6. Effect of applied force on recorded signals. (A) Schematic illustrating displacement (Z) of the cell substrate with respect to a NWFET. (B) Comparison of single peaks recorded with ΔZ values of 0 (purple), 8.2 (blue), 13.1 (green), and 18.0 μm (red). (C) Summary of the recorded conductance signals and calibrated voltages vs. ΔZ, where the open red circles (filled blue triangles) were recorded for increasing (decreasing) ΔZ. Adapted from Cohen-Karni et al. (2009) (Proceedings of the National Academy of Sciences of the United States of America, copyright 2009).

Color image of this figure appears in the color plate section at the end of the book.

shows the systematic 2.3-fold increase in peak amplitude and, moreover, demonstrates these amplitude changes are reversible. Recent studies of *Aplysia* neurons cultured on planar FET devices have also reported an increase in peak amplitude when the cell body was displaced (Cohen et al. 2008). In both cases, the enhanced signal amplitudes can be attributed to a decrease in gap between the cell membrane and devices, although future studies will be needed to quantify such junction changes.

Multiplexed Recording with NWFETs

Multiplexed measurements from NWFET arrays have been reported previously in the context of other biological systems (Zheng et al. 2005; Patolsky et al. 2006; Timko et al. 2009). We carried out multiplexed measurements using the NWFET arrays interfaced with cardiomyocytes. The devices used in these studies were configured in a linear array (Fig. 7A) so that signal propagation within cardiomyocyte monolayers could be characterized. Simultaneous recording from NW1, NW2 and NW3 devices in contact with spontaneously beating monolayer (Fig. 7B) yield stable and high S/N ~10 peaks. Time differences between the signals recorded

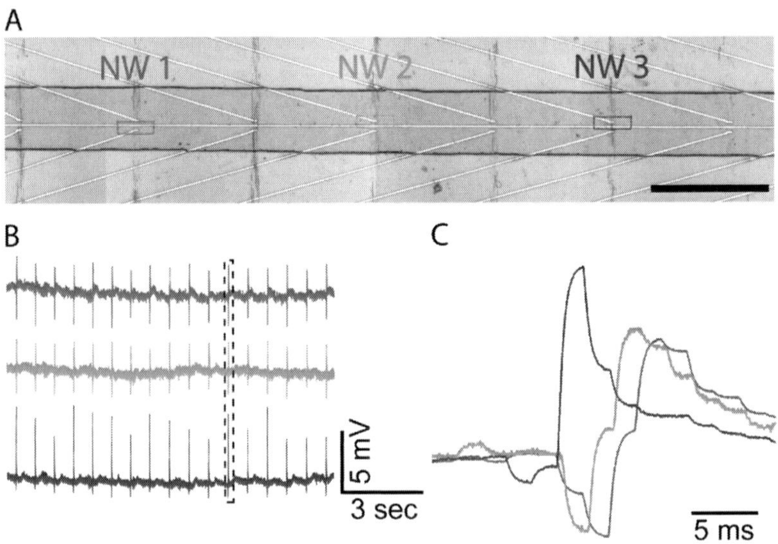

Fig. 7. Multiplexed NWFET recording. (A) Optical micrograph showing 3 NWFET devices (NW1, NW2, NW3) in a linear array; scale bar is 150 μm. (B) Conductance vs. time signals recorded simultaneously from NW1, NW2, and NW3. (C) Comparison of temporally correlated peaks highlighted by the black dashed box in B. Adapted from Cohen-Karni et al. (2009) (Proceedings of the National Academy of Sciences of the United States of America, copyright 2009).

Color image of this figure appears in the color plate section at the end of the book.

by the three devices can be estimated from comparison of single peaks (Fig. 7C), although a cross-correlation methodology (Cohen-Karni et al. 2009) provides a quantitative measure of the time delays. These results and device separations yield propagation speeds of 0.07–0.21 m/s that are consistent with other measurements on monolayers of neonatal rat cardiomyocytes (Fast and Kleber 1994). The variation in propagation speeds is not surprising given the monolayer inhomogeneity and suggests an important future direction. Specifically, high-resolution multiplexed NWFET recording combined with optical imaging will enable details of both intra- and intercellular propagation for well-defined cellular assemblies.

INTRACELLULAR INTERFACES OF NWFET AND CELLS

Localized and tunable 3D sensing and recording using the prototypical nanoelectronic device, a nanoscale field-effect transistor (nanoFET) (Lu and Lieber 2007), have not been previously demonstrated because almost all transistors are created on planar substrates. Ideally, one would like to move the nanoFET with necessary source (S) and drain (D) electrical connections into contact with the cell and probe within the cell membrane. However, minimally invasive insertion of a nanoFET into the confined 3D space of single cells or even 3D cellular networks is a significant challenge because the S and D typically dominate the overall device size and define a planar and rigid structure regardless of whether the nanoFET is on or suspended above a planar substrate (Lu and Lieber 2007; Heller et al. 2009). In this section, we will review the design and synthesis of a revolutionary 3D, flexible nanoFET and its implementation as localized intracellular probe.

NW Probe Synthesis

Recently, we demonstrated a breakthrough in nanostructure synthesis with the introduction of 120° kinks along the axis of a growing nanowire (Tian et al. 2009); moreover, we showed that the junction regions could be doped to create nanoFETs. We have extended this methodology to create a two-terminal FET probe that could be inserted into single cells (Tian et al. 2010). First, we incorporated two or three *cis*-linked kinked units to yield probe tip angles of 60° or 0°, respectively (Fig. 8A). Because two *trans*-linked units (Fig. 8A) would be unsuitable for a probe, the selective and well-controlled synthesis of *cis*-linked units is central to our probe design. A representative scanning electron microscopy (SEM) image (Fig. 8B) shows a doubly kinked structure with well-defined *cis*-linkage, and overall 60° tip angle. Second, we used selective doping during synthesis to localize

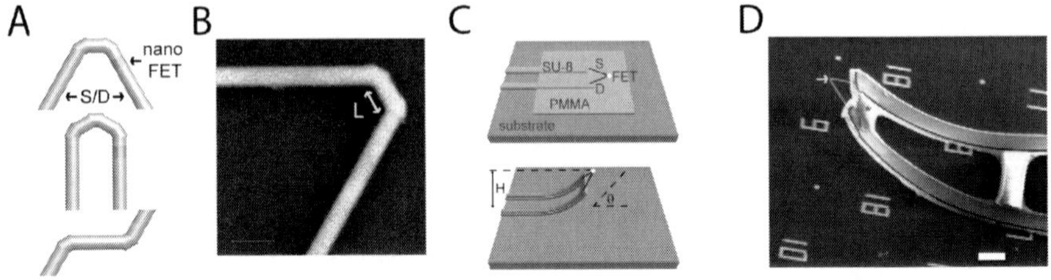

Fig. 8. Synthesis of kinked silicon nanowire probes. (A) Schematics of 60° (top) and 0° (middle) multikinked nanowires, and *'cis'* (top) and *'trans'* (bottom) configurations. The blue and magenta regions designate source/drain (S/D) and nanoFET, respectively. (B) SEM image of a double kinked nanowire with *'cis'* configuration. (C) Schematics of device fabrication. H and θ are tip height and orientation, respectively, and the white dot designates the nanoFET. (D) SEM image of a 3D nanoprobe. The yellow arrow and magenta star mark the nanoFET and SU-8, respectively. Scale bar is 5 μm. Adapted from Tian et al. (2010) (American Association for the Advancement of Science, copyright 2010).

Color image of this figure appears in the color plate section at the end of the book.

and self-label the nanoscale FET element adjacent to the topologically defined probe tip (Fig. 8A, magenta segments), and simultaneously "wire up" the FET channel with heavily doped nanowire S/D components (Fig. 8A, blue segments).

Device Fabrication

We next designed an unconventional fabrication approach to present the nanoFET in 3D for use as an intracellular probe. Remote electrical interconnects were made to the S/D nanowire arms using a multi-layer structure whose interfacial stress bent the probe upward in a controllable manner (Fig. 8C). A representative SEM image of one free-standing device (Fig. 8D) demonstrates that the 60° kinked probe remains intact after fabrication with the two nanowire arm terminals connected to larger metallic electrodes leading that are used for connections to measurement instruments. Our nanoprobes are unique because (1) the nanoFET and S/ are integrated synthetically at the nanoscale, (2) the nanoFET is free-standing, and (3) the acute-angle kinked nanowire geometry and the extended S/D arms spatially separate the functional nanoFET from the bulky allowing intracellular interrogation with minimum interference from ancillary parts of the nanoFET.

Surface Modification and Cellular Entry

The 3D nanoFET probes were coated with phospholipid bilayers (Misra et al. 2009), which can fuse with cell membranes, to enable intracellular recoring (Fig. 9A). Fluorescence microscopy images of dye-labeled phospholipid-modified probes (Fig. 9B) show that the lipid forms a

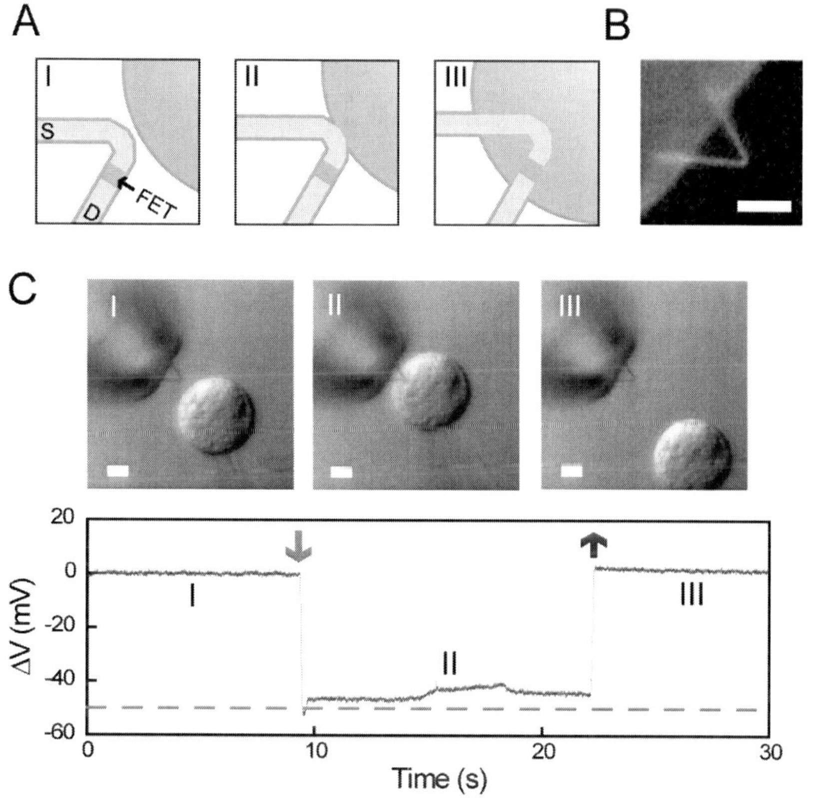

Fig. 9. Surface modification and cellular entry. (A) Schematics of nanowire probe entrance into a cell. Purple, light purple, magenta and blue colors denote the phospholipid bilayers, doped nanowire arms, active sensor segment and cytosol, respectively. (B) Fluorescence image of a lipids-coated nanowire probe. (C) Differential interference contrast (DIC) microscopy images (upper panels) and electrical recording (lower panel) of an HL-1 cell and nanowire probe as the cell approaches (I), contacts and internalizes (II), and is retracted from (III) the nanoprobe. A micropipette was used to manipulate and voltage-clamp (broken green line) the cell. Scale bars: 5 μm. Adapted from Tian et al. (2010) (American Association for the Advancement of Science, copyright 2010).

Color image of this figure appears in the color plate section at the end of the book.

continuous shell on nanoprobes. We then monitored the calibrated potential change of phospholipid-modified nanoFET probe while an isolated HL-1 cell was moved into contact and then away from the nanoprobe using a glass micropipette under microscopy visualization (Fig. 9C, top). The micropipette was also used to clamp the intracellular potential at –50 mV (Sakmann and Neher 1984). Notably, measurement of the potential versus time from the nanoFET probe shows a sharp ~50 mV drop within 250 ms after cell/tip contact, which is maintained until the cell was detached from the nanoprobe end. Interestingly, nanoFET probes of similar sensitivity that were not coated with a phospholipid bilayer modification exhibited only baseline fluctuations as HL-1 cells were brought into contact and then retracted. These results suggest the biochemical state of the nanowire probe surfaces is critical for assisting access to the intracellular region, and is distinct from larger, more rigid probes commonly used for intracellular electrical recording (Tian et al. 2010).

Intracellular Recordings

We have also investigated the formation of intracellular interfaces between our 3D nanoFET probes and spontaneously firing cardiomyocytes (Fig. 10A). Representative conductance versus time data recorded from a 3D nanoFET probe initially in gentle contact with a spontaneously beating cardiomyocyte cell showed a sequence of distinct features (Fig. 10B). Initially, we observed regularly spaced ~1 ms duration spikes with a frequency of ca. 2.3 Hz consistent with extracellular recordings made with nanowire devices on substrates (Cohen-Karni et al. 2009). After a relatively brief (~40 s) period of extracellular signals, we observed several pronounced changes with the initial extracellular signals gradually disappearing (Figs. 10B and C, II, magenta stars) and a concomitant decrease in baseline potential and new peaks emerging that had an opposite sign, similar frequency, much greater amplitude, and longer duration (Fig. 10B, II, green stars). These new peaks, which are coincident with cell beating, rapidly reached a steady state (Fig. 10B, III) with an average calibrated peak amplitude of ~80 mV and duration of ~200 ms. The amplitude, sign, and duration are near those reported for whole-cell patch clamp recordings from cardiomyocytes (Zipes et al. 2004), and thus we conclude that these data represent a transition to steady-state intracellular recording with the 3D nanowire probe (Tian et al. 2010).

Detailed analysis of the latter steady-state peaks (Fig. 10C) shows five characteristic phases of a cardiac intracellular potential (Zipes et al. 2004): resting state, rapid depolarization, plateau, rapid repolarization and hyperpolarization (a-e respectively). In addition, a sharp transient peak (blue star) and the notch (orange star) possibly associated with the inward sodium and outward potassium currents (Zipes et al. 2004) can be resolved. When the PDMS/cell substrate was mechanically retracted from the 3D kinked nanowire devices, the intracellular peaks disappeared, but they reappeared when the cell substrate was brought back into gentle contact with the device. These results confirm that electrical recording arises from the highly localized, point-like nanoFET near the probe tip, which (1) initially records only extracellular potential, (2) simultaneously records both extracellular and intracellular signals as the nanoFET spans the cell membrane, and (3) records only intracellular signals when fully inside the cell (Tian et al. 2010).

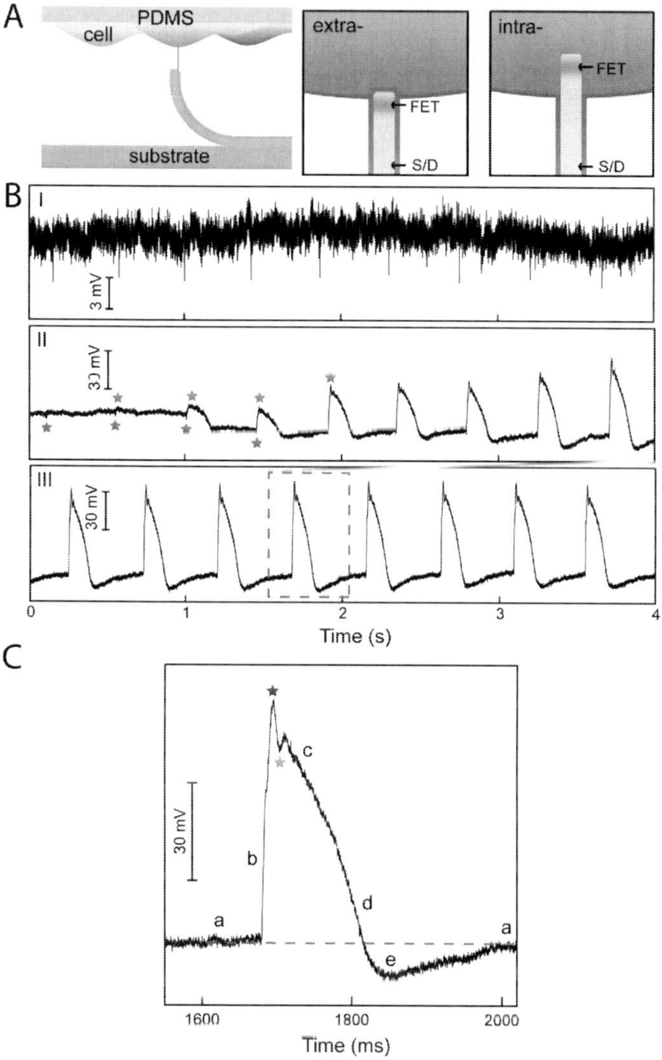

Fig. 10. Electrical recording from beating cardiomyocytes. (A) Schematics of cellular recording from cardiomyocyte monolayer (left panel), and extracellular and intracellular nanowire/cell interfaces. The cell membrane and nanowire lipids coatings are marked as purple lines. (B) Recording from beating cardiomyocytes: (I) extracellular recording; (II) transition from extracellular to intracellular recording; and (III) steady-state intracellular recording. Green and magenta stars mark peak positions of intracellular and extracellular signal components, respectively. (C) Zoom-in signal from the corresponding red dashed region in (B). Blue and orange stars designate features associated with inward sodium and outward potassium currents, respectively. a-e are defined in the text. Adapted from Tian et al. (2010) (American Association for the Advancement of Science, copyright 2010).

Color image of this figure appears in the color plate section at the end of the book.

CONCLUSIONS AND PROSPECTS

In recent years, NW devices have been implemented as a broad platform for electronic interfaces with cells and tissue, demonstrating several key points (Patolsky et al. 2006; Timko et al. 2009; Cohen-Karni et al. 2009; Qing et al. 2010; Tian et al. 2010). First, NWFETs having orders of magnitude smaller recording area than previous techniques were used to measure extracellular signals from individual cells and tissue. These signals, in the mV range, are considerably larger than those measured using planar, lithographically patterned FETs or MEAs (Halbach et al. 2003; Reppel et al. 2004; Ingebrandt et al. 2001; Yeung et al. 2001), possibly because of enhanced coupling between the nanoscale device and cell membrane. The signal magnitude is directly correlated to (1) device sensitivity (transconductance) (Timko et al. 2009; Cohen-Karni et al. 2009) and (2) interface quality between cells and substrate (Cohen-Karni et al. 2009). Second, NW devices were used to connect with individual neurons at the level of individual axons or dendrites (Patolsky et al. 2006), and with heart tissue (Timko et al. 2009) or cardiomyocytes at the subcellular level. Moreover, multiplexed recording from NWFETs has subsequently enabled signal mapping in the context of neurons, acute brain slices, heart tissue, and cardiomyocytes (Patolsky et al. 2006; Timko et al. 2009; Cohen-Karni et al. 2009; Qing et al. 2010). Third, using our well-controlled synthesis method and the advantages of the bottom-up approach we demonstrated intracellular recording with a nanoFET probe (Tian et al. 2010).

These methods represent a general and flexible approach for creating hybrid nanoelectronic-biological devices that serve as the foundation for new, fundamental studies as well as novel directions in biomedical research and applications (Fig. 11). A remarkable property of NWs is that their material composition and corresponding properties can be tuned at

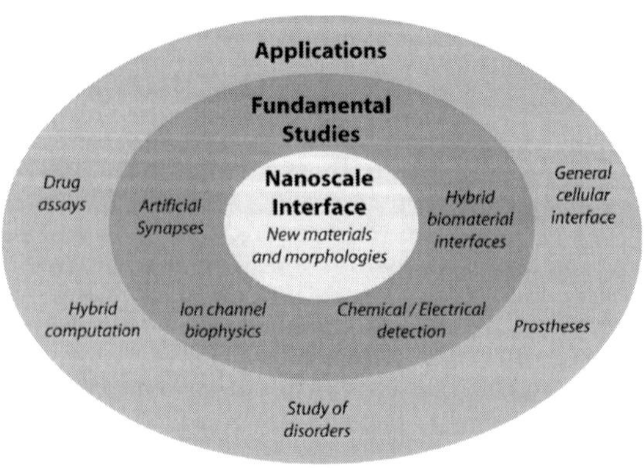

Fig. 11. Bottom-up paradigm for nanowire nanobioelectronics.

the time of synthesis. This opens up future opportunities in the design of ultra-high sensitivity transistors through band structure engineering (Lu and Lieber 2007), incorporation of photonic materials for light-addressable interfaces with photoactive cells (Pappas et al. 2007), or use of additional novel geometries to further improve device-membrane coupling.

Finally, in the future NW device arrays may be used to correlate electronic signaling with chemical release or to simultaneously detect a matrix of biologically relevant species. These techniques, pushing the boundaries of present technologies, could lead to new drug assays or breakthroughs in fundamental bioscience. NWs could also serve as a foundation for new and powerful prosthetic devices. Significantly, NW devices assembled on biocompatible substrates that form intimate extracellular or intracellular connections with excitable cells and tissue may be used as functional prosthetics, which could complement current biomedical technologies, by blurring the distinction of biological systems and electronic systems.

Key Facts

- Silicon nanowires are synthesized via vapour-liquid-solid method, which was first introduced in 1964 at Bell Labs by R.S. Wagner and W.C. Ellis.
- Nanostructures such as nanotubes, nanowires and nanocrystals are excellent bottom-up building blocks that show enhanced coupling to artificial membranes, cells, and tissue.
- These nano-bio interfaces exist at a length scale natural for biological systems and offer better sensitivity and spatial resolution than conventional planar structures.
- Nanowires can readily be assembled on nearly any type of surface, including those that are typically not compatible with standard semiconductor processing, such as flexible plastic substrates.
- Sensing of chemical or biological analytes with high sensitivity and specificity can be done by chemically modifying the native oxide surface of silicon nanowires by various functional groups.

Definitions

Action potential: The flow of information between, for example, nerves and muscles in the human body is mediated by electrical activity. A fundamental event that makes the electrical conduction of information is the action potential. The action potential is a process by which the membrane depolarizes from its resting potential, followed by a rapid potential swing ending with hyperpolarization and final return to the rest potential. These events are due to ionic currents flowing through the cell membrane via voltage gated ion channels.

Extracellular electrical recording: the recording of electrical activity of cells, e.g., the extracellular field potentials of excitable cells, outside of the cell membrane.

Field effect transistor (FET): In a standard FET, a semiconductor such as p-type silicon (p-Si) is connected to source and drain electrodes, through which a current is injected and controlled. The conductance of the semiconductor between source and drain is switched on and off by a third gate electrode capacitively coupled through thin dielectric layer. In the case of p-Si, applying negative (positive) gate leads to accumulation of negative (positive) charges at the interface between the dielectric and the semiconductor and leads to accumulation (depletion) of carriers and increase (decrease) in conductance.

Intracellular electrical recording: A technique used to measure the potential across, or electrical currents passing through, cellular membranes by, for example, inserting a sharp glass microelectrode electrode inside a cell.

Vapor-liquid-solid growth: Semiconductor NWs are synthesized by employing metal nanoclusters as catalysts via a vapor-liquid-solid process. In this process, the metal nanoclusters are heated above the eutectic temperature for the metal–semiconductor system of choice in the presence of a vapor-phase source of the semiconductor, resulting in a liquid droplet of the metal/semiconductor alloy. Supersaturation of the eutectic leads to nucleation of the solid semiconductor, and the solid-liquid interface acts as a sink causing the continued semiconductor incorporation into the lattice and, thereby, growth of the nanowire with the alloy droplet riding on the top. In the case of SiNW growth, silane (SiH_4) and Au nanoparticles are normally used as the precursor and catalysts, respectively.

Summary Points

- NWFETs with orders of magnitude smaller recording area than previous planar FETs and MEAs were used to measure robust electrical signals from individual cells and tissue.
- The measured extracellular signals, in the mV range, are considerably larger than those measured using planar, lithographically patterned FETs or MEAs, possibly because of enhanced coupling between the nanoscale device and cell membrane.
- The signal magnitude is directly correlated to (a) device sensitivity (transconductance) and (b) interface quality between cells and substrate.

- Multiplexed recording from NWFETs has subsequently enabled signal mapping in the context of heart tissue and cardiomyocytes.
- "Stereocontrolled" synthesis was used to create a new class of hairpin-like nanoscale 3D FET probes.
- Intracellular electrical signals were recorded non-invasively for the first time using 3D nanoFET hairpin probes.

Abbreviations

3D	:	three dimensional
CNT	:	carbon nanotubes
FET	:	field effect transistor
FWHM	:	full width at half maximum
MEA	:	multielectrode arrays
nanoFET	:	nanoscale field effect transistor
NW	:	nanowire
NWFET	:	nanowire field effect transistor
PDMS	:	polydimethylsiloxane
S/D	:	source/drain
SEM	:	scanning electron microscope
SiNW	:	silicon nanowire

References

Banks, D.J., W. Balachandran, P.R. Richards and D. Ewins. 2002. Instrumentation to evaluate neural signal recording properties of micromachined microelectrodes inserted in invertebrate nerve. Physiol. Meas. 23(2): 437–448.

Cohen, A., J. Shappir, S. Yitzchaik and M.E. Spira. 2008. Reversible transition of extracellular field potential recordings to intracellular recordings of action potentials generated by neurons grown on transistors. Biosens. Bioelectron. 23(6): 811–819.

Cohen, M.L., R.H. Hoyt, J.E. Saffitz and P.B. Corr. 1989. A high density *in vitro* extracellular electrode array: description and implementation. Am. J. Physiol. 257 (2 Pt 2): H681–9.

Cohen-Karni, T., B.P. Timko, L.E. Weiss and C.M. Lieber. 2009. Flexible electrical recording from cells using nanowire transistor arrays. Proc. Natl. Acad. Sci. USA 106(18): 7309–7313.

Cui, Y. and C.M. Lieber. 2001. Functional nanoscale electronic devices assembled using silicon nanowire building blocks. Science 291(5505): 851–853.

Cui, Y., Q. Wei, H. Park and C.M. Lieber. 2001. Nanowire nanosensors for highly sensitive and selective detection of biological and chemical species. Science 293(5533): 1289–92.

Efimov, I.R., V.P. Nikolski and G. Salama. 2004. Optical imaging of the heart. Circ. Res. 95(1): 21–33.

Halbach, M., U. Egert, J. Hescheler and K. Banach. 2003. Estimation of action potential changes from field potential recordings in multicellular mouse cardiac myocyte cultures. Cell. Physiol. Biochem. 13(5): 271–284.

Heller, I., W.T. Smaal, S.G. Lemay and C. Dekker. 2009. Probing macrophage activity with carbon-nanotube sensors. Small 5(22): 2528–2532.

Hu, J.T., T.W. Odom and C.M. Lieber. 1999. Chemistry and physics in one dimension: Synthesis and properties of nanowires and nanotubes. Acc. Chem. Res. 32(5): 435–445.

Ingebrandt, S., C.K. Yeung, M. Krause and A. Offenhausser. 2001. Cardiomyocyte-transistor-hybrids for sensor application. Biosens. Bioelectron. 16(7–8): 565–570.

Javey, A., S. Nam, R.S. Friedman, H. Yan and C.M. Lieber. 2007. Layer-by-layer assembly of nanowires for three-dimensional, multifunctional electronics. Nano Lett. 7(3): 773–777.

Kotov, N.A., J.O. Winter, I.P. Clements, E. Jan, B.P. Timko, S. Campidelli, S. Pathak, A. Mazzatenta, C.M. Lieber, M. Prato, R.V. Bellamkonda, G.A. Silva, N.W.S. Kam, F. Patolsky and L. Ballerini. 2009. Nanomaterials for neural interfaces. Adv. Mater. 21(40): 3970–4004.

Lu, W., and C.M. Lieber. 2007. Nanoelectronics from the bottom up. Nat. Mater. 6(11): 841–850.

McAlpine, M.C., R.S. Friedman, S. Jin, K.H. Lin, W.U. Wang, and C.M. Lieber. 2003. High-performance nanowire electronics and photonics on glass and plastic substrates. Nano Lett. 3(11): 1531–1535.

Misra, N., J.A. Martinez, S.C.J. Huang, Y.M. Wang, P. Stroeve, C.P. Grigoropoulos and A. Noy. 2009. Bioelectronic silicon nanowire devices using functional membrane proteins. Proc. Natl. Acad. Sci. USA 106(33): 13780–13784.

Morales, A.M. and C.M. Lieber. 1998. A laser ablation method for the synthesis of crystalline semiconductor nanowires. Science 279 (5348): 208–211.

Pappas, T.C., W.M. Wickramanyake, E. Jan, M. Motamedi, M. Brodwick and N.A. Kotov. 2007. Nanoscale engineering of a cellular interface with semiconductor nanoparticle films for photoelectric stimulation of neurons. Nano Lett. 7(2): 513–519.

Patolsky, F., B.P. Timko, G. Yu, Y. Fang, A.B. Greytak, G. Zheng and C.M. Lieber. 2006. Detection, stimulation, and inhibition of neuronal signals with high-density nanowire transistor arrays. Science 313(5790): 1100–1104.

Patolsky, F., G. Zheng, O. Hayden, M. Lakadamyali, X. Zhuang and C.M. Lieber. 2004. Electrical detection of single viruses. Proc. Natl. Acad. Sci. USA 101(39): 14017–14022.

Pui, T.S., A. Agarwal, F. Ye, N. Balasubramanian and P. Chen. 2009. CMOS-Compatible nanowire sensor arrays for detection of cellular bioelectricity. Small 5(2): 208–212.

Qing, Q., S.K. Pal, B. Tian, X. Duan, B.P. Timko, T. Cohen-Karni, V.N. Murthy and C.M. Lieber. 2010. Nanowire transistor arrays for mapping neural circuits in acute brain slices. Proc. Natl. Acad. Sci. USA 107(5): 1882–1887.

Ramgir, N.S., Y. Yang and M. Zacharias. 2010. Nanowire-based sensors. Small 6(16): 1705–1722.

Reppel, M., F. Pillekamp, Z.J. Lu, M. Halbach, K. Brockmeier, B.K. Fleischmann and J. Hescheler. 2004. Microelectrode arrays: a new tool to measure embryonic heart activity. J. Electrocardiol. 37 Suppl: 104–109.

Sakmann, B. and E. Neher. 1984. Patch clamp techniques for studying ionic channels in excitable membranes. Annu. Rev. Physiol. 46: 455–472.

Scanziani, M. and M. Hausser. 2009. Electrophysiology in the age of light. Nature 461(7266): 930–939.

Schoning, M.J., and A. Poghossian. 2002. Recent advances in biologically sensitive field-effect transistors (BioFETs). Analyst 127(9): 1137–1151.

Sniadecki, N.J., R.A. Desai, S.A. Ruiz and C.S. Chen. 2006. Nanotechnology for cell-substrate interactions. Ann. Biomed. Eng. 34(1): 59–74.

Stern, E., J.F. Klemic, D.A. Routenberg, P.N. Wyrembak, D.B. Turner-Evans, A.D. Hamilton, D.A. LaVan, T.M. Fahmy and M.A. Reed. 2007. Label-free immunodetection with CMOS-compatible semiconducting nanowires. Nature 445(7127): 519–522.

Sze, S.M. and K.K. Ng. 2007. Physics of Semiconductor Devices. 3rd ed. Wiley-Interscience, Hoboken, N.J.

Taccardi, B., B.B. Punske, E. Macchi, R.S. Macleod and P.R. Ershler. 2008. Epicardial and intramural excitation during ventricular pacing: effect of myocardial structure. Am. J. Physiol. Heart. Circ. Physiol. 294 (4): H1753–66.

Tian, B., T. Cohen-Karni, Q. Qing, X. Duan, P. Xie and C.M. Lieber. 2010. Three-dimensional, flexible nanoscale field-effect transistors as localized bioprobes. Science 329(5993): 830–834.

Tian, B., P. Xie, T.J. Kempa, D.C. Bell and C.M. Lieber. 2009. Single-crystalline kinked semiconductor nanowire superstructures. Nat. Nanotechnol. 4(12): 824–829.

Timko, B.P., T. Cohen-Karni, G. Yu, Q. Qing, B. Tian and C.M. Lieber. 2009. Electrical recording from hearts with flexible nanowire device arrays. Nano Lett. 9(2): 914–918.

Yeung, C.K., S. Ingebrandt, M. Krause, A. Offenhausser and W. Knoll. 2001. Validation of the use of field effect transistors for extracellular signal recording in pharmacological bioassays. J. Pharmacol. Toxicol. Methods 45(3): 207–214.

Zheng, G.F., F. Patolsky, Y. Cui, W.U. Wang and C.M. Lieber. 2005. Multiplexed electrical detection of cancer markers with nanowire sensor arrays. Nat. Biotech. 23(10): 1294–1301.

Zipes, D.P. and J. Jalife. 2004. Cardiac Electrophysiology: From Cell to Bedside. 4th ed. Saunders, Philadelphia.

9

Nanospheres and Applications to Cardiology. Multifunctionality: The Key to Future Success

Andrea Masotti

ABSTRACT

Nanobiotechnology, the joining of nanotechnology with molecular biology and medicine, recently emerged as a new scientific research area. One of the most interesting aspects of this new field concerns the preparation and use of multifunctional nanospheres. Multifunctionality results from the contemporaneous exploitation of single functionalities (i.e., fluorescence, magnetism, targeting moieties) of the nanostructured system. Nanostructured systems have their own advantages and provide many exciting opportunities in biomedicine. It is now well acknowledged that nanoparticles and nanospheres have controllable sizes ranging from a few up to tens of nanometers matching the common needs of biomedical applications. An important area of application is the diagnosis and treatment of cardiovascular diseases, recognized among the leading causes of death in industrialized nations. According to the World Health Organization, cardiovascular diseases are the world's largest killers, claiming 17.1 million

Gene Expression—Microarrays Laboratory, IRCCS-Children's Hospital Bambino Gesù, P.za S.Onofrio 4, 00165 Rome, ITALY; E-mail: andrea.masotti@opbg.net; andrea_maso@yahoo.it

List of abbreviations after the text.

lives a year. Tobacco use, an unhealthy diet, physical inactivity and harmful use of alcohol increase the risk of heart attacks and strokes. Therefore, techniques that can suggest novel diagnostic and therapeutic approaches in the field of cardiology are highly desirable. In this chapter, some major applications linking multifunctional nanotechnology to the field of cardiology are reviewed and discussed.

INTRODUCTION

Nanobiotechnology, the joining of nanotechnology with molecular biology and medicine, recently emerged as a new scientific research area (Whitesides 2003). One of the most interesting aspects of this new field concerns the preparation and use of multifunctional nanospheres. Multifunctionality results from the contemporaneous exploitation of single functionalities (i.e., fluorescence, magnetism, targeting moieties) of the nanostructured system. Nanostructured systems have their own advantages and provide many exciting opportunities in biomedicine. In general, nanoparticles and nanospheres have controllable sizes ranging from a few up to tens of nanometers matching the common needs of biomedical applications. Recently, we reported the use of a novel patented osmosis- based method to obtain well-defined chitosan and chitosan/DNA nanospheres without the use of surfactants (Masotti et al. 2008b). The novelty of the reported method resides in the general applicability to various synthetic or natural biopolymers. To prepare nanospheres, we employed one of the most well-known polysaccharides: chitosan (Fig. 1).

Chitosan is a biodegradable polysaccharide obtained from deacetylated chitin (66% to 95% deacetylation) and the commercial product has an average molecular weight ranging between 4 and 20 kDa. It contains several amino groups (Figs. 1 and 2) that in acidic pH may undergo protonation leading to its solubilization in water.

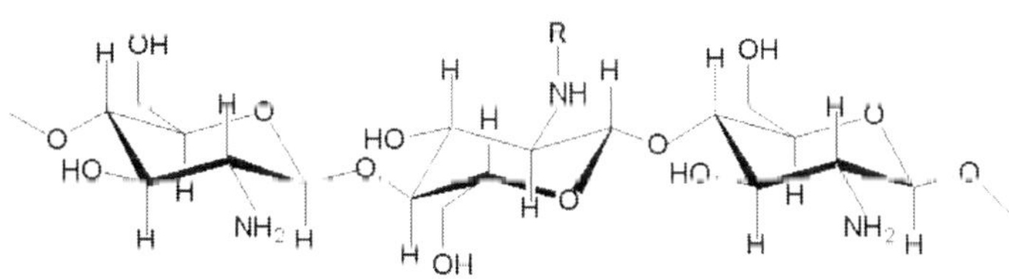

R = H or $COCH_3$

Fig. 1. Chemical structure of chitosan. Chitosan is a polysaccharide that we employed to prepare nanospheres. (Unpublished material of the author.)

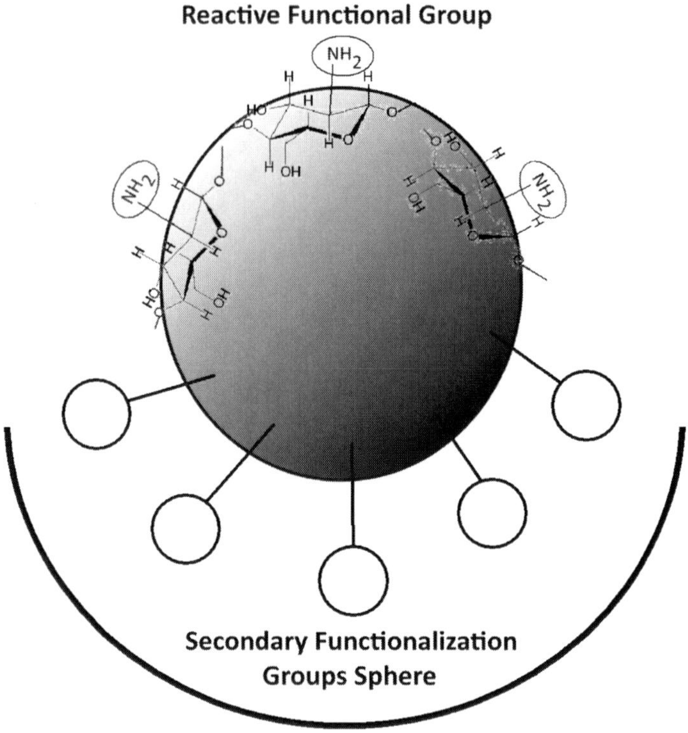

Fig. 2. Schematic representation of a nanosphere employed for secondary functionalization. Chitosan nanospheres may be employed for secondary (multifunctional) functionalization thanks to the numerous amino groups present on the surface. (Unpublished material of the author.)

Chitosan may also establish electrostatic interactions with the negatively charged DNA to form complexes (polyplexes). Being non-toxic, chitosan is also widely used in pharmaceutical research and in industry for the controlled release of antibiotics, DNA, proteins, peptide drugs or vaccines. The size and shape of these systems are important factors for several medical applications: to improve bioavailability (i.e., overcoming enzymatic or adsorption barriers and in the case of nasal administration the mucociliary clearance) and to prolong the residence time of drug delivery systems at the site of drug absorption. Solvent, temperature and membrane cut-off are the physicochemical parameters that can control the overall osmotic process, resulting in several nanostructured systems with different size and shape that may be used in several biotechnological applications. We prepared chitosan nanospheres with an average diameter of 45 ± 9 nm (measured by dynamic light scattering) (Fig. 3) that might be employed for secondary multifunctional functionalization (Fig. 2).

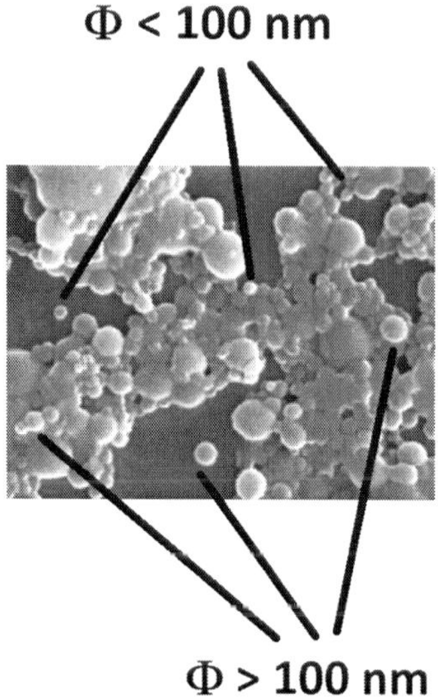

Fig. 3. Micrograph of a chitosan nanosphere. Nanospheres of different dimensions can be obtained with the method we recently reported. (Unpublished material of the author.)

The patented method that we employed to produce chitosan nanospheres (Masotti et al. 2008b) can be eventually employed also for other kind of polymers to produce similar nanostructures. One of the most intensively studied polyamine polymers in biomedicine is the high molecular weight branched polyethylenimine (Fig. 4).

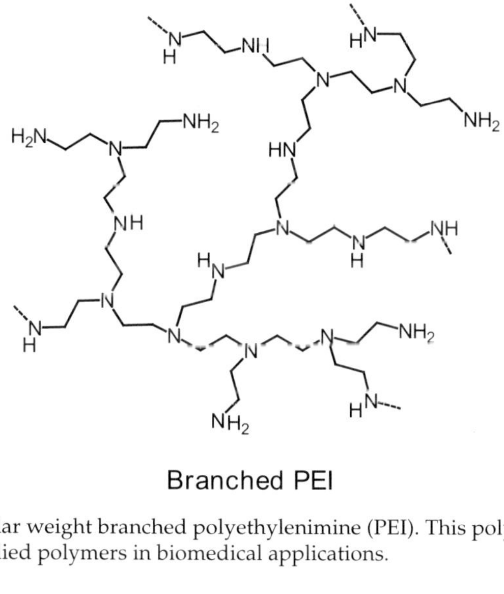

Branched PEI

Fig. 4. High molecular weight branched polyethylenimine (PEI). This polyamine polymer is one of the most studied polymers in biomedical applications.

We reported in the last few years that polyethylenimine is a versatile polymer because of its numerous amino groups on the surface that can be easily functionalized. We also reported that this polymer can be functionalized with near-infrared probes (i.e., IR-820 dye) (Masotti et al. 2008a), can incorporate iron oxide nanoparticles (Masotti et al. 2009) and can bind to DNA and other nucleic acids at the same time. Therefore, this functionalized polymer, once "moulded" into nanospheres, can be potentially employed as a powerful multifunctional vector.

The introduction of magnetic ability (i.e., superparamagnetism) within nanostructured systems leads to a further advantage represented by the ability to manipulate these systems by an external magnetic force (Lauterbur 1989). Moreover, the superparamagnetism is also responsible for their responsiveness to magnetic resonance imaging (MRI), leading to the preparation of innovative contrast agents (Pankhurst 2003). Techniques and procedures for producing monodispersed and size-controllable magnetic nanoparticles (e.g., FePt, Fe_3O_4, and γ-Fe_2O_3) with (Masotti et al. 2009) or without additional coatings (Park et al. 2007) have been published. Additionally, the conjugation with other "biofunctional" molecules resulted in "multifunctional systems" (Masotti and Ortaggi 2008). In fact, when magnetic nanoparticles are coated with biofunctional molecules (e.g., antibodies, ligands, or receptors), they are able to interact with biological entities with high affinity, providing a controlled "tagging" (Fig. 5). The further combination of magnetic nanomaterials with other functional nanostructures (by sequential growth or coating) can produce a single complex entity with multiple functions in nanoscale dimension.

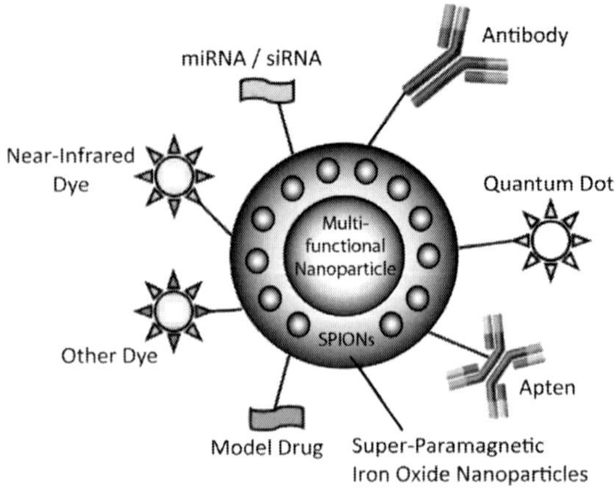

Fig. 5. Schematic representation of a multifunctional nanoparticle having different properties due to its numerous derivatization with antibodies, fluorophores, contrast agents and other small biologically relevant molecules. (Unpublished material of the author.)

Color image of this figure appears in the color plate section at the end of the book.

An important area of application is the diagnosis and treatment of cardiovascular diseases (CVDs), which are recognized among the leading causes of death in industrialized nations. The role of nanotechnology is steadily increasing in this area. For example, nanoparticles are being widely accepted as biosensors and imaging tools for detection and monitoring of the progression of the disease, as well as drug delivery systems for therapeutic purposes (Iverson et al. 2008). Some major applications linking multifunctional nanotechnology to the field of cardiology are reviewed and discussed in this chapter.

MOLECULAR IMAGING IN ATHEROSCLEROSIS

The first reported approach aiming at identifying suspect lesions (McCarthy and Weissleder 2008) consisted in the use of magnetic nanoparticles for targeted imaging macrophages and therapy. Macrophages, key components of atherosclerotic vessels, can constitute 10–20% of the cells present within a lesion. In particular, macrophages contribute crucially to all stages of atherogenesis and are involved in plaque rupture, ultimately leading to lethal thrombosis in advanced lesions (Hansson 2005; Hansson and Libby 2006). In light of the morbidity and mortality associated with atherosclerotic vascular disease, it is not surprising that novel methods to identify and treat inflamed atheromas are urgently needed.

Jaffer and co-workers used fluorophore-derived cross-linked iron oxide nanoparticles (CLIO) to investigate the localization within atherosclerotic lesions, and were able to visualize inflamed plaques both by MRI and optical imaging (Jaffer et al. 2006; Pande et al. 2006). The cellular localization of CLIO labeled with both a fluorophore and radionuclide (^{64}Cu) within these lesions was also investigated (Nahrendorf et al. 2008). Examination of excised aortas obtained from animal models injected with the nanoparticle preparation revealed that about 75% of the nanoparticles within the atherosclerotic lesions were associated with macrophages, 20% with neutrophils, and the remainder with endothelial cells, lymphocytes and smooth muscle cells.

Vascular Cell Adhesion Molecule 1

A potentially useful imaging target is represented by vascular cell adhesion molecule 1 (VCAM-1). VCAM-1 expressed on activated endothelial cells, macrophages, and smooth muscle cells participates in the onset of the inflammatory process, and in the progression of atherosclerotic lesions. Magneto-fluorescent nanoparticles bioconjugated with antibodies (Tsourkas et al. 2005) have been investigated to test the preferential targeting ability *in vitro* against primary cultures of murine heart endothelial cells (MHEC) expressing high levels of VCAM-1, and murine dermal endothelial cells (MDMEC) with low VCAM-1 expression. Flow cytometry demonstrated a preferential binding of nanoparticles to VCAM-1 in MHEC, and no binding to the MDMEC. Therefore, the authors tested the *in vivo* targeting ability of the conjugate in a murine model of inflammation. In this model, a severe local inflammatory response was induced by injection of tumor necrosis factor α (TNF-α). After injection, the VCAM-1–targeted particles demonstrated a specific binding on the surface of the vasculature in the TNF-α–treated model, with peak labeling 6 h post-injection.

However, one of the major drawbacks of the use of antibody-modified nanoparticles is the lack of cellular internalization and limited loading of the surface of the nanoparticles. For this reason, peptide-based targeting nanoparticles were investigated by Kelly and co-workers that identified two potentially useful peptide sequences, CVHSPNKKC (VHS) (Kelly et al. 2005) and VHPKQHR (termed VCAM-1 internalizing peptide) targeting VCAM-1 (Kelly et al. 2006; Nahrendorf et al. 2006). The first peptide, when fluorescently labeled, exhibits rapid internalization by MHEC expressing high levels of VCAM-1 with respect to a scrambled control peptide. To demonstrate the efficacy of this peptide, when the cells were pre-incubated with an anti-VCAM-1 antibody, actively blocking the VCAM-1 binding site, neither the VHS nor control peptides were internalized. When conjugated to magnetofluorescent nanoparticles modified with Cy5.5, the peptide-targeted particles exhibited a 10-fold increase in uptake versus the control particle. The targeting ability of this peptide was also tested *in vivo* in a murine TNF-α inflammation model. Four hours after injection of the VHS peptide–targeted nanoparticle the targeted agent remained bound to the inflamed vasculature, as observed by intravital fluorescence microscopy. Twenty-four hours post-injection, the nanoagent was still observed with both non-invasive techniques and histology. The VCAM-1–targeted nanoparticles are also able to localize within atherosclerotic lesions and visualized by MRI in apolipoprotein E knockout mice. This mouse model spontaneously develops atheromas, especially when fed a high-fat diet. Following intravenous injection of the VHS nanoparticle, significant signal decrease was observed in the aortic wall. This result was not observed in mice injected with an untargeted control nanoparticle, or in C57BL/6 wild-type mice bearing no atherosclerotic lesions.

The second linear peptide reported by Kelly demonstrated a 20-fold increase in binding to MHECs, as compared to the VHS peptide and an excellent uptake both *in vitro* and *in vivo* (animal models and human endarterectomy specimens) (Kelly et al. 2006; Nahrendorf et al. 2006).

Detection of Intravascular Thrombi

The synthesis and characterization of two multimodal thrombus-targeted multifunctional nanoparticles exhibiting either covalent or noncovalent binding to thrombi have been recently reported by McCarthy et al. (2009). Thrombosis is characterized by the accumulation of platelets and fibrin induced by the activation of coagulation factors. Thrombosis underlies several life-threatening cardiovascular syndromes including myocardial infarction, stroke, and pulmonary embolism, and is one of the leading causes of death in the developed world (Murray and Lopez 1997). Therefore, the localization and characterization of thrombi *in vivo* is highly desirable in order to determine treatment options and their consequent efficacy. One of the strategies adopted by McCarthy et al. was to focus the study on activated factor XIII (FXIIIa). FXIIIa is responsible for the cross-linking of fibrin α- and γ-chains, increases fibrinolytic resistance, and is a hallmark of biologically acute thrombi, since its activity diminishes over time. The peptide-based fluorescent imaging agent targeted to FXIIIa (Jaffer et al. 2004) is tightly cross-linked into clots undergoing thrombogenesis, exhibiting a time-dependent increase in the fluorescence signal. This characteristic renders this system useful in distinguishing acute from sub-acute thrombi *in vivo*. The functionalization of fluorescently labeled CLIO nanoparticles with targeting peptides for fibrin or FXIIIa yielded thrombus-targeted nanoparticles detectable by both MRI and optical imaging. These systems are, in a broader sense, "multifunctional" since they can allow for the simultaneous monitoring of thrombus formation, the determination of biological thrombus age, and also the efficacy of thrombolytic therapies.

Multimodal Theranostic Probes: Targeted Imaging and Therapy of Atherosclerosis

Integrated diagnostic imaging and therapeutic systems, or "theranostic" systems, are increasingly appreciated as a valuable advance for targeted therapy. Addition of a diagnostic imaging moiety to a targeted therapeutic enables temporal and spatial monitoring of the therapeutic agent. Imaging information in theranostics can be used to confirm delivery of therapy at the desired target, identify the correct clinical dose/response strategy, quantify and track the "molecular efficacy" of the therapeutic *in vivo*, and identify likely responders and non-responders at the onset of therapy. Targeting atherosclerosis with theranostic probes is still in the early stages and only a few examples have been reported: photosensitizers with intrinsic NIR fluorescence capabilities (Tawakol et al. 2008) that could offer the ability to characterize inflamed atheromata before photodynamic therapy or the NIR fluorescent magnetic nanoparticles carrying therapeutic siRNA molecules (Medarova et al. 2007).

Small Multifunctional Nanoclusters (Nanoroses) for Targeted Cellular Imaging and Therapy

Small multifunctional nanoclusters, called "nanoroses" (Ma et al. 2009), are composed of iron oxide nanoparticles with thin gold coatings assembled into very stable ~30 nm nanoclusters. These nanoparticles incorporating two inorganic materials, iron oxide and gold, display intense NIR (700–850 nm) absorbance and *r*2 magnetic relaxivity (>200 $mM^{-1}s^{-1}$) in solution and strong optical contrast in targeted macrophages. The small average size and relatively low polydispersity of the nanoclusters are achieved by kinetic control of self-assembly, as a function of the iron oxide nanoparticle concentration and gold/iron oxide ratio, in the presence of dextran stabilizer. Nanoclusters, or nanoroses, were formed by the reduction of $HAuCl_4$ onto the surfaces of 5 nm iron oxide nanoparticles (Ofir et al. 2008) with hydroxylamine as a seeding agent that adsorbed on the iron oxide particle surface and favored selective formation of gold on the iron oxide surface rather than in bulk solution (Lyon et al. 2004). The reaction resulted in ~30 nm (from 23 ± 3 and 34 ± 2 nm) relatively open clusters with a low polydispersity shown by dynamic light scattering, composed of much smaller primary particle domains. The small size of the nanoclusters and the dextran surface coating are shown to facilitate high uptake by macrophages, which contain dextran receptors (Jaffer et al. 2006). The uptake of nanoroses by macrophages, endothelial cells, and smooth muscle cells in the aorta was evaluated to quantify the specificity of targeting of macrophage cells associated with atherosclerosis. Macrophages exhibited the highest uptake with respect to other cell lines and this result was exploited to enhance diagnosis and therapy of atherosclerosis *in vivo* on an animal model. The plaque-based macrophages engulf nanoroses following intravenous injection, as demonstrated via co-localization of macrophage immunohistochemistry and nanorose identified with both dark field microscopy and hyperspectral imaging. The high nanorose uptake and absorbance provided high contrast between macrophages and surrounding tissues. The properties of these systems have been also exploited for photothermal destruction of nanocluster-laden macrophages, rendering these systems potentially useful in multifunctional theranostics.

Iron Oxide Core High-Density Lipoprotein Nanoparticles

Another multifunctional nanostructured system is represented by iron oxide core high-density lipoprotein nanoparticles employed in an experimental model of atherosclerosis (Skajaa et al. 2011). Lipoproteins are a family of plasma nanoparticles responsible for the transportation of lipids throughout the body. High-density lipoprotein (HDL), the smallest of the lipoprotein family, measures 7–13 nm in diameter is initially synthesized in the liver and the intestine and consists of a cholesteryl ester and triglyceride core that is covered with a monolayer of phospholipids and apolipoproteins. The most abundant protein components are apolipoprotein A-I (apoA-I) and apolipoprotein A-II (apoA-II). Most important, HDL is the main actor in the process known as reverse cholesterol transportation, in which HDL promotes the transportation of excess cholesterol from extra-hepatic or peripheral tissue to the liver for elimination through the biliary system. Reverse efflux of cholesterol from plaque macrophages has an important protective effect in atherosclerosis.

Skajaa et al. recently published a study on a novel nanocrystal HDL-based nanoparticle platform (Cormode et al. 2008), which allowed multimodal imaging of macrophages in atherosclerotic plaques of the apoE knockout (apoE KO) mouse model of atherosclerosis (Fig. 6).

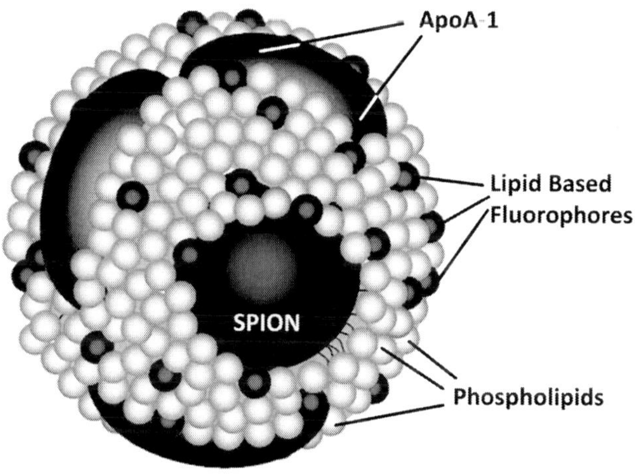

Fig. 6. Schematic representation of FeO-HDL multifunctional nanoparticles based on high-density lipoprotein (HDL), fluorophores, superparamagnetic iron oxide nanoparticles (SPION) and apolipoproteinA-1 (ApoA-1). (Unpublished material of the author.)

Excised aortas of apoE KO mice injected with FeO-HDL nanoparticles were examined on sub-cellular level with transmission electron microscopy and iron oxide particles were found in large quantities within atherosclerotic plaques.

The iron oxide core HDL nanoparticle (FeO-HDL), which has a lipid-based fluorophore incorporated in the phospholipid layer, provides contrast for optical imaging, MRI and transmission electron microscopy and can be visualized on the anatomical, cellular and sub-cellular level. The authors suggest that FeO-HDL can be used as a probe to investigate HDL biology and as an imaging agent or vector to deliver drugs or proteins in disease settings where HDL is key.

APPLICATIONS TO OTHER AREAS OF HEALTH AND DISEASE

Multifunctional nanoparticles have several potential applications in many clinical fields and in biomedicine (Masotti 2010). The great number of papers and patents demonstrate that nanotechnology coupled with biomedicine, namely nanobiotechnology, is a field that holds great promise in the near future. The principle of coupling different functionalities into the same nanostructured moiety will result in modular systems with targeting, delivery, imaging and other intrinsic abilities that might be exploited in all biomedical fields in which these properties are of paramount importance. It is impossible to delineate other fields of applications since advances in medicine, biology, chemistry and physics will open new horizons and allow for applications of these systems in fields that are now unimaginable. At present, there are almost a hundred records in Pubmed (http://www.ncbi.nlm.nih.gov/pubmed) for "multifunctional nanoparticles" and "medicine", but only a few are about cardiology and cardiovascular diseases. This is a clear indication of how innovative these systems are and how far we are from current clinical applications. The future of the exciting field of nanobiotechnology depends on the creativity and imagination of researchers to find ever novel systems and the ability of clinicians to translate these concepts from "bench to bedside".

Key Facts

- The word "nanotechnology" was first used by the physicist Richard Feynman in a talk entitled "There's Plenty of Room at the Bottom" given at an American Physical Society meeting at Caltech on December 29, 1959.
- Multifunctional nanoparticles applied to cardiology and related fields have received, so far, little attention in the scientific literature. In a search of Pubmed for "multifunctional nanoparticles" and "medicine", 82 papers can be retrieved, and only 25 for "nanotechnology" and "cardiology".
- Of the 24,540 papers on nanotechnology, 4695 were published in the last three years and 1369 in 2010 (at the time of the submission of this chapter).
- According to the World Health Organization, cardiovascular diseases (CVDs) are the world's largest killers, claiming 17.1 million lives a year.
- Tobacco use, an unhealthy diet, physical inactivity and harmful use of alcohol increase the risk of heart attacks and strokes. Low- and middle-income countries are disproportionally affected: 82% of CVD

deaths take place in low- and middle-income countries and occur almost equally in men and women.

- By 2030, almost 23.6 million people will die from CVDs. The largest percentage increase will occur in the Eastern Mediterranean Region and the largest increase of deaths will occur in South-East Asia.

Definitions

Nanobiotechnology: The joining of two disciplines: nanotechnology and biotechnology.

Nanoclusters: Aggregates (or clusters) of nanoparticles of different forms and dimensions.

Photodynamic therapy: Therapeutic procedure involving the use of a photosensitizer, a delivery vector and a laser light. The photosensitizer is excited by light of a specific wavelength and produces singlet oxygen that is very reactive toward nearby biomolecules inducing cell death through apoptosis or necrosis.

Superparamagnetism: A form of magnetism appearing in ferromagnetic or ferrimagnetic small particles having nanometer dimensions.

Theranostic: A system or a nanoparticle having both therapeutic and diagnostic properties.

Summary Points

- Nanobiotechnology representing the joining of nanotechnology with molecular biology and medicine, has recently emerged as a new scientific research area.
- Multifunctional nanoparticles are nanostructured systems. Multifunctionality is a result of the contemporaneous exploitation of single functionalities (i.e., fluorescence, magnetism, targeting moieties) of the nanostructured system.
- Magnetic nanoparticles have been used for targeted imaging of macrophages and atherosclerosis therapy.
- Vascular cell adhesion molecule 1 is expressed on activated endothelial cells, macrophages, and smooth muscle cells, and participates in the onset of the inflammatory process, and in the progression of atherosclerotic lesions. It can be efficiently monitored with magnetofluorescent nanoparticles bioconjugated with antibodies.
- Multifunctional nanoparticles exhibiting either covalent or noncovalent binding to thrombi have been recently reported: the study focused on activated factor XIII, which is responsible for the cross-linking of

fibrin α- and γ-chains and for the increase of fibrinolytic resistance, and is a hallmark of biologically acute thrombi.

- Dextran-coated crystalline iron oxide magnetic nanoparticles have been used for multimodality molecular imaging of lesional macrophages.
- Theranostics are innovative systems that can be used to confirm delivery of therapy at the desired target, identify the correct clinical dose/response strategy, quantify and track the "molecular efficacy" of the therapeutic *in vivo*, and identify likely responders and non-responders at the onset of therapy.
- Iron oxide core high-density lipoprotein nanoparticles have been employed in an experimental model of atherosclerosis. High-density lipoprotein, the smallest of the lipoprotein family, measures 7–13 nm.

Abbreviations

(apoA-I)	:	apolipoprotein A-I
(apoA-II)	:	apolipoprotein A-II
CLIO	:	cross-linked iron oxide nanoparticles
CVD	:	cardiovascular diseases
FeO-HDL	:	nanoparticle
FXIIIa	:	activated factor XIII
HDL	:	high-density lipoprotein
HDL	:	iron oxide core
MDMEC	:	murine dermal endothelial cells
MHEC	:	murine heart endothelial cells
MRI	:	magnetic resonance imaging
NIR	:	near-infrared
TNF-α	:	tumor necrosis factor-α:
VCAM-1	:	vascular cell adhesion molecule 1
VHS	:	CVHSPNKKC peptide

References

Cormode, D.P., T. Skajaa, M.M. van Schooneveld, R. Koole, P. Jarzyna, M.E. Lobatto, C. Calcagno, A. Barazza, R.E. Gordon, P. Zanzonico, E.A. Fisher, Z.A. Fayad and W.J. Mulder. 2008. Nanocrystal core high-density lipoproteins: a multimodality contrast agent platform. Nano Lett. 8: 3715–3723.

Hansson, G.K. 2005. Inflammation, atherosclerosis, and coronary artery disease. N. Engl. J. Med. 352: 1685–1695.

Hansson, G.K. and P. Libby. 2006. The immune response in atherosclerosis: a double-edged sword. Nat. Rev. Immunol. 6: 508–519.

Iverson, N., N. Plourde, E. Chnari, G.B. Nackman and P.V. Moghe. 2008. Convergence of nanotechnology and cardiovascular medicine: progress and emerging prospects. BioDrugs 22: 1–10.

Jaffer, F.A., P. Libby and R. Weissleder. 2006. Molecular and cellular imaging of atherosclerosis: emerging applications. J. Am. Coll. Cardiol. 47: 1328–1338.

Jaffer, F.A., C.H. Tung, J.J. Wykrzykowska, N.H. Ho, A.K. Houng, G.L. Reed and R. Weissleder. 2004. Molecular imaging of factor XIIIa activity in thrombosis using a novel, near-infrared fluorescent contrast agent that covalently links to thrombi. Circulation 110: 170–176.

Kelly, K.A., J.R. Allport, A. Tsourkas, V.R. Shinde-Patil, L. Josephson and R. Weissleder. 2005. Detection of vascular adhesion molecule-1 expression using a novel multimodal nanoparticle. Circ. Res. 96: 327–336.

Kelly, K.A., M. Nahrendorf, A.M. Yu, F. Reynolds and R. Weissleder. 2006. *In vivo* phage display selection yields atherosclerotic plaque targeted peptides for imaging. Mol. Imaging Biol. 8: 201–207.

Lauterbur, P.C. 1989. Image formation by induced local interactions. Examples employing nuclear magnetic resonance. 1973. Clin. Orthop. Relat. Res. (244): 3–6.

Lyon, J.L., D.A. Fleming, M.B. Stone, P. Schiffer and M.E. Williams. 2004. Synthesis of Fe oxide core/Au shell nanoparticles by iterative hydroxylamine seeding. 4: 719–723.

Ma, L.L., M.D. Feldman, J.M. Tam, A.S. Paranjape, K.K. Cheruku, T.A. Larson, J.O. Tam, D.R. Ingram, V. Paramita, J.W. Villard, J.T. Jenkins, T. Wang, G.D. Clarke, R. Asmis, K. Sokolov, B. Chandrasekar, T.E. Milner and K.P. Johnston. 2009. Small multifunctional nanoclusters (nanoroses) for targeted cellular imaging and therapy. ACS Nano 3: 2686–2696.

Masotti, A. 2010. Multifuntional nanoparticles: preparation and applications in biomedicine and in non-invasive bioimaging. Recent. Pat. Nanotechnol. 4: 53–62.

Masotti, A. and G. Ortaggi. 2008. Peptide nucleic acid-polyethylenimine conjugates promising multifunctional therapeutic tools for the future. Oligonucleotides 18: 301–303.

Masotti, A., A. Pitta, G. Ortaggi, M. Corti, C. Innocenti, A. Lascialfari, M. Marinone, P. Marzola, A. Daducci, A. Sbarbati, E. Micotti, F. Orsini, G. Poletti and C. Sangregorio. 2009. Synthesis and characterization of polyethylenimine-based iron oxide composites as novel contrast agents for MRI. MAGMA 22: 77–87.

Masotti, A., P. Vicennati, F. Boschi, L. Calderan, A. Sbarbati and G. Ortaggi. 2008a. A novel near-infrared indocyanine dye-polyethylenimine conjugate allows DNA delivery imaging *in vivo*. Bioconjug. Chem. 19: 983–987.

Masotti, A., F. Bordi, G. Ortaggi, F. Marino and C. Palocci. 2008b. A novel method to obtain chitosan/DNA nanospheres and a study of their release properties. Nanotechnology 19: 055302 (6 pp.).

McCarthy, J.R., P. Patel, I. Botnaru, P. Haghayeghi, R. Weissleder and F.A. Jaffer. 2009. Multimodal nanoagents for the detection of intravascular thrombi. Bioconjug. Chem. 20: 1251–1255.

McCarthy, J.R. and R. Weissleder. 2008. Multifunctional magnetic nanoparticles for targeted imaging and therapy. Adv. Drug Deliv. Rev. 60: 1241–1251.

Medarova, Z., W. Pham, C. Farrar, V. Petkova and A. Moore. 2007. *In vivo* imaging of siRNA delivery and silencing in tumors. Nat. Med. 13: 372–377.

Murray, C.J. and A.D. Lopez. 1997. Alternative projections of mortality and disability by cause 1990–2020: Global Burden of Disease Study. Lancet 349: 1498–1504.

Nahrendorf, M., F.A. Jaffer, K.A. Kelly, D.E. Sosnovik, E. Aikawa, P. Libby and R. Weissleder. 2006 Noninvasive vascular cell adhesion molecule-1 imaging identifies inflammatory activation of cells in atherosclerosis. Circulation 114: 1504–1511.

Nahrendorf, M., H. Zhang, S. Hembrador, P. Panizzi, D.E. Sosnovik, E. Aikawa, P. Libby, F.K. Swirski and R. Weissleder. 2008 Nanoparticle PET-CT imaging of macrophages in inflammatory atherosclerosis. Circulation 117: 379–387.

Ofir, Y., B. Samanta and V.M. Rotello. 2008. Polymer and biopolymer mediated self-assembly of gold nanoparticles. Chem. Soc. Rev. 37: 1814–1825.

Pande, A.N., R.H. Kohler, E. Aikawa, R. Weissleder and F.A. Jaffer. 2006. Detection of macrophage activity in atherosclerosis *in vivo* using multichannel, high-resolution laser scanning fluorescence microscopy. J. Biomed. Opt. 11: 021009.

Pankhurst, Q.A., et al. 2003. Applications of magnetic nanoparticles in biomedicine. J. Phys. D: Appl. Phys. 36, no. 13: R167.

Park, J., J. Joo, S.G. Kwon, Y. Jang and T. Hyeon. 2007 Synthesis of monodisperse spherical nanocrystals. Angew.Chem.Int.Ed Engl. 46: 4630–4660.

Skajaa, T., D.P. Cormode, P.A. Jarzyna, A. Delshad, C. Blachford, A. Barazza, E.A. Fisher, R.E. Gordon, Z.A. Fayad and W.J. Mulder. 2011. The biological properties of iron oxide core high-density lipoprotein in experimental atherosclerosis. Biomaterials 32: 206–213.

Tawakol, A., A.P. Castano, F. Gad, T. Zahra, G. Bashian, R.Q. Migrino, A. Ahmadi, J. Stern, F. Anatelli, S. Chirico, A. Shirazi, S. Syed, A. J. Fischman, J.E. Muller and M.R. Hamblin. 2008. Intravascular detection of inflamed atherosclerotic plaques using a fluorescent photosensitizer targeted to the scavenger receptor. Photochem. Photobiol. Sci. 7: 33–39.

Tsourkas, A., V.R. Shinde-Patil, K.A. Kelly, P. Patel, A. Wolley, J.R. Allport and R. Weissleder. 2005. *In vivo* imaging of activated endothelium using an anti-VCAM-1 magnetooptical probe. Bioconjug. Chem. 16: 576–581.

Whitesides, G.M. 2003. The 'right' size in nanobiotechnology. Nat. Biotechnol. 21: 1161–1165.

10

Nanomonitor Technology and Its Applicability to Diagnosis of Cardiac Disease

Shalini Prasad,[1,2,*] *Manish Bothara,*[3,a] *Ravikiran K. Reddy,*[3,b] *Thomas Barrett*[4] and *John Carruthers*[5]

ABSTRACT

In this chapter we present the nanomonitor technology as a 'point-of-care' device for early diagnosis of protein biomarkers that have association with vulnerable coronary vascular plaque rupture. Here, we present label-free, highly sensitive detection of proteins with the use of electrical impedance spectroscopy through a single capture immunoassay methodology. The principle of the

[1]Department of Electrical Engineering and Computer Science, Wichita State University, Wichita, KS, 67260; E-mail: shalini.prasad@wichita.edu

[2]Bioengineering Program, Wichita State University, Wichita, KS, 67260.

[3]Department of Electrical and Computer Engineering, Portland State University, Portland, OR, 97201.

[a]E-mail: mbothara@pdx.edu

[b]E-mail: ravikk@pdx.edu

[4]Department of Veteran Affairs, Oregon Health Sciences University, Portland, OR 97209; E-mail: barreth@ohsu.edu

[5]Department of Physics, Portland State University, Portland, OR 97201; E-mail: caruthe@pdx.edu

*Corresponding author

List of abbreviations after the text.

detection is based on the formation of an electrical double layer and its perturbations caused by proteins trapped in a nanoporous alumina membrane over a microelectrode array platform. High sensitivity and rapid detection of study protein biomarkers for vulnerable coronary vascular plaque rupture, C-reactive protein and myeloperoxidase in pure and clinical samples through label-free electrical detection were achieved. C-reactive protein and myeloperoxidase were detected in pure solutions and in human serum samples at picogram/ml and nanogram/ml sensitivities respectively. These two study proteins were also detected from multiplexed samples containing a mixture of proteins and human serum samples. The performance parameters of the nanomonitors—detection within minutes, volume of reagents of a few microliters and low cost per assay—are potential advantages of the nanomonitor device in the context of developing a "lab-on-a-chip" assay for rapid clinical diagnostics. Nanomonitors also provide a large linear dynamic range of label-free detection and allow a significant reduction in the size of assay, thus proving an ideal candidate for a clinical diagnostic lateral flow device for protein biomarker profiling and hence early disease diagnosis.

INTRODUCTION

Proteomics research has elucidated many new biomarkers that have the potential to greatly improve disease diagnosis (Darain et al. 2004; Hahm et al. 2004; Nam et al. 2004; Niwa et al. 1996). Combination of multiple biomarkers has been determined to provide the information necessary for robust diagnosis of a disease (Abeloff et al. 2000; Chou et al. 2004; Danesh et al. 1998). In addition, detection of biomarkers associated with different stages of disease pathogenesis could facilitate early detection. Widespread use of protein biomarkers in health care will ultimately depend upon the development of techniques that allow rapid and multiplexed detection of a wide range of biomarkers with high selectivity and sensitivity. This goal has not been attained with any existing immunoassay method, including ELISA (Chou et al. 2004; Prasad et al. 2004a, b). On the other hand, electrical biosensors can be extremely sensitive and can be modified to function as immunoassays meeting the improved performance metric requirements.

There are currently three electrical biosensor technologies for biomolecule detection: (1) electrochemical conductance measurements, (2) field effect transistor (FET)-based transconductance measurements (Etzioni et al. 2003) and (3) electrochemical capacitance measurements (DeSilva et al. 1995). Lieber's group at Harvard has demonstrated biomolecule (DNA) detection limits of as low as 10-femtomolar sensitivity (Zheng et al. 2002) by employing the electrochemical conductance detection technology. There are, however, issues regarding the specificity of detection in non-DNA–based assays. The second technique using transconductance measurements has been used to demonstrate the detection of prostate-specific antigen using silicon nanowire–based FET arrays up to 0.9 pg/ml sensitivity (Zheng et al. 2002). The FET method based on transconductance measurements is rapid and extremely sensitive but very complex to implement, whereas the conductance and capacitive change methods have sensitivities equal to standard immunoassay methods but are much simpler to implement and therefore more likely to find clinical applications sooner.

The electrochemical capacitance technique has demonstrated specificity in protein biomarker detection. This technique has been used to distinguish between monoclonal and polyclonal antibodies (Berney et al. 1998). Berggren and co-workers demonstrated interleukin (IL 6) antigen detection sensitivity of 0.5 pg/ml using this technique (Berggren et al. 2001).

Traditional assay methods for detection of protein biomarkers for diseases are primarily gel electrophoresis, western blot and enzyme-linked immunosorbant assay (ELISA). The limitations and experimental disadvantages of such biomarker detection methods are (1) the use of reagents up to milliliter in volume with multiple wash cycles, (2) the lack of ability to simultaneous identify multiple biomarkers from a test sample, (3) increased cost associated with multiple assays for one result, (4) the greater requirement of laboratory staff and time for identifying the detection event on the assay, and (5) the inability to perform rapid detection of the biomarker (resolution of hours). Unlike the traditional assays, label-free electrical immunoassays offer possibilities of comparable sensitivity and specificity at improved speed and decreased cost with reagent minimization along with the integration of the detection and measurement capabilities onto a single platform. Thus, they have the potential to enable the development of a proteomic test device with reduced volume per assay, reduced cost per assay and the scope for simultaneous multiple biomarker detection, also known as multiplexed detection. This device can be applied for rapid and early detection of protein biomarkers that are indicative of a specific disease state.

The clinical application for this emerging technology is the identification of the disease state: presence of vulnerable coronary plaque, which is the cause of acute coronary syndromes (ACS) such as stroke, peripheral vascular limb ischemia, and other end-organ ischemic diseases, and thus is the leading cause of death worldwide. Inflammation and thrombosis are key mediators of vulnerable coronary plaque. It has been determined that this disease state can only be identified in the pre-operative condition by the presence of a set of certain inflammatory protein biomarkers.

Currently, there is no clinical immunoassay technique to simultaneously detect the protein biomarkers responsible for the identification of vulnerable coronary plaque with the same robustness as the traditional immunoassay methods. Vulnerable coronary plaque is believed to be present in the blood vessels of a patient by the simultaneous detection of at least five proteins: C-reactive protein (CRP), myeloperoxidase (MPO), CD40 ligand, oxidized low-density lipoproteins (LDL), and tissue factor. The pre-operative identification of the disease state is expected to improve the post-surgical outcomes of the patient population. Thus, one possible solution to identify vulnerable coronary plaque is to develop an immunoassay device that can detect these proteins biomarkers in a rapid and multiplexed manner.

The principle of electrochemical capacitance measurements has been adopted as the mechanism of detection in the electrical immunoassay system developed as a part of this research, because of the simplicity and improved signal amplification offered by this technique within the immunoassay configuration. The capability of the electrical immunoassay in detecting two of the target five protein biomarkers in the purified form and in human serum is demonstrated. It has been hypothesized by a number of research groups that the multiple protein biomarkers of vulnerable coronary plaque in the pre-operative state can be detected by an electrical immunoassay device fabricated from nanoporous alumina membranes operating on the principle of electrochemical signal transduction. These devices are referred to henceforth as "nanomonitors" (NM). The rationale for this study is the demonstration of the performance metrics of the NM to be comparable to ELISA, with reduced cost and volume per assay, and improved detection speed. In addition, multiplexed protein detection for two of the five protein biomarkers in a robust manner from purified as well as spiked human serum samples has been demonstrated.

There is a strong interest in applying the capabilities of miniaturization in electrical engineering and biophysics to important problems in the biomedical sciences, and the focus is on a clinical issue with high impact for patients. The proposed study is innovative because it establishes a novel proteomic immunoassay device that integrates the measurement capability into a detection platform, which the current techniques lack. The NM are a miniature version of the microtiter plate technology with reduced cost per assay and decreased volume per assay. The application of the electrical biosensing technique has resulted in a simpler, cost-effective device with performance metrics comparable to ELISA. This device will be attractive for multiplexed protein biomarker detection.

STANDARD IMMUNOASSAY TECHNIQUES AND THEIR LIMITATIONS IN PROTEIN BIOMARKER DETECTION

Standard immunoassay techniques that are currently used for protein biomarker detection in clinical diagnostics can be classified into three categories: gel electrophoresis, western blot and ELISA. While the first technique uses a combination of electrical and fluorescence-based optical methods, the other two are essentially fluorescent optical detection mechanisms. Of the three techniques, ELISA is the most robust and is the "gold standard" for clinical diagnostics.

ELISA-based protein biomarker detection occurs in a microtiter well plate. Despite the robustness of this technique it has certain inherent limitations: it requires a large sample volume (microliter-milliliter) per well, it costs more per assay (each ELISA kit costs $200–$1000 depending on the protein that is being identified), lab-based sample preparation techniques are complex, and the time of detection varies from a few hours to a couple of days. These factors result in additional complexity when applied to multiplexed protein biomarker detection. The scope of this application is the demonstration of the NM technology that overcomes these limitations.

APPLICATION OF NANOTECHNOLOGY IN THE ELECTRICAL DETECTION OF PROTEIN BIOMARKERS

Electrical detection of protein biomarkers is a relatively new approach that monitors a specific electrical parameter that undergoes change during the detection event. With the advent of nanotechnology in the field of clinical diagnostics, material with nanometer physical dimensions has been used for its improved surface area in developing detectors with enhanced sensitivity and reduced use of reagents. The major classes of nanomaterials that have been used for protein biomarker detection are nanotubes, nanowires, nanoparticles and nanotemplates. Dai and co-workers at Stanford demonstrated the use of carbon nanotubes with sensitivity in the range of lower nanogram/ml (ng/ml) to a few hundred picogram/ml (pg/ml). Detection was achieved using the electrochemical conductance technique (Chen et al. 2004). Lieber and his group at Harvard implemented the FET-based transconductance technique using silicon nanowires with detection sensitivity in the range of picogram/ml to femtogram/ml (Zheng et al. 2002). Willner et al. (2003) demonstrated the use of antibody-coated nanoparticles as electrical biosensors operating on electrochemical conductance technique with sensitivities in the pg/ml range.

In all the nanomaterial-based electrical protein biosensors, detection response is in the order of minutes with very low sample volume (in the order of nanoliters). But, of the four nanomaterial classes, nanotemplates enable protein isolation and localization that in turn results in the development of robust sensors with improved specificity at lower concentrations. Hence, for this application we incorporated the use of nanotemplates for the development of the NM.

SELECTION OF THE NANOPOROUS TEMPLATE

A wide range of nanoporous material has been used as templates for scaffolding as well as trapping biomolecules. Nanoporous silica as well as polymer-based templates have been used for the development of protein biosensors (Dancil et al. 1999; Norman et al. 2006). While controlling the pore dimensions in silica is a complex fabrication process, polymer-based pores require complex biochemical functionalization and surface treatment for biosensing. In this application, nanoporous alumina has been identified as a suitable material for localizing and isolating the protein biomarkers. There has been a precedent in the use of this material for biosensing. Takhistov used nanotemplates for the detection of penicillin up to 0.1 µg/ml by electrochemical conductance measurement (Takhistov et al. 2004). Also, the physical dimensions as well as the pore density can be precisely controlled via a cost-effective wet bench–based electrochemical technique known as two-step anodization. The pore dimensions can be matched to the dimensions of the specific protein biomarkers to ensure effective physical adsorption as well as trapping of the proteins into the pores. This in turn ensures improved sensitivity and lowers the limit of detection. Finally, nanoporous alumina is an electrical insulator; this ensures electrical isolation of individual nanopores. This in turn reduces electrical signal interference between the pores, which is crucial during multiplexed detection.

SELECTION OF INFLAMMATION-RELATED PROTEIN BIOMARKERS FOR DETECTION OF VULNERABLE CORONARY PLAQUE

The inflammatory proteins were selected from a review of the literature for ACS and correspond to the conceptual construct of selecting proteins from each of the histologic participants that constitute a vulnerable plaque (see Fig. 1 and Table 1). The constituent elements of a vulnerable plaque include vascular cells (endothelial and smooth muscle cells) (soluble CD40 ligand), leukocytes (monocytes/macrophages, lymphocytes, T, B, and natural

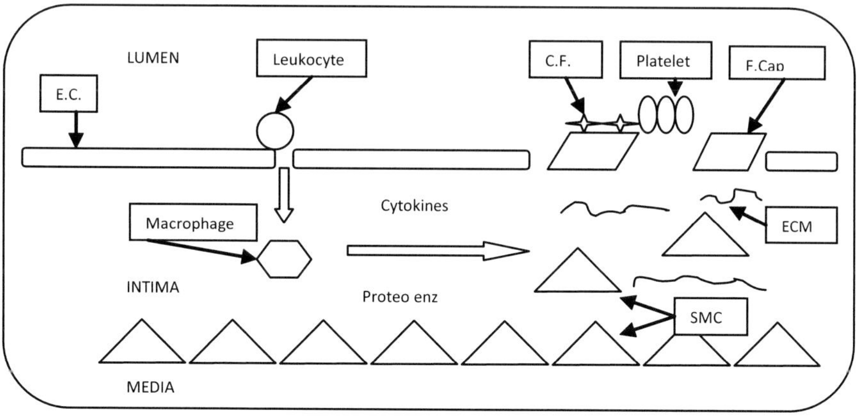

Fig. 1. Conceptual construct of proteins that contribute to vulnerable plaque showing the distribution of the various proteins that contribute to the formation of vulnerable plaque. E.C., endothelial cells; C.F., coagulation factors; F.Cap, fibrous cap; ECM, extracellular matrix; SMC, smooth muscle cells; Proteo enz, proteolytic enzymes.

Table 1. Proteins associated with vulnerable coronary vascular plaque rupture.

Vascular cells	Leukocytes		Platelets	Coagulation factor
CD40 Ligand	CRP	Tissue factor	CD40 Ligand	Tissue factor
	CD40 Ligand	Ox-LDL		
	MPO			

These proteins are classified by the sources that generate them.

killer T cells) (CRP, oxidized LDL, MPO), platelets (soluble CD40 ligand), as well as coagulation factors (tissue factor), the extracellular matrix and the fibrous cap. (Norman et al. 2006; Hyun et al. 2001; Block et al. 2006). In this chapter the detection of two protein markers, namely C-reactive protein and myeloperoxidase, is presented. It is important that the NM measure low levels of multiple protein biomarkers simultaneously. This is especially relevant in the clinical environment as it will provide a method to detect levels of multiple biomarkers simultaneously to assess activation of the endothelium, leukocytes, platelets, the coagulation system, and plaque turnover to identify patients who may already be primed for an exaggerated inflammatory response to surgery.

These proteins have been extensively associated with myocardial ischemia, which is the clinical outcome in which we are most interested for our application of NM to vulnerable plaque rupture. The rationale for selecting each protein relates to its role in the histology of vulnerable plaque rupture as above, and to associations with ACS (Apple et al. 2005; Ferri et al. 2006; Jaffe et al. 2006; Libby et al. 2005; Nagahavi et al. 2003; Vasan et al. 2006).

The scope of the current application is the demonstration of sensitivity in identification of two of the five protein biomarkers using NM to be comparable to ELISA both individually and in the multiplexed format. The proteins have been detected in both the purified form and from spiked serum samples.

APPLICATIONS TO AREAS OF HEALTH AND DISEASE

The NM technology, which is a portable, label-free diagnostics platform operating on the principle of electrical impedance spectroscopy, can be applied to early detection of a wide range of diseases. Most of the grave diseases can be identified by detecting proteins that are expressed during the inception and progress of the disease. These proteins are thought to be biomarkers of the disease. But they are present in ultralow concentrations and it is essential to accurately detect them from complex solutions such as human blood and serum. The use of nanotextured surfaces in diagnostics has enhanced the process of biomolecule binding onto sensor surfaces. In this chapter we have leveraged the use of nanoporous alumina membranes with microelectronic platforms to design a portable diagnostics assay that can be used to detect a wider range of protein biomarkers associated with a range of diseases including cardiovascular disease, neurodegenerative disease and cancer.

NANOMONITOR ASSAY ARCHITECTURE

The NM has a multi-scale architecture whose dimensions span from the micro scale to the nanoscale. It comprises three parts: the microelectrode array base platform, a nanoporous alumina membrane over-layer, which forms the nanowells, and the microfluidic encapsulant to enable the lateral flow of reagents over and into the nanowells. The microelectrode base platform comprises an array of circular, gold measurement/sensing sites where the binding of the protein molecules occurs in a controlled manner (Fig. 2a). Each sensing site comprises a working electrode (WE) and a counter electrode (CE) (Fig. 2b); the surface area ratio of the CE to WE is 225:1 (Fig. 2b), which ensures an increase in the differential surface area between the electrodes. Additionally, the distance of separation between the circular portions of the WE and CE is very small (2 μm). The capacitance change due to biomolecule binding is measured across the WE and CE. Based on the fundamental principles of electrical engineering, the capacitance change is maximized by an increase in the differential surface area and a decrease in the distance of separation between WE and CE. The metal electrodes

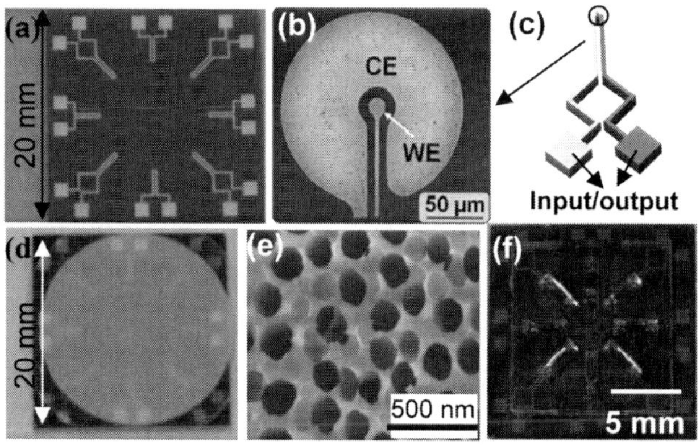

Fig. 2. Optical micrograph representing the manufacturing of a nanomonitor platform. (a) Nanomonitor (NM) device. (b) Sensing site—working electrodes (WE) and counter electrodes (CE). (c) Input/output pads connected to sensing site through interconnects. (d) Nanoporous membrane overlying on NM. (e) SEM of membrane. (f) Microfluidic device encapsulating the NM.

Color image of this figure appears in the color plate section at the end of the book.

function as the base for multiple nanowells once the nanoporous alumina membrane is overlaid. Protein binding and capacitance changes occur at these electrodes (sensing sites). These electrodes are connected to input/output measurement leads. An input low-voltage alternating current (AC) signal is provided to the sensing site through these leads and the output capacitance signal is measured across the same leads (Fig. 2c). Both the electrodes are designed to be circular to attain maximum surface area of interaction and to avoid any possible edge effects. The microelectrode is designed to get eight such sensing sites on a 20x20 mm^2 silicon chip using the standard process of photolithography used for fabricating microelectronics.

The second part of the NM is the nanoporous alumina membrane (Fig. 2d and 2e). A commercially available nanoporous alumina membrane with nanopore diameter of 200 nm and membrane thickness of 250 nm (Anodisc, Walman, NJ, USA) is overlaid and adhered on to the base microelectrode array. The pore diameter is selected such that the pore is large enough to function as a scaffold for one or more antibody-antigen binding complexes. At 200 nm diameter there are approximately a quarter million nanowells on a single sensing site. The third part is the microfluidic encapsulant fabricated out of biocompatible, plasma-treated, polydimethylsiloxane (Fig. 2f) to regulate the lateral flow of reagents onto the sensing sites. The encapsulant is interfaced to the NM device through a pressure-sensitive adhesive. It consists of eight microchannels, each encapsulating one sensing site and consisting of an input and output port. Each micro-channel supports the flow of 8 μl onto the sensing site.

PRINCIPLE OF OPERATION: ELECTRICAL IMPEDANCE SPECTROSCOPY

The protein binding and detection process is achieved by incorporating a single capture immunoassay into each nanowell. All the biomolecules (linker molecules, antibodies and antigens) have surface charges. The binding of these molecules to the base of each nanowell perturbs the charge distribution in the electrical double layer that forms at the solid/liquid interface. This charge perturbation produces a capacitance change in the electrical double layer. The biomolecule binding–induced capacitance change can be measured by the electrical impedance spectroscopy technique. This technique is widely used in electrical biosensors for detecting surface-charged analytes. In this technique a low AC voltage is provided as the input to a sensing site, which functions as the electrical stimulus to direct the surface-charged biomolecules onto a sensing site. The binding of the biomolecules produces a change in the measured output impedance across a sensing site. The measured impedance is the sum total of two components, the resistance and capacitance of the sensing site. It is only the capacitive component that is a measure of the biomolecule binding event, as the capacitive component indicates the surface charge differential in the electrical double layer as a function of biomolecule binding. Hence the input voltage parameters (frequency and voltage amplitude) need to be optimized to measure this capacitive component.

CAPACITANCE CHANGES WITHIN A NANOWELL AND ACROSS A SENSING SITE

In the NM configuration, there are approximately a quarter million nanowells on one sensing site. We determined that an AC voltage of 100 mV at a frequency of 1 kHz when applied between the WE and CE of a sensing site enabled the measurement of the changes to the capacitive component associated with the protein binding event.

In this section the single capture assay–induced capacitance change within an individual nanowell is described. At these input AC voltage conditions, the ions in phosphate buffered saline (PBS), which is the ionic buffer used for aliquoting the proteins, get attracted to the voltage-biased base of the nanowell to form a charge-neutral layer, which is also known as the inner Helmholtz plane (iHp) and is the fixed end of the electrical double layer. The remainder of the free ions of the dissociated PBS buffer functions as electrostatic cues to guide the linker molecules towards the charge neutral nanowell base. The linker molecules are eventually tethered to the base of the well through chemical conjugation. This forms the outer Helmholtz plane (oHP) and is the movable end of the electrical double layer (EDL). The capacitance is measured across the iHP and oHP. When the antibodies are added to the NM device, they are tethered to the linkers by biochemical conjugation causing the oHP to move upwards, to encompass the linker molecules as well as the antibodies. Because of this perturbation of the EDL, the capacitance across the EDL changes. Capacitance further changes during the addition of the antigen as antibody antigen conjugation further perturbs the EDL by shifting the oHP. Hence, monitoring the capacitance change in the EDL is a measure of the protein binding event. An increase in antigen concentration will in turn increase the capacitance change in the EDL up to an upper bound that corresponds to the absence of free antibody sites. The EDL is approximately 20–50 nm in height, hence capacitance changes only due to biomolecule binding occurs in this region. The region outside the oHP is the bulk region and contributes to the resistive component of the impedance measurements (Fig. 3a). The capacitance change due to EDL perturbation is replicated in all the nanowells during each step of the single capture assay. As capacitance is measured across the WE and CE, in essence the quarter million nanowells behave such that they are connected in parallel, i.e., across the same two points. Based on capacitor principles, the equivalent capacitance across the sensing site is the sum of the individual capacitances of the nanowells. The equivalent electrical circuit of a sensing site (Fig. 3b)

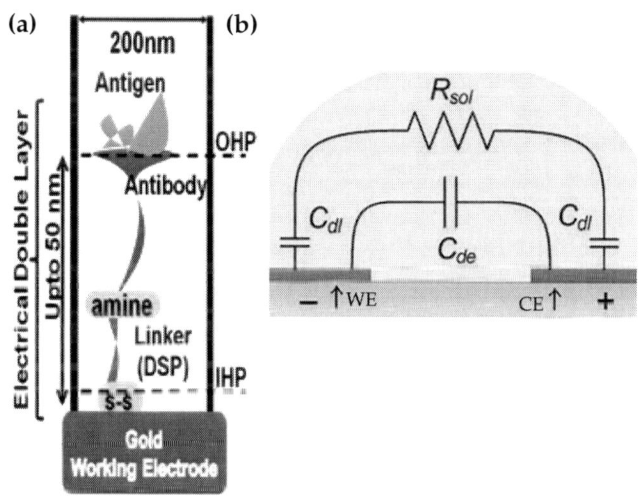

Fig. 3. Representation of biomolecule binding within the nanowells of the nanomonitor. (a) Schematic showing the antibody-antigen binding inside a nanowell. (b) Equivalent circuit of a sensing site showing a resistive capacitive circuit, where the capacitor correlates to the double layer capacitance.

Color image of this figure appears in the color plate section at the end of the book.

comprises a resistor and two capacitors. The resistor is a measure of the bulk resistance of the buffer solution, Rsol, due to the ions present in the bulk regions of the nanowells (> 50 nm from base of the nanowell). The dielectric capacitance between the two electrodes due to the base substrate is represented by C_{de}. This capacitance does not change for varying input AC voltage frequencies and hence this capacitance is accounted for when the NM is calibrated during baseline/control measurements. The EDL capacitance is represented by C_{dl}, which undergoes a change due to the biomolecule-based perturbation to the EDL. The measured Cdl capacitance is a sum of the contributions from the WE and CE nanowells. Based on electrochemistry, it has been determined that the bulk resistive component dominates the impedance measurements at high frequencies (> 10 kHz) and the capacitive double layer component dominates the impedance measurements at low frequencies (1 kHz and lower) (Wang 2000; Wang 2005; Zimmerman and Minton 1993). We experimentally identified 1 kHz to be the optimum frequency for operating the NM as we observed maximum change in the EDL capacitance with minimum background electrical noise.

NANOMONITOR SURFACE FUNCTIONALIZATION

Antibodies were immobilized into the nanowells using dithiobis (succinimidyl propionate) (DSP) as the covalent linker. DSP (Fig. 3a) has a thiol (sulfur) end group that has great affinity to gold (Reddy et al. 2008a; Reddy et al. 2008b). Hence, the thiol end of the linker bound to the gold base of the nanowell. The NHS end group at the opposite epitopes of the linker bound to the antibody. The entire gold base surface of the nanowell was saturated with DSP linker molecules prior to injecting antibodies. The NM was cleaned using 90% ethanol for 2 min followed by a rinse in de-ionized water for 2 min and blow-dried under dry nitrogen to maintain device sterility. AC input voltage (100 mV, 1 kHz) was applied across WE and CE of each sensing site. The capacitance from each sensing site was obtained, which were the Cde measurements for calibrating the NM device. Eight microliter of PBS (Sigma Aldrich, MO, USA, at a concentration of 1 X PBS, stock solution, with bulk ionic conductivity of 0.13 S/cm) was injected onto individual sensing site. Capacitance measurement from each sensing site was obtained after the capacitance reading stabilized. This occurred within 5 min of injecting PBS onto the sensing site. These measurements indicated the baseline or control measurements for the device. Eight microliters of DSP linker at 10 µg/ml aliquoted in PBS was injected onto each sensing site to saturate the sensing sites. Capacitance measurements were obtained once stability in the recording was achieved, which typically occurred within 2–5 min post-injection of the linker molecules. These measurements when subtracted from the Cde measurements indicated the Cdl due to the DSP linker molecules.

ANTIBODY SATURATION

The antibodies monoclonal anti-CRP and monoclonal anti-MPO were purchased from Calbiochem (San Diego, CA, USA). A serial dilution of both the antibodies was prepared using PBS. The antibody concentrations ranged from 1 ng/ml to 10 μg/ml. A number of DSP linker saturated NM devices were prepared to perform five replicated capacitance measurements for each concentration dose of the antibody. Eight microliter antibody aliquot was injected per sensing site and the device was incubated at room temperature for 15 min. AC voltage (100 mV, 1 kHz) was applied across the WE and the CE of each sensing site. The capacitance was measured across the device and stable measurements were obtained within 5 min of measurement. With an increase in the antibody concentration an increase in the measured capacitance was observed until for a particular concentration of the antibody, the capacitance measurement saturated. The concentration of antibody corresponding to the saturation in capacitance measurements was determined to be the antibody saturation concentration. It was determined to be 10 μg/ml for anti-CRP and 5 μg/ml for anti-MPO (Fig. 4a). We expect that at these concentrations the antibodies would have saturated the nanowell surface (Bothara et al. 2008; Nagaraj et al. 2010). After the antibody saturation step we injected 8 μl of 1 μg/ml bovine serum albumin (BSA, Sigma Aldrich, MO, USA) as a blocker into each sensing site, after room temperature incubation for 5 min we washed the blocker using PBS. We measured the EDL capacitance and it was determined to be within 1% variation of the antibody saturation capacitances as indicated in Fig. 4a. Figure 4b shows the change in cap acitance measured continuously over the span of the experiment. The capacitance increased in a stepwise manner as the successive reagents are added.

ANTIGEN DETECTION

The antigens CRP and MPO were purchased from Calbiochem (San Diego, CA). Two sets of samples were prepared. The first set was a serial dilution of the antigens from 100 pg/ml to 100 μg/ml using PBS. After saturating the sensing sites with antibodies, the antigen to be detected was injected onto the sensing site. The capacitance across the WE and CE for a sensing site was monitored. We observed an increase in the capacitance from the previous steady-state capacitance value associated with antibody

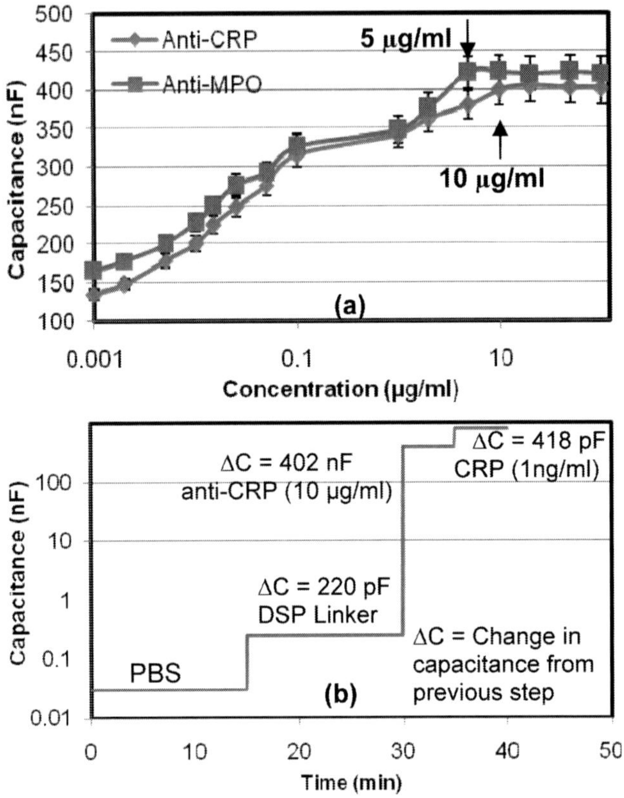

Fig. 4. Antibody saturation cure and representation of a time series for the dose response. (a) Antibody saturation curves for Anti-C-reactive protein and Anti-Myeloperoxidase. (b) "Real- time" detection showing stepwise change in capacitance with time.

Color image of this figure appears in the color plate section at the end of the book.

saturation. The addition of antigens perturbed the EDL and we observed an increase in the Cdl capacitance. Each concentration dose was measured in five replicates. Steady-state capacitance values were observed 5 min after the injection of the antigen. In order to determine the specificity/ cross-reactivity of the assay we injected the serial dilution aliquots of CRP onto an anti-MPO-saturated device and vice-versa. Throughout the antigen dose response detection the sensing sites were biased with an input AC voltage of 100 mV and 1 kHz. The cross-reactivity specifications as well as the linear dynamic range of detection of the two proteins are indicated in Table 3.

The second set of antigen samples was prepared by serially diluting human serum (U.S. Biological, MA, USA) with PBS. The human serum comprises many competing proteins including CRP and MPO. The concentration of CRP and MPO already present in human serum was recorded by testing in standard ELISA kits for CRP and MPO. Aliquots of CRP and MPO in human serum were then prepared by diluting human serum such that the concentrations were similar to the pure samples tested previously. Figure 5 shows the dose response curves for the proteins CRP and MPO detected in pure form and from serum samples in a multiplexed manner. The performance parameters of NM for both pure and serum samples of CRP and MPO are given in Table 3 and are comparable to ELISA.

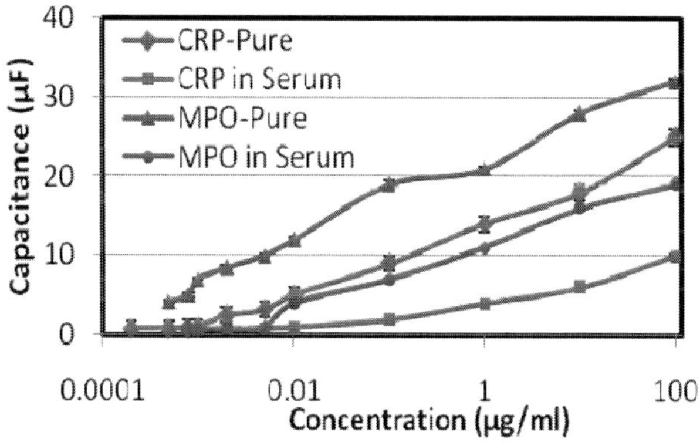

Fig. 5. Dose response curves for the inflammatory proteins. Dose response for C-reactive protein and myeloperoxidase detection from pure and serum samples.

Color image of this figure appears in the color plate section at the end of the book.

VALIDATION OF NANOMONITOR PERFORMANCE

The key technical idea that required validation was the demonstration that the antibody bound on the gold surface at the base of each nanowell. A validation experiment was designed by placing small sections of the NM base gold micro platform into individual wells of a 48-well ELISA plate. In this section we present the validation results in detecting anti-CRP binding on the NM surface. The NM incorporated into individual wells of the ELISA plate were treated with the DSP linker (4 μg/ml of dimethyl sulfoxide for 30 min at room temperature) and anti-CRP solution at different concentrations. A horseradish peroxidase (HSP) conjugated detection antibody was then added followed by the TMB (3, 3', 5, 5'-tetramethylbenzidine) substrate to produce the change in color, which was a visual indicator of the antibody binding event. The plate was washed three times after each step with 3% BSA/PBS solution. A total of eight samples consisting of four controls, two test samples and two blank samples were tested in duplicates. Table 2 gives the details of the testing conditions. The entries of each cell in Table 2 indicate the various reagents that were used to treat the NM base micro platform in that cell. Rows A-D in columns 1 and 2 were set as four negative controls (cells A1, A2, B1, B2, C1, C2, D1 and D2) where one or more of the reagents such as DSP, anti-CRP and HRP-antibody (the HRP conjugated antibody) were not added. The next two samples (cells E1, E2, F1, and F2) were test samples where increasing concentrations of anti-CRP were added to the gold (Au) substrate functionalized with DSP linker. The binding of anti-CRP antibody on Au was confirmed by adding the HRP-antibody followed by TMB substrate. The TMB substrate on reaction with the HRP-antibody changed to blue. The increase in strength of the blue color indicated that with increased concentration of the anti-CRP, the binding on the Au increased, thereby capturing more HRP-antibody. Figure 6 shows a digital photograph of the gold NM platform at the base of each well in the ELISA plate. The cells in which the TMB substrate solution turned blue indicated that HRP-conjugated antibody had been captured by the antibody attached to the Au in that cell. This in turn validated the binding of the antibody on the NM base micro platform.

Table 2. Digital micrograph of the 48 well gold Nanomonitor-ELISA plate experiment.

	1	2	3	4
A	Au, HRP-Ab, TMB **(No DSP, No anti-CRP)**		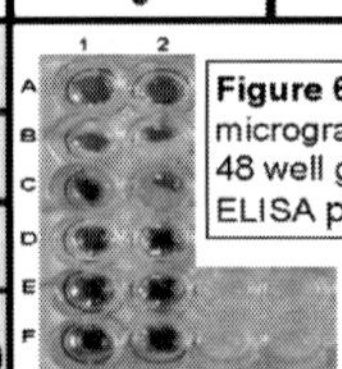	**Figure 6:** Digital micrograph of 48 well gold NM ELISA plate.
B	Au, DSP, HRP-Ab, TMB **(No anti-CRP)**			
C	Au, anti-CRP (50 µg/ml), HRP-Ab, TMB **(No DSP)**			
D	Au, DSP, anti-CRP (50 µg/ml), TMB **(No HRP-Ab)**			
E	Au, DSP, **anti-CRP (10 µg/ml)**, HRP-Ab, TMB		Blank	
F	Au, DSP, **anti-CRP (50 µg/ml)**, HRP-Ab, TMB		Blank	

Details of the solutions added for the 12 samples in the validation experiment. The important steps in each cell are in bold type. Purple cells represent the controls, red cells represent the anti-c-reactive protein tests, and white cells are blank.

Color image of this figure appears in the color plate section at the end of the book.

The accuracy of the concentrations of CRP and MPO was validated using commercial ELISA kits (Immunology Consultants, OR, USA) with a coefficient of variation of 3% for detection range of 3.125 ng/ml to 100 ng/ml.

We evaluated the NM device as a proteomic assay through two study inflammatory proteins, CRP and MPO, and our findings are summarized in Table 3a. We demonstrated performance metrics comparable to ELISA, as indicated in Table 3b.

Table 3a. Summary of key performance metrics of NM in detection of two inflammatory proteins.

CRP	**Pure samples in PBS**	**Human serum samples**
Limit of detection (LoD)	200 pg/ml	1 ng/ml
Capacitance value at LoD	820 nF	620 nF
Linear dynamic range	200 pg/ml to 100 µg/ml	500 pg/ml to 100 µg/ml
Protein binding response time	150–180 sec	150–180 sec
Concurrent detection	CRP and MPO at 10 ng/ml	CRP and MPO at 10 ng/ml
Specificity/cross-reactivity	2% at ng/ml of CRP on anti-MPO	3% at ng/ml of CRP on anti-MPO
Validation	NM ELISA	Standard ELISA
MPO	**Pure samples in PBS**	**Human serum samples**
Limit of detection (LoD)	500 pg/ml	800 pg/ml
Capacitance value at LoD	4.2 □F	700nF
Linear dynamic range	500 pg/ml to 100 µg/ml	800 ng/ml to 100 µg/ml
Protein binding response time	120-150 sec	120–150 sec
Concurrent detection	CRP and MPO at 1 ng/ml	CRP and MPO at 10 ng/ml
Specificity/cross-reactivity	2% at ng/ml of MPO on anti-CRP	3% at ng/ml of MPO on anti-CRP
Validation	NM ELISA	Standard ELISA

C-reactive protein and myeloperoxidase from pure phosphate-buffered saline samples.

Table 3b. Performance metrics for CRP and MPO detection in human serum.

Performance metrics	**CRP in serum**	**MPO in serum**
Limit of detection (LoD)	100 pg/ml	400 pg/ml
Linear dynamic range	500 pg/ml to 8 ng/ml	400 ng/ml to 100 ng/ml
Assay time	3.5 h	3.5 h
Interassay variation	7.6%	7%
Sample volume	50 µl/well	100 µl/well

C-reactive protein and myeloperoxidase from human serum samples.

CONCLUSIONS

A highly sensitive device for label-free protein detection was designed and fabricated. The technique has several advantages over current existing immunoassay methods:

- It is label free and hence no external tags or chemicals are needed and the risk of contamination is minimized.
- It has high potential to be developed into a portable device.
- It has lower cost for manufacturing and lower infrastructure needs than a standard immunoassay.

This technology has potential to be expanded to be of clinical significance in the near future.

Key Facts

- Progressive atherosclerotic disease and its clinical sequel remain the number one cause of mortality in the U.S. The majority of patients with acute coronary syndromes present with unstable angina, acute myocardial infarction, or sudden cardiac death secondary to sudden luminal thrombosis.
- The immobilization of biomolecules on a solid substrate and their localization in "small" size-matched regions are major requirements for a variety of biomedical diagnostic applications, where rapid and accurate identification of multiple biomolecules is essential.
- In this specific application we have fabricated nanomonitors for identifying specific protein biomarkers based on the electrical detection of antibody-antigen binding events.
- The nanomonitor, lab-on-a-chip device technology is based on electrical detection of protein biomarkers. It is based on developing high-density, low-volume multi-well plate devices.
- The scientific core of this technology lies in integrating nanomaterial with micro fabricated chip platforms and exploiting the improved surface area to volume to improve the detection.
- The devices that have been developed use electrical detection mechanisms where capacitance and conductance changes due to protein binding are used as "signatures" for biomarker profiling.

Definitions

Electrical double layer: A structure that describes the variation of electric potential near a surface and has a significant influence on the behavior of colloids and other surfaces in contact with solutions or solid-state fast ion conductors.

Electrical impedance spectroscopy: A technique for measuring changes to the impedances associated with the electrical double layer at a solid-liquid interface over a range of frequencies.

"Lab-on-a-Chip": A device that integrates one or several laboratory functions on a single chip as small as some millimeters to a few square centimeters. The device is used in handling extremely small fluid volumes.

Label-free detection: Detection of target biomolecules without attaching a fluorescent dye.

Macromolecular crowding: The phenomenon of macromolecular crowding alters the properties of molecules in a solution when high concentrations of macromolecules such as proteins are present.

Nanomonitor technology: The technology is based onmeasuring the capacitance changes at the electrical double layer interface as a result of binding of protein molecules on to the diagnostics platform.

Summary Points

- The nanomonitor technology is a heterogeneous technology comprising nanoporous alumina membrane integrated onto a microelectronic platform for ultra-sensitive detection of protein biomarkers.
- Detection of two cardiovascular protein markers, C-reactive protein and myeloperoxidase, is demonstrated associated with pre-operative detection of vulnerable coronary vascular plaque rupture.
- Detection is achieved from phosphate-buffered saline samples as well as human serum samples. Detection limit is obtained in the range of lower picogram/ml from PBS and in the ng/ml regime from human serum samples.
- Detection of the two proteins is achieved in a multiplexed manner with detection time in the order of minutes, which is considerably faster than ELISA.
- Nanomonitor performance is also validated using a nanomonitor ELISA technique and the nanomonitor has been calibrated to demonstrate effective detection in the clinically relevant detection regime of ng/ml.

Abbreviations

ACS	:	acute coronary syndrome
CF	:	coagulation factors
CRP	:	C-reactive protein
DSP	:	dithiobis (succinimidyl propionate)
EC	:	endothelial cells
ECM	:	extracellular matrix
EDL	:	electrical double layer
F.Cap	:	fibrous cap
FET	:	field effect transistor
HRP	:	(horseradish peroxidase)
LDL	:	low-density lipoproteins
MPO	:	myeloperoxidase
NM	:	nanomonitors
Proteo enz	:	proteolytic enzymes
SMC	:	smooth muscle cells

References

Abeloff, M.D., J.O. Armitage, A.S. Lichter and J.E. Niederbuber. 2000. Clinical Oncology. Churchill Livingstone, New York. 4th edn. 528–625.

Apple, F.S., et al., 2005. Future biomarkers for detection of ischemia and risk stratification in acute coronary syndrome. Clin. Chem. 51(5): 810–824.

Berney, C.H., J. Alderman, W.A. Lane and J.K. Collins. 1998. Development of capacitive immunosensor: a comparison of monoclonal and polyclonal capture antibodies as the primary layer. J. Mol. Recog. 11: 175–177.

Berggren, C., B. Bjarnason and G. Johansson. 2001. Capacitive biosensors. Electroanalysis 13(3): 173–179.

Block, I.D., L.L. Chan and B.T. Cunningham. 2006. Photonic crystal optical biosensor incorporating structured low-index porous dielectric, sensors and actuators. B: Chemical 120: 187–193.

Bothara, M., V. Venkatraman, R.K.K. Reddy, T. Barrett, J. Carruthers, S. Prasad. 2008. "Nanomonitors: electrical immunoassays for protein biomarker profiling", Nanomedicine. 3(4): 423–436.

Chen R.J., H.C. Choi, S. Bangsaruntip, E. Yenilmez, X. Tang, Q. Qian Wang, Y.L. Chang and H. Dai. 2004. An Investigation of the mechanisms of electronic sensing of protein adsorption on carbon nanotube devices. J. Am. Chem. Soc. 126(5): 1563–1568.

Chou, S.F., W.L. Hsu, J.M. Hwang and C.Y. Chen. 2004. Development of an immunosensor for human ferritin, a nonspecific tumor marker, based on surface plasmon resonance. Biosens. Bioelectron. 19: 999–1005.

Dancil, K., D.P. Greiner and M.J. Sailor. 1999. A porous silicon optical biosensor: detection of reversible binding of IgG to a protein A-modified surface. J. Am. Chem. Soc. 121(34): 7925–7930.

Danesh, J., R. Collins, P. Appleby and R. Peto. 1998. Association of fibrinogen, C-reactive protein, albumin, or leukocyte count with coronary heart disease: metaanalyses of prospective studies. JAMA 279(18): 1477–1482.

Darain, F., D.S. Park, J.S. Park and Y.B. Shim. 2004. Development of an immunosensor for the detection of vitellogenin using impedance spectroscopy. Biosens. Bioelectron. 19: 1245–1252.

DeSilva, M.S., Y. Zhang, P.J. Hesketh, G.J. Maclay, S.M. Gendel and J.R. Stetter. 1995. Impedance based sensing of the specific binding reaction between Staphylococcus enterotoxin B and its antibody on an ultra-thin platinum film. Biosens. Bioelectron. 10(8): 675–682.

Etzioni, R., et al., 2003. The case for early detection. Nat. Rev. Cancer 3: 243–252.

Ferri, N., R. Paoletti and A. Corsini. 2006. Biomarkers for atherosclerosis: pathophysiological role and pharmacological modulation. Curr. Opin. Lipidol. 17(5): 495–501.

Hahm, J. and C.M. Lieber. 2004. C. Direct ultrasensitive electrical detection of DNA and DNA sequence variations using nanowire nanosensors. Nano Lett. 4(1): 51–54.

Hyun, Y.Z., M. Heidaran, A. Liebmann-Vinson, T.P. Beebe and A. Chilkoti. 2001. Microstamping on an activated polymer surface: Patterning biotin and streptavidin on common polymeric biomaterials, Langmuir 17: 6358–6367.

Jaffe, A.S., L. Babuin and F.S. Apple. 2006. Biomarkers in acute cardiac disease: the present and the future. J. Am. Coll. Cardiol. 48(1): 1–11.

Libby, P., and P. Theroux. 2005. Pathophysiology of coronary artery disease. Circulation 111(25): 3481–3488.

Naghavi, M., et al. 2003. From vulnerable plaque to vulnerable patient: a call for new definitions and risk assessment strategies: Part II. Circulation 108(15): 1772–1778.

Nagaraj, V.J., S. Aithal, S. Eaton, M. Bothara, P. Wiktor and S. Prasad. 2010. Nanomonitors: miniature electronic biosensors for glycan biomarker detection, Nanomedicine. 5(3): 369–378.

Nam, J.M., C.S. Thaxton and A. Mirkin. 2004. Nanoparticle-based bio-bar codes for the ultrasensitive detection of proteins. Science 301: 1884–1886.

Niwa, O., M. Morita and H. Takei. 1996. Electrochemical behavior of reversible redox species at interdigitated array electrodes with different geometries: consideration of redox cycling and collection efficiency. Anal.Chem. 62: 447–456.

Norman, J.J. and T.A. Desai. 2006. Methods for fabrication of nanoscale topography for tissue engineering scaffolds. Ann. Biomed. Eng. 34: 89–101.

Prasad, S., X. Zhang, C.S. Ozkan and M. Ozkan. 2004. Neuron-based microarray sensors for environmental sensing. Electrophoresis 25(21–22): 3746–60.

Prasad, S. and J. Quiano. 2004. Development of nanostructured biomedical micro drug testing device based on *in-situ* cellular activity monitoring. Biosens. Bioelectron. 21(7): 1219–1229.

Reddy, R.K., S. Prasad, T. Barrett, J. Carruthers. 2008a. Electrical Immunoassays towards clinical diagnostics: Identification of Vulnerable cardiovascular plaque, Journal of the Association for Laboratory Automation. 13(1): 33–39.

Reddy, R.K., M.G. Bothara, T.W. Barrett, J. Carruthers and S. Prasad. 2008b. Nanomonitors: Protein Biosensors for Rapid Analyte Analysis, IEEE Sensors Journal. 8(6): 720–723.

Takhistov, P. 2004. Electrochemical synthesis and impedance characterization of nanopatterned biosensor substrate. Biosens. Bioelectron. 9(11): 1445–1456.

Vasan, R.S. 2006. Biomarkers of cardiovascular disease: molecular basis and practical considerations. Circulation 113(19): 2335–2362.

Wang, J. 2000. Analytical electrochemistry. Wiley-VCH New York.

Wang, J. 2005. Nanomaterial-based electrochemical biosensors. Analyst, 130(4): 421{6.

Willner, I., A.N. Shipway and B. Willner. 2003. Layered nanoparticle architectures on surfaces for sensing and electronic functionalities. In: M. Lieberman (ed.). Molecules as Components of Electronic Devices. ACS Symposium Series. 844, Washington, DC, Chapter 8: 88–105.

Zimmerman, S.B. and A.P. Minton. 1993. Macromolecular crowding: biochemical, biophysical, and physiological consequences. Annu Rev Biophys Biomol Struct. 22: 27{65.

Zheng, G., F. Patolsky, Y. Cui, W.U. Wang and C.M. Lieber. 2002. Multiplexed electrical detection of cancer markers with nanowire sensor arrays. Nature 23(10): 1294–1301.

11

Nanospin Probes and Applications to Cardiology

Valery V. Khramtsov[1,a,*] and *Denis A. Komarov*[1,b]

ABSTRACT

Ischemia in a healthy myocardium produces a complex cascade of events resulting in alteration of the myocardial tissue microenvironment including low oxygen tension, severe tissue acidosis, and a higher reducing capacity. Monitoring the parameters of the myocardial microenvironment may provide important insight in the mechanisms of ischemia-reperfusion (I/R) injury, preconditioning and pharmacological interventions, and an opportunity to design new therapeutic approaches. Electron paramagnetic resonance (EPR) allows for non-invasive *in vivo* measurements of the parameters of myocardial microenvironment in animal I/R models and in isolated hearts. These functional EPR applications rely on advances in the design

[1]Dorothy M. Davis Heart & Lung Research Institute and Division of Pulmonary, Allergy, Critical Care & Sleep Medicine, Department of Internal Medicine, College of Medicine, The Ohio State University, 201 DHLRI, 473 W. 12th Avenue, Columbus, OH 43210, USA;
[a]E-mail: Valery.Khramtsov@osumc.edu
[b]E-mail: Denis.Komarov@osumc.edu
*Corresponding author

List of abbreviations after the text.

of specific paramagnetic probes, here termed "nanospins", which include soluble molecular spin probes, nano- and micro-crystalline particulate paramagnetic materials, and spin probes encapsulated in nano-sized structures. More specifically, we recently used the term "nanoSPINs" to abbreviate EPR sensors represented by nano-sized particles with incorporated nitroxides such as liposome-encapsulated pH-sensitive spin probes. This chapter reviews recent progress in the development of nanospin probes with functional sensitivity to oxygen, pH, tissue redox state and glutathione, the parameters critically affected in myocardial tissue during ischemia and reperfusion. The EPR capacity for non-invasive real time monitoring of these parameters is illustrated by applications to animal models of I/R and in isolated hearts.

INTRODUCTION

With the development of low-field EPR techniques aimed to increase penetration depth of the microwaves in aqueous samples, *in vivo* biomedical EPR applications became feasible. However, *in vivo* EPR exclusively relies on the use of exogenous paramagnetic probes (Khramtsov and Zweier 2010). In the last decade the progress in design of the spin probes with functional sensitivity and enhanced stability in living tissues resulted in a number of *in vivo* functional EPR applications for cardiology. These functional spin probes, here termed "nanospins", include (1) soluble molecular spin probes, such as nitroxide radicals (NRs) and triarylmethyl radicals (TAMs), (2) particulate paramagnetic probes, and (3) spin-labeled macromolecules and encapsulated spin probes. Specifically, we recently used the term "nanoSPINs" to abbreviate EPR sensors represented by nano-sized particles with incorporated nitroxides such as liposome-encapsulated pH-sensitive spin probes (Woldman et al. 2009). In this chapter the application of nanospins for EPR monitoring of myocardial oxygenation, redox, pH and glutathione is overviewed.

MYOCARDIAL EPR APPLICATIONS OF NANOSPIN OXYGEN-SENSITIVE PROBES

After the invention of magnetic resonance imaging (MRI), oxygen was the primary candidate to be considered as potential contrast agent, but this possibility was ruled out because of the small effect of oxygen on the relaxation times of water protons. Several nuclear magnetic resonance (NMR) oximetric approaches were developed that use effects dependent on blood oxygen level, or use exogenous probes such as perfluorocarbon emulsions and fluorinated nitroimidazoles. However, these methods still suffer from low intrinsic NMR sensitivity and difficulties with quantitative data interpretation. EPR oximetry is one of the most promising and rapidly developing techniques for measurement of oxygen in living tissues (Swartz 2004). The EPR oximetry is based on Heisenberg exchange between the diradical molecule of oxygen and a spin probe. The exchange-induced EPR line broadening of the probe is proportional to oxygen concentration. An advantage of EPR oximetry is that it does not interfere with the oxygen metabolism, therefore providing a basis for non-invasive oxygen measurements in biological systems, including those *in vivo*. The distribution of the spin probes in living tissue can be measured using EPR imaging (EPRI) techniques. Moreover, spectral-spatial EPRI allows for extracting an oxygen concentration map and spin probe distribution. To date, to obtain the highest sensitivity and best quality images, minutes to hours of time are required with common continuous wave EPRI. Because of technical restrictions, pulsed EPR approaches are limited to paramagnetic probes with long relaxation times. Recent technical advances made *in vivo* pulsed EPRI at 300 MHz feasible using nitroxides with narrow EPR lines (Hyodo et al. 2009). An alternative imaging modality that employs unpaired electrons is proton-electron double-resonance imaging (PEDRI). Using the PEDRI approach, the EPR signal spatial distribution is reconstructed from the enhanced NMR signal of water protons after EPR irradiation of the paramagnetic probe resulting in a transfer of polarization from the electrons to protons by Overhauser effect. Since PEDRI is based on the proton MRI, it inherently offers high spatial resolution and rapid image data acquisition. PEDRI with variable saturation power provides the method for mapping of oxygen concentrations *in vivo* (Krishna et al. 2002).

A number of non-invasive EPR applications to myocardial oxygenation studies in animals and in isolated hearts were developed and are systemized below on the basis of type of oxygen-sensitive nanospin probes.

Nitroxides

NRs were the first paramagnetic probes used for EPR oximetry (Backer et al. 1977). The line-broadening effect of oxygen on the peak-to-peak linewidth, ΔH_{pp}, of the NRs varies slightly with the structure and on average is about 500 mG/100% oxygen. In general, NRs have low toxicity and can be administered to an animal by infusion or injection.

Figure 1 represents L-band (1.2 GHz) EPR oximetry application using 2,2,6,6-tetramethylpiperidine-1-oxyl nitroxide (TEMPO) for monitoring ischemia-induced oxygen depletion in isolated rat hearts

(Zweier and Kuppusamy 1988; Kuppusamy et al. 1994). The spectral data showed a gradual decrease in the linewidth over the duration of ischemia approaching the linewidth observed in the absence of oxygen. Oxygen consumption observed in cardiopleged hearts subjected to global ischemia (Kuppusamy et al. 1994) was significantly slower than in non-cardiopleged hearts (Zweier and Kuppusamy 1988). This would be expected since contractile function of cardiopleged hearts is arrested. The oxygen consumption data for cardiopleged heart were obtained from the spectral-spatial EPRI, which allowed for spatially resolved oxygen mapping but required long acquisition times from 12 to 16 min (Fig. 1a). PEDRI provides a faster alternative for oxygen mapping. Application of PEDRI for myocardium oxygen mapping was first demonstrated in perfused sheep heart using a high concentration, 4 mM, of Fremy's salt nitroxide (Grucker and Chambron 1993).

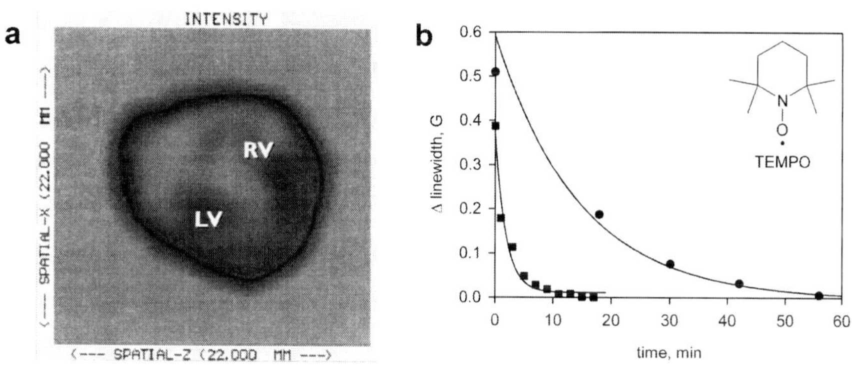

Fig. 1. EPR-measured 2D-spatial image of the rat heart and kinetics of myocardial oxygen depletion during ischemia. (a) Cross-sectional transverse 2D-spatial EPR image of the isolated rat heart preloaded with 1 mM TEMPO (field of view, 22×22 mm^2; acquisition time, 16 min). The structure of the left ventricle (LV) and right ventricle (RV) are observed. (b) Changes in L-band EPR linewidth of TEMPO spin label induced by oxygen depletion during ischemia in cardiopleged (•) and non-cardiopleged (■) hearts. The hearts were perfused by the method of Langendorff with Krebs buffer equilibrated with 95% O_2 and 5% CO_2. (Adapted from Zweier and Kuppusamy 1988; Kuppusamy et al. 1994; copyright National Academy of Sciences, U.S.A.)

Insensitivity of the NRs to oxygen tension below 50 mmHg and limited stability in living tissues makes NR-based quantitative measurements of myocardial oxygenation difficult. Furthermore, concentration-induced line-broadening effects complicate quantitative oximetric applications of the NRs. Typically, the self-broadening effect is about 100–200 mG/mM of NRs and may interfere with accurate oxygen measurements at NR concentrations above 100 μM.

Application of deuterated NRs provides significant enhancement in sensitivity to low oxygen concentrations due to their narrow linewidths (Kuppusamy et al. 1994). However, deuterated NRs still suffer from the concentration-induced line broadening, which is difficult to disentangle from oxygen-induced broadening. To overcome this problem, Halpern et al. proposed selectively deuterated nitroxide with only one hydrogen hyperfine splitting. Increasing NR concentration but not oxygen results in narrowing of hydrogen hyperfine splitting. This allows for discrimination between oxygen- and concentration-induced line broadening and, therefore, for quantitative oxygen detection in living tissues with sensitivity about 10 mmHg (Halpern et al. 1994).

Encapsulated Nitroxides

Stability of nitroxides in living systems can be increased by their encapsulation in semipermeable capsules. Baker et al. (1997) applied encapsulated TEMPO derivatives for real time measurement of oxygen depletion in isolated immature rabbit hearts during global no-flow ischemia at 20°C. The hearts were perfused with a buffer containing lipid-soluble ^{14}N-TEMPO laurate encapsulated in perfluorocarbon micelles and positively charged ^{15}N-TEMPO choline encapsulated in the aqueous inner volume of phospholipid liposomes. The EPR spectra measured at high modulation amplitude were found to be oxygen-sensitive only for ^{14}N-TEMPO laurate due to the high oxygen solubility in hydrophobic micelles. This allowed for calculation of extracellular O_2 concentration from the ratio of peak intensities of oxygen-sensitive ^{14}N-TEMPO and oxygen-insensitive ^{15}N-TEMPO. Authors observed a significant drop in myocardial oxygenation from 1195 μM to 393 μM in 40 min and 61 μM in 60 min of ischemia, respectively.

TAM Oximetric Probes

TAM radicals were developed for biomedical applications by Nycomed Innovation AB (Anderson et al. 1996). TAMs are extraordinarily stable *in vivo* and have a very narrow single EPR line of about 100 mG or less (see Fig. 2 for the representative structures). The oxygen-induced line broadening of the TAMs in water is about 500 mG/mM of oxygen, similar to that for NRs; however, the concentration line broadening of the TAMs is about 10–30 mG/mM, which is one order of magnitude less than that for the NRs. These properties make TAMs superior oximetric probes for *in vivo* EPR/EPRI and PEDRI applications.

Recently we reported (Dhimitruka et al. 2010) a synthetic procedure for large-scale synthesis of deuterated triarylmethyl radical, dTAM. The

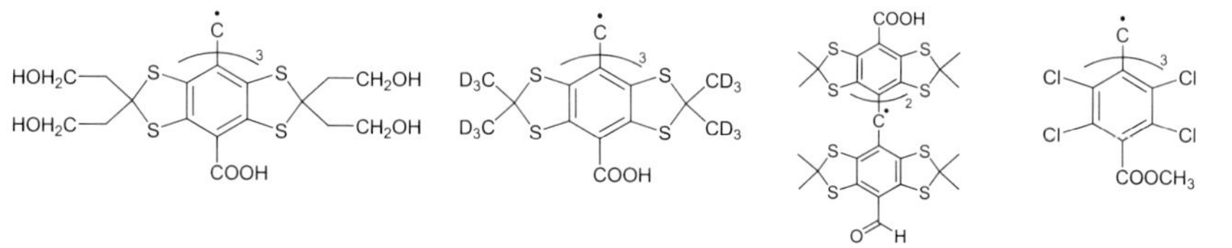

Fig. 2. Representative structures of oxygen-sensitive TAM probes.

probe has an EPR linewidth about 50 mG and high oxygen sensitivity, 0.53 mG/mmHg of O_2, allowing for detection of oxygen tensions as low as 1 mmHg. We used the probe to study myocardial tissue oxygenation in isolated perfused rat hearts. The hearts were placed directly in the resonator of an L-band EPR spectrometer equipped with automatic coupling and tuning controls as shown in Fig. 3a. For EPR oximetric measurements the hearts were perfused for 15 min to achieve stabilization of contractile parameters followed by addition of 100 μM of dTAM to the perfusate buffer. The presence of dTAM in the perfusate buffer did not affect contractile parameters of the hearts, suggesting low if any toxicity of the probe. The measured myocardial oxygen concentration was about 160 mmHg before ischemia and dropped to 2 ± 1 mmHg in less than one minute of global no-flow ischemia. The myocardial oxygen concentration only partially recovered to about 50 mmHg in 5 min of reperfusion. Apparently a major factor in lower oxygen concentration after ischemia was decreased oxygen supply due to reduction in perfusate flow rate from 16 ml/min before ischemia to 8 ml/min after ischemia as a consequence of I/R damage to the myocardium.

Recently we synthesized triarylmethyl radical with only one resolved hydrogen hyperfine splitting, TAM-H (Fig. 2) (Bobko et al. 2009). This probe demonstrates enhanced oxygen sensitivity (see Fig. 4) similar to that described by Halpern et al. (1994) for selectively deuterated nitroxide. TAM-H has a much higher sensitivity to oxygen (about 1 mmHg) compared with that for the nitroxide (about 10 mmHg) due to more narrow linewidth.

As with NRs, encapsulation of TAMs may be useful to improve their oxygen sensitivity, aqueous solubility and biocompatibility. Sostaric et al. (2007) used high-power ultrasound (20 kHz) to encapsulate a hexamethyldisiloxane solution of water-insoluble perchlorinated TAM (Fig. 2) into a polymerized shell of bovine serum albumin. The formed microspheres of 0.5–3 μm diameter were water-soluble and possessed high oxygen sensitivity. The microspheres were employed to measure oxygen tension in the muscle and tumor tissue of living mice.

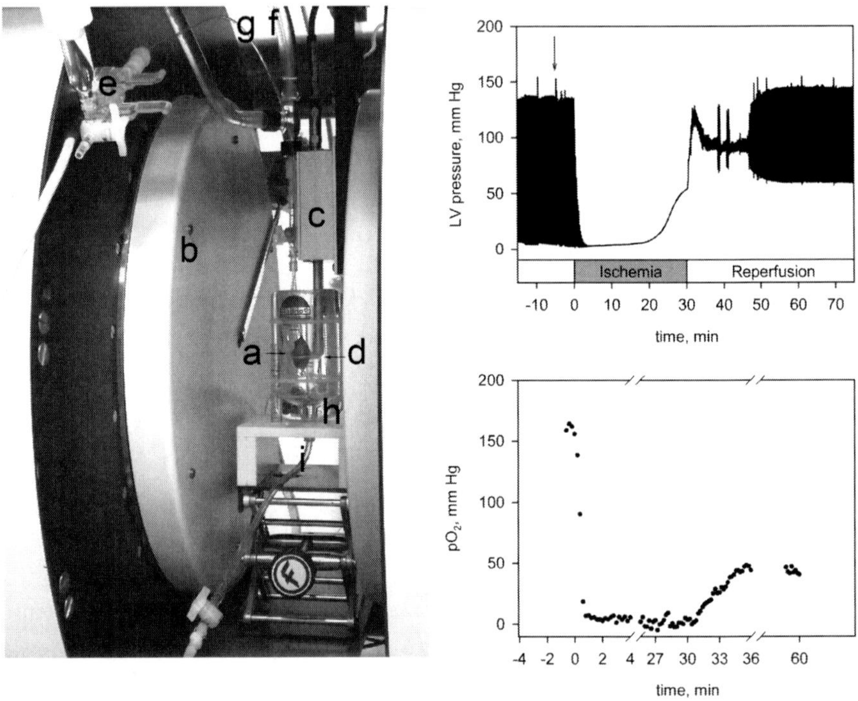

Fig. 3. L-band EPR measurements of oxygen in isolated perfused rat heart. Experimental setup allowing for heart perfusion, contractile parameter measurements and EPR data acquisition: (a) rat heart; (b) EPR magnet; (c) resonator; (d) resonator loop; (e) pressure transducer; (f) perfusate line; (g) hydraulic line; (h) temperature-controlled bath; (i) perfusate drain. Isolated rat hearts were perfused with Krebs buffer equilibrated with 95% O_2 and 5% CO_2. The hearts were maintained at 37°C and relative humidity of approximately 100%. Contractile parameters of the hearts were continuously monitored using a water-filled latex balloon inserted into the left ventricle and attached via hydraulic line to a pressure transducer. The top graph shows the left ventricular pressure of the isolated perfused rat heart. The heart was subjected for 30 min of global no-flow ischemia followed by reperfusion. Arrow indicates an addition of oxygen-sensitive probe to the perfusate (dTAM, 100 μM). The EPR spectra were collected (acquisition time, 10 sec) and fitted with Voigt function. Myocardial tissue oxygen tensions were calculated from Lorentz linewidth using a calibration curve. Unpublished data.

Particulate Oxygen Probes

Several types of particulate probes have been used for EPR oximetry, among them naturally occurring coal and its derivatives, synthetic carbon-based amorphous materials, and crystalline materials. The last type is represented by crystals and microcrystalline powders of lithium phthalocyanine (LiPc) and its derivatives (see Fig. 5). Following the first report by Liu et al. in

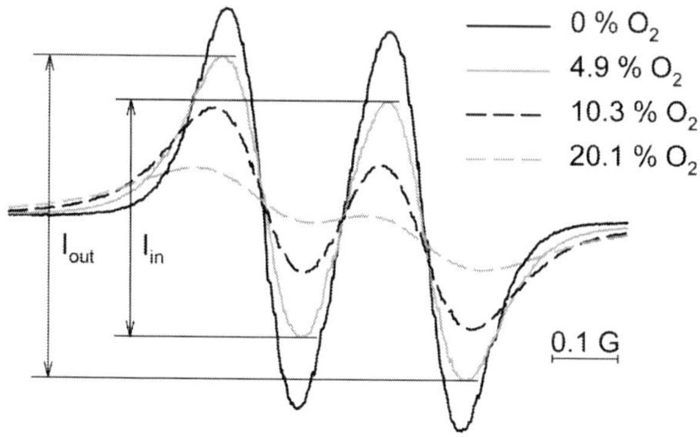

Fig. 4. Dependence of EPR spectrum of TAM-H on concentration of oxygen. EPR spectra of 50 μM TAM-H in aqueous solution at various oxygen concentrations are shown. The ratio of peak intensities, I_{out}/I_{in}, represents a highly oxygen-sensitive parameter. (From Bobko et al. 2009, with permission from Elsevier.)

1993 the LiPc crystals were widely recognized as important EPR oximetric probes. The LiPc probes possess an extremely narrow single line EPR spectrum (5–26 mG) highly sensitive to partial pressure of oxygen in the environment (6.1–9.56 mG/mmHg of O_2). However, their use is limited because of EPR saturation at microwave power above 1 mW, requirement of low modulation frequency, and dependence of the oxygen sensitivity on crystal size. Kuppusamy and colleagues developed several derivatives of LiPc and naphthalocyanine, namely LiPc-OPh (Pandian et al. 2007), LiNc (Pandian et al. 2009) and LiNc-OBu (Khan et al. 2009) (see Fig. 5). These radicals showed certain advantages over LiPc, especially with respect to oxygen sensitivity, power saturation, and nanoparticle formation.

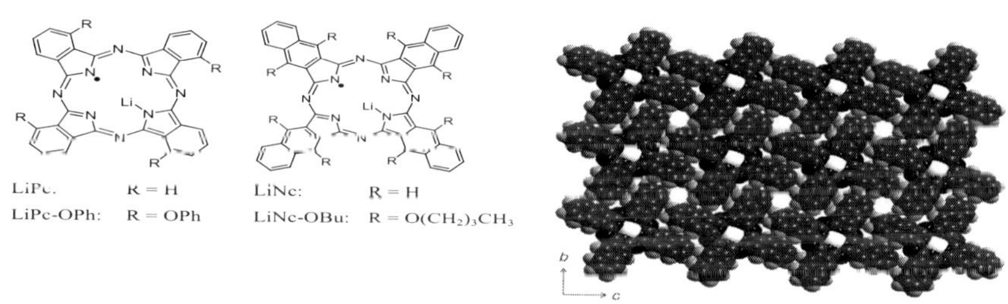

Fig. 5. Molecular structures of LiPc, LiPc-OPh, LiNc and LiNc-OBu, and three-dimensional representation of the packing of LiNc molecules. A structural framework possesses long, large channels of 5.0×5.4 Å^2 running parallel to the a-axis of packing and enabling diffusion of oxygen molecules through the channel. The white spheres represent lithium, the black spheres nitrogen, the dark grey carbon and the light grey hydrogen. (From Pandian et al. 2009, reproduced by permission of the Royal Society of Chemistry.)

The microcrystalline powders of LiPc-OPh, LiNc and LiNc-OBu exhibit a single, sharp anoxic EPR line (210–630 mG) with extraordinarily high oxygen-dependent line broadening of 8.5–31.2 mG/mmHg of O_2. Figure 5 exemplifies the three-dimensional packing of LiNc molecules in a crystal revealed by X-ray diffraction studies. The crystals of LiNc possess long channels enabling free oxygen diffusion. Intermolecular spin exchange between the LiNc probe and molecular oxygen results in shortening of the spin-spin relaxation time and broadening of the EPR spectrum. Note that microcrystalline preparations showed variations in anoxic linewidth and oxygen sensitivity and, therefore, calibration is required for each batch. A reduction in the size of particles from the raw preparations by sonication provides particles of uniform size and oxygen sensitivity. Thus suspension of LiNc sonicated at 22.5 kHz showed particle size distribution from 120 to 500 nm with a maximum at approximately 330 nm, and linear oxygen sensitivity about 27 mG/mmHg of O_2 (Pandian et al. 2009).

Particulate probes have been used in numerous I/R model studies of myocardial oxygenation both in isolated hearts and *in vivo*. Figure 6 demonstrates typical pO_2 measurements performed in living rats using L-band EPR spectrometer (Khan et al. 2009). The regional ischemia was induced by ligating the left anterior descending coronary artery for 30 min, followed by reperfusion for 1 or 24 h by releasing the ligation.

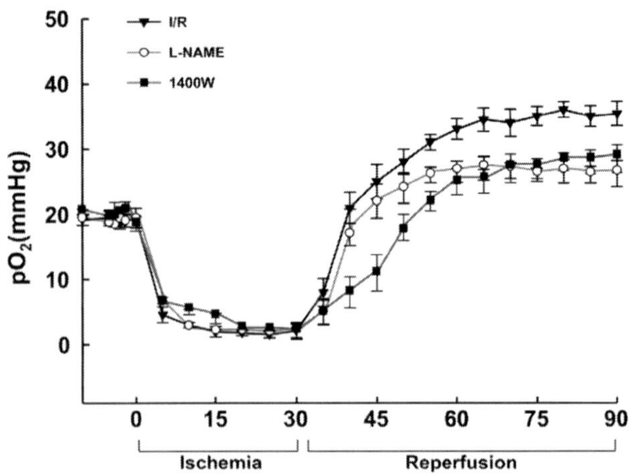

Fig. 6. Myocardial pO_2 monitoring using EPR oximetry in a model of regional ischemia in living rats. The animals pretreated with vehicle (I/R), L-NAME, or 1400W were anesthetized with isoflurane, and microcrystals of LiNc-OBu were implanted in the left ventricular mid-myocardium. Regional ischemia was induced by temporarily ligating the left anterior descending coronary artery (LAD). Data represent mean ± S.D., obtained from six animals/group. (Adapted from Khan et al. 2009, with permission from Mary Ann Liebert, Inc. publishers.)

Microcrystals of LiNc-OBu were implanted in the ischemic site, left ventricular mid-myocardium, using a 25-gauge needle. The EPR measured myocardial pO_2 of 19 ± 2 mmHg fell to about 2 mmHg after initiation of ischemia. The reperfusion resulted in an overshoot of pO_2 to about 35 mmHg. The observed overshoot of myocardial pO_2 was attributed to lower oxygen demand by the damaged myocardial tissue. Additionally, authors demonstrated involvement of nitric oxide in the oxygen overshoot phenomenon. Oral administration of NO synthase inhibitor, L-NAME, and specific inhibitor of inducible NO synthase, 1400W, in drinking water for 3 d prior to induction of I/R resulted in a significantly lower level of hyperoxygenation (27 ± 3 mmHg) during reperfusion.

The advantages of solid or particulate probes for EPR oximetry are high functional sensitivity, stability in living tissue and minimal toxicity. The particulate probe may remain in the implanted location for long time periods, enabling repeated measurements of tissue pO_2 for days and weeks after implantation. On the other hand, soluble probes provide an opportunity for spatial-resolved pO_2 measurements using EPR-based imaging techniques.

MYOCARDIAL EPR APPLICATION OF NANOSPIN pH-SENSITIVE PROBES

The crucial role of tissue acidosis in myocardial I/R is well documented but is still far from understood and contains many areas of controversy. ^{31}P-NMR of inorganic phosphate is the most widely used approach for non-invasive myocardial pH detection. However, this method has certain limitations, which are rarely discussed, including lack of resolution (about 0.2–0.3 units and even less at lower pH), variability of phosphate concentrations with metabolism, and dependence of its chemical shift on ionic strength. In our lab we developed a large number of soluble pH-sensitive nitroxide (Khramtsov and Volodarsky 1998; Khramtsov 2005) and triarymethyl (Bobko et al. 2007; Dhimitruka et al. 2008) probes for EPR spectroscopy and imaging applications in living tissue.

Figure 7 shows representative structures of pH-sensitive nitroxide and triarymethyl radicals. In aqueous solution these probes exist in two ionization states, R and RH^+ (or RH and R^-), because of the presence of ionizable group in their structure. The ratio of concentrations of the ionization states is described by the Henderson-Hasselbalch equation, $[H^+] = K_a[RH^+]/[R]$, where K_a is an equilibrium constant. This provides the basis for EPR measurements of pH around the probe pK_a. In the case of imidazoline NRs, protonation of nitrogen atom N-3 of heterocycle results in a significant decrease in nitrogen hyperfine splitting ($\Delta a_N \approx 0.8$ G), which

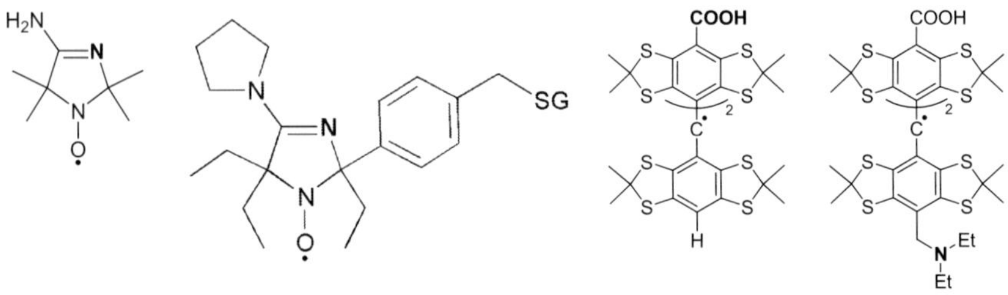

ATI R-SG (GSH=glutathione) cTAM-H aTAM

Fig. 7. Representative structures of pH-sensitive nitroxide and TAM probes. The presence of ionizable groups in the structures of the probes (shown in bold) is responsible for pH sensitivity of their EPR spectra.

has been used as a pH-sensitive parameter in numerous applications (Khramtsov 2005). Recently we developed several pH-sensitive TAM probes (Fig. 7). Deprotonation of cTAM-H probe carboxyl group ($pK_a \approx 2.6$) results in a decrease of the doublet splitting from a single hydrogen atom (Δa_H = 23 mG) of the aromatic ring (Bobko et al. 2007). The other probe, aTAM, exhibits an EPR spectrum that is sensitive to pH in physiologically relevant range from 6.8 to 8.8 (Dhimitruka et al. 2008). Protonation of aTAM amino group results in a large decrease of hyperfine splitting from one of the methylene hydrogens ($\Delta a_H \approx 1$ G). However, a disadvantage of the aTAM probe is its comparatively complex multiplet EPR spectrum.

We used the nitroxide pH probes, ATI (pK_a 6.1) and R-SG (pK_a 6.6, see Fig. 8a), to monitor ischemia-induced acidosis in isolated rat hearts. R-SG probe demonstrated excellent EPR signal stability, while 90% of the ATI probe was reduced in myocardial tissue within 5 min. The enhanced stability of the R-SG probe is due to the bulky substitutes introduced around the paramagnetic NO fragment. Previous studies showed that this substitution resulted in about 30-fold increase of the nitroxide radical lifetime when compared with unsubstituted analogs (Kirilyuk et al. 2004). Both probes reported similar myocardial acidification during ischemia but the ATI probe was informative only for the first 5 min (see Fig. 8b). As seen in Fig. 8b, ischemic preconditioning improved pH homeostasis during the global no-flow ischemia. Note that R-SG probe is bound to the highly hydrophilic membrane-impermeable tripeptide, glutathione, and reports extracellular pH value. The ATI probe easily penetrates the phospholipid membrane barrier and apparently reports averaged extra- and intracellular pH. Similar kinetics of myocardial acidification observed by the two probes support fast pH equilibration between extra- and intracellular spaces in agreement with data obtained previously using glass microelectrodes and ^{31}P-NMR. We did not observe any influence of the pH probes on the contractile function of the hearts supporting low toxicity of the NRs.

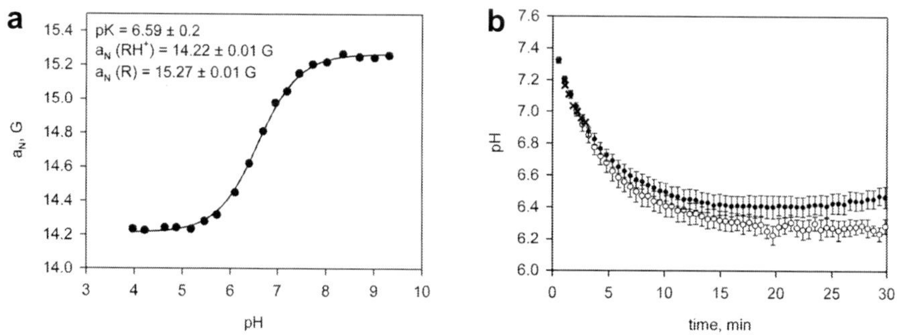

Fig. 8. L-band EPR measurements of ischemia-induced acidosis in isolated rat hearts. (a) Titration curve of the hyperfine splitting, a_N, of the nitroxide pH probe, R-SG. (b) Changes of myocardial pH during global no-flow ischemia in control (○) and ischemic preconditioned hearts (●) as measured using R-SG spin probe. The data obtained during first five minutes of ischemia using ATI probe are denoted by (×) symbols. Isolated rat hearts were perfused for 30 min before the onset of ischemia. Preconditioned hearts were subjected to two episodes of 5 min ischemia followed by 5 min of reperfusion. The probes (1.5 mM) were infused into the hearts through side arm of the perfusate line upon onset of global ischemia and EPR spectra were continuously recorded; acquisition time, 20 sec. Data are mean ± S.E.; n = 6. Unpublished data.

The stability of the NRs in the biological milieu is critical for *in vivo* applications. To protect the NRs against bioreduction we designed nano-sized particles with incorporated nitroxides, or nanoSPINs (Woldman et al. 2009). Figure 9a illustrates the concept of nanoSPINs based on phospholipid liposomes. Application of the pH-sensitive nanoSPINs in a rat heart homogenate shows protection of the encapsulated NR against bioreduction (Fig. 9b) and allows for monitoring of ischemia-induced acidosis (Fig. 9c). The extraordinary stability of the nanoSPINs may be of significant advantage for pH monitoring in living animals, e.g., in the model of regional ischemia.

Spatially resolved *in vivo* pH measurement is critically important in allowing identification of the acidosis-affected area, particularly in the case of regional myocardial ischemia. Continuous wave EPRI allows for pH mapping but requires long acquisition times from several minutes to hours in the case of 4D (3-spatial, 1-spectral) imaging. Recently we made significant progress towards practical pH mapping of living tissue using new functional PEDRI approaches. Variable field (Khramtsov et al. 2010) and variable radio frequency (Efimova et al. 2010) PEDRI measure the enhanced NMR signal of water protons after EPR irradiation of the paramagnetic probe at pre-selected EPR fields or frequencies, correspondingly. The latter approach allowed for pH mapping of the mouse-sized phantom with only 6.8 sec of total acquisition time (Efimova et al. 2010). This approach may find applications for pH mapping of living tissues, e.g., animal myocardium, and it has potential for applications to humans.

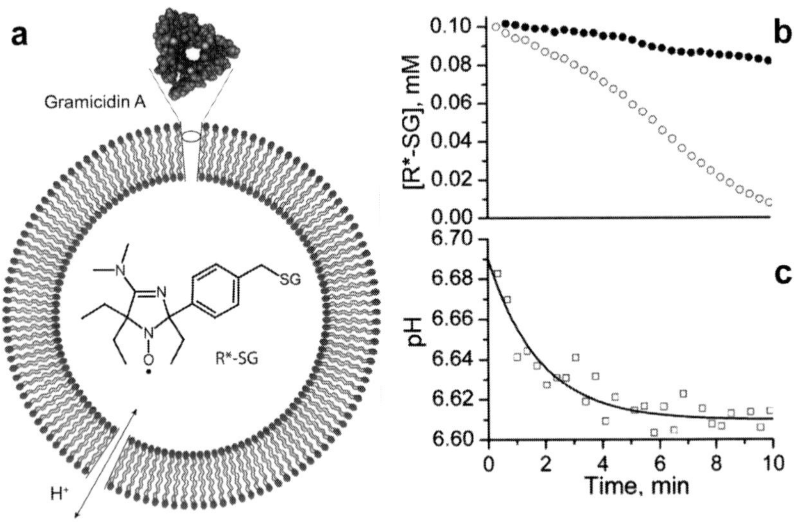

Fig. 9. NanoSPINs and their application for EPR monitoring of ischemia-induced acidosis in rat heart homogenate. (a) Schematic design of pH-sensitive nanoSPIN: the phospholipid membrane of the liposome with incorporated gramicidin A provides selective permeability to H^+, while protecting encapsulated nitroxide, R*-SG, from biological reductants. (b) Kinetics of the reduction of free (○) and encapsulated (•) R*-SG probe in the rat heart homogenate measured from its EPR signal intensity decay. (c) Ischemia-induced pH decrease calculated from the EPR spectra of the encapsulated R*-SG. (From Woldman et al. 2009, reproduced by permission of The Royal Society of Chemistry.)

Color image of this figure appears in the color plate section at the end of the book.

EPR ASSESSMENT OF MYOCARDIAL REDOX AND INTRACELLULAR GLUTATHIONE

Regulation of tissue redox status is important for maintenance of normal physiological conditions in the living body. The redox couple of glutathione, GSH, and its disulfide form, GSSG, is considered to be a major intracellular redox buffer (Schafer and Buettner 2001). Ischemic heart disease represents a pathological condition with compromised redox state. Alterations of tissue redox status, and particularly of intracellular GSH, are of central importance in the pathogenesis of I/R injury. Several available techniques for measuring tissue redox status, such as cytochrome c reduction and lucigenin-enhanced chemiluminescence, can not be applied *in vivo*. EPR spectroscopy using specifically designed nanospin probes provides an opportunity for *in vivo* assessment of both tissue redox status and intracellular GSH content.

Nitroxide Redox Probes

In general, EPR-measured rates of NR reduction to EPR-silent hydroxylamines provide information on overall tissue redox status (Kuppusamy and Krishna 2002). The reduction of the membrane-permeable NRs *in vivo* is primarily intracellular and is determined by intracellular redox state.

Figure 10 demonstrates the application of EPR spectroscopy using the NR probe to measure *in vivo* myocardial tissue redox status (Zhu et al. 2007) in a mouse model of regional ischemia. Authors observed about two-fold ischemia-induced increase of the overall myocardial reducing capacity in the area of risk. On the other side, the NR reduction rates in reperfused hearts were about 30% lower than in pre-ischemic hearts. The latter correlated with oxygen overshoot of 26.2 mmHg in post-ischemic hearts over pre-ischemic value of 16.3 mmHg observed by EPR using LiPc probe (Zhu et al. 2007). Both lower redox potential and higher oxygen may result in elevated production of reactive oxygen species and related myocardial damage. Authors observed that ischemic preconditioning normalizes both post-ischemic myocardial oxygenation and tissue redox status, which may contribute to the mechanisms of myocardial protection.

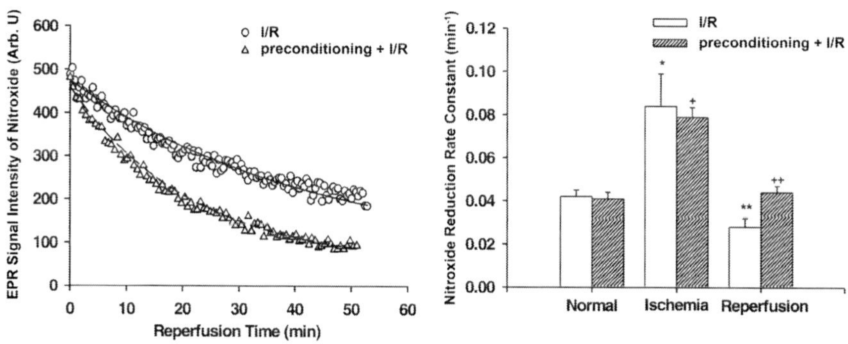

Fig. 10. *In vivo* EPR measurements of myocardial tissue redox status in a mouse model of regional ischemia and reperfusion. Regional ischemia was initiated by LAD occlusion. Animals from control I/R group were subjected to 30 min ischemia and 60 min reperfusion; animals from preconditioned I/R group were subjected to three cycles of 5 min LAD occlusion followed by 5 min reperfusion with the last reperfusion prolonged to 15 min before main course of ischemia. The probe, 2, 2, 5, 5-tetramethyl-3-carboxylpyrrolidine-1-oxyl (5 μl, 10 mM), in phosphate buffer was injected in the area at risk. Left: Representative EPR signal intensity decays for the control (○) and preconditioned groups measured during reperfusion (Δ). Right. Reduction rate constants of the nitroxide before, during, and after LAD occlusion in control I/R and preconditioned I/R mice. Data are means ± SE, n = 7/group. (From Zhu et al. 2007, with permission from the American Physiological Society.)

Specific GSH-sensitive Spin Probes

GSH redox status might be a useful marker of possible oxidative damage. To date, EPR represents the only reported *in vivo* GSH assay, first applied in our lab, for detection of GSH content in tumor tissue in living mice (Roshchupkina et al. 2008). The method is based on use of the disulfide nitroxide biradical labels, $R^{\bullet}SSR^{\bullet}$, where $R^{\bullet}$ is imidazolidine nitroxide radical. Being a paramagnetic analog of widely used Ellman's reagent, $R^{\bullet}SSR^{\bullet}$ reacts with GSH via the reaction of thiol-disulfide exchange, $R^{\bullet}SSR^{\bullet} + GSH \rightarrow R^{\bullet}SSG + R^{\bullet}SH$. The splitting of the disulfide bond results in the formation of two monoradicals. EPR spectrum of the biradical $R^{\bullet}SSR^{\bullet}$ significantly differs from the triplet pattern of the mononitroxides due to the appearance of additional biradical components as consequence of intramolecular spin exchange between two radical fragments. Therefore, reaction with GSH results in decrease of EPR intensity of biradical spectral components and simultaneous increase of the monoradical ones (see Fig. 11). The rate of initial increase of monoradical spectral component is proportional to GSH concentration. The EPR approach for *in vivo* intracellular GSH measurement is based on fast diffusion of the lipophilic $R^{\bullet}SSR^{\bullet}$ reagent across cellular membranes and predominant contribution of the GSH in fast-reacting low-molecular-weight intracellular thiols (Roshchupkina et al. 2008).

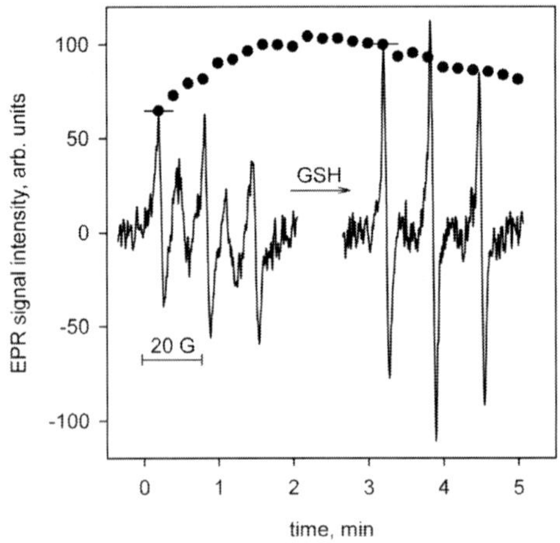

Fig. 11. L-band EPR measurement of myocardial tissue GSH using $R^{\bullet}SSR^{\bullet}$ probe. Isolated rat heart was perfused for 20 min with Krebs buffer equilibrated with 95% O_2 and 5% CO_2. Then the heart was cardiopleged, and $R^{\bullet}SSR^{\bullet}$ probe (10 µl, 30 mM) was injected intramuscularly. EPR spectra were acquired for each 12 sec. The spectra measured immediately and 3.2 min after injection of the probe demonstrate the increase of the monoradical signal intensity. The corresponding kinetics of the low-field component intensity is presented. Tissue GSH concentration evaluated from the initial rate of signal increase was found to be equal to 2.2 ± 0.4 mM. Unpublished data.

Recently we applied this approach for quantitative *in vivo* assessment of myocardial GSH. Figure 11 represents the kinetics of the EPR spectrum of $R^{\bullet}SSR^{\bullet}$ low-field monoradical component measured in isolated rat heart. The increase in spectral intensity during the first minutes of the kinetics as a result of the probe reaction with GSH was followed by signal decrease due to bioreduction of the probe. The calculated GSH concentration from *in vivo* measurements, 2.2 ± 0.4 mM, is in good agreement with GSH concentration, 2.0 ± 0.2 mM, measured *in vitro* using a tissue biopsy sample from the same heart. Injection of the probe into cardiac muscle did not affect significantly its contractile parameters, suggesting low toxicity of the probe in concentrations used in the experiments. Therefore, the approach represents a specific non-invasive method for *in vivo* assessment of tissue GSH concentration.

APPLICATIONS TO AREAS OF HEALTH AND DISEASE

Ischemic heart disease caused by insufficient blood supply to regions of the myocardium is a major form of cardiovascular disease and the leading cause of mortality in the United States, and in the world. Clinical interventions used to reintroduce blood flow to an ischemic region of the myocardium are accompanied by additional damage to the myocardium, or reperfusion injury. Myocardial ischemia produces a complex cascade of events resulting in alteration of the tissue microenvironment including low oxygen tension, severe tissue acidosis, and higher reducing capacity. The ischemia-induced alterations of myocardial microenvironment parameters and their restoration during reperfusion correlate with I/R injury to myocardium. Therefore, non-invasive *in vivo* monitoring of the parameters of myocardial microenvironment may provide important insight in the mechanisms of I/R injury, preconditioning and pharmacological interventions, and an opportunity for the design of new therapeutic approaches. EPR spectroscopy and imaging of specially designed paramagnetic probes, here termed "nanospins", provide unique opportunities for non-invasive real-time monitoring of these parameters in animal models of I/R and in isolated hearts.

Key Facts

- Good penetration of low-frequency electromagnetic waves in the living tissues makes low-field EPR technique feasible for non-invasive *in vivo* functional measurements.
- Cardiological applications of functional EPR spectroscopy and imaging include monitoring of the myocardial tissue microenvironment in

animal models of cardiovascular diseases and have potential for applications in humans.

- EPR measurements of the parameters of tissue microenviroment are based on application of specially designed paramagnetic probes, termed "nanospins", which include soluble molecular spin probes, nano- and micro-crystalline paramagnetic materials and spin probes encapsulated in nano-sized structures.
- More specifically, the term "nanoSPINs" abbreviates EPR sensors represented by nano-sized particles with incorporated nitroxides such as liposome-encapsulated pH-sensitive spin probes.
- Ischemic heart disease caused by insufficient blood supply to regions of the myocardium is a major form of cardiovascular disease and the leading cause of mortality in the United States, and in the world.
- *In vivo* EPR studies of the parameters of myocardial microenvironment during ischemia-reperfusion and pharmocological interventions provide an opportunity for the design of new therapeutic approaches.

Definitions

Electron paramagnetic resonance: A technique for spectroscopy and imaging of the paramagnetic species that have one or more unpaired electrons. Low-field EPR techniques allow sufficient penetration depth of the microwaves in living tissues, making biomedical EPR applications feasible. Since most stable molecules have all their electrons paired, *in vivo* EPR applications rely on exogenously added spin probes.

Ischemia-reperfusion model of ischemic heart disease: Models designed to study mechanisms of injury to myocardium caused by insufficient blood supply, followed by reintroduction of the blood flow to the ischemic region. The models include *in vivo* animal models of ischemia-reperfusion and isolated perfused hearts.

Nanospin: Paramagnetic probes designed for functional EPR applications. Those include soluble molecular spin probes, nano- and micro-crystalline paramagnetic materials and spin probes encapsulated in nano-sized structures.

Nitroxide radical: A stable organic radical with unpaired electron localized at the sterically protected NO group. The well-developed chemistry of the nitroxide radicals allows for the design of structures that vary in solubility, functionality and ability to be targeted.

Particulate paramagnetic probes: Several types of particulate probes have been used for EPR oximetry, among them naturally occurring coal and its derivatives, synthetic carbon-based amorphous materials, and crystalline materials. Particulate probes have the advantages of high spectral sensitivity to oxygen, stability in living tissue, and minimal toxicity. The particulate probe may remain in the implanted location for long time periods, enabling repeated measurements of tissue oxygenation days and weeks after implantation.

Triarylmethyl radical: Stable organic radicals with approximately 60% of electron spin density localized at the sterically protected central carbon atom and 40% delocalized over three adjacent aryl groups. Triarylmethyl radicals have certain advantages over nitroxides in extraordinary stability towards bioreduction and superior spectral properties. This makes triarylmethyl probes particularly attractive for imaging applications. Nitroxide and triarymethyl radicals are two major classes of soluble molecular spin probes.

Summary Points

- Electron paramagnetic resonance (EPR) allows for non-invasive *in vivo* monitoring of myocardial microenvironment parameters, such as tissue oxygenation, pH, redox state and glutathione.
- Functional EPR applications rely on the advances in design of specific paramagnetic probes, termed nanospins, which include soluble molecular spin probes, nano- and micro-crystalline paramagnetic materials, and spin probes encapsulated in nano-sized structures.
- EPR oximetric probes include soluble nitroxide and triarymethyl probes, and particulate probes such as naturally occurring coals, synthetic, carbon-based amorphous materials, and crystalline materials represented by nano- and micro-sized particles of lithium phthalocyanine and its derivatives.
- EPR pH-sensitive probes are represented by soluble triarymethyl and nitroxide probes, including liposome-based nanospins.
- Redox- and GSH-sensitive probes are represented by nitroxides and biradical disulfide nitroxides, respectively. The latter are paramagnetic analogs of Ellman's reagent allowing for non-invasive *in vivo* assessment of intracellular GSH.
- EPR oximetry using nanospin probes in *in vivo* ischemia-reperfusion models demonstrated the kinetics of ischemia-induced myocardial oxygen depletion. The reperfusion results in an overshoot in the oxygen concentration that correlates with the degree of myocardium recovery.

- By application of pH-sensitive spin probes it was shown that oxygen depletion during ischemia is followed by the myocardial tissue acidosis, which is partially improved by ischemic preconditioning.
- EPR redox measurements show significant increase in myocardial reducing capacity during ischemia. Reperfusion results in the decrease of the myocardium reducing capacity, supporting oxidative stress-related myocardial damage after reintroduction of oxygen to the ischemic regions.
- Measurements of intracellular GSH in perfused isolated rat hearts using disulfide biradical nitroxide have been demonstrated, therefore extending the functionality of EPR applications to living tissues.
- In conclusion, *in vivo* EPR using nanospin probes allows for multifunctional parameterization of the myocardial tissue microenvironment, therefore providing an experimental tool for the study of pathological conditions and the development of new therapeutic approaches.

Abbreviations

ATI	:	4-amino-2,2,5,5-tetramethylimidazoline-1-oxyl
EPR	:	electron paramagnetic resonance
EPRI	:	electron paramagnetic resonance imaging
GSH	:	glutathione
I/R	:	ischemia-reperfusion
LAD	:	left anterior descending coronary artery
LiNc	:	lithium naphtalocyanine
LiPc	:	lithium phtalocyanine
MRI	:	magnetic resonance imaging
nanoSPIN	:	nano-sized particles with incorporated nitroxide
NMR	:	nuclear magnetic resonance
NR	:	nitroxide radical
PEDRI	:	proton-electron double-resonance imaging
TAM	:	triarylmethyl radical
TEMPO	:	2,2,6,6-tetramethylpiperidine-1-oxyl

References

Andersson, S., F. Radner, A. Rydbeek, R. Servin and L.-G. Wistrand. 1996. Free radicals. US Patent: 5, 530, 140.

Backer, J.M., V.G. Budker, S.I. Eremenko and Y.N. Molin. 1977. Detection of the kinetics of biochemical reactions with oxygen using exchange broadening in the ESR spectra of nitroxide radicals. Biochim. Biophys. Acta 460: 152–156.

Baker, J.E., W. Froncisz, J. Joseph and B. Kalyanaraman. 1997. Spin label oximetry to assess extracellular oxygen during myocardial ischemia. Free Radic. Biol. Med. 22: 109–115.

Bobko, A.A., I. Dhimitruka, T.B. Eubank, C.B. Marsh, J.L. Zweier and V.V. Khramtsov. 2009. Trityl-based EPR probe with enhanced sensitivity to oxygen. Free Radic. Biol. Med. 47: 654–658.

Bobko, A.A., I. Dhimitruka, J.L. Zweier and V.V. Khramtsov. 2007. Trityl radicals as persistent dual function pH and oxygen probes for in vivo electron paramagnetic resonance spectroscopy and imaging: concept and experiment. J. Am. Chem. Soc. 129: 7240–7241.

Dhimitruka, I., A.A. Bobko, C.M. Hadad, J.L. Zweier and V.V. Khramtsov. 2008. Synthesis and characterization of amino derivatives of persistent trityl radicals as dual function pH and oxygen paramagnetic probes. J. Am. Chem. Soc. 130: 10780–10787.

Dhimitruka, I., O. Grigorieva, J.L. Zweier and V.V. Khramtsov. 2010. Synthesis, structure and EPR characterization of deuterated derivatives of Finland trityl radical. Bioorg. Med. Chem. Lett. 20: 3946–3949.

Efimova, O.V., Z. Sun, G.L Caia, E. Kesselring, S. Petryakov, D. Johnson, J.L. Zweier, V.V. Khramtsov and A. Samouilov. 2011. Variable radio frequency proton-electron double-resonance imaging; application to pH mapping of aqueous samples. J. Magn. Reson. 209: 227–232.

Grucker, D. and J. Chambron. 1993. Oxygen imaging in perfused hearts by dynamic nuclear polarization. Magn. Reson. Imaging 11: 691–696.

Halpern, H.J., C. Yu, M. Peric, E. Barth, D.J. Grdina and B.A. Teicher. 1994. Oxymetry deep in tissues with low-frequency electron paramagnetic resonance. Proc. Natl. Acad. Sci. USA 91: 13047–13051.

Hyodo, F., S. Matsumoto, N. Devasahayam, C. Dharmaraj, S. Subramanian, J.B. Mitchel and M.C. Krishna. 2009. Pulsed EPR imaging of nitroxides in mice. J. Magn. Reson. 197: 181–185.

Khan, M., I.K. Mohan, V.K. Kutala, S.R. Kotha, N.L. Parinandi, R.L. Hamlin and P. Kuppusamy. 2009. Sulfaphenazole protects heart against ischemia-reperfusion injury and cardiac dysfunction by overexpression of iNOS, leading to enhancement of nitric oxide bioavailability and tissue oxygenation. Antioxid. Redox Signal. 11: 725–38.

Khramtsov, V.V. 2005. Biological imaging and spectroscopy of pH. Curr. Org. Chem. 9: 909–923.

Khramtsov, V.V., G.L. Caia, K. Shet, E. Kesselring, S. Petryakov, J.L. Zweier and A. Samouilov. 2010. Variable field proton-electron double-resonance imaging: application to pH mapping of aqueous samples. J. Magn. Reson. 202: 267–273.

Khramtsov, V.V. and L.B. Volodarsky. 1998. Use of imidazoline nitroxides in studies of chemical reactions. ESR measurements of the concentration and reactivity of protons, thiols and nitric oxide. In: L.J. Berliner (ed.). Spin Labeling. The Next Millennium. Plenum Press, New York, pp. 109–180.

Khramtsov, V.V. and J.L. Zweier. 2010. Functional *in vivo* EPR spectroscopy and imaging using nitroxide and trityl radicals. In: R. Hicks (ed.). Stable Radicals: Fundamentals and Applied Aspects of Odd-Electron Compounds. John Wiley & Sons, Ltd., pp. 537–566

Kirilyuk, I.A., A.A. Bobko, I.A. Grigor'ev and V.V. Khramtsov. 2004. Synthesis of the tetraethyl substituted pH-sensitive nitroxides of imidazole series with enhanced stability towards reduction. Org. Biomol. Chem. 2: 1025–1030.

Krishna, M.C., S. English, K. Yamada, J. Yoo, R. Murugesan, N. Devasahayam, J.A. Cook, K. Golman, J.H. Ardenkjaer-Larsen, S. Subramanian and J.B. Mitchel. 2002. Overhauser enhanced magnetic resonance imaging for tumor oximetry: coregistration of tumor anatomy and tissue oxygen concentration. Proc. Natl. Acad. Sci. USA 99: 2216–2221

Kuppusamy, P., M. Chzhan, K. Vij, M. Shteynbuk, D.J. Lefer, E. Giannella and J.L. Zweier. 1994. Three-dimensional spectral-spatial EPR imaging of free radicals in the heart: a technique for imaging tissue metabolism and oxygenation. Proc. Natl. Acad. Sci. USA 91: 3388–3392.

Kuppusamy, P. and M.C. Krishna. 2002. EPR Imaging of tissue redox status. Curr. Top. Biophys. 26: 29–34.

Liu, K.J., P. Gast, M. Moussavi, S.W. Norby, N. Vahidi, T. Walczak, M. Wu and H.M. Swartz. 1993. Lithium phthalocyanine: a probe for electron paramagnetic resonance oximetry in viable biological systems. Proc. Natl. Acad. Sci. USA 90: 5438–5442.

Pandian, R.P., M. Dolgos, V. Dang, J.Z. Sostaric, P.M. Woodward and P. Kuppusamy. 2007. Structure and oxygen-sensing paramagnetic properties of a new lithium 1, 8, 15, 22-tetraphenoxyphthalocyanine radical probe for biological oximetry. Chem. Mater. 19: 3545–3552.

Pandian, R.P., M. Dolgos, C. Marginean, P.M. Woodward, P.C. Hammel, P.T. Manoharan and P. Kuppusamy. 2009. Molecular packing and magnetic properties of lithium naphthalocyanine crystals: hollow channels enabling permeability and paramagnetic sensitivity to molecular oxygen. J. Mater. Chem. 19: 4138–4147.

Roshchupkina, G.I., A.A. Bobko, A. Bratasz, V.A. Reznikov, P. Kuppusamy and V.V. Khramtsov. 2008. *In vivo* EPR measurement of glutathione in tumor-bearing mice using improved disulfide biradical probe. Free Rad. Biol. Med. 45: 312–320.

Schafer, F.Q. and G.R. Buettner. 2001. Redox environment of the cell as viewed through the redox state of the glutathione disulfide/glutathione couple. Free Rad. Biol. Med. 30: 1191–1212.

Sostaric, J.Z., R.P. Pandian, A. Bratasz and P. Kuppusamy. 2007. Encapsulation of a highly sensitive EPR active oxygen probe into sonochemically prepared microspheres. J. Phys. Chem. B. 111: 3298–3303.

Swartz, H.M. 2004. Using EPR to measure a critical but often unmeasured component of oxidative damage: oxygen. Antioxid. Redox Signal. 6: 677–686.

Woldman, Y.Y., S.V. Semenov, A.A. Bobko, I.A. Kirilyuk, J.F. Polienko, M.A. Voinov, E.G. Bagryanskaya and V.V. Khramtsov. 2009. Design of liposome-based pH sensitive nanoSPIN probes: nano-sized particles with incorporated nitroxides. Analyst 134: 904–910.

Zhu, X., B. Liu, S. Zhou, Y.R. Chen, Y. Deng, J.L. Zweier and G. He. 2007. Ischemic preconditioning prevents *in vivo* hyperoxygenation in postischemic myocardium with preservation of mitochondrial oxygen consumption. Am. J. Physiol. Heart Circ. Physiol. 293: H1442–H1450.

Zweier, J.L. and P. Kuppusamy. 1988. Electron paramagnetic resonance measurements of free radicals in the intact beating heart: a technique for detection and characterization of free radicals in whole biological tissues. Proc. Natl. Acad. Sci. USA 85: 5703–5707.

12

Native Endothelium Mimicking Nanomatrices and Applications

Adinarayana Andukuri,[1,a] *Chidinma P. Anakwenze,*[1,b] *Bryan A. Blakeney*[1,c] and *Ho-Wook Jun*[1,d,*]

ABSTRACT

Cardiovascular disease is a major cause of death worldwide and, to improve vascular grafting and stents, scientists are turning to nanotechnology. While synthetic grafts are useful in many artery bypass surgery cases, grafts are prone to thrombosis due to platelet adhesion and intimal hyperplasia due to smooth muscle proliferation. Stenting closed arteries is also a widely used treatment; however, stents also cause injury to the artery and can cause similar types of thrombosis and hyperplasia as grafts. The lack of endothelialization of grafts and stents is a major limitation of these implants, and nanotechnology may provide several strategies to overcome this deficit. Since the native extracellular matrix consists of nanofibers, it is not surprising to find that studies

[1]Department of Biomedical Engineering, University of Alabama at Birmingham, Birmingham, AL 35294.
[a]E-mail: andukuri@uab.edu
[b]E-mail: chidi89@uab.edu
[c]E-mail: adamblak@uab.edu
[d]E-mail: hwjun@uab.edu
*Corresponding author

List of abbreviations after the text.

show nanotopographical cues such as grooves smaller than 800 nm are more favorable to endothelial cell growth than larger features. For vascular grafting, electrospinning shows great promise as a method that produces strong nanofibrous material from a number of different polymers that can support numerous types of cells. Bioactivity can be increased by functionalizing materials with peptide ligands such as fibronectin-derived Arg-Gly-Asp (RGD), and laminin-derived Tyr-Ile-Gly-Ser-Arg (YIGSR), and by releasing "endothelium-derived relaxation factor" nitric oxide. Bare metal stents have been modified by coating with polymers that release anti-proliferative drugs to inhibit smooth muscle growth causing intimal hyperplasia, such as sirolimus and paclitaxel , to prevent restenosis of the artery. One coating, polyvinylidenefluoro–hexafluoropropylene, also supports the growth of endothelial cells. A recently introduced strategy uses peptide amphiphile self-assembling nanofibers to coat electrospun nanofibers or bare metal stents with nanofibers that include the fibronectin ligand RGD, laminin ligand YIGSR, or poly-lysine peptides reacted to NO, to provide nanotopographical cues, bioactive ligands, and drug-elution to the cardiovascular implants. Multi-functional approaches are possible to creating cardiovascular grafts and stents that mimic vascular native endothelium, and the next generation of devices have great potential to solve the problems of current cardiovascular implants.

INTRODUCTION

The limitations faced by current cardiovascular therapies have encouraged scientists to increasingly look at nanotechnologies as a viable alternative for developing better treatments. Cardiovascular diseases are a major cause of death worldwide. The underlying cause of most cardiovascular diseases is atherosclerosis, the build-up of fatty plaque in the arteries. Major conventional treatment techniques for these diseases are grafting and stenting.

CLINICAL SIGNIFICANCE

Grafting is commonly used to replace a blocked artery in surgeries including coronary artery bypass graft (CABG). Synthetic vascular grafts are increasingly being used in such cases. There are several commercially available synthetic vascular grafts. These grafts typically suffer from characteristic limitations, including lack of endothelialization, thrombosis due to adhesion and activation of platelets especially at small diameters (< 5 mm), intimal hyperplasia due to proliferation of smooth muscle cells, which eventually leads to restenosis (Greisler 1990). The inability of these vascular grafts to promote re-endothelialization is thought to be a major factor in their lack of clinical patency and eventual failure (Mitchell and Nicklason 2003). Stenting is a common procedure used for treatment of blocked arteries. It involves the deployment of an expandable metallic scaffold to hold the artery open after using a balloon to expand the clogged blood vessel. Originally, bare metal stents (BMS) were used to prevent the elastic recoil of the artery. However, these BMS suffered from restenosis (Sheiban et al. 2002), the re-closure of the artery due to a cascade of events resulting from the injury to the blood vessel. The injury results in denudation of the endothelial layer. This activates thrombotic factors in the blood, leading to thrombosis. It also promotes the invasion and proliferation of smooth muscle cells, leading to neointimal hyperplasia, and eventually restenosis. In response to these limitations, drug-eluting stents (DES) were developed. DES is typically coated with a polymer matrix that contains and releases an anti-proliferative drug to prevent the invasion and proliferation of smooth muscle cells. These stents are able to limit the problem of neointimal hyperplasia (Kolodgie et al. 2002), but there is a concern of late-stent thrombosis (Camenzind et al. 2007). This has ensured that there is no significant difference in the mortality rates of BMS and DES (Babapulle et al. 2004). Therefore, there is a need for next generation stents to comprehensively combat all the limitations of the current stents as shown in Fig. 1.

It is therefore evident that the problems faced by these devices can be outlined as lack of endothelialization, thrombosis, and restenosis. Biological compatibility of implants can be improved by mimicking the native biological systems. Therefore, a solution to these problems can be obtained by mimicking the native endothelium in the blood vessels. The endothelium consists of a layer of endothelial cells and underlying layers of smooth muscle cells and viscoelastic, nanofibrillar extracellular matrix (ECM) proteins. It produces factors such as nitric oxide (NO), which regulates cellular homeostasis. NO promotes the migration and proliferation of endothelial cells, limits the proliferation of smooth muscles and prevents the adhesion and activation of platelets. With the recent advances in the field of nanotechnologies, scientists have been looking to incorporate elements of the native endothelium in the development of next generation cardiovascular implants. To mimic the chemical and the biological complexity of the native endothelium, the following features need to be incorporated: (1) nanofibrillar structure; (2) endothelial cell adhesive ligands to promote endothelialization; (3) enzyme-mediated degradable sites for promoting cell migration, and (4) release of factors such as NO for regulation of cell behavior. The next few sections will provide an account of approaches that incorporate the aforementioned features of nanotechnology in the development of cardiovascular implants, vascular grafts and stents.

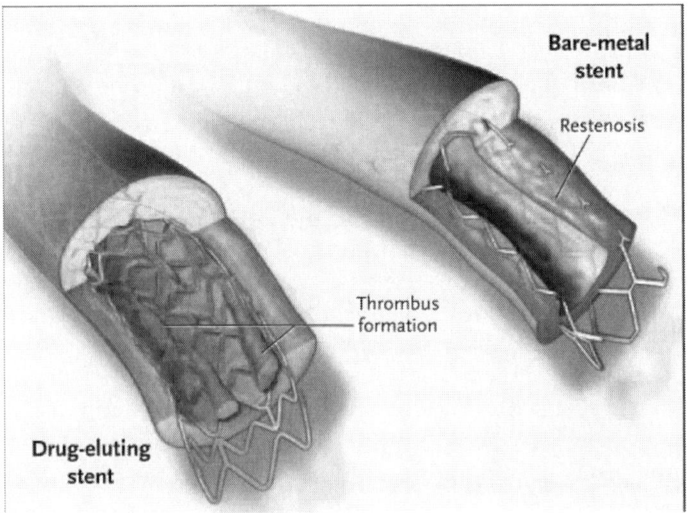

Fig. 1. Contrasting mechanisms of obstruction of bare metal stents and drug-eluting stents. Bare metal stents may be narrowed or obstructed by ingrowth of tissue. With drug-eluting stents, this process is inhibited, but since the struts remain uncovered, they may be prone to thrombosis after antiplatelet therapy is discontinued. Reproduced with permission (Shuchman 2006).

Color image of this figure appears in the color plate section at the end of the book.

VASCULAR GRAFTS

Vascular grafts are essentially conduits that are designed to replace diseased arteries. Several synthetic vascular grafts are commercially available, and these are mostly made of Dacron, polyurethane or expanded polytetrafluoroethylene (ePTFE). While these grafts perform satisfactorily for large vessels, small vessels are plagued by thrombosis, intimal hyperplasia and eventually restenosis. It is therefore critical to develop novel vascular graft materials by mimicking the native endothelium.

Mechanical Properties

Structurally, the native endothelial ECM consists of nanofibrillar proteins such as laminin, elastin and collagen. In order to promote endothelialization, it is essential to replicate the nanoscale topographies that these proteins present to the cells. In one study, it was reported that vascular cell adhesion and proliferation were greatly improved on nanostructured Ti when compared to conventional Ti (Choudhary et al. 2007). Furthermore, nanostructured Ti showed increased ECM production when compared to conventional Ti. In another study, vascular cell responses to nanostructures were studied by growing them on PLGA films cast on 100 nm, 200 nm and 500 nm polystyrene films. It was found that the 200 nm films promoted vascular cell responses and fibronectin interconnectivity (Miller et al. 2007). In another study, polymers with anisotropically ordered grooves ranging from 200 nm to 2000 nm were used to study the cell behaviors of endothelial cells (Liliensiek et al. 2010). Cellular response was found to be favorable only when the grooves were smaller than 800 nm. These studies show that endothelial cells respond to nanotopographical cues.

The success of nano-topographies led to an increased interest in the use of electrospun nanofibrous materials for vascular grafts as shown in Fig. 2 (Thomas et al. 2009). Electrospinning is the process of generating nanoscale fibers from a polymer solution by using a high voltage electric field. The fibers produced are of similar dimensions to ECM proteins and the process can be tuned to vary the dimensions of the fibers. The attractiveness of electrospinning lies in its versatility. Several synthetic polymers have been

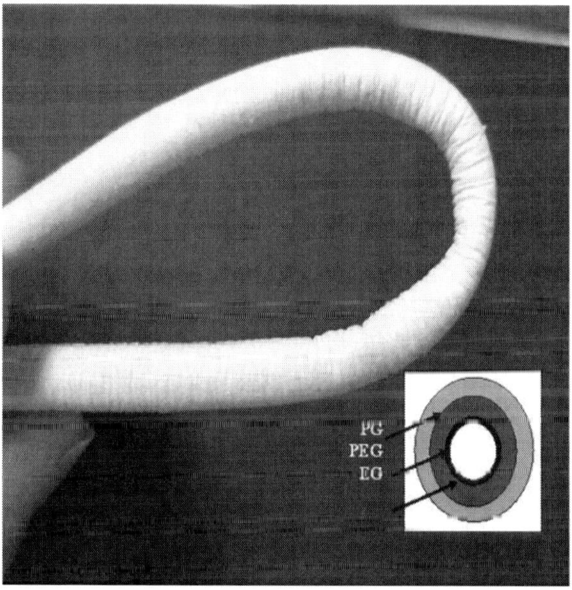

Fig. 2. Tubular electrospun conduit of 20 cm length and 4 mm inner diameter. Inset, a schematic showing the trilayered structure of scaffold. The lumen layer is rich in protein and outer layers are rich in PDS. Reproduced with permission (Thomas et al. 2009).

electrospun (Stitzel et al. 2006) and the process has also been successfully extended to include natural polymers such as collagen and elastin (He et al. 2009). In addition, these polymers can be mixed in varying ratios to generate fibers of required biomechanical properties similar to native arteries. These fibers can also be deposited on a rotating mandrel to produce a conduit. Electrospun fibers can withstand high pressure and pulsatile blood flow. Therefore, electrospinning has great potential for generation of nanostructured vascular grafts. However, the fibers are limited by the lack of the ability to regulate specific cell behavior.

Directing Cell Behavior via Biomimetic Approaches

It is critical to functionalize the surfaces to selectively promote the adhesion and proliferation of endothelial cells. Surface functionalization confers the ability to direct cell behavior without changing the mechanical properties of the bulk material. The key parameters in controlling cell behaviors are ligand specificity, density, material architecture and mechanical properties (Saha et al. 2007). Native ECM proteins possess sequences of amino acids that show specificity for certain cell types. In order to mimic this feature, several studies have attempted the incorporation of endothelial cell–specific ligands for improving the patency of vascular grafts. Initially, common fibronectin-derived cell adhesive ligand Arg-Gly-Asp (RGD) was used to functionalize surfaces to improve endothelialization with limited success but RGD also promotes the adhesion of platelets, which can lead to thrombosis. A commonly used ligand for endothelial cell specificity is Tyr-Ile-Gly-Ser-Arg (YIGSR) ligand (Massia and Hubbell 1991). It is derived from laminin, the major glycoprotein found in the underlying basement membrane in the endothelium. YIGSR is known to promote the adhesion and growth of endothelial cells. When used to modify surfaces of polymers such as polyethylene terephthalate and polytetrafluoroethylene, it has been shown to selectively enhance endothelial cell adhesion, proliferation and migration (Massia and Hubbell 1991). These ligands have also been incorporated in self-assembling peptide systems (Andukuri et al. 2010) and the peptide amphiphiles promoted endothelial cell adhesion and growth. In one study of functionalized peptide sequences from collagen IV (TAGSCLRKFSTM) and laminin (YIGSR, RYVVLPR) onto RAD16-I peptide, these self-assembling peptides improved endothelial cell growth and also improved their function as evidenced by increased ECM production and NO release (Genove et al. 2005).

The most important aspect of mimicking the native endothelium is conferring anti-thrombogenicity on the graft material. This can be achieved by immobilization of anti-thrombogenic molecules along with the acceleration of endothelialization. The sustained release of immobilized anti-thrombogenic molecules prevents the adhesion and activation of platelets. Therefore, proteins such as heparin (Seifert et al. 1995) and thrombomodulin (Li et al. 2002) have been incorporated. In one study, an anti-thrombogenic material was developed by incorporating polyhedral oligomeric silsesquioxane in nanocomposite polyurethane (Wilson et al. 2007). Recently, NO has been gaining attention for its unique functions in the cardiovascular system (Verma and Marsden 2005). NO is released by the healthy endothelium and is also known as "endothelium-derived relaxation factor" (EDRF). NO helps in maintaining vascular cell homeostasis. It promotes the proliferation of endothelial cells (Ziche et al. 1994), but limits the proliferation of smooth muscles cells. Importantly, it also prevents the adhesion and activation of platelets (Bohl and West 2000). NO causes these cell behaviors by acting through cGMP pathways. An increase in cGMP activates cGMP-dependent kinases that reduce intracellular calcium. This leads to relaxation of smooth muscle cells and inactivation of platelets. These properties make NO an attractive therapeutic candidate for cardiovascular diseases. The hurdle in using NO lies in its short half-life. Therefore, it is necessary to identify potential donors that can release NO in a controlled manner. In this regard, diazeniumdiolates or NONOates have been used as NO donors (Keefer et al. 1996). Vascular grafts with NO-releasing microspheres using NONOates have been studied (Pulfer et al. 1997). NO-releasing polyethyleneimine microspheres (PEIX) were developed using water-in-oil emulsion technique and cross-linked using *bis*-epoxide. NO release was successfully observed when these microspheres were incorporated in a gore-tex–based vascular graft, as shown in Fig. 3. Diazeniumdiolates can also be synthesized by reacting nucleophilic amino acids such as lysine with NO at high pressure. The reaction is reversed and NO is released when the molecule is exposed to buffers such as blood or cell culture media. Release of NO from lysine-based polymers has been shown to reduce smooth muscle cell proliferation and platelet adhesion (Bohl and West 2000). This was incorporated into peptide amphiphiles (PAs) in a study by Andukuri et al. (2011). PAs are molecules that consist of hydrophilic peptide sequences attached to hydrophobic alkyl tails. The presence of these opposing groups drives self-assembly of PAs in aqueous media to form nanofibers with diameters ranging from 8 to 10 nm. In this study, the peptide sequence comprised two parts: an MMP-2 enzyme-degradable sequence (Gly-Thr-Ala-Gly-Leu-Ile-Gly-Gln) that was attached to either a YIGSR moiety or a polylysine (KKKKK) moiety. The polylysine moiety acts as an NO donor. These two PAs were mixed in an optimum ratio to develop PA-YK-NO, an NO-releasing PA mixture. This mixture was self-assembled onto electrospun polycaprolactone nanofibers (diameters 300–500 nm) to functionalize the surface and thereby direct cellular behaviors.

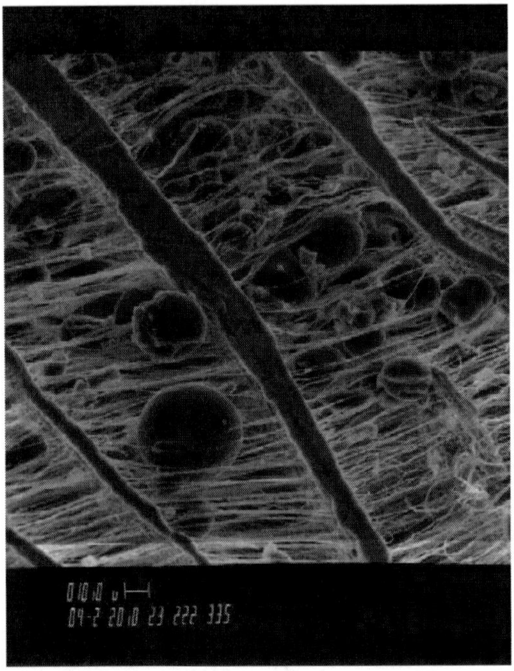

Fig. 3. Scanning electron micrograph of FITC-PEIX/NO and PEIX/ NO microspheres in a vascular graft. FITC-PEIX/NO and PEIX/NO microspheres were incorporated with vacuum into the pores of a 60 mm, 4 mm ID, non-reinforced Gore-tex vascular graft. Indicator bar = 10 mm. Used with permission (Pulfer et al. 1997).

This system replicates the biological and the chemical complexity of the native endothelium by providing the cells with nanofibrous topography, cell adhesive ligands, enzyme-degradable sequences and NO donors. The study reported that the material promoted the adhesion and proliferation of endothelial cells but limited the proliferation of smooth muscles and the adhesion of platelets. This material, therefore, has great potential for application in vascular grafts.

Development of the ideal vascular graft has long been the pursuit of several researchers. The key to this ideal graft lies in the native endothelium and is therefore limited by the inability to recreate its biochemical complexity. The recent advances discussed above represent biomimetic approaches of nanotechnology to specifically target the problems faced by current clinically used vascular grafts resulting in the development of novel cell-responsive materials. This shows that there is great potential in the field of nanotechnology with regard to vascular grafts and the application of these technologies increases the likelihood of overcoming the current clinical limitations of vascular grafts.

STENTS

Stents are lattice-shaped expandable metallic scaffolds that were introduced in the 1990s as an improvement over balloon angioplasty. Balloon angioplasty provided immediate relief to the patient by reopening the clogged artery; however, it leads to cases of re-closure due to elastic recoil of the artery walls. The stent in a collapsed form was inserted into the balloon and then inflated. BMS solved the problem of immediate re-closure of arteries by preventing the elastic recoil of the walls (Ozaki et al. 1996). However, a new set of problems arose due to the presence of a metallic structure in the artery. The major cause of concern was in-stent restenosis, which is eventual re-closure of the artery due to negative vessel remodeling and neointimal hyperplasia (Sheiban et al. 2002). The BMS damaged the endothelial lining of the artery, thereby exposing it to the thrombotic cascade. This also led to the invasion and proliferation of smooth muscle cells. These smooth muscle cells change from a contractile state to a synthetic state, which leads to increased migration and production of ECM proteins resulting in neointimal hyperplasia and eventually in-stent restenosis. This situation necessitated the use of drugs to limit the proliferation of smooth muscle cells and led to the development of DES.

Drug-eluting Stents

First generation DES is a modified BMS coated with a polymer matrix that releases one or a mixture of anti-proliferative drugs in a controlled manner after implantation. Several drug release approaches, such as diffusion, degradation, dissolution and ion-exchange, have been studied for DES (Acharya and Park 2006). Commercially available DES typically uses drugs such as sirolimus and paclitaxel or their analogs/derivatives. A list of current DES is shown in Table 1. Both sirolimus and paclitaxel limit proliferation of cells by inhibiting the cell cycle. Sirolimus is an immunosuppressive drug that promotes kinase activation, leading to the inhibition of the cellular growth phase. Paclitaxel binds to microtubules in dividing cells and causes them to assemble, thereby preventing mitosis (Wessely et al. 2006). While these DES reduce the rates of in-stent restenosis (Kolodgie et al. 2002), it is interesting to note that there is no significant difference in the mortality rates of DES and BMS (Babapulle et al. 2004). This can be explained by the occurrence of late-stent thrombosis in DES. In addition, the use of anti-proliferative drugs also causes reduced endothelialization when compared to BMS. This coupled with increased fibrin deposition leads to delayed arterial healing (Finn et al. 2007).

Table 1. Commercially available drug-eluting stents.

Stent	Drug (Concentration [μ/cm²])	Drug Mechanism	Polymer	Polymer Thickness, μm	Release Kinetics, 28 Days	Metal	Geometry	Strut Thickness, μm	Crimped Profile, mm	Maximum Cell C/D, mm
CYPHER	Sirolimus (140)	Inhibits mTOR, cytostatic	Polyethelyne co-vinyl acetate and poly-n-butyl methacrylate	12.6	80%	SS	Closed cell	140	1.198	9.5/3.0
TAXUS Express	Paclitaxel (100)	Microtubule inhibitor, cell cycle arrest in G0/G1 and G2/M	Poly(styrene-b-isobutylene-b-styrene)	16.0	<10%	SS	Open cell	132	1.245	11.9/3.8
TAXUS Liberté	Paclitaxel (100)	Microtubule inhibitor, cell cycle arrest in G0/G1 and G2/M	Poly(styrene-b-isobutylene-b-styrene)	16.0	<10%	SS	Hybrid	97	1.219	13.7/4.4
Endeavor	Zotarolimus (100)	Inhibits mTOR, cytostatic	Phosphorylcholine	4.1	95%†	CoCr	Open cell	91	1.130	19.8/6.3
Xience V	Everolimus (100)	Inhibits mTOR, cytostatic	Polyvinylidene fluoride co-hexafluoropropylene and poly-n-butyl methacrylate	7.6	80%	CoCr	Open cell	81	1.055	12.6/4.0

Specifications of drug-eluting stents approved by the Food and Drug Administration. C/D, circumference/diameter; CoCr, cobalt chromium; mTOR, mammalian target of rapamycin; SS, stainless steel. Based on a 3 mm stent. †Drug release at 14 d. Used with permission (Garg and Serruys 2010).

With vascular compatibility in mind, new DES (ENDEAVOR and Xience-V) have been developed. ENDEAVOR (Medtronic) uses an anti-thrombogenic phosphorylcholine coating (Lewis et al. 2001) to deliver Zotarolimus drug. Xience-V uses a copolymeric coating of polyvinylidenefluoro–hexafluoropropylene (PVDF-HFP) to release an anti-proliferative drug (Everolimus) (Sheiban et al. 2008). While these stents have shown improved characteristics compared to earlier devices in clinical trials, it is important to understand that the current DES still uses anti-proliferative drugs to prevent neointimal growth. These drugs are non-specific and therefore also affect endothelial cells. It is therefore critical to develop new coatings that support endothelial cell retention and proliferation but limit the proliferation of smooth muscle cells to prevent in-stent restenosis. It is also critical for the coating to limit the adhesion and activation of platelets, thereby preventing late stent thrombosis. Therefore, there is a need for development of next generation DES by using biomimetic coatings.

Directing Cell Behavior via Biomimetic Approaches

There are several approaches to develop biomimetic coatings for DES. An interesting concept is the use of phosphorylcholine (PC) based coatings to mimic the cell membranes of red blood cells, thereby conferring vascular compatibility on the surfaces. While this coating has been shown to have no effect on restenosis rates, its anti-thrombogenicity, long-term stability and ability to release drugs has generated interest (Lewis et al. 2001). Another study performed by Meyers et al. detailed the development of a peptide-based interfacial biomaterial (IFBM) that was capable of functionalizing inert surfaces (Meyers et al. 2009). IFBMs can simultaneously bind the biomaterial surface while endowing it with desired biological functionalities at the interface between the material and biological realms, as shown in Fig. 4. The authors define IFBMs as modular peptides containing at least one material-binding domain and at least one bioactive functionality. Sargeant et al. developed a functional surface by covalently linking NiTi (nickel-titanium) surfaces with PA nanofibers. In their study, PA nanofibers containing RGDS moiety were linked to the NiTi surface. This was found to increase the adhesion and spreading of endothelial cells (Sargeant et al. 2008).

A multifunctional approach for developing a biomimetic DES coating was studied by Kushwaha et al. (2010). It involves the release of NO. As previously mentioned, NO plays a unique role in the cardiovascular system. It maintains vascular cell homeostasis by promoting endothelialization but limits smooth muscle cell proliferation and platelet adhesion (Fig. 5). In this study, two PAs containing enzyme-degradable sequences linked to

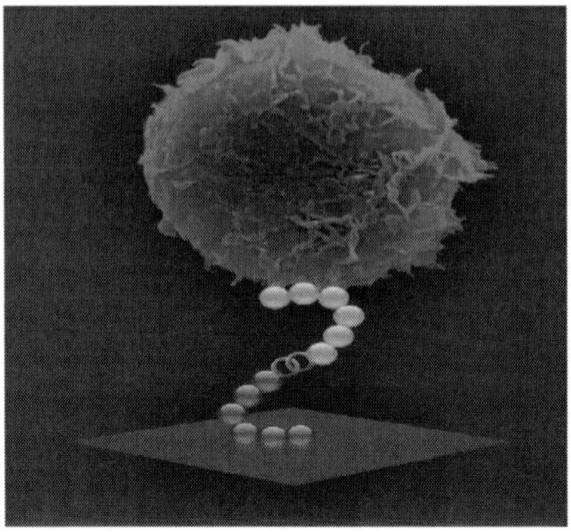

Fig. 4. Schematic diagram of an IFBM at the interface between the biologic and material realms. Used with permission (Meyers et al. 2009).

Color image of this figure appears in the color plate section at the end of the book.

either endothelial cell adhesive ligand (YIGSR) or polylysine NO donor (KKKKK) were synthesized and mixed in an optimum ratio designed to maximize endothelial cell adhesion. This PA mixture self-assembles into nanofibers with diameters ranging from 8 to 10 nm by the process of solvent (water) evaporation. This system attempts to incorporate all the complexities of the native endothelium by possessing cell adhesive ligands, NO donors, enzyme-degradable sequences and self-assembled nanofibers. The PA nanofiber system was shown to enhance the adhesion and proliferation of endothelial cells but limit the proliferation of smooth muscle cells. More importantly, it greatly reduces the adhesion of platelets. This system therefore has great potential as a coating for next generation DES.

There is a critical need for development of a new generation of DES to overcome the limitations of the current DES. It is evident from the studies discussed here that sufficient advances are being made towards using multifunctional approaches to develop biomimetic stent coatings. Recently, there has been great interest in accelerating the endothelialization of stent surfaces by recruiting endothelial progenitor cells (EPCs). EPCs are immature bone marrow–derived cells that circulate in the blood. They possess the ability to proliferate, migrate, and differentiate into endothelial cells but have not yet acquired the characteristics of mature endothelial cells. This has led to the development of EPC capture stents. The most prominent model uses immobilized anti-CD34 antibodies (Aoki et al. 2005) or anti-VEGFR2 antibodies (Markway et al. 2008). Other studies have used high-specificity DNA aptamers (Hoffmann et al. 2008) and superparamagnetically labeled cells (Pislaru et al. 2006). These technologies

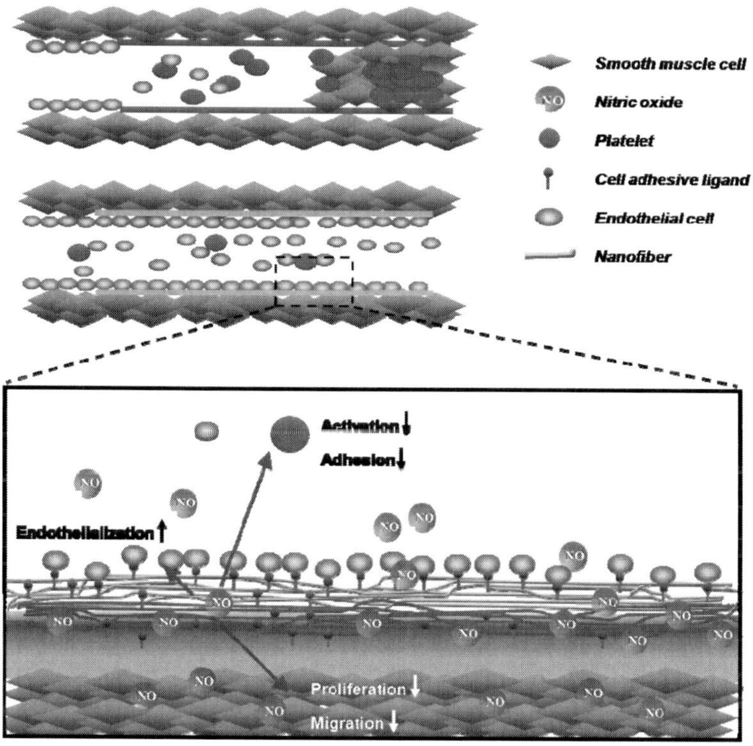

Fig. 5. Schematic comparison of current stents and endothelial ECM-mimicking PA coated stents. (a) Schematic comparison of current stent/vascular grafts (blue) and endothelial ECM-mimicking nanofibrous matrix (green). (1) Endothelial disruption exposes vessel wall, leading to restenosis, thrombosis and incomplete endothelialization. (2) Native endothelial ECM-mimic nanofibrous matrix will prevent restenosis and thrombosis while promoting endothelialization. (b) Components of the native endothelial ECM-mimic nanofibrous matrix. NO released from the nanofibrous matrix prevents platelet activation, adhesion, smooth muscle cell proliferation and migration and promotes endothelialization. Used with permission (Kushwaha et al. 2010)

Color image of this figure appears in the color plate section at the end of the book.

have met with reasonable success in clinical trials but there have been some concerns. EPC capture technologies are limited by small population of EPCs which cannot be defined by one specific marker (Wendel et al. 2009). In addition, there are also concerns of recruiting pro-inflammatory cells (Wendel et al. 2009). Some EPC capture stents have seen increased neointimal formation (Kaul et al. 2008), which may be due to certain smooth muscle progenitor cells expressing CD34 as well (Simper et al. 2002). Therefore, a greater progressive understanding of EPCs is required. It will then be possible to develop technologies that incorporate them in improving the clinical patency of DES.

CONCLUSION

Cardiovascular implants such as vascular grafts and stents suffer from certain characteristic limitations such as lack of re-endothelialization, intimal hyperplasia, thrombosis, and restenosis. The key to solving these issues is to develop nanostructured materials capable of providing the native endothelium-mimicking environment. As described in this chapter, several approaches have been studied to give cardiovascular implant surfaces an endothelium-mimicking nature. In order to completely mimic the native endothelium, a multifunctional approach that combines all the unique properties of the native tissue is required. With emerging nanotechnologies, it is possible to recreate these properties in a bottom-up manner. This is particularly evident in the highly studied field of vascular grafts, where the use of techniques such as electrospinning and self-assembling peptides are being investigated with great potential. Stents have seen great advancements from BMS to the current clinical trials involving EPC capture stents. It is also interesting to note that the techniques developed are not limited to vascular grafts and stents. Because of their anti-thrombotic nature, they may have great potential in other cardiovascular implant devices including mechanical heart valves, which require continued use of anti-coagulants. In conclusion, the native endothelium-mimicking nanotechnologies described in this chapter have great potential to fight the limitations of current cardiovascular implants and can lead to the development of next generation of devices.

APPLICATIONS TO AREAS OF HEALTH AND DISEASE

Cardiovascular diseases are the leading cause of death in the United States. In 2006, one third of all deaths in the United States were caused by cardiovascular diseases. The underlying cause of most cardiovascular diseases is atherosclerosis, which is the build-up of fatty plaque in the arteries. This leads to eventual narrowing and obstruction of the artery. Stenting and vascular grafting are two common therapies currently used in the treatment of cardiovascular diseases. They suffer from lack of endothelialization, thrombosis, neointimal hyperplasia and restenosis. Recent developments in nanotechnology can aid in the design and development of endothelium-mimicking nanomatrices that offer an opportunity to increase the patency of these cardiovascular implants.

Key Facts

- Balloon angioplasty was originally used to open clogged arteries. Stents were later introduced to keep the artery open after balloon angioplasty. Vascular grafts are used to replace diseased blood vessels in surgeries including CABG.
- Cardiovascular implants such as stents and vascular grafts suffer from characteristic limitations such as lack of endothelialization, neointimal hyperplasia, thrombosis and restenosis.
- The endothelium is a monolayer of endothelial cells and endothelial ECM. The proteins that form endothelial ECM contain endothelial cell adhesive sites.
- Nitric oxide helps maintain vascular cell homeostasis. It promotes endothelial cell proliferation, limits smooth muscle cell migration and proliferation and prevents platelet adhesion.
- EPCs are circulating mononuclear cells that eventually differentiate into endothelial cells. Recruitment of EPCs by cardiovascular implants is thought to quicken the process of endothelialization.

Definitions

Coronary stent: A lattice-shaped metallic scaffold that helps keep arteries open after balloon angioplasty.

Endothelium: A monolayer of endothelial cells that forms the interface between the blood vessel lumen and blood.

Neointimal hyperplasia: Thickening of the tunica intima layer of a blood vessel by proliferation of smooth muscle cells. This is a response of the blood vessel to injury.

Restenosis: Re-narrowing of an artery following a cardiovascular procedure such as balloon angioplasty and stenting.

Thrombosis: Formation of a blood clot inside a blood vessel leading to obstruction of blood flow.

Summary Points

- Cardiovascular diseases are a major cause of death all over the world.
- Two major therapeutic methods for cardiovascular diseases are grafts and stents.
- Grafts are used to replace diseased blood vessels in surgeries such as CABG, while stents are used to open clogged arteries.

- Both these cardiovascular implant devices suffer from similar problems: lack of endothelialization, thrombosis, intimal hyperplasia and restenosis.
- There is a requirement for an endothelium-mimicking coating on these implants to overcome the aforementioned limitations.

Abbreviations

BMS	:	bare metal stent
CABG	:	coronary artery bypass graft
cGMP	:	cyclic guanosine monophosphate
CoCr	:	cobalt-chromium
DES	:	drug eluting stent
DNA	:	deoxyribonucleic acid
ECM	:	extracellular matrix
EDRF	:	endothelium-derived relaxing factor
EPC	:	endothelial progenitor cell
ePTFE	:	expanded polytetrafluoroethylene
IFBM	:	interfacial biomaterial
KKKKK	:	polylysine peptide sequence
mTOR	:	mammalian target of Rapamycin
NiTi	:	nickel-titanium alloy
NO	:	nitric oxide
PA	:	peptide amphiphile
PC	:	phosphorylcholine
PEIX	:	polyethyleneimine microspheres
PVDF-HFP	:	polyvinylidelenefluoro hexafluoropropylene
RGD	:	arginine-glycine-aspartate peptide sequence
SS	:	stainless steel
Ti	:	titanium
YIGSR	:	tyrosine-isoleucine-glycine-serine-arginine

References

Acharya, G. and K. Park. 2006. Mechanisms of controlled drug release from drug-eluting stents. Adv. Drug. Deliv. Rev. 58: 387–401.

Andukuri, A., M. Kushwaha, A. Tambralli, J. Anderson, D. Dean, J. Berry, Y. Sohn, Y. Yoon, B. Brott and Jun HW. 2011. A hybrid biomimetic nanomatrix composed of electrospun polycaprolactone and bioactive peptide amphiphiles for cardiovascular implants. Acta Biomaterialia 7: 225–233.

Andukuri, A., W. Minor, M. Kushwaha, J. Anderson and H.W. Jun. 2010. Effect of endothelium mimicking self-assembled nanomatrices on cell adhesion and spreading of human endothelial cells and smooth muscle cells. Nanomedicine 6: 289–297.

Aoki, J., P.W. Serruys, H. van Beusekom, A.T. Ong, E.P. McFadden, G. Sianos, W.J. van der Giessen, E. Regar, P.J. de Feyter, H.R. Davis et al. 2005. Endothelial progenitor cell capture by stents coated with antibody against CD34: the HEALING-FIM (Healthy

Endothelial Accelerated Lining Inhibits Neointimal Growth-First In Man) Registry. J. Am. Coll. Cardiol. 45(10): 1574–1579.

Babapulle, M.N., L. Joseph, P. Belisle, J.M. Brophy and M.J. Eisenberg. 2004. A hierarchical Bayesian meta-analysis of randomised clinical trials of drug-eluting stents. Lancet 364(9434): 583–591.

Bohl, K.S. and J.L. West. 2000. Nitric oxide-generating polymers reduce platelet adhesion and smooth muscle cell proliferation. Biomaterials 21: 2273–2278.

Camenzind, E., P.G. Steg and W. Wijns. 2007. Stent thrombosis late after implantation of first-generation drug-eluting stents. Circulation 7(115): 1440–1455.

Choudhary, S., K.M. Haberstroh and T.J. Webster. 2007. Enhanced functions of vascular cells on nanostructured Ti for improved stent applications. Tissue Eng. 13(7): 1421–1430.

Finn, A.V., G. Nakazawa, M. Joner, F.D. Kolodgie, E.K. Mont, H.K. Gold and R. Virmani. 2007. Vascular responses to drug eluting stents: importance of delayed healing. Arterioscler. Thromb. Vasc. Biol. 27(7): 1500–1510.

Garg, S. and P.W. Serruys. 2010. Coronary stents: current status. J. Am. Coll. Cardiol. 56(10 Suppl): S1–42.

Genove, E., C. Shen, S. Zhang and C.E. Semino. 2005. The effect of functionalized self-assembling peptide scaffolds on human aortic endothelial cell function. Biomaterials 26(16): 3341–3351.

Greisler, H.P. 1990. Interactions at the blood/material interface. Ann. Vasc. Surg. 4(1): 98–103.

He, W., Z. Ma, W.E. Teo, Y.X. Dong, P.A. Robless, T.C. Lim and S. Ramakrishna. 2009. Tubular nanofiber scaffolds for tissue engineered small-diameter vascular grafts. J. Biomed. Mater. Res. A 90(1): 205–216.

Hoffmann, J., A. Paul, M. Harwardt, J. Groll, T. Reeswinkel, D. Klee, M. Moeller, H. Fischer, T. Walker, T. Greiner et al. 2008. Immobilized DNA aptamers used as potent attractors for porcine endothelial precursor cells. J. Biomed. Mater. Res. A. 84: 614–621.

Kaul, U., V. Bhatia, T. Ghose, R. Gupta, R. Kachru and G. Singh. 2008. Angiographic follow-up of genous bioengineered stent in acute myocardial infarction (GENAMI)-a pilot study. Indian Heart J. 60(6): 532–535.

Keefer, L.K., R.W. Nims, K.M Davies and D.A. Wink. 1996. "NONOates" (1-substituted diazen-1-ium-1, 2-diolates) as nitric oxide donors: convenient nitric oxide dosage forms. Methods Enzymol. 268: 281–293.

Kolodgie, F.D., M. John, C. Khurana, A. Farb, P.S. Wilson, E. Acampado, N. Desai, P. Soon-Shiong and R. Virmani. 2002. Sustained reduction of in-stent neointimal growth with the use of a novel systemic nanoparticle paclitaxel. Circulation 106: 1195–1198.

Kushwaha, M., J. Anderson, W. Minor, A. Andukuri, C. Bosworth, J. Lancaster, B. Brott, P. Anderson and H.W. Jun. 2010. Native endothelium mimicking self-assembled nanomatrix for cardiovascular devices. Biomaterials 31: 1502–1508.

Lewis, A.L., T.A. Vick, A.C. Collias, L.G. Hughes, R.R. Palmer, S.W. Leppard, J.D. Furze, A.S. Taylor and P.W. Stratford. 2001. Phosphorylcholine-based polymer coatings for stent drug delivery. J. Mater. Sci. Mater. Med. 12(10–12): 865–870.

Li, J.M., M.J. Singh, P.R. Nelson, G.M. Hendricks, M. Itani, M.J. Rohrer and B.S. Cutler. 2002. Immobilization of human thrombomodulin to expanded polytetrafluoroethylene. J. Surg. Res. 105(2): 200–208.

Liliensiek, S.J., J.A. Wood, J. Yong, R. Auerbach, P.F. Nealey and C.J. Murphy. 2010. Modulation of human vascular endothelial cell behaviors by nanotopographic cues. Biomaterials 31(20). 5418–5826.

Markway, B.D., O.J. McCarty, U.M. Marzec, D.W. Courtman, S.R. Hanson and M.T. Hinds. 2008. Capture of flowing endothelial cells using surface-immobilized anti-kinase insert domain receptor antibody. Tissue Eng. Part C Methods.

Massia S.P. and J.A. Hubbell. 1991. Human endothelial cell interactions with surface-coupled adhesion peptides on a nonadhesive glass substrate and two polymeric biomaterials. J. Biomed. Mater. Res. 25(2): 223–242.

Meyers, S.R., X. Khoo, X. Huang, E.B. Walsh, M.W. Grinstaff and D.J. Kenan. 2009. The development of peptide-based interfacial biomaterials for generating biological functionality on the surface of bioinert materials. Biomaterials 30(3): 277–286.

Miller, D.C., K.M. Haberstroh and T.J. Webster. 2007. PLGA nanometer surface features manipulate fibronectin interactions for improved vascular cell adhesion. J. Biomed. Mater. Res. A 81(3): 678–684.

Mitchell, S.L. and L.E. Nicklason. 2003. Requirements for growing tissue-engineered vascular grafts. Cardiovasc. Pathol. 12: 59–64.

Ozaki, Y., A.G. Violaris and P.W. Serruys. 1996. New stent technologies. Prog. Cardiovasc. Dis. 39(2):129–40.

Pislaru, S.V., A. Harbuzariu, G. Agarwal, T. Witt, R. Gulati, N.P. Sandhu, C. Mueske, M. Kalra, R.D. Simari and G.S. Sandhu. 2006. Magnetic forces enable rapid endothelialization of synthetic vascular grafts. Circulation 114: I314–I318.

Pulfer, S.K., D. Ott and D.J. Smith. 1997. Incorporation of nitric oxide-releasing crosslinked polyethyleneimine microspheres into vascular grafts. J. Biomed. Mater. Res. 37(2): 182–189.

Saha, K., J.F. Pollock, D.V. Schaffer and K.E. Healy. 2007. Designing synthetic materials to control stem cell phenotype. Curr. Opin. Chem. Biol. 11(4): 381–387.

Sargeant, T.D., M.S. Rao, C.Y. Koh and S.I. Stupp. 2008. Covalent functionalization of NiTi surfaces with bioactive peptide amphiphile nanofibers. Biomaterials 29(8): 1085–1098.

Seifert, B., T. Groth, K. Herrmann and P. Romaniuk. 1995. Immobilization of heparin on polylactide for application to degradable biomaterials in contact with blood. J. Biomater. Sci. Polym. Ed. 7(3): 277–287.

Sheiban, I., L. Carrieri, B. Catuzzo, P. Destefanis, E. Oliaro, C. Moretti and G.P. Trevi. 2002. Drug-eluting stent: the emerging technique for the prevention of restenosis. Minerva Cardioangiol. 50(5): 443–453.

Sheiban, I., G. Villata, M. Bollati, D. Sillano, M. Lotrionte and G. Biondi-Zoccai. 2008. Next-generation drug-eluting stents in coronary artery disease: focus on everolimus-eluting stent (Xience V). Vasc. Health Risk Manag. 4(1): 31–38.

Shuchman, M. 2006. Trading restenosis for thrombosis? New questions about drug-eluting stents. N. Engl. J. Med. 355(19): 1949–1952.

Simper, D., P.G. Stalboerger, C.J. Panetta, S. Wang and N.M. Caplice. 2002. Smooth muscle progenitor cells in human blood. Circulation 106(10): 1199–1204.

Stitzel, J., J. Liu, S.J. Lee, M. Komura, J. Berry, S. Soker, G. Lim, M. Van Dyke, R. Czerw, J.J. Yoo et al. 2006. Controlled fabrication of a biological vascular substitute. Biomaterials 27(7): 1088–1094.

Thomas, V., X. Zhang and Y.K. Vohra. 2009. A biomimetic tubular scaffold with spatially designed nanofibers of protein/PDS bio-blends. Biotechnol. Bioeng. 104(5): 1025–1033.

Verma, S. and P. Marsden. 2005. Nitric oxide-eluting polyurethanes—Vascular grafts of the future? N. Engl. J. Med. 353(7): 730–731.

Wendel, H.P., M. Avci-Adali and G. Ziemer. 2009. Endothelial progenitor cell capture stents—hype or hope? Int. J. Cardiol.: doi:10.1016/j.ijcard.2009.06.020.

Wessely, R., A. Schomig and A. Kastrati. 2006. Sirolimus and Paclitaxel on polymer-based drug-eluting stents: similar but different. J. Am. Coll. Cardiol. 47(4): 708–714.

Wilson, J.T., W. Cui, X.L. Sun, C. Tucker-Burden, C.J. Weber and E.L. Chaikof. 2007. *In vivo* biocompatibility and stability of a substrate-supported polymerizable membrane-mimetic film. Biomaterials 28(4): 609–617.

Ziche, M., L. Morbidelli, E. Masini, S. Amerini, H.J. Granger, C.A. Maggi, P. Geppetti and F. Ledda. 1994. Nitric oxide mediates angiogenesis *in vivo* and endothelial cell growth and migration in vitro promoted by substance P. J. Clin. Invest. 94(5): 2036–2044.

13

Nanofibre-based Vascular Grafts

Sarra de Valence,[1,a] *Beat H. Walpoth*[2] and *Michael Möller*[1,b,*]

ABSTRACT

While 500,000 coronary artery bypass surgeries are performed each year in the U.S., there are no clinically acceptable alternatives to autologous venous or arterial grafts for these small diameter vessels (< 6 mm). Therefore, much research is currently being conducted to develop efficient small diameter vascular grafts. Nanotechnologies have proven to be a promising tool in taking up such a challenge. Microfibre and nanofibre-based polymeric vascular grafts can be fabricated with electrospinning to create tubular scaffolds with a good biocompatibility and adequate mechanical strength. Biodegradable polymers of both natural and synthetic origin have been employed to promote vascular tissue regeneration, which should ideally lead to better long-term clinical performance. To tailor optimal tissue response to electrospun vascular grafts, the micro-architecture of the fibrous

[1]School of Pharmaceutical Sciences, University of Geneva, University of Lausanne, Pharmaceutics, Quai Ernest Ansermet, 30, CH-1211 Geneva 4, Switzerland.
[a]E-mail: sarra.devalence@unige.ch
[b]E-mail: michael.moeller@unige.ch
[2]Cardiovascular Research, Service of Cardiovascular Surgery, Faculty of Medicine, Geneva University Hospital, Rue Gabrielle-Perret-Gentil 4, CH-1211 Geneva 14, Switzerland.
E-mail: beat.walpoth@hcuge.ch
*Corresponding author

List of abbreviations after the text.

scaffold can be adapted, for example, with fibre size or pore size, or by creating different layers in the construct. In addition, fibres can be functionalized by drug incorporation, physical and chemical modifications, or cell seeding. Nanotechnologies offer promising solutions to problems often encountered in the vascular environment such as thrombosis, improper tissue responses, and calcification. Although active research in this field is still needed, the next generation of small diameter vascular grafts may emerge conjointly from the fields of nanotechnologies and regenerative medicine.

INTRODUCTION

The quest for the ideal vascular prosthesis has been going on for more than a century. Carrel and Guthrie first explored vascular replacement in the beginning of the 20th century when they successfully inserted a vein into the arterial circulation (Carrel and Guthrie 1906). They won the Nobel Prize in Medicine and Physiology in 1912. The first synthetic vascular graft was hand sewn by Voorhees out of Vinyon-N cloth in 1952 and implanted into dogs (Voorhees et al. 1952). Many other textiles and materials were evaluated for vascular applications in the following years including kitted polyethylene terephthalate (Dacron) in 1958 by DeBakey (Debakey et al. 1958) and expanded polytetrafluoroethylene (ePTFE) tubes in the early 1970s by Eiseman (Kelly and Eiseman 1982; Soyer et al. 1972), both of which proved very successful in the clinic and still remain the standard synthetic materials for revascularization procedures. The development of synthetic vascular prostheses was pivotal in the practice of vascular reconstruction, but while modern grafting materials have near excellent clinical results for large diameter arterial replacements, small diameter synthetic repairs (< 6 mm) are prone to failure. Small diameter arteries differ from larger diameters in terms of hemodynamics. Subject to lower blood flow velocities and lower wall shear stress, small diameter grafts are inclined to thrombosis, making inappropriate the synthetic ePTFE and Dacron grafts because of lack of endothelium (Berger et al. 1972). Autologous vessels such as the saphenous vein or the internal mammary artery remain the standard for small diameter revascularization procedures, but autologous material is not always available because of vascular disease or previous surgery. A new generation of man-made vascular grafts is therefore required to fill the clinical need for high-performing small diameter arterial prosthesis. The ideal small diameter vascular graft should mimic to the greatest extent the natural biological and physical properties of the artery it is meant to replace. Two main strategies can be employed to create such grafts: using a biostable material with excellent blood biocompatibility or designing a transient scaffold aimed for degradation and remodelling to allow the artery to regenerate in the long term. Because of the unique mechanical and biological environment in arteries, there are several requirements that must be satisfied for clinically acceptable small diameter vascular grafts: they must have good hemocompatibility, tissue biocompatibility, infection resistance, durable mechanical strength, kink resistance, no blood leakage, and suture retention, and they must be appropriate for surgical handling.

Nanotechologies are finding an increasing number of applications in modern medicine. Applying nanotechnologies to regenerative medicine brings a world of possibilities and may create solutions to the long-lasting challenge of small diameter vascular grafts. Nanofibre-based vascular grafts have many interesting advantages such as good mechanical properties, various porosities allowing cell ingrowth, ease of fabrication, and possibilities for advanced functionalization.

This chapter reviews the different fabrication techniques, materials, and advanced functionalization possibilities for nanofibre-based vascular grafts and discusses the potential of such constructs as next generation medical devices.

NANOFABRICATION TECHNIQUES FOR VASCULAR GRAFTS

Electrospinning

General Principles

Electrospinning is a remarkably simple and versatile technique to produce polymeric nano- and microfibres ranging from tens of nanometres to several micrometres, which can be easily collected to form a non-woven fabric. This technique has been known for more than 100 years and was patented in 1934 (Formhals 1934), but did not receive much attention until the end of the century. In the 1990s, in parallel to important advances in biomaterials science, electrospinning found much interest in academic circles for biomedical applications. Currently, the main biomedical applications for such fibres include scaffolds for tissue engineering, dressings to improve wound healing, and drug delivery. Vascular grafts have emerged as a promising application for such fibres (Greiner and Wendorff 2007).

Electrospinning consists of producing fine polymeric fibres with the help of a strong electrical field between a charged polymer solution and a grounded or oppositely charged collector. The minimum equipment required for an electrospinning setup is a syringe pump to eject a viscous polymer solution, an electrode to charge this solution, a high-voltage DC generator, and a grounded or oppositely charged surface to collect the fibres. The electrically charged polymer solution is fed through a nozzle forming a droplet. As the electrostatic forces overcome the droplet's surface tension, the fluid is stretched into the Taylor cone, resulting in a fluid jet (Fig. 1). As the polymer solution travels through the air, bending instabilities cause the jet to whip around, stretching the fluid stream to form solid fibers as the solvent evaporates. This process results in the deposition of solid fibres with diameters ranging from nanometres to micrometres onto the collector. In order to produce seamless vascular grafts with this method (Fig. 2), a cylindrical rotating and translating mandrel is used as a collector (Fig. 1).

Several parameters affect fibre morphology and quality:

- Polymer solution: solvent, concentration, viscosity, conductivity, surface tension, dielectric constant, volatility.
- Solution feed rate.
- Applied potential: voltage, polarity of the polymer solution and the collector.
- Collector: material, geometry, needle-collector distance, spin and translation speed.
- Environment: humidity, temperature.

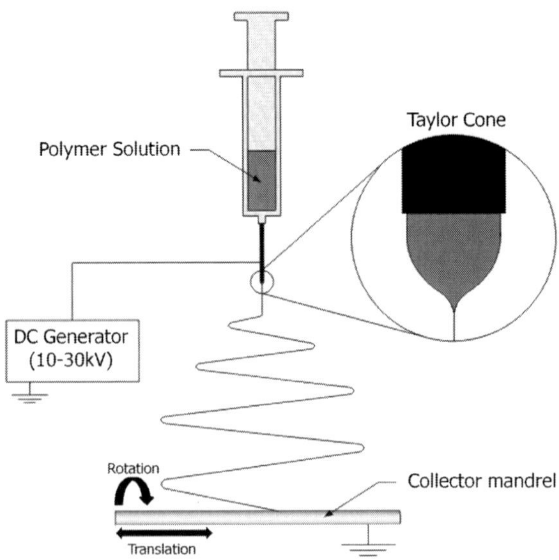

Fig. 1. Electrospinning. The basic setup is made of a viscous polymer solution that can be ejected at a controlled flow rate, an electrode to charge this solution, and a grounded collector. As the solution travels through the air, the solvent evaporates and fibres accumulate on the collector. Unpublished material from the author.

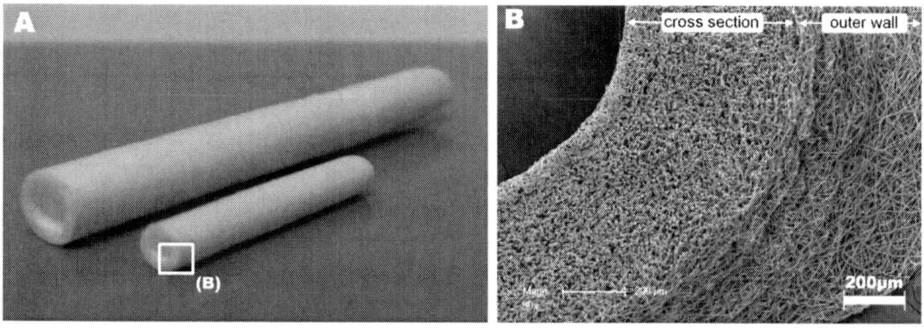

Fig. 2. Electrospun vascular graft. Macroscopic and microscopic view of an electrospun PCL vascular graft (2 mm and 4 mm inner diameter grafts in A). Electrospinning makes it possible to create homogeneous and highly porous scaffolds with sufficient mechanical properties to withstand pulsatile arterial blood flow. Unpublished material from the authors.

At first view, the simplicity of the technique may suggest that it is easily controllable, but fine-tuning the many parameters to obtain the desired fibre size, size distribution and aspect is a tedious task that have been addressed through complex factorial designs (Nottelet et al. 2009). Furthermore, optimal parameters are different for each electrospinning apparatus, so they must be redefined for every slight change in the setup. Poor control of these parameters can lead to imperfections such as merged fibres, bead formation (Fig. 3), or delamination, which can modify cell response and mechanical properties of the construct. A detailed description of the electrospinning process and the importance of the different parameters can be found in a book on polymer nanofibres (Andrady 2008).

Electrospinning clearly has many advantages, but its main drawback for tissue engineering scaffold applications is its mainly two-dimensional porosity. In an electrospun scaffold, the pore size depends directly on the fibre size. Nanoscale fibres will create very small pores that may impede cellular ingrowth if they are smaller than cell dimensions.

Advanced Electrospinning

Improvements to the basic electrospinning method described above have led to more advanced fibre structures, tailored for specific needs. Some of these features include fibre orientation, or core-shell fibre spinning.

Generally, electrospun fibres are randomly oriented, but achieving aligned fibres is also possible and has its advantages (Fig. 4a). This can be achieved in two ways: either by increasing the collector rotation speed and/or by modifying the electrical field with charged electrodes placed between the needle and the collector. Fibre alignment has been shown to increase the radial mechanical strength of an electrospun graft, therefore increasing its pressure resistance (Drilling et al. 2009), and also to enhance cell migration along the direction of orientation (He et al. 2006).

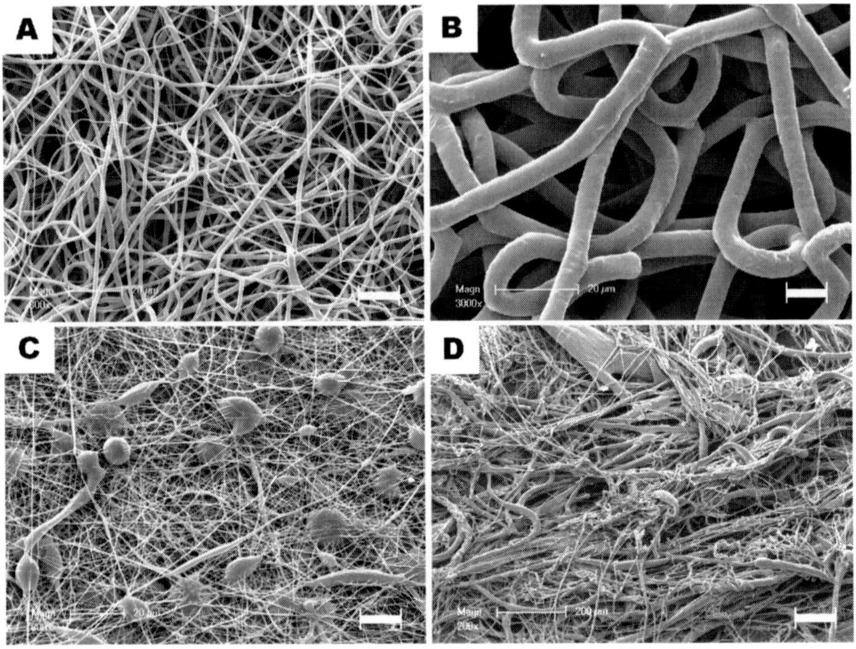

Fig. 3. Electrospun polymeric fibres. Electrospun fibres can have a variety of aspects and sizes (A and B). Imperfections such as polymer beads (C) or non-homogeneous surfaces (D) can be created if suboptimal spinning parameters are used. Added scale bars are 10 μm. Unpublished material from the authors.

Coaxial electrospinning is a modified form of electrospinning where two polymer solutions are fed through concentric feeder needles. When proper voltage and flow rate parameters are applied, the polymer solutions do not mix in the Taylor cone, and result in core-shell fibres (Fig. 4b). These types of fibres are interesting for several applications such as long-term drug release, mechanical support of low molecular weight polymers that cannot be electrospun on their own, or incorporation of biological molecules into electrospun fibres without altering their functionality with harsh solvents (Greiner and Wendorff 2007).

Other advanced features of electrospinning such as surface nanostructured fibres have been described in the literature, but have not yet been applied in the field of vascular grafts. These features are not discussed further in this chapter, but deserve attention from researchers interested in electrospinning technology (Andrady 2008).

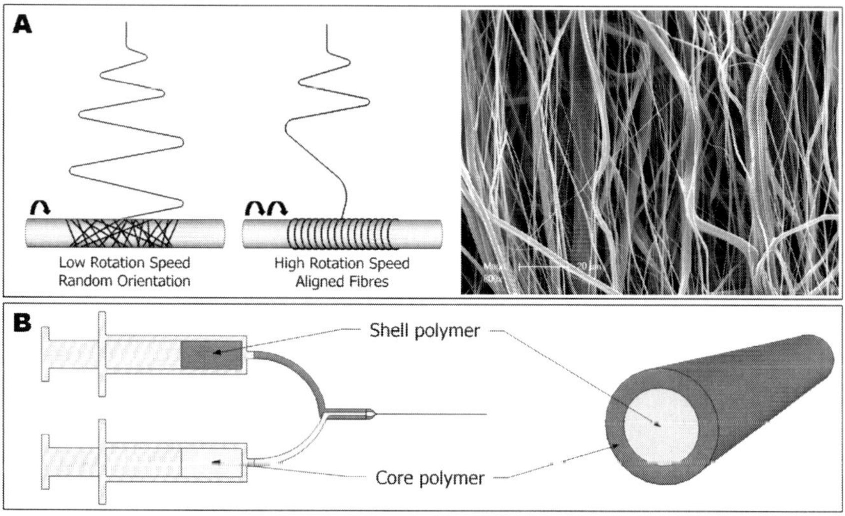

Fig. 4. Advanced electrospinning. Fibre alignment can be achieved by several techniques including increasing mandrel rotation speed (A). Coaxial electrospinning will create fibres with two distinct core and shell layers coming from different polymer solutions that are jointly electrospun (B). Unpublished material from the author.

Other Techniques

Although electrospinning remains the gold standard for the fabrication of nanofibre vascular grafts, there are two other established techniques to produce nanofibres: self-assembly and phase separation.

Self-assembly involves the spontaneous arrangement of molecules into patterns or structures through non-covalent interactions (Hartgerink et al. 1996). This is mostly applied to peptide formation, although some studies have been carried out with synthetic polymers. Self-assembly offers a three-dimensional pore arrangement, it can take place after injection for *in vivo* assembly, and cells can easily be integrated in the scaffold. However, it is a complex procedure with a limited range for fibre diameter, size (typically 2–30 nm diameter and 10 µm length), and orientation and three-dimensional arrangement cannot be controlled (Beachley and Wen 2010).

Scaffolds can also be formed by thermally induced liquid-liquid phase separation (Ma and Zhang 1999). This technique involves first dissolving the polymer in a solvent, then thermally inducing phase separation and gelation; the solvent can then be removed by water extraction, and the polymer is freeze-dried to form a three-dimensional nanofibrous scaffold with fibre diameters ranging from 50 to 500 nm. Phase separation does not require specialized equipment and is reproducible, but can only be applied to a limited number of polymers and is difficult to apply beyond the laboratory scale.

In the remainder of this chapter, fabrication of nanofibres is referred to as electrospinning or spinning of a polymer, since this is the most commonly used technique.

NANOFIBRE-BASED VASCULAR GRAFTS

Materials for Vascular Grafts

Biostable Materials

The requirements for a biostable vascular graft are somewhat different from those for a biodegradable graft. Since the material will remain in the body in the long term, it must have the right inherent properties to resist thrombosis, intimal hyperplasia, calcification, and aneurysms. To reduce the risk of intimal hyperplasia and calcification, the graft must have a good compliance match with the native artery. To resist aneurysms, the material must have a minimal hysteresis and creep at physiological pressure levels so that it does not distend over time. Thrombosis can be most efficiently prevented by a healthy luminal layer of endothelial cells so the material must provide a good surface for endothelial cell migration and proliferation, or it must be treated in order to durably prevent thrombosis.

As an example, polyurethanes have been found to be a promising material for biostable vascular grafts because of their superior tensile and elastic properties and their good biocompatibility. Vascular grafts have been electrospun from thermoplastic polyurethane (Grasl et al. 2010; Kang et al. 2008) and segmented polyurethane (Matsuda et al. 2005) and show unmatched compliance and mechanical properties for synthetic vascular grafts. Nevertheless, it is important to mention that very slow degradation by oxidation occurs in polyurethanes *in vivo*. This has been observed in several other biomedical devices such as pacemaker leads (Wiggins et al. 2001) and could be an issue for vascular graft design if not addressed properly.

Biodegradable Materials

Biodegradable materials act as transient scaffolds for regeneration of the vessel. A biodegradable scaffold must therefore have excellent biocompatibility and must be favourable to cell migration, infiltration, and proliferation. In addition, it must degrade fast enough to not impede the regeneration process and slow enough to not induce an aneurysm. This balance is very difficult to achieve and slowly degrading polymers are generally preferred because of the catastrophic outcome of aneurysms (Walpoth et al. 2007). Two types of biodegradable polymers are being evaluated for vascular grafts: synthetic polymers and natural polymers. While synthetic polymers have excellent mechanical properties, natural polymers have superior cell affinity.

Synthetic Polymers

The most frequently used synthetic biodegradable polymers used for vascular grafts are polycaprolactone, poly(lactic acid), poly(glycolic acid) , and their copolymers. Polydioxanone and other polymers are also occasionally employed, but will not be discussed in detail in this section.

Polycaprolactone. PCL is a biocompatible aliphatic polyester that degrades slowly (about 2 y) and possesses high tensile and elongation properties. Its excellent mechanical properties make it a good candidate for use in vascular grafts and it is probably the most frequently used polymer in nanofibre-based vascular graft studies. In addition, some PCL-based medical products have been approved by the U.S. Food and Drug Administration, which would significantly decrease the time to market if such a graft were to be used in the clinic. PCL has been electrospun with a variety of common solvents including acetone, chloroform, dimethylformamide, tetrahydrofuran, and 1, 1, 1, 3, 3, 3 hexafluoro-2-propanol (HFP). Uniform electrospun PCL fibres with a wide range of diameters can easily be obtained to form flexible, but poorly compliant scaffolds. This material is highly hydrophobic, which can lead to poor cell affinity, but makes it a good polymer for long-term applications and drug delivery. Electrospun PCL vascular grafts have been evaluated in the rat model up to 6 months and show rapid endothelialization, good mechanical stability, and favourable healing characteristics (Pektok et al. 2008).

Poly(glycolic acid). PGA is a hydrophilic, degrading polymer with better than average biocompatibility. Electrospun PGA fibres will be completely bioabsorbed *in vivo* after 6 to 12 mon, depending on the molecular weight, but they rapidly lose their mechanical properties since hydrolytic degradation accounts for nearly 60% loss in tensile strength after 2 wk implantation. Degradation of PGA also creates a sharp increase in local pH, which can be detrimental to tissue response (Barnes et al. 2007). Vascular grafts made of PGA must therefore be carefully designed to stimulate rapid tissue regeneration to compensate for the fast loss in mechanical strength of the scaffold.

Poly(lactic acid). PLA is a relatively hydrophobic polymer, which degrades more slowly than PGA. The homopolymer (PLLA) of the naturally occurring isomer L-lactide is often used for medical applications. It has a high strength and degrades slowly, taking more than 2 y to clear *in vivo.* Few studies have investigated pure PLA or PLLA as a vascular graft, most probably since this material has little elasticity and a too long residence time, but it is frequently used as a copolymer (Andrady 2008).

Common copolymers. Copolymers of common synthetic biodegradable polymers have been evaluated for the vascular graft application. By designing specific copolymers, it is possible to tailor degradation rate and mechanical properties to better meet the needs of the arterial environment.

One commonly used copolymer is poly(L-lactide-co-ε-caprolactone). Several groups have investigated these copolymers with different arrangements and different proportions of PLLA and PCL (Chung et al. 2010; He et al. 2009; Inoguchi et al. 2006; Qiu et al. 2009). These constructs have good mechanical properties and a compliance comparable to native arteries. They have been implanted *in vivo* and have shown good structural integrity and patency up to 7 wk (He et al. 2009).

Natural Polymers

In arteries, the extracellular matrix plays an essential role in arterial physiology. This complex multilayered organ must withstand pulsatile blood flow and high flow rates and pressures. The three main layers—the intima, the media, and the adventitia—play different physiological roles, shaped by different matrix compositions. The intimal layer is in direct contact with the bloodstream. It is made of a single layer of endothelial cells on a thin subendothelial layer of collagen type IV and elastin. The thick media is populated by smooth muscle cells embedded in collagen type I and III, elastin, and proteoglycans. The outermost adventitial layer is made of fibroblasts and collagen type I. In order to closer mimic natural artery physiology and improve nanofibre-graft results, there is a strong trend to incorporate biopolymers such as collagen and elastin in vascular graft constructs.

Collagen. Collagen is one of the most plentiful proteins found in the native artery, and with smooth muscle cells, it confers strength and resistance to rupture of blood vessels. Although there are more than 20 different types of collagen throughout the body, the main collagens of interest for cardiovascular engineering are types I, III, and IV. Type I collagen has excellent tensile strength, type III collagen contributes to both tensile strength and flexibility of the blood vessel, and type IV is a non-fibrillar collagen that becomes a sheet-like structure to form the major component of the basal lamina. In addition to its essential role for mechanical support, collagen has an important biological role. Collagen has very good affinity with cells and provides biological signals to promote tissue regeneration. Collagen has been successfully electrospun by several groups using HFP or PBS/ethanol as solvents, and these fibres closely mimic the biological and structural properties of natural collagen. However, electrospun collagen is generally not stable and has low mechanical strength, but cross-linking improves its structural stability. These features make collagen an excellent candidate for nanofibre-based vascular grafts aimed towards vessel regeneration (Sell et al. 2009).

Elastin. Elastin is an insoluble, but very stable polymer composed of massive arrays of covalently cross-linked molecules of tropoelastin, which provides the necessary elasticity for the artery to withstand pulsatile blood flow. In the arterial wall, elastogenic cells (fibroblasts, smooth muscle cells, and endothelial cells) secrete tropoelastin, which then binds together to form elastin fibres. While elastin is insoluble, the soluble forms of this molecule include alpha, beta or kappa elastin obtained by hydrolysing insoluble elastin, and naturally occurring tropoelastin. These forms can interact with elastin receptors on cells and mediate elastin synthesis, chemotaxis, inhibition of calcification, proliferation of fibroblasts and arterial smooth muscle cells, vasorelaxant activity, and matrix protease activity. Elastin and its derivatives can therefore influence the regenerative process in vascular grafts (Daamen et al. 2007). Soluble forms of elastin have been successfully electrospun into microfibres using HFP as a solvent (Li et al. 2005). However, these fibres of pure soluble elastin will rapidly disintegrate upon hydration if not cross-linked and they have poor inherent mechanical properties (Sell et al. 2009).

Gelatin. Gelatin is an amorphous biopolymer obtained by partial hydrolysis of collagen. Similarly to collagen, it is biodegradable and biocompatible, but available at a much lower cost, making it an attractive component for engineered scaffolds. However, gelatin forms a colloidal sol at temperatures above 37°, making it necessary to use cross-linking for making suitable grafts. Gelatin has been successfully electrospun with 2, 2, 2-trifluoroethanol as a solvent.

Fibrin. Fibrin is a large structural protein made from the polymerization of blood-borne fibrinogen after cleavage by thrombin. It is a central constituent of blood clots and plays a major role in wound healing where it acts as a natural scaffold for tissue regeneration. It contains many cell binding sites and is known to promote extracellular matrix secretion by cells. Fibrin gel, reinforced with a poly(D/L)lactide mesh, has recently been used to fabricate a tissue engineered vascular graft and shows promising results (Koch et al. 2010; Tschoeke et al. 2009). Fibrinogen can be electrospun (Wnek et al. 2003) and could also be an interesting component for nanofibre-based vascular grafts.

Silk. Silk fibroin is a naturally occurring polymer which has gained much attention in the biomedical fields mainly due to its exceptionally good mechanical properties. The silk is harvested from silkworm *Bombyx mori* cocoons and commercially available in large quantities. Silk is slowly degraded by enzymatic degradation after implantation *in vivo* and degradation rates can be tailored from months to years depending on the processing technique. Silk is also very stable to high temperatures allowing it to be autoclaved for sterilization. Silk can be electrospun in an aqueous solution, making it an excellent matrix for the incorporation of bioactive components such as proteins, growth factors, or cells that are sensitive to the harsh solvents used to dissolve other polymers (Zhang et al. 2009). *In vitro* studies have been conducted to demonstrate the favourable interactions between silk electrospun fibres and arterial smooth muscle cells and endothelial cells, showing the potential of this material for vascular graft applications (Zhang et al. 2008).

Mix of Natural and Synthetic Polymers

As described above, natural and synthetic polymers both have their advantages. While natural polymers have interesting biocompatibility properties, synthetic polymers confer good mechanical properties. Mixing natural and synthetic polymers may overcome the shortcomings of each type. Many studies have been published with different mixes of natural and synthetic polymers and are not described in detail here. Generally, a blend solution of a synthetic polymer and one or several biological polymers is electrospun to form a vascular graft. The overall mechanical strength of the graft is reduced compared to a purely synthetic graft, but the bioaffinity is increased.

For clinical application, several important aspects must be taken into account, including material source and sterilization. Synthetic polymers are usually available at medical grade, but natural polymers are often from animal sources, which could cause undesirable immunogenic side effects (Ratner et al. 2004). This problem would be solved by using human-derived or recombinant polymers, but these can be difficult to acquire. There are three main methods for sterilizing medical devices: thermal sterilization, radiation sterilization, and chemical sterilization. Depending on the materials used, some methods may denature or significantly affect mechanical properties of the construct. It is therefore important to consider this aspect while designing new vascular grafts.

Advanced Functionalization of Vascular Grafts

Drug Release

In order to increase the biocompatibility of materials used for vascular grafts, bioactive molecules can be directly incorporated into the structure for local drug release and action. There are several ways to incorporate drugs in a biomaterial polymer/drug mixing, covalent bonding of the drug onto the polymer, or core-shell electrospinning of the drug and the polymer.

The drug can be incorporated directly into the polymer solution before electrospinning, creating a drug-releasing scaffold. However, there are several conditions that must be satisfied for this approach to work. First of all, the drug must be stable in the solvent system used. Since harsh solvents are used in most cases for electrospinning, this can be an important limiting factor, especially for biological molecules such as proteins or growth factors, which can easily be denatured and lose their functionality in such solvents. The drug must also be stable to other stresses applied during graft fabrication such as high electrical fields and sterilization. Second, for a drug to play for instance a significant role in thrombosis or in the vascular remodelling process, it must be delivered for several days to weeks. In terms of drug delivery technology, this can only be achieved if the drug has a very high affinity with the polymer, generally through hydrophobic interactions. In the vascular graft literature, this approach has been used for the local delivery of heparin (Luong-Van et al. 2006), nitric oxide (Kapadia et al. 2008), paclitaxel (Innocente et al. 2009), and FGF2 (Ye et al. 2010b).

Covalent bonding of a drug to the polymer is another approach for timely controlled delivery. The bonding will allow long-term local activity of the drug with a much slower loss of activity. This is interesting for vascular graft applications since regeneration occurs on long time spans, typically months. Moreover, covalent bonding to the surface of the polymer can be done directly on the electrospun scaffold, allowing the drug to not be exposed to the solvent system. This approach has been applied for example to covalently link gelatin (Ma et al. 2005a, b), RGD peptides (Choi et al. 2008), vascular endothelial growth factor (Miyazu et al. 2010), and heparin (Ye et al. 2010a).

Physical and Chemical Modifications

Several studies describe physical and chemical methods to increase mechanical properties and/or chemical properties of nanofibre grafts. Mechanical properties can for example be increased by thermal treatments consisting of heating the graft to a temperature near the melting temperature of the polymer to increase connection points between the fibres (Lee et al. 2008). Chemical treatment with NaOH or HCL (Boland et al. 2004), or plasma treatment (He et al. 2006; Sanders et al. 2005) will increase the hydrophilicity of the polymer surface, improving its affinity to cells.

Cell Seeding

The importance of cell seeding into vascular grafts prior to implantation is a subject of much discussion among the scientific community. Some support cell-free vascular grafts, which are shelf-ready and rely entirely on the scaffold's ability to attract the necessary cells for tissue regeneration. Others endorse endothelial cell, smooth muscle cell, or stem cell seeding and maturation before implantation. A comprehensive review of the different techniques of cell seeding has been published by Villalona et al. (2010). Although pre-seeding will enhance biocompatibility of vascular grafts, it raises issues such as cell source, seeded cell survival, delayed availability of grafts in the clinic, and added complexity with regulatory aspects. However, if the benefit of pre-seeding vascular grafts is clearly demonstrated (Zilla et al. 2007b), this technology could be used in the future for high-performance vascular grafts. This may prove to be important for revascularization procedures of elderly patients, since their tissue regeneration potential is greatly reduced by their age (Zilla et al. 2007a). Obtaining a confluent and healthy endothelium in such patients *in situ* may be too slow and expose the patient to high risks of thrombosis. Cell seeding is therefore important to consider, since it has shown positive clinical results (Deutsch et al. 1999).

ADDRESSING VASCULAR GRAFT PROBLEMS WITH NANOFIBRE TECHNOLOGY

Vascular Regeneration

Since biostable materials have proven to be susceptible to many issues in the long term, such as infection, intimal hyperplasia, calcification, or thrombosis, much research is being carried out on biodegradable vascular grafts to serve as transient scaffolds for tissue regeneration. Vascular regeneration involves two main areas: regeneration of the luminal endothelial lining and regeneration of the vessel wall, which is rich in smooth muscle cells. A healthy endothelium will prevent thrombosis and inhibit intimal hyperplasia, and smooth muscle cells will secrete the extracellular matrix to confer elasticity and mechanical resistance of the construct. Biodegradable vascular grafts must therefore be remodelled for proper biological and mechanical matching of the regenerated vessel and the adjacent native vessel in order to assure good long-term performance.

The layer-by-layer approach can address the specific problems of the lumen and the adventitia separately (Ju et al. 2010). In terms of architecture, a layer of small diameter fibres on the inside of the graft will allow rapid proliferation and migration of endothelial cells. However, scaffolds with nano-scale fibre diameters have small pore sizes, which can inhibit cellular infiltration. In the graft wall, cellular infiltration should be maximal to promote proper tissue regeneration and timely polymer degradation. Therefore, a structure with larger fibres and larger pores is needed. Different layers can also be loaded with different drugs in order to further enhance the overall performance. Although the layer-by-layer approach has a strong potential for vascular grafts, it is also important to consider the risk of delamination between layers. If the layers delaminate, the structural integrity of the graft is lost, which will affect overall mechanical properties and cellular response.

Thrombosis

Thrombosis remains a major issue for vascular grafts and is the main cause of small diameter ePTFE graft failure in the clinic. The best anti-thrombogenic surface known today remains the natural endothelial cell lining and maintaining a confluent, adherent, and functional lining is the best protection against thrombosis. In order to avoid thrombosis on vascular grafts, one option is to pre-seed the lumen with endothelial cells (Ma et al. 2005b). Although this strategy shows good results for resisting thrombosis, it is not discussed further here. Instead, we focus on nanotechnology strategies to prevent thrombosis.

Surface architecture is an important aspect of endothelial cell affinity with nanofibres. Fibre size and pore size influence endothelial cell proliferation and migration. Fibre architectures that most favour endothelialization have small fibres and small pore sizes (Ju et al. 2010).

In addition to surface architecture, there are three main strategies to create thromboresistant luminal surfaces of nanofibre vascular grafts:

1. The immobilization of anti-thrombotic molecules on the surface to create an acellular thromboresistant lumen.
2. The immobilization of molecules to increase population speed and stability of endothelial cells.
3. The enhancement of endothelialization via nitric oxide production.

An interesting review by Jordan et al. has been published on modifications of materials to create thromboresistant surfaces for vascular grafts (Jordan and Chaikof 2007). It lists the different molecules that have been incorporated in or bound to vascular grafts to improve their thromboresistance.

Calcification

Calcification of biomaterials in humans, and particularly in children, is still incompletely understood. For example, calcification of ePTFE vascular grafts is frequently observed in the clinic (Mehta et al. 2010), but has been only lightly reported on compared to other issues with these grafts. Calcifications greatly reduce the overall compliance of the construct, which increases the risk of intimal hyperplasia leading to patency problems. Moreover, calcified nodules are very stable and degrade with difficulty, so it is important to prevent their occurrence rather than treat them. Of the many risk factors for calcification, three probable ones are inflammation (Aikawa et al. 2007), hypoxia, and compliance mismatch. Hypoxia could be reduced by better scaffold architecture and composition to enhance neoangiogenesis in the graft wall. Compliance mismatch between the graft and the native artery can be addressed by fibre organization and choice of materials. Very few studies mention calcifications in nanofibre-based vascular grafts and mechanisms of this calcification are poorly understood, but this is probably because few constructs have been evaluated *in vivo* for a long enough time. Nevertheless, graft calcification remains an important problem to be addressed early in the process of designing new vascular grafts and deserves more attention from researchers in this field.

CONCLUSION

The creation of clinically acceptable vascular grafts with nanofibre technologies is highly promising. Current technologies offer many possibilities to modify and functionalize nanofibre-based vascular grafts, which may lead to the next generation of small diameter vascular grafts. While currently used grafts are mostly biologically inert, the next generation of grafts will be biofunctional and guide vascular regeneration to form a new vessel, rather than replacing it in the long term.

APPLICATIONS TO AREA OF HEALTH AND DISEASE

Myocardial and peripheral revasularization procedures are performed millions of times each year. These procedures require small-calibre vascular grafts and most often autologous arterial or venous grafts are used. However, autologous material is not always available because of previous surgery or generalized vascular disease. Also, vein removal from the leg creates discomfort for the patient. Currently approved synthetic vascular grafts materials (ePTFE and Dacron) show good long-term results for large diameter vessel replacements (> 6 mm), but are prone to failure by thrombosis and intimal hyperplasia for smaller diameter vessels. There is therefore an important clinical need for shelf-ready small diameter vascular grafts. Past experiences have shown that it is difficult to design durable biostable vascular conduits: no material can work as well as natural tissues. Nanotechnology-based methods have therefore been used to find solutions to this problem. Nanofibre-based electrospun biodegradable vascular grafts show a good potential for guiding vascular regeneration and providing the clinically needed long-lasting small diameter vascular replacement.

Definitions

Aneurysm: A localized dilation of a blood vessel. In the case of vascular grafts, this can occur if the mechanical strength of the graft is not sufficient to withstand blood pressure and pulsatile flow. Aneurysms can lead to the rupture of the vessel, which can be catastrophic for the patient.

Compliance: Vascular compliance is a measure of elasticity. It reflects the ability of a vessel to expand, then recoil in response to pressure waves.

Electrospinning: A method to produce micro and nano polymeric fibres with the help of a strong electrical field.

Intimal hyperplasia: An uncontrolled smooth muscle cell proliferation under the endothelial layer of vascular grafts resulting in a narrowing of the vessel lumen, which can eventually lead to occlusion.

Patency: A patent vessel or vascular graft has no stenosis, or obstruction to flow.

Key Facts

- Small diameter vascular grafts are used mainly for coronary and peripheral bypass and access surgery. In coronary bypass surgery, a 15- to 20 cm-long conduit is grafted to the ascending aorta and to the obstructed coronary artery, downstream of the obstructions in order to maintain proper perfusion of the heart muscle.
- More than 1,000,000 m of small diameter vascular grafts are implanted each year as coronary bypasses, the majority of which are of autologous arterial or venous source, but clinically approved synthetic vascular grafts are prone to failure by thrombosis and intimal hyperplasia when used at diameters less than 6 mm.

- In addition to a good tissue regeneration potential, biodegradable nanofibre vascular grafts must have hemocompatibility and tissue biocompatibility, infection resistance, durable mechanical strength, kink resistance, no blood leakage, and suture retention, and they must be appropriate for surgical handling.
- The main issues that need to be avoided in vascular grafts are thrombosis, aneurysms, intimal hyperplasia, infections, and calcification. Nanotechnologies offer promising solutions to all these problems.
- More than 100,000 times each day, the heart pumps pressurized blood into the arteries. For a vascular graft to withstand such an environment, it must retain excellent mechanical strength throughout the patient's lifetime since a rupture of the graft could have fatal consequences.

Summary Points

- The main method for nanofibre-based vascular graft fabrication is electrospinning.
- Electrospinning creates tubular scaffolds with varying porosities suitable for cell invasion and good mechanical properties.
- Polymers used for vascular grafts can be synthetic or natural and biostable or biodegradable.
- Synthetic polymers offer superior mechanical properties, whereas biological polymers have a better bioaffinity. The current trend is to create vascular grafts from a blend of synthetic and natural polymers in order to benefit from the advantages of both types.
- Functionalizing nanofibre-based vascular grafts is possible by drug incorporation in the fibres, drug covalent bonding to the surface of the fibres, physical or chemical modifications of the polymer, and cell pre-seeding into or on the scaffold.
- Nanotechnologies can help address vascular-specific problems such as vascular tissue regeneration, thrombosis, and calcification.

Abbreviations

ePTFE	:	expanded polytetrafluoroethylene
HFP	:	1, 1, 1, 3, 3, 3-hexafluoro-2-propanol
PBS	:	phosphate buffered saline
PCL	:	polycaprolactone
PGA	:	poly(glycolic acid)
PLA	:	poly(lactic acid)
PLLA	:	poly-L-lactide acid

References

Aikawa, E., M. Nahrendorf, J.L. Figueiredo, F.K. Swirski, T. Shtatland, R.H. Kohler, F.A. Jaffer, M. Aikawa and R. Weissleder. 2007. Osteogenesis associates with inflammation in early-stage atherosclerosis evaluated by molecular imaging in vivo. Circulation 24: 2841–2850.

Andrady, A.L. 2008. Science and Techology of Polymer Nanofibers. John Wiley & Sons, Inc., New Jersey.

Barnes, C.P., S.A. Sell, E.D. Boland, D.G. Simpson and G.L. Bowlin. 2007. Nanofiber technology: Designing the next generation of tissue engineering scaffolds. Adv. Drug Delivery Rev. 14: 1413–1433.

Beachley, V. and X.J. Wen. 2010. Polymer nanofibrous structures: Fabrication, biofunctionalization, and cell interactions. Prog. Polymer Sci. 7: 868–892.

Berger, K., S.J. Wood, L.R. Sauvage and A.M. Rao. 1972. Healing of arterial prostheses in man—its incompleteness. Ann. Surg. 1: 118 ff.

Boland, E.D., T.A. Telemeco, D.G. Simpson, G.E. Wnek and G.L. Bowlin. 2004. Utilizing acid pretreatment and electrospinning to improve biocompatibility of poly(glycolic acid) for tissue engineering. J. Biomed. Mater. Res. B-Appl. Biomater. 1: 144–152.

Carrel, A. and C. Guthrie. 1906. Uniterminal and Biterminal Venous Transplantations. Surg. Gynecol. Obstet. 2: 266–286.

Choi, W.S., J.W. Bae, H.R. Lim, Y.K. Joung, J.C. Park, I.K. Kwon and K.D. Park. 2008. RGD peptide-immobilized electrospun matrix of polyurethane for enhanced endothelial cell affinity. Biomed. Mater. 3(4): 044104.

Chung, S.W., N.P. Ingle, G.A. Montero, S.H. Kim and M.W. King. 2010. Bioresorbable elastomeric vascular tissue engineering scaffolds via melt spinning and electrospinning. Acta Biomaterialia 6: 1958–1967.

Daamen, W.F., J.H. Veerkamp, J.C.M. van Hest and T.H. van Kuppevelt. 2007. Elastin as a biomaterial for tissue engineering. Biomaterials 30: 4378–4398.

Debakey, M.E., D.A. Cooley, E.S. Crawford and G.C. Morris. 1958. Clinical application of a new flexible knitted Dacron arterial substitute. Arch. Surg. 5: 713–724.

Deutsch, M., J. Meinhart, T. Fischlein, P. Preiss and P. Zilla. 1999. Clinical autologous *in vitro* endothelialization of infrainguinal ePTFE grafts in 100 patients: A 9-year experience. Surgery 5: 847–855.

Drilling, S., J. Gaumer and J. Lannutti. 2009. Fabrication of burst pressure competent vascular grafts via electrospinning: Effects of microstructure. J. Biomed. Mater. Res. A. 4: 923–934.

Formhals, A. 1934. Process and apparatus for preparing artificial threads. U.S. Patent 1975504.

Grasl, C., H. Bergmeister, M. Stoiber, H. Schima and G. Weigel. 2010. Electrospun polyurethane vascular grafts: *In vitro* mechanical behavior and endothelial adhesion molecule expression. J. Biomed. Mater. Res. A. 2: 716–723.

Greiner, A. and J.H. Wendorff. 2007. Electrospinning: A fascinating method for the preparation of ultrathin fibres. Angewandte Chemie-Intl. Ed. 30: 5670–5703.

Hartgerink, J.D., J.R. Granja, R.A. Milligan and M.R. Ghadiri. 1996. Self-assembling peptide nanotubes. J. Am. Chem. Soc. 1: 43–50.

He, W., Z.W. Ma, W.E. Teo, Y.X. Dong, P.A. Robless, T.C. Lim and S. Ramakrishna. 2009. Tubular nanofiber scaffolds for tissue engineered small-diameter vascular grafts. J. Biomed. Mater. Res. A. 1: 205–216.

He, W., T. Yong, Z.W. Ma, R. Inai, W.E. Teo and S. Ramakrishna. 2006. Biodegradable polymer nanofiber mesh to maintain functions of endothelial cells. Tissue Eng. 9: 2457–2466.

Innocente, F., D. Mandracchia, E. Pektok, B. Nottelet, J.C. Tille, S. de Valence, G. Faggian, A. Mazzucco, A. Kalangos, R. Gurny, M. Moeller and B.H. Walpoth. 2009. Paclitaxel-eluting biodegradable synthetic vascular prostheses—a step towards reduction of neointima formation? Circulation 11: S37–S45.

Inoguchi, H., I.K. Kwon, E. Inoue, K. Takamizawa, Y. Maehara and T. Matsuda. 2006. Mechanical responses of a compliant electrospun poly(L-lactide-co-epsilon-caprolactone) small-diameter vascular graft. Biomaterials 8: 1470–1478.

Jordan, S.W. and E.L. Chaikof. 2007. Novel thromboresistant materials. J. Vasc. Surg. 104A–115A.

Ju, Y.M., J.S. Choi, A. Atala, J.J. Yoo and S.J. Lee. 2010. Bilayered scaffold for engineering cellularized blood vessels. Biomaterials 15: 4313–4321.

Kang, Y.K., C.H. Park, H. Chang, K. Minn and C.Y. Park. 2008. Development of thermoplastic polyurethane vascular prostheses. J. Appl. Polymer Sci. 5: 3267–3274.

Kapadia, M.R., L.W. Chow, N.D. Tsihlis, S.S. Ahanchi, J.A. Hrabie, J. Murar, J. Martinez, D.A. Popowich, Q. Jiang, J.A. Hrabie, J.E. Saavedra, L.K. Keefer, J.F. Hulvat, S.I. Stupp and M.P. Kibbe. 2008. Nitric oxide and nanotechnology: A novel approach to inhibit neointimal hyperplasia. J. Vasc. Surg. 1: 173–182.

Kelly, G.L. and B. Eiseman. 1982. Development of a new vascular prosthetic—lessons learned. Arch. Surg. 10: 1367–1370.

Koch, S., T.C. Flanagan, J.S. Sachweh, F. Tanios, H. Schnoering, T. Deichmann, V. Ella, M. Kellomaki, N. Gronloh, T. Gries, R. Tolba, T. Schmitz-Rode and S. Jockenhoevel. 2010. Fibrin-polylactide-based tissue-engineered vascular graft in the arterial circulation. Biomaterials 17: 4731–4739.

Lee, S.J., S.H. Oh, J. Liu, S. Soker, A. Atala and J.J. Yoo. 2008. The use of thermal treatments to enhance the mechanical properties of electrospun poly (epsilon-caprolactone) scaffolds. Biomaterials 10: 1422–1430.

Li, M.Y., M.J. Mondrinos, M.R. Gandhi, F.K. Ko, A.S. Weiss and P.I. Lelkes. 2005. Electrospun protein fibers as matrices for tissue engineering. Biomaterials 30: 5999–6008.

Luong-Van, E., L. Grondahl, K.N. Chua, K.W. Leong, V. Nurcombe and S.M. Cool. 2006. Controlled release of heparin from poly(epsilon-caprolactone) electrospun fibers. Biomaterials 9: 2042–2050.

Ma, P.X., and R.Y. Zhang. 1999. Synthetic nano-scale fibrous extracellular matrix. J. Biomed. Mater. Res. 1: 60–72.

Ma, Z.W., W. He, T. Yong and S. Ramakrishna. 2005a. Grafting of gelatin on electrospun poly(caprolactone) nanofibers to improve endothelial cell spreading and proliferation and to control cell orientation. Tissue Eng. 7–8: 1149–1158.

Ma, Z.W., M. Kotaki, T. Yong, W. He and S. Ramakrishna. 2005b. Surface engineering of electrospun polyethylene terephthalate (PET) nanofibers towards development of a new material for blood vessel engineering. Biomaterials 15: 2527–2536.

Matsuda, T., M. Ihara, H. Inoguchi, K. Kwon, K. Takamizawa and S. Kidoaki. 2005. Mechano-active scaffold design of small-diameter artificial graft made of electrospun segmented polyurethane fabrics. J. Biomed. Mater. Res. A. 1: 125–131.

Mehta, R.I., A.K. Mukherjee, T.D. Patterson and M.C. Fishbein. 2010. Pathology of explanted polytetrafluoroethylene vascular grafts. Cardiovasc. Pathol. In Press:

Miyazu, K., D. Kawahara, H. Ohtake, G. Watanabe and T. Matsuda. 2010. Luminal surface design of electrospun small-diameter graft aiming at in situ capture of endothelial progenitor cell. J. Biomed. Mater. Res. B-Appl. Biomater. 1: 53–63.

Nottelet, B., E. Pektok, D. Mandracchia, J.C. Tille, B. Walpoth, R. Gurny and M. Moller. 2009. Factorial design optimization and in vivo feasibility of poly(epsilon-caprolactone)-micro- and narofiber-based small diameter vascular grafts. J. Biomed. Mater. Res. A. 4: 865–875.

Pektok, E., B. Nottelet, J.C. Tille, R. Gurny, A. Kalangos, M. Moeller and B.H. Walpoth. 2008. Degradation and healing characteristics of small-diameter poly(epsilon-caprolactone) vascular grafts in the rat systemic arterial circulation. Circulation 24: 2563–2570.

Qiu, R.X., C.M. Li, L. Ye, J.D. Dong, A.Y. Zhang, Y.Q. Gu and Z.G. Feng. 2009. Electrospinning of synthesized triblock copolymers of epsilon-caprolactone and L-lactide for the application of vascular tissue engineering. Biomed. Mater. 4(4): 044105.

Ratner, B.D., A.S. Hoffman, F.J. Schoen and J.E. Lemons. 2004. Biomaterials Science: An Introduction to Materials in Medicine. Academic Press.
Sanders, J.E., S.E. Lamont, A. Karchin, S.L. Golledge and B.D. Ratner. 2005. Fibro-porous meshes made from polyurethane micro-fibers: effects of surface charge on tissue response. Biomaterials 7: 813–818.
Sell, S.A., M.J. McClure, K. Garg, P.S. Wolfe and G.L. Bowlin. 2009. Electrospinning of collagen/biopolymers for regenerative medicine and cardiovascular tissue engineering. Adv. Drug Delivery Rev. 12: 1007–1019.
Soyer, T., P. Cooper, L. Norton, M. Lempinen and B. Eiseman. 1972. New venous prosthesis. Surgery 6: 864.
Tschoeke, B., T.C. Flanagan, S. Koch, M.S. Harwoko, T. Deichmann, V. Ella, J.S. Sachweh, M. Kellomaki, T. Gries, T. Schmitz-Rode and S. Jockenhoevel. 2009. Tissue-engineered small-caliber vascular graft based on a novel biodegradable composite fibrin-polylactide scaffold. Tissue Eng. A. 8: 1909–1918.
Villalona, G.A., B. Udelsman, D.R. Duncan, E. McGillicuddy, R.F. Sawh Martinez, N. Hibino, C. Painter, T. Mirensky, B. Erickson, T. Shinoka and C.K. Breuer. 2010. Cell-seeding techniques in vascular tissue engineering. Tissue Eng. B-Rev. 3: 341–350.
Voorhees, A.B., A. Jaretzki and A.H. Blakemore. 1952. The use of tubes constructed from Vinyon-N cloth in bridging arterial defects—a preliminary report. Ann. Surg. 3: 332–336.
Walpoth, B.H., M. Cikirikcioglu, D. Mugnai, J.C. Tiller, E. Pektok, R. Pfister, D. Palombo, A. Kalangos and G. Bowlin. 2007. Biodegradable small calibre polydioxanone-based vascular prostheses: Potential as coronary bypass grafts. Circulation 16: 444–444.
Wiggins, M.J., B. Wilkoff, J.M. Anderson and A. Hiltner. 2001. Biodegradation of polyether polyurethane inner insulation in bipolar pacemaker leads. J. Biomed. Mater. Res. 3: 302–307.
Wnek, G.E., M.E. Carr, D.G. Simpson and G.L. Bowlin. 2003. Electrospinning of nanofiber fibrinogen structures. Nano Lett. 2: 213–216.
Ye, L., X. Wu, Q. Mu, B. Chen, Y. Duan, X. Geng, Y. Gu, A. Zhang, J. Zhang and Z.G. Feng. 2010a. Heparin-conjugated PCL scaffolds fabricated by electrospinning and loaded with fibroblast growth factor 2. J. Biomater. Sci. Polym. Ed.
Zhang, X.H., C.B. Baughman and D.L. Kaplan. 2008. *In vitro* evaluation of electrospun silk fibroin scaffolds for vascular cell growth. Biomaterials 14: 2217–2227.
Zhang, X.H., M.R. Reagan and D.L. Kaplan. 2009. Electrospun silk biomaterial scaffolds for regenerative medicine. Adv. Drug Delivery Rev. 12: 988–1006.
Zilla, P., D. Bezuidenhout and P. Human. 2007a. Prosthetic vascular grafts: Wrong models, wrong questions and no healing. Biomaterials 34: 5009–5027.

Section 2: Focused Areas, Treatments and Diseases

14

Nanoparticle Processing of Cholesterol-Lowering Drug

Toshiro Fukami,[1,*] *Toyofumi Suzuki,*[2,a]
Ayyalusamy Ramamoorthy[3] and *Kazuo Tomono*[2,b]

ABSTRACT

High cholesterol levels (or hypercholesterolemia) are associated with many diseases, particularly with the risk of coronary heart diseases. Probucol is commonly used to reduce cholesterol in blood; however, the effectiveness of this drug is highly dependent on its solubility and it is almost insoluble (solubility is about 5 ng/ml in water). Therefore, it is essential to develop approaches to increase the solubility, and thus the bioavailability, to enhance the efficiency of this drug. We reviewed new methods for increasing the solubility of probucol in water and to enhance

[1]Laboratory of Pharmaceutics, School of Pharmacy, Nihon University, 7-7-1 Narashinodai, Funabashi, Chiba 274-8555, Japan.
E-mail: fukami.toshiro@nihon-u.ac.jp

[2]Laboratory of Pharmaceutics, School of Pharmacy, Nihon University.
[a]E-mail: suzuki.toyofumi@nihon-u.ac.jp
[b]E-mail: tomono.kazuo@nihon-u.ac.jp

[3]Department of Chemistry and Biophysics, University of Michigan, 930 North University Avenue, Ann Arbor, MI 48109-1055, USA.
E-mail: ramamoor@umich.edu

*Corresponding author

List of abbreviations after the text.

drug absorption *in vivo*. These new methods for processing the drug nanoparticles use the grinding of probucol together with sodium dodecylsulfate and aqueous polymers. Particle size measurement of the nanoparticles prepared from ternary components indicated that probucol is present as colloidal nanoparticles in water. Microscopic techniques, such as scanning electron microscopy and atomic force microscopy, confirmed the presence of primary nanoparticles in the powder of a ternary ground mixture. In addition, atomic force micrographs indicated that the nanoparticles have a core-shell structure and the rigid core originates from probucol nanocrystals covered with a soft substance. Solid-state nuclear magnetic resonance experiments were performed to elucidate the underlying molecular interactions in the formation of nanoparticles with atomic-level resolution. 2D $^{1}H/^{1}H$ correlation experiments suggest that the hydrophobic interaction between probucol and the polymer plays an important role in their binding. Caco-2 cells were used to test the amounts of probucol permeated from a ternary mixture. The experimental results suggest that the nanoparticles permeate the cell membrane, while the probucol alone or a physical mixture of non-ground 1:5:1 probucol/methacrylic copolymer/surfactant did not permeate the cell membrane, even after 1 h. Shudo et al. (2008) reported that *in vivo* absorption was significantly improved using probucol/polyvinylpyrrolidone/surfactant ternary ground mixtures, compared with those of ternary physical mixtures and unprocessed probucol. Although a powder mixture containing drug nanoparticles was successfully prepared, the powder was sticky and difficult to handle, rendering it unsuitable for practical use by patients. Therefore, care was taken to prepare a pharmaceutical dosage form using the drug nanoparticles available for oral administration. Spraying of the suspended ternary ground mixture provided pharmaceutical granules or powders that contained probucol nanoparticles in the polymeric structure.

INTRODUCTION

High cholesterol levels (or hypercholesterolemia) are associated with numerous diseases, particularly with an increased risk of coronary heart diseases (Kathiresan et al. 2008). Anti-hypolipidemic drugs based on statin are commonly used to inhibit cholesterol synthesis and lower cholesterol serum levels. However, a number of side effects, such as hepatic failure, thrombocytopenia, and skeletal muscle abnormalities including myopathy and rhabdomyolysis, have reportedly been caused by these drugs (Kobayashi et al. 2008). There is therefore significant interest in developing alternative compounds for treating high cholesterol levels.

Probucol (PBC) is commonly used to reduce cholesterol in blood. The main advantages of using PBC over other drugs are its improved acceptance and tolerance by patients, and ease of administration. It is also considerably less expensive than other anti-hypolipidemic drugs available on the market. PBC lowers serum cholesterol levels by increasing the fractional rate of low-density lipoprotein (LDL) catabolism in the final metabolic pathway for cholesterol elimination from the body (Sawayama et al. 2002). Studies have shown that PBC inhibits dietary cholesterol absorption, biosynthesis of cholesterol, and oxidation and tissue deposition of LDL cholesterol (Braun et al. 2003). However, since the extremely poor solubility of PBC markedly decreases the drug's ability to suppress cholesterol levels (Betge et al. 2007), development of new approaches to increase the solubility of PBC is necessary.

PBC functions via a different mechanism from that of conventional anti-hyperlipidemic agents based on statin, and such new approaches would enable PBC and its derivatives to suppress the level of pathogenic degenerate LDL in blood. The focus of this chapter is a nanoparticle processing approach that enhances the bioavailability of PBC. In addition, the design of solid dosage forms for practical use, which was achieved by the integration of a cogrinding and a spraying technique, will be described. The approach will enable hydrophobic pharmaceutical compounds to be maintained as nanosized particles.

ENHANCEMENT IN SOLUBILITY OF POORLY WATER-SOLUBLE DRUGS

Most of the newly developed active pharmaceutical ingredients (API) that have been produced in recent years are poorly water-soluble (Timpe 2007). APIs often show low bioavailability when administered orally, because the dissolution rate of the drugs in the gastrointestinal tract is usually a rate-limiting step (Blagden et al. 2007). As a result, these hydrophobic drugs must be taken in large doses in order to reach the therapeutic plasma concentration, which may result in undesired side effects. Therefore, it is essential to develop novel methods to enhance the effectiveness of hydrophobic drugs. Various techniques have been attempted to improve the dissolution behavior of insoluble APIs, including the use of solubilizing excipients (Avdeef et al. 2007), micronization of drug particles (Kecka and Müller 2006), preparation of solid dispersions (Hasegawa et al. 2007), formation of molecular complexes (Umeda et al. 2007) and inclusion complexes with cyclodextrins (CDs) (Uekama et al. 2006; Loftsson and Duchêne 2007).

The reduction of particle size leads to a significant increase in the dissolution rate of the API, which can in turn lead to a marked increase in drug absorption. Generally, particles with sizes less than 0.1 μm are referred to as nanoparticles, while particles measuring hundreds of nanometers are defined as submicron-region particles. Several techniques can be employed to reduce particle size to nano-order, including precipitation, supercritical fluid processing (Pathak et al. 2004), freeze-drying (Zhang et al. 2008), high pressure homogenization (Kecka and Müller 2006) and cogrinding (Itoh et al. 2003; Fukami et al. 2009). The molecular state of a drug substance prepared as nanoscale particles is usually one of two forms, crystalline or amorphous. Although the solubility of amorphous drugs is generally higher than that of the crystalline form, an amorphous drug has an intrinsic problem of crystallization due to its high energy state during storage. Therefore, the crystalline form has been selected as the preferred solid form for the development of commercial pharmaceutics. The processing of nanoparticles consisting of nanocrystals is expected to improve the dissolution behavior of insoluble drugs and impart thermodynamic stability during storage under ambient conditions.

NANOPARTICLE PROCESSING BY SOLID-STATE REACTION

Micronization of drug substances by milling is a popular technique to enhance drug solubility. Recent studies have shown that nanoparticle formation by wet-grinding (Kecka and Müller 2006) or cogrinding (Itoh et al. 2003; Fukami et al. 2009) is a promising method to improve the solubility of hydrophobic drugs. However, the wet-grinding approach typically results in submicron-size drug particles that aggregate when dried during the preparation of solid dosage forms (Chaubal and Popescu 2008). Due to the simplicity of the procedure and the lack of organic solvents, the dry-cogrinding approach was used to produce nanoparticles (< 100 nm). In addition, the method also has environmental and cost-effective advantages for the production of solid pharmaceutical applications.

Wongmekiat et al. reported noticeable improvements in the formation of drug nanoparticles by cogrinding with cyclodextrins (Wongmekiat et al. 2002). Complexation with cyclodextrins, in addition to particle size reduction, may modify the dissolution properties of some drugs (Wongmekiat et al. 2006). We have also reported an improvement in the dissolution behavior of certain organic compounds by cogrinding with several types of cyclic glucan in the solid state (Fukami et al. 2004, 2006).

Thus, application of this technique is expected to enhance the reactivity between poorly water-soluble drugs and solubilizing agents.

Furthermore, the cogrinding of ternary components, consisting of API, aqueous polymer and surfactant, has also been attempted to prepare drug nanoparticles (Ito et al. 2003). These pharmaceutical ingredients may enhance grinding efficiency and maintain the size of the generated drug nanoparticles during storage. Each of four poorly water-soluble drugs, N-5159, griseofulvin, glibenclamide and nifedipine, was ground with polyvinylpyrrolidone (PVP) and sodium dodecyl sulfate (SDS) using a vibrational rod mill according to the procedure shown in Fig. 1A. Very few drug nanoparticles were obtained when the drug was ground by itself, or from the binary ground mixtures (GMs) and ternary physical mixtures (PMs). However, when a ternary PM was ground, most of the coarse drug

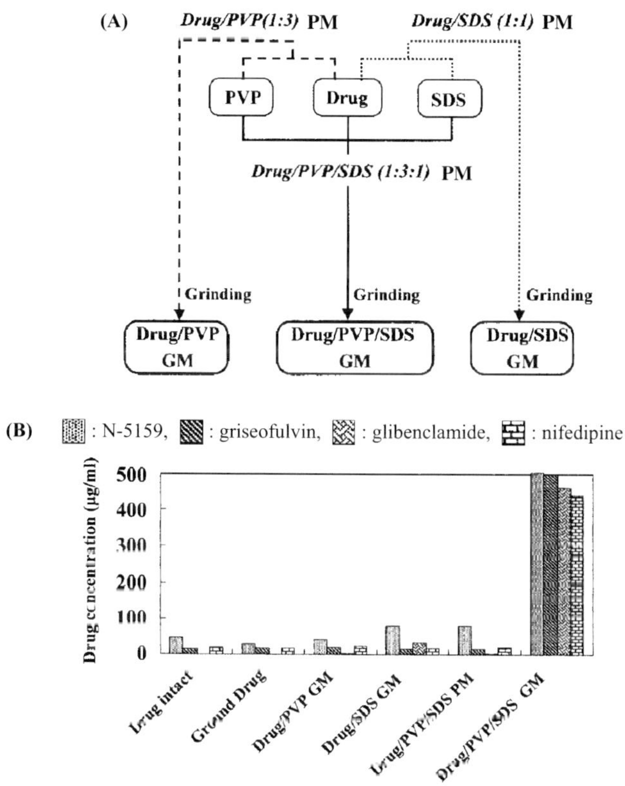

Fig. 1. Processing of nanoparticles by the ternary cogrinding method. (A) Procedures for the preparation of binary and ternary ground mixtures (GMs). (B) Variation in the amount of fine drug particles (< 0.2 μm) after dispersion of various GMs and physical mixtures (PMs) into distilled water at 37°C. (Reproduced with permission from the Pharmaceutical Society of Japan; Itoh et al. 2003.)

particles were processed into fine particles that passed through a 0.2 μm filter (Fig. 1B). A suitable mixing ratio of the ternary GM was concluded to be 1:3:1 (drug/PVP/SDS, by weight), which showed the highest yield of drug nanoparticles while minimizing the quantity of additives.

We previously used a methyl methacrylate copolymer (MMC) instead of PVP as an aqueous polymer in the ternary cogrinding system (Fukami et al. 2009) and found that the maximum number of nanoparticles was obtained using a combination of 1:5:1 (PBC/MMC/SDS) after cogrinding for 10 min. Although the difference in the optimum weight ratio of ternary components between PVP and MMC is not completely understood, the intermolecular interaction among the components and/or the bulk powder fluidity during grinding may affect the formation of drug nanoparticles.

EVALUATION OF THE PHYSICO-CHEMICAL PROPERTIES OF A NANONIZED DRUG

Measurement of Particle Size Distribution

The solubility, bioavailability and efficiency of a drug can be enhanced by reducing the particle size, or by formulating the drug in lipids or detergents (Jiang et al. 2008). Dynamic light scattering (DLS) analysis, which is one of the most useful techniques for determining the size distribution profile of small particles in suspension, revealed the existence of drug nanoparticles. The particle size distributions of GMs prepared by cogrinding PBC, MMC and SDS indicated that particle size depends on the weight ratios of these three compounds (Fig. 2).

Two different particle size distributions are indicated by the two peaks around 72 and 290 nm in Fig. 2C. The larger peak at about 290 nm was similar to the mean particle size of MMC, as shown in Fig. 2A, which meant that the smaller size particles (about 72 nm) were considered to be drug nanoparticles generated by the cogrinding process. These results suggest that PBC exists as an approximately 72-nm-sized colloidal nanoparticle in water. It has also been shown that a two-component ground mixture of PBC and MMC does not produce nanoparticles (Fukami et al. 2009).

Imaging of Nanoparticles by Scanning Electron Microscopy

Imaging analysis is also an important method for assessing the particle size distribution results from DLS and to provide visual confirmation of the presence of nanoparticles. Nanoparticles entrapped by the 0.05 μm pore size membrane filter were investigated by scanning electron microscopy (SEM) as shown in Fig. 3A. These nanoparticles were prepared from curcumin, which was used as a model compound, by cogrinding with

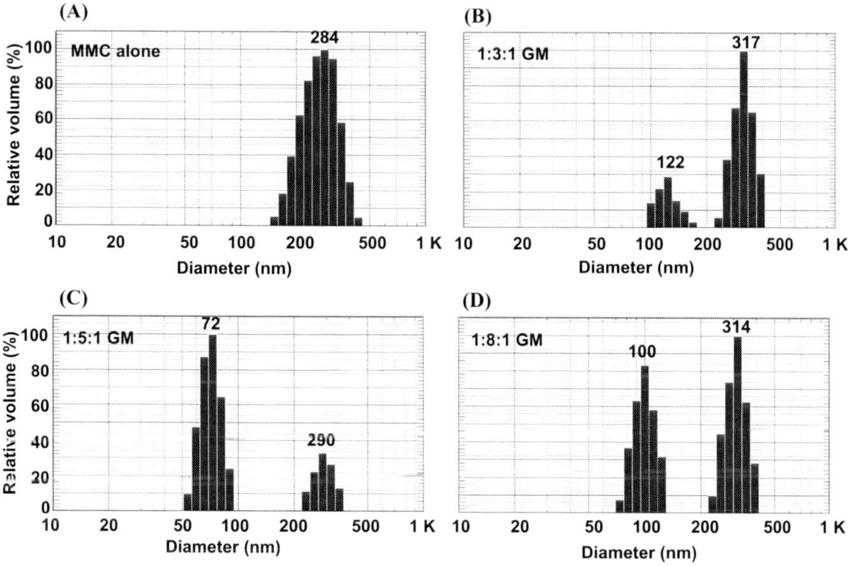

Fig. 2. Particle size distributions measured using the DLS method. Particle size distributions of (A) MMC, and PBC/MMC/SDS GMs in ratios of (B) 1:3:1, (C) 1:5:1 and (D) 1:8:1. The GMs exhibited two peak distributions corresponding to small drug nanoparticles and the larger MMC particles. The mean particle size of the smaller particle peaks was consistent with nanoparticle formation, and the smallest mean particle size was observed in 1:5:1 GM.

MMC and SDS under the same conditions employed for the processing of PBC. Nanoparticles with approximate diameters of 100 nm were sparsely distributed on the membrane surface. The morphology of the PBC nanoparticles was then examined more closely using high resolution SEM (Fig. 3B), which confirmed the presence of primary nanoparticles in a ternary GM powder consisting of PBC/PVP/SDS. The agglomerates were composed of spherical nanoparticles with diameters of approximately 20 nm (Pongpeerapat et al. 2006).

Imaging of Nanoparticles by Atomic Force Microscopy

Recently, atomic force microscopy (AFM) was used to evaluate the morphology of nanostructures. AFM has various analytical modes, which can be applied to the measurement of surface shape and/or the stiffness and rigidity of nanoparticles. AFM topography and phase image observations were performed to evaluate the morphology and surface states of colloidal PBC nanoparticles (Moribe et al. 2008). The AFM topographical image indicated that the particle was covered with a less dense substance (Fig. 4A left). The AFM phase image, another analytical mode of AFM, provides information regarding the physicochemical properties, such as

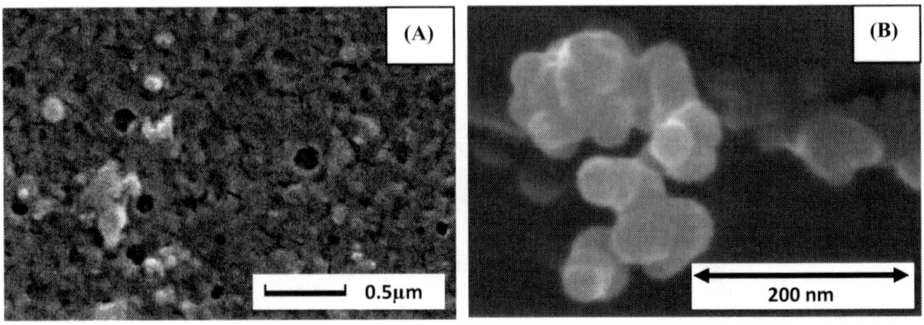

Fig. 3. SEM images of drug nanoparticles. (A) SEM micrograph of CCM/MMC/SDS (unpublished work). (B) High resolution SEM micrograph of PBC nanoparticles obtained from the PBC/PVP K17/SDS GM powder. (Reprinted with permission from Elsevier; Pongpeerapat et al. 2008.)

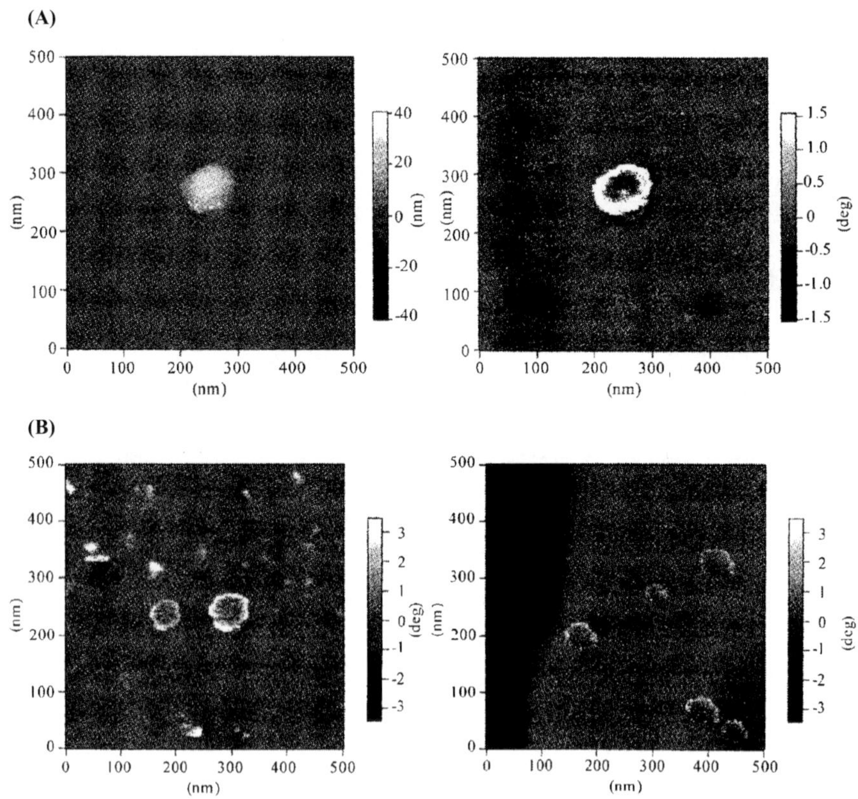

Fig. 4. AFM images of drug nanoparticles. (A) AFM images of colloidal PBC nanoparticles with PVP K17. Left, AFM topography image. Right, AFM phase image. (B) Comparison of AFM phase images of colloidal PBC nanoparticles with PVP K17 (left) and those with PVP K12 (right). (Reproduced with permission from the Pharmaceutical Society of Japan; Moribe et al. 2008.)

stiffness and rigidity. The phase image indicated a difference between the core and the surface of the particle, as shown in the right micrograph of Fig. 4A. Thus, Moribe et al. speculated that the particle had a core-shell structure and the stiff core originated from PBC nanocrystals covered with a softer substance, such as a PVP-SDS complex. From the comparison of AFM phase images of colloidal PBC nanoparticles with PVP K12 and with PVP K17 (where K12 and K17 indicate the molecular weights of PVP and are 12,000 and 17,000, respectively) are shown in Fig. 4B. The outline of PBC nanoparticles with PVP K17 was more pronounced than that of PBC nanocrystals covered with PVP K12, indicating that the surface structure and states of the nanoparticles differed depending on the molecular weight of PVP. These results confirm that AFM is a useful tool for characterizing the physical properties of nanoparticles.

Solid-state Nuclear Magnetic Resonance Spectroscopy

It is important to understand the underlying molecular interactions associated with the formation of nanoparticles. Solid-state nuclear magnetic resonance spectroscopy (NMR) (Varian 400 and 600 MHz NMR spectrometers in Ann Arbor, and Bruker 900 MHz in Lansing, USA) experiments were performed on nanoparticles in GMs, PMs and the pure forms of all three individual components in the mixture of PBC, MMC and SDS (Fukami et al. 2009). The spectra of PBC-I and PBC-II were different, and the differences in the spectra of the GMs and PMs were noted (Fig. 5). In the figure, PBC-I and PBC-II represent the polymorphic crystalline forms I and II, respectively. The ^{1}H and ^{13}C spectra suggested that the grinding of pure PBC-I results in the formation of PBC-II. A combination of ^{1}H and ^{13}C chemical shift spectra (Fig. 5), 2D $^{1}H/^{1}H$ chemical shift correlation under magic angle spinning (MAS) (Fig. 6), and 2D $^{1}H/^{13}C$ chemical shift correlation under MAS (data not shown) were used to assign spectral lines from the individual chemical components and from the nanoparticles. Our assignment of the NMR spectra was in agreement with previously reported results (Pongpeerapat et al. 2006). The spectra of the GM and PM samples were dominated by PBC and MMC components, as shown in Fig. 5A. ^{1}H and ^{13}C spectra revealed that PBC is present in the PBC-I form in PM, while it is converted to the PBC-II form in the GM. 2D $^{1}H/^{1}H$ correlation experiments (similar to 2D nuclear Overhauser effect spectroscopy (NOESY)) were performed under MAS using the pulse sequence shown in Fig. 6B. A spin echo was applied prior to signal acquisition to filter out peaks with short T_2 values. 2D $^{1}H/^{1}H$ correlation spectra of the PM and GM samples are given in Fig. 6 and 2D correlation spectra of pure PBC-I and PBC-II are also shown for comparison. All spectra exhibited relatively high resolutions in both frequency dimensions. Peaks in 2D spectra were

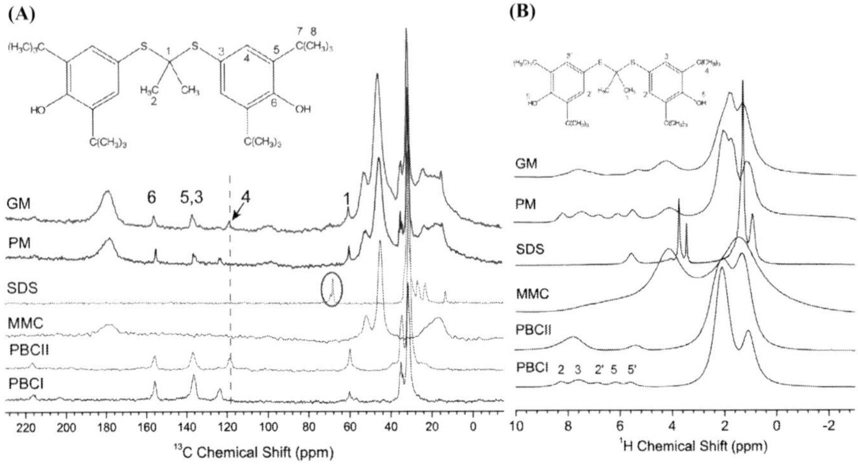

Fig. 5. Solid-state NMR spectra of ternary components. ^1H and ^{13}C NMR spectra: (A) ^{13}C NMR spectra of SDS, MMC, PBC-II and PBC-I were recorded using a ramped-cross-polarization pulse sequence under 18 kHz MAS. (B) ^1H NMR spectra were recorded using a solid-echo pulse sequence under 18 kHz MAS at room temperature. Different refocusing times were used for different samples: 1 ms (SDS), 40 μs (MMC), 2 ms (PM), 2 ms (GM), 1 ms (PBC-I) and 1 ms (PBC-II). (Reprinted with permission from the American Chemical Society; Fukami et al. 2009.)

assigned on the basis of the connectivity of the proton resonances shown in the spectra (Fig. 6). The presence of cross peaks between PBC and MMC in the GM 2D $^1H/^1H$ spectrum (indicated by a circle) and their absence in the PM spectrum suggest that these two molecules are nearer in space in the GM than in the PM. Specifically, CH_3 of PBC and CH_2 of MMC are within a distance of around 6 Å, suggesting that hydrophobic interactions between the drug and the polymer play an important role in binding between MMC and PBC, and thus in the formation of nanoparticles. An increase in the line width observed (particularly in ^1H spectra) in the spectrum of the GM suggests that cogrinding of the PM changes the crystalline PBC to an amorphous state. This observation is in agreement with powder X-ray diffraction and Fourier transform infrared (FT-IR) spectroscopy measurements (Fukami et al. 2009).

These results also demonstrate that solid-state NMR experiments can be used to examine the quality of nanoparticles and provide insights into the role of intermolecular interactions with atomic-level resolution that can contribute to the design of nanoparticles. In addition, these nanoparticles are not well suited to studies using the most commonly used high-resolution physical techniques, including X-ray crystallography and solution NMR spectroscopy.

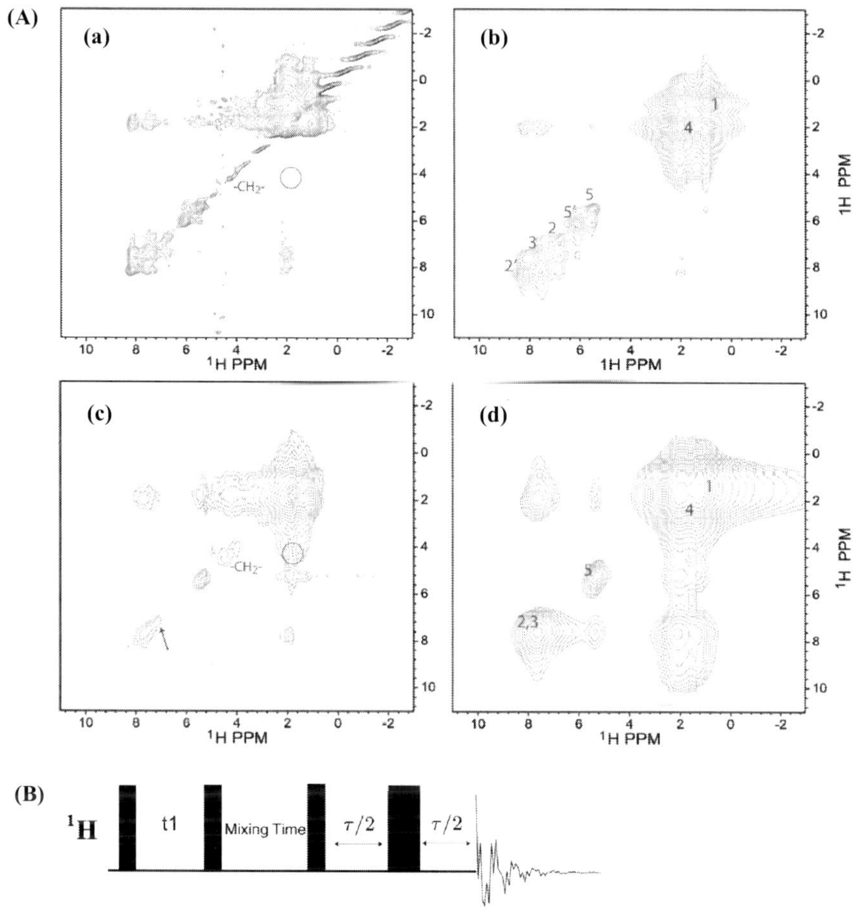

Fig. 6. 2D correlation spectra of solid-state NMR of ternary components. (A) 2D H/H correlation spectra of PM (a), PBC-I (b), GM (c) and PBC-II (d) obtained using the pulse sequence given in (B). τ = 1 ms, an 11.11 ms mixing time, 16 scans and 62 t1 increments were used. The spinning frequency of the sample was synchronized with the mixing time of the pulse sequence. The hypercomplex data collection method was applied to achieve quadrature detection in the indirect frequency dimension. (Reprinted with permission from the American Chemical Society; Fukami et al. 2009.)

ENHANCED ABSORPTION OF PBC VIA ORAL ROUTE

Since the cholesterol-lowering efficiency of PBC is directly correlated to its ability to permeate membranes, it is important to examine the ability of the newly developed nanoparticles to permeate cell membranes. Caco-2 cells are commonly used to test the absorption of drugs in humans; although the Caco-2 cells used in this study are of colonic origin, unlike normal colonocytes, they exhibit drug transport similar to the human small intestine (Shah et al. 2006). The amounts of PBC that permeated from a mixture of 1:5:1 PBC/MMC/SDS with (i.e., around 72 nm nanoparticles) and without grinding were measured. The experimental results suggest that nanoparticles permeate the cell membrane, whereas solely PBC or a PM of 1:5:1 PBC/MMC/SDS without grinding did not permeate the cell membrane, even after 1 h (Fukami et al. 2009). The surfactant SDS may open the junction between Caco-2 cells and allow the drug to transit between cells. The transepithelial electrical resistance value decreased to about 70% from the initial condition after the permeation experiment. Studies are currently being conducted in our laboratories to address the effect of SDS on the transmembrane permeation of PBC.

Shudo et al. (2008) reported *in vivo* drug absorption for the PBC/PVP/SDS ternary GM. PBC formulations were administered by oral gavages to male Sprague-Dawley rats (Charles River Laboratories Inc., Wilmington, MA). The PBC formulations were dispersed into distilled water with sonication to prepare suspensions for administration to the rats at 200 mg/kg. In the case of unprocessed PBC, 0.5% hydroxypropylcellulose solution was used instead of distilled water to facilitate dispersion, due to its low wettability.

Plasma concentration—time profiles of PBC after oral administration in rats are shown in Fig. 7. *In vivo* absorption was significantly improved for the PBC/PVP/SDS ternary GMs compared to that observed using the ternary PMs and unprocessed PBC (Fig. 7A). There were no significant differences in the pharmacokinetic parameters, such as area under the blood concentration time curve and maximum drug concentration values, between the ternary PM and the unprocessed PBC. Additionally,

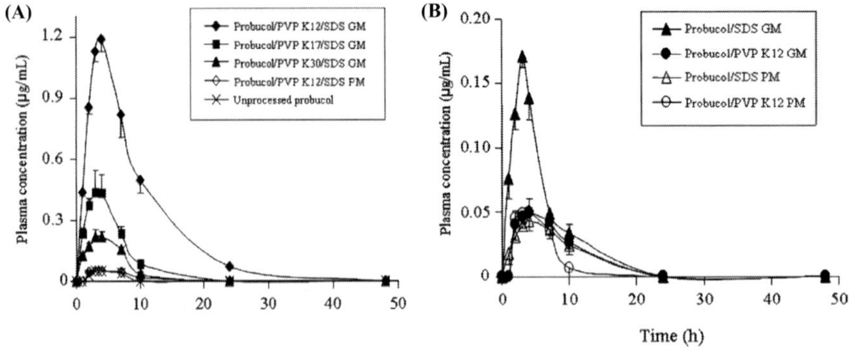

Fig. 7. *In vivo* drug absorption study. Plasma concentration of probucol following oral administration of (A) Probucol/PVP/SDS ternary and (B) Probucol/PVP or Probucol/SDS binary mixture suspensions results are expressed as mean ± S.E. (n=3). (Reproduced with permission from the Pharmaceutical Society of Japan; Shudo et al. 2008.)

the PBC/PVP K12 binary GM and the binary PMs demonstrated similar concentration-time profiles to that obtained using the unprocessed PBC, although a slight improvement in absorption was observed for the PBC/SDS binary GM (Fig. 7B). Shudo et al. concluded that not only the reduced particle size of PBC, but also modification of the particle surface conditions by PVP and SDS, appear to influence the *in vivo* absorption of PBC. This estimation was reconfirmed by *in vivo* assays in which the mean particle size of PBC was controlled to be approximately equal, and the surface properties of the nanoparticles were distinguished by AFM measurement (Moribe et al. 2008).

FORMULATION FOR A SOLID DOSAGE FORM OF PBC NANOPARTICLES

When ternary components including PBC were coground, a powder mixture containing drug nanoparticles was successfully prepared. However, the powder was sticky and difficult to handle, and thus unsuitable for practical use by patients. Therefore, formulating a solid dosage form, such as tablets or granules with or without capsules, is important for improving the convenience associated with oral administration. Considerable care was therefore taken to uniformly coat a carrier particle with the nanoparticle mixture.

The ternary GM consisting of PBC/MMC/SDS (1:5:1) was suspended in distilled water and 1.7% ammonium hydroxide was gradually added and equilibrated with slow agitation using a magnetic stirrer for 60 min. Polysorbate 80, triethyl citrate, and glycerol monostearate were dispersed in hot water (80°C) and poured into the ternary GM suspension, which was then agitated overnight. A fluidized bed coater (FL-Mini, Freund Corporation, Tokyo, Japan) was used to prepare fine film-coated granules. Core particles (Nonpareil 103, 355–500 μm, Freund Corporation, Tokyo, Japan) were used as carrier particles to produce a solid dosage form.

SEM micrographs confirmed the formation of a polymeric nanoparticle film on the surface of the carrier particle (Fig. 8), and the bright areas caused by refraction of polarized light in the polarized micrographs indicated a crystalline component consisting of PBC and/or SDS in the film. In addition, excellent fluidity of fine carrier particles has been achieved by preparing pharmaceutical granules coated with polymer film containing drug nanoparticles (Fukami et al. 2009).

Pharmaceutical granules were prepared by spray-coating of a ternary GM suspension onto carrier particles, although the drug content was only around 1%. In order to increase the drug content in the formulated solid dosage form, a spray-drying method was used. A suspension containing

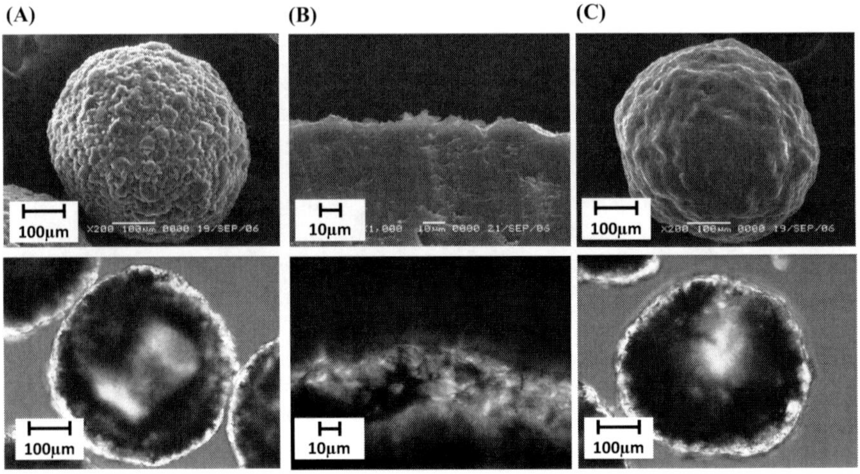

Fig. 8. Images of solid dosage form coated with drug nanoparticles and polymeric film. Micrographs of intact and coated carrier particles. Upper row: SEM photographs; bottom row: polarized photographs; (A) an intact carrier particle, (B) a cross-sectional view of (A), and (C) a carrier particle coated with a polymer layer containing drug nanoparticles.

PBC nanoparticles was prepared by the same procedure for the spray-coating method as described above. A desktop spray-dryer (SD-1, Tokyo Rika Corp., Tokyo, Japan) was used to prepare the spray-dried particles. SEM photographs of spray-dried powders prepared at 75°C (SD75), 90°C (SD90) and 105°C (SD105) are shown in Fig. 9. These powder samples are composed of spherical particles with smooth surfaces. Although the mean diameters of these spherical particles were similar (around 3 μm) among the SD75, SD90 and SD105 samples, smaller particles were more abundant in the SD75 samples than in the SD105 samples. DLS analysis of the particle size distribution in the spherical spray-dried powder revealed the presence of around 47-nm-sized nanoparticles in the SD75 and SD90 powder samples. On the other hand, relatively large nanoparticles (around 75 nm diameter) were observed in the SD105 sample, with the particle size increasing with measuring time due to aggregation (Io et al. 2010).

The percentages of PBC released from the SD75, SD90 and SD105 powders were examined at pH 1.2 and pH 6.8 to simulate the pH environments of the stomach and small intestine, respectively. The percentages of PBC released from all three samples were less than 2% at pH 1.2, even after 240 min. On the other hand, SD75, SD90 and SD105 immediately released 80, 60 and 50% of PBC at pH 6.8, respectively. These results imply that SD75 has the best enteric characteristics and that most of the SD75 drug nanoparticles were released within 15 min. These results also suggest that (1) the agglomeration of drug nanoparticles occurs

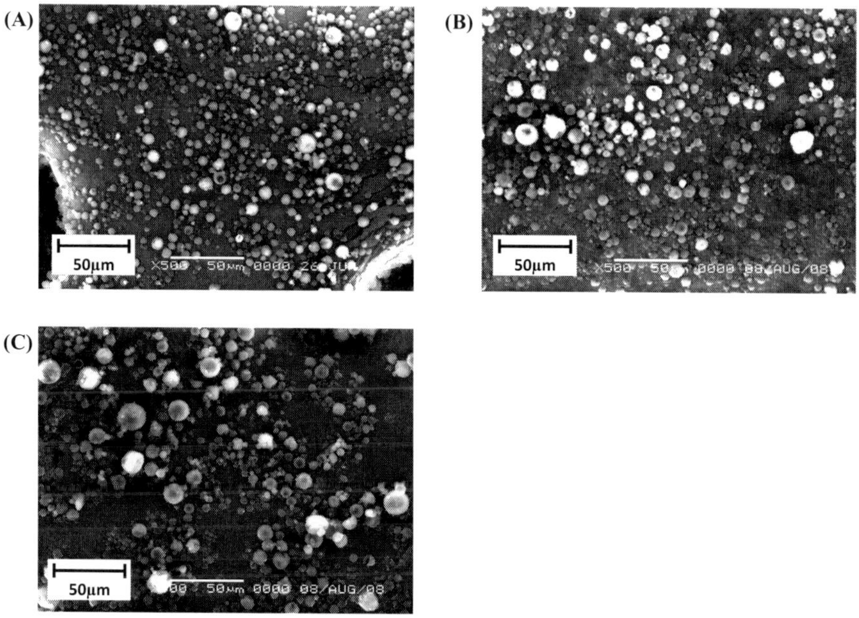

Fig. 9. SEM Images of spray-dried powders containing drug nanoparticles. SEM photomicrographs of (A) SD75, (B) SD90 and (C) SD105 spray-dried powders. (Reprinted with permission from the American Chemical Society; Io et al. 2010.)

during spray drying at higher temperatures, and (2) these nanoparticles exhibit pH-dependent functional properties, implying that hydrophobic drugs could be effective in the alkaline environment of the small intestine. Therefore, MMC was selected as the aqueous polymer instead of PVP because of its enteric properties, which means that MMC dissolves in the neutral pH region but not under acidic conditions, and that MMC appears to protect nanoparticles and increase the solubility of hydrophobic drugs. A Caco-2 cell monolayer was used to examine the cell membrane permeation of PBC nanoparticles released from the spray-dried powder. The best permeation was observed for SD75; around 8 μg of PBC was transported from the apical to the basal side of the monolayer over 120 min. The permeation rate of the SD75 nanoparticles, which was twice that of nanoparticles prepared without the spray-drying process (Fukami et al. 2009), suggests that the spray-drying process can significantly improve the efficiency of hydrophobic drugs. The results of the Caco-2 cell permeation test were consistent with the dissolution study and particle size analysis of cogrinding-spray-drying processed drug nanoparticles (Io et al. 2010). Since the solubility, fraction, and size of nanoparticles are correlated with the dissolution rate according to the Noyes-Whitney equation, the smaller nanoparticles would be more efficient for cell membrane permeation. To the best of our knowledge, this is the first study that successfully combined dry-cogrinding and spraying technique to prepare a polymeric dosage form of drug nanoparticles.

CONCLUSION

Recent research in this field has focused on the diversity of the pharmacological action of PBC, such as anti-oxidation and anti-inflammatory effects (Braun et al. 2003). These effects are expected to be important in treating myocardial infarction by inhibiting the oxidation of LDL and its accumulation in the coronary artery. This is an important requirement of cholesterol-lowering drugs; therefore, the approach presented in this chapter will facilitate the efficient application of PBC to multiple routes in the human body. In addition, PBC nanoparticles prepared by cogrinding exhibited enhanced oral bioavailability from the *in vivo* study and were effectively absorbed through the mucous membranes of the human small intestine. We believe that this novel approach of cogrinding and spraying could have significant advantages for the effective use and oral dosage of PBC in patients who require higher and faster rates of mucosal absorption. This production technology could also be applied on an industrial scale to other poorly soluble cholesterol-lowering drugs.

Key Facts about Magnetic Resonance Imaging

- Nuclear magnetic resonance (NMR) techniques and data analyses are used to reveal the intermolecular interaction between drug and ingredients with atomic-level resolution.
- NMR has a rich history and several Nobel prizes have been awarded to scientists in this field: Felix Block (1952), Edward Purcell (1952), Richard Ernst (1991), Kurt Wuthrich (2002), Paul Lauterbur (2003), and Peter Mansfield (2003).
- NMR technology similar to magnetic resonance imaging (MRI) has been employed to image human internal structures.
- MRI uses an external magnetic field to polarize the nuclear spin magnetization of hydrogen atoms in the water molecules of the body.
- The first cross-sectional MRI of a living mouse was published in 1974 (Lauterbur 1974).
- Paul Lauterbur and Peter Mansfield were awarded the 2003 Nobel Prize in Physiology or Medicine for their "discoveries concerning MRI".

Definitions

Amorphous: An amorphous material is a noncrystalline solid with a disordered structure. Amorphous materials are often prepared by rapidly cooling liquid materials, as well as by using additives that interfere with the ability of the primary constituent to crystallize. In the field of pharmaceutical sciences, solid or molecular dispersions are developed by combining the drug substance with a polymer to improve the dissolution behavior of poorly water-soluble drugs. Although solid dispersions are commonly prepared from organic solutions, cogrinding and hot-melt extrusion methods have also been extensively studied.

Caco-2 cells: Caco-2 cells are derived from human colon carcinoma. When cultured on a porous membrane filter, the cells differentiate to form a confluent monolayer having tight junctions, producing P-glycoprotein and a number of enzymes, and resemble the enterocytes lining the small intestine. The *in vitro* apparent permeability coefficient (Papp) across Caco-2 monolayers correlates with the *in vivo* fraction absorbed (Fa) and bioavailability. Therefore, Caco-2 monolayers are widely used in the pharmaceutical industry as an *in vitro* screening model for the permeability of orally administered drugs through the human small intestine. Recently, the Madin-Darby canine kidney (MDCK) epithelial cell line and the parallel artificial membrane permeability assay (PAMPA) were also used for the same purpose.

Crystal: A material with its constituent atoms, molecules, or ions arranged in an orderly repeating pattern that extends in all three spatial dimensions. Drugs can potentially have more than one crystalline structure, which is known as polymorphism, where the molecular arrangements and/or conformations of the drug are different. Each crystalline form can exhibit different physico-chemical properties, particularly solubility, so that control of the crystalline form is considered important during drug production and storage.

Cyclodextrin: Substances produced from starch by enzymatic conversion. Typical cyclodextrins consisting of six, seven and eight glucopyranoside units are defined as α-, β- and γ-cyclodextrin, respectively. The characteristic toroidal shape with a hydrophobic interior and hydrophilic exterior can accommodate other hydrophobic molecules and can be solubilized in water as an inclusion complex. Cyclodextrins have attracted considerable interest in numerous fields, especially in pharmaceutical applications, because an inclusion complex of cyclodextrin with an insoluble drug can be more readily absorbed orally than when the drug is administrated alone.

Solid dosage forms: Powders, granules, tablets, and capsules. Pharmaceutical powder is a mixture of finely divided drugs and/or ingredients in a dry form. Granules are prepared aggregates of powdered materials to form larger particles (0.5–2 mm). The sizes of granules are defined in the pharmacopoeia of each country. Because granules flow better than powders, they are more commonly used to prepare tablets and capsules. Pharmaceutical functions, including modified drug release, enteric release, and taste- or odor-masking, can be added to conventional solid dosage forms through the addition of various polymer ingredients.

Summary Points

- In recent years, most newly developed active pharmaceutical ingredients (APIs) are poorly water-soluble and often have low bioavailability when administered orally.
- Reduction of particle size is a promising method for significantly increasing the dissolution rate of the API.
- Nanoparticle processing method by cogrinding of ternary components, consisting of API, aqueous polymer and surfactant, was developed for improving the solubility of drugs.
- Nanoparticles prepared from ternary components indicated that probucol (PBC), which is a cholesterol-lowering drug, exists as colloidal nanoparticles in water by dynamic light scattering method.
- Atomic force microscopy measurements revealed that the observed particle had a core-shell structure, and that the rigid and stiff core originated from PBC nanocrystals covered by a soft substance.
- 2D $^{1}H/^{1}H$ correlation spectra of solid-state nucleic magnetic resonance suggested that hydrophobic interactions between the drug and the polymer play an important role in the binding between methacrylic copolymer and PBC, and thus in the formation of nanoparticles.
- *In vivo* absorption was significantly improved for PBC/polymer/surfactant ternary ground mixtures compared to those of the ternary physical mixtures and unprocessed PBC.
- Care was taken to prepare a pharmaceutical dosage form with the drug nanoparticles available for practical use by patients via oral administration route.
- Spray-coating of the suspended ternary ground mixture provided pharmaceutical granules that were covered with polymeric film containing PBC nanoparticles.
- To increase the drug content in the dosage form, the spray-drying method was employed and spherical particles containing about 10% of PBC nanoparticles were obtained.

Acknowledgements

This work was supported in part by a grant from the "High-Tech Research Center" Project for Private Universities: matching fund subsidy from the Ministry of Education, Culture, Sports, Science, and Technology (MEXT) 2007–2011 of Japan (to T.S.) , Grant-in-Aid for Scientific Research (22590149 to T.S. and T.F., 20366628 to T.F.), and the National Institute of Health (RR023597 to A.R.). The authors thank Dr. Jiadi Xu and Kazutoshi Yamamoto (University of Michigan, Ann Arbor) for performing the solid-state NMR experiments, and the 900 MHz Biomolecular NMR facility in East Lansing.

Abbreviations

AFM	:	atomic force microscopy
API	:	active pharmaceutical ingredient
CD	:	cyclodextrin
DLS	:	dynamic light scattering
FT-IR	:	Fourier transform-infrared
GM	:	ground mixture
LDL	:	low-density lipoprotein
MAS	:	magic-angle spinning
MDCK	:	Madin-Darby canine kidney
MMC	:	methyl methacrylate copolymer
MRI	:	magnetic resonance imaging
NMR	:	nuclear magnetic resonance
PAMPA	:	parallel artificial membrane permeability assay
PBC	:	probucol
PM	:	physical mixture
PVP	:	polyvinylpyrrolidone
SDS	:	sodium dodecyl sulfate
SEM	:	scanning electron microscopy

References

Avdeef, A., S. Bendels, O. Tsinman, K. Tsinman and M. Kansy. 2007. Solubility-excipient classification gradient maps. Pharm. Res. 24: 530–545.

Betge, S., K. Lutz, M. Roskos and H.R. Figulla. 2007. Oral treatment with probucol in a pharmacological dose has no beneficial effects on mortality in chronic ischemic heart failure after large myocardial infarction in rats. Eur. J. Pharmacol. 558. 119–127.

Blagden, N., M. de Matas, P.T. Gavan and P. York. 2007. Crystal engineering of active pharmaceutical ingredients to improve solubility and dissolution rates. Adv. Drug Deliv. Rev. 59: 617–630.

Braun, A., S. Zhang, H.E. Miettinen, S. Ebrahim, T.M. Holm, E. Vasile, M.J. Post, D.M. Yoerger, M.H. Picard, J.L. Krieger, N.C. Andrews, M. Simons and M. Krieger. 2003. Probucol prevents early coronary heart disease and death in the high-density lipoprotein receptor SR-BI/apolipoprotein E double knockout mouse. Proc. Natl. Acad. Sci. USA 100: 7283–7288.

Chaubal, M.V., C. Popescu. 2008. Conversion of nanosuspensions into dry powders by spray drying: A case study. Pharm. Res. 25: 2302–2308.

Fukami, T., A. Mugishima, T. Suzuki, S. Hidaka, T. Endo, H. Ueda and K. Tomono. 2004. Enhancement of water solubility of fullerene by cogrinding with mixture of cycloamyloses, novel cyclic α-1,4-glucans, via solid-solid mechanochemical reaction. Chem. Pharm. Bull. 52: 961–964.

Fukami, T., T. Furuishi, T. Suzuki, S. Hidaka, H. Ueda and K. Tomono. 2006. Improvement in solubility of poorly water soluble drug by cogrinding with highly branched cyclic dextrin. J. Incl. Phenom. Macrocycl. Chem. 56: 61–64.

Fukami, T., T. Ishii, T. Io, N. Suzuki, T. Suzuki, K. Yamamoto, J. Xu, A. Ramamoorthy and K. Tomono. 2009. Nanoparticle processing in the solid state dramatically increases the cell membrane permeation of a cholesterol-lowering drug, probucol. Mol. Pharmaceutics 6: 1029–1035.

Hasegawa, S., N. Furuyama, S. Yada, T. Hamaura, A. Kusai, E. Yonemochi and K. Terada. 2007. Effect of physical properties of troglitazone crystal on the molecular interaction with PVP during heating. Int. J. Pharm. 336: 82–89.

Io, T., T. Fukami, K. Yamamoto, T. Suzuki, J. Xu, K. Tomono and A. Ramamoorthy. 2010. Homogeneous nanoparticles to enhance the efficiency of a hydrophobic drug, antihyperlipidemic probucol, characterized by solid-state NMR. Mol. Pharmaceutics 7: 299–305.

Itoh, K., A. Pongpeerapat, Y. Tozuka, T. Oguchi and K. Yamamoto. 2003. Nanoparticle formation of poorly water-soluble drugs from ternary GMs with PVP and SDS. Chem. Pharm. Bull. 51: 171–174.

Jiang, W., B.Y.S. Kim, J.T. Rutka and W.C.W. Chan. 2008. Nanoparticle-mediated cellular response is size-dependent. Nat. Nanotech. 3: 145–150.

Kathiresan, S., O. Melander, D. Anevski, C. Guiducci, N.P. Burtt, C. Roos, J.N. Hirschhorn, G. Berglund, B. Hedblad, L. Groop, D.M. Altshuler, C. Newton-Cheh and M.N. Orho-Melander. 2008. Polymorphisms associated with cholesterol and risk of cardiovascular events. Engl. J. Med. 358: 1240–1249.

Kecka, C.M. and R.H. Müller. 2006. Drug nanocrystals of poorly soluble drugs produced by high pressure homogenization. Eur. J. Pharm. Biopharm. 62: 3–16.

Kobayashi, M., I. Chisaki, K. Narumi, K. Hidaka, T. Kagawa, S. Itagaki, T. Hirano and K. Iseki. 2008. Association between risk of myopathy and cholesterol-lowering effect: A comparison of all statins. Life Sci. 82: 969–975.

Lauterbur, P.C. 1974. Magnetic resonance zeugmatography. Pure Appl. Chem. 40: 149–157.

Loftsson, T. and D. Duchêne. 2007. Cyclodextrins and their pharmaceutical applications. Int. J. Pharm. 329: 1–11.

Moribe, K., C. Wanawongthai, J. Shudo, K. Higashi and K. Yamamoto. 2008. Morphology and surface states of colloidal probucol nanoparticles evaluated by atomic force microscopy. Chem. Pharm. Bull. 56: 878–880.

Pathak, P., M.J. Meziani, T. Desai and Y.P. Sun. 2004. Nanosizing drug particles in supercritical fluid processing. J. Am. Chem. Soc. 126: 10842–10843.

Pongpeerapat, A., K. Higashi, Y. Tozuka, K. Moribe and K. Yamamoto. 2006. Molecular interaction among probucol/PVP/SDS multicomponent system. Pharm. Res. 23: 2566–2574.

Pongpeerapat, A., C. Wanawongthai, Y. Tozuka, K. Moribe and K. Yamamoto. 2008. Formation mechanism of colloidal nanoparticles obtained from probucol/PVP/SDS ternary ground mixture. Int. J. Pharm. 352: 309–316.

Sawayama, Y., C. Shimizu, N. Maeda, M. Tatsukawa, N. Kinukawa, S. Koyanagi, S. Kashiwagi and J. Hayashi. 2002. Effects of probucol and pravastatin on common carotid atherosclerosis in patients with asymptomatic hypercholesterolemia—Fukuoka Atherosclerosis Trial (FAST). J. Am. Coll. Cardiol. 39: 610–616.

Shah, P., V. Jogani, T. Bagchi and A. Misra. 2006. Role of Caco-2 cell monolayers in prediction of intestinal drug absorption. Biotechnol. Prog. 22: 186–198.

Shudo, J., A. Pongpeerapat, C. Wanawongthai, K. Moribe and K. Yamamoto. 2008. *In vivo* assessment of oral administration of probucol nanoparticles in rats. Biol. Pharm. Bull. 31: 321–325.

Timpe, C. 2007. Strategies for formulation development of poorly water-soluble drug candidates—A recent perspective. Am. Pharm. Rev. 10: 104–109.

Uekama, K., F. Hirayama and H. Arima. 2006. Recent aspect of cyclodextrin-based drug delivery system. J. Incl. Phenom. Macrocycl. Chem. 56: 3–8.

Umeda, Y., T. Fukami, T. Furuishi, T. Suzuki, M. Makimura and K. Tomono. 2007. Molecular complex consisting of two typical external medicines: Intermolecular interaction between indomethacin and lidocaine. Chem. Pharm. Bull. 55: 832–836.

Wongmekiat, A., Y. Tozuka, T. Oguchi and K. Yamamoto. 2002. Formation of fine drug particles by cogrinding with cyclodextrins. I. The use of β-cyclodextrin anhydrate and hydrate. Pharm. Res. 19: 1867–1872.

Wongmekiat, A., S. Yoshimatsu, Y. Tozuka, K. Moribe and K. Yamamoto. 2006. Investigation of drug nanoparticle formation by co-grinding with cyclodextrins: Studies for indomethacin, furosemide and naproxen. J. Incl. Phenom. Macrocycl. Chem. 56: 29–32.

Zhang, H., D. Wang, R. Butler, N.L. Campbell, J. Long, B. Tan, D.J. Duncalf, A.J. Foster, A. Hopkinson, D. Taylor, D. Angus, A.I. Cooper and S.P. Rannard. 2008. Formation and enhanced biocidal activity of water-dispersible organic nanoparticles. Nat. Nanotech. 3: 506–511.

15

Intratracheal Gene Transfer Using Polyplex Nanomicelles and Their Application to Cardiology

Noriyuki Iwamoto[1] and *Mariko Hrada-Shiba*[2,*]

ABSTRACT

Advances in nanotechnology have led to its application in the field of biomaterials, including drug delivery systems (DDS). To carry nucleic acid drugs, other drugs, and genes to the targeted organs, devices that can carry them intact all the way from the site of administration to the target organs are needed. The development of DDS materials is a challenging and attractive research area whose goal is to treat intractable diseases such as metabolic, cardiovascular, and cancer diseases including genetic diseases with least adverse effects.

Nanocarrier systems based on non-viral DDS, such as lipoplexes and polyplexes, including our polymeric nanomicelles from

[1]Division of Endocrinology and Metabolism, National Cerebral and Cardiovascular Center, 5-7-1 Fujishirodai, Suita, Osaka 565-8565, Japan; E-mail: niwamoto@hsp.ncvc.go.jp

[2]Department of Molecular Innovation in Lipidology, National Cerebral and Cardiovascular Center Research Institute, 5-7-1 Fujishirodai, Suita, Osaka 565-8565, Japan; E-mail: mshiba@ri.ncvc.go.jp

*Corresponding author

List of abbreviations after the text.

poly(ethylene glycol)-poly(amino acid) block copolymer, are among the most functional devices for gene delivery. To deliver genes successfully to target tissues, three primary factors must be clearly addressed: the gene carrier, the therapeutic gene, and the route of administration. The intratracheal administration of therapeutic genes by using polyplex or lipoplex is a promising strategy for delivering genes because it has many advantages over intravenous administration, which has problems such as the degradation of the gene by the nuclease, binding and aggregation with other proteins or erythrocytes in the blood, and difficulty breaking through endothelial barriers. We recently reported that in polyplex nanomicelles consisting of poly(ethylene glycol)-poly(amino acid) block copolymer and adrenomedullin, a therapeutic gene, could attenuate animal models of pulmonary arterial hypertension. This chapter focuses on challenges in the development of gene carriers, therapeutic genes currently used, and the characteristics of the administration route of successful gene therapy for cardiovascular disease. We also discuss our recent report of intratracheal administration of the adrenomedullin gene by using our delivery systems.

INTRODUCTION

Gene therapy is one of the strategic approaches to continuously supplying therapeutic peptides or proteins to target tissues. Gene therapy using viral vectors is reported to deliver therapeutic genes effectively, gain access to host cells, and exploit the cellular machinery to facilitate their own replication. However, viral vectors proved to have adverse effects of toxicity, immunogenicity, inflammatory properties, and high cost (Thomas et al. 2003). In addition, an overwhelming immune reaction against adenovirus occurred in a patient at Pennsylvania University in 1999 (Marshall 2000), and a leukemia-like disease was reported in a French patient in 2002 (Marshall 2002).

Non-viral vectors show certain advantages over viral ones in terms of safety, immunogenicity, and ease of manufacture. To obtain successful gene delivery to the target tissue, three primary factors must be clearly addressed: (1) the development of an appropriate gene carrier, (2) the selection of a therapeutic gene, and (3) the selection of an administration route. (1) The development of advanced carrier systems is essential for protecting therapeutic plasmids or drugs in the environment, for providing site-specific targeting, and for releasing effectively the plasmids or drugs for the desired pharmacological effect. (2) After the genome project, many genes related to the pathogenesis and pathophysiology of diseases were identified and their functions clarified. The appropriate genes that show therapeutic effects in the target tissue should be selected. (3) Selection of a route depending on the target tissue and the gene carrier is essential for successful delivery. Intravenous gene delivery is extremely difficult because of the propensity for rapid nuclease degradation in the blood.

Non-viral delivery systems used to present low levels of transfection and expression efficiency of the gene. Recent advances in the technology of non-viral vectors have yielded techniques with transfection efficiencies similar to those of viruses (Bae and Kataoka 2009). One of the intelligent non-viral gene delivery systems (Fig. 1) is poly(ethylene glycol)(PEG)-based block catiomers, polyplex nanomicelles, composed of a hydrophilic shell and a hydrophobic core. These are well dispersed even in aqueous media containing serum proteins, and they protect plasmid DNA from degradation by nuclease *in vitro* and *in vivo* (Katayose and Kataoka 1998; Wakebayashi et al. 2004; Harada-Shiba et al. 2002). This chapter focuses on the characteristics of non-viral vectors including lipoplexes, polyplexes, and nanomicelle carriers and their application to cardiovascular disease.

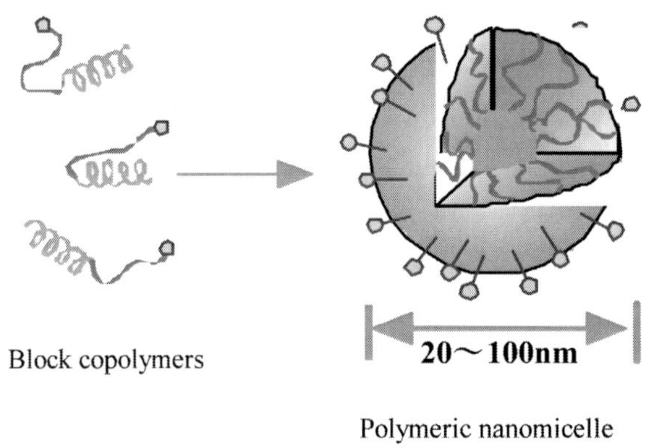

Fig. 1. The schema of the structure of polymeric nanomicelle. Polymeric nanomicelle consists of DNA and block copolymer showing a core shell structure.

Color image of this figure appears in the color plate section at the end of the book.

DEVELOPMENT OF GENE CARRIERS

To achieve safe and effective gene therapy by using a non-viral vector, the construction of a reliable carrier with minimal toxicity is essential. For successful gene delivery to the target cells, the carrier should offer the following: (1) protection of plasmids from the environment during the route to the target cells, (2) incorporation into the target cells, (3) endosomal escape, and (4) transportation to the nucleus.

From Administered Site to Target Cells

The first step of the targeting process is the compartmentalization of plasmid DNA into a carrier that is resistant to the circumstances during the route to the target cells. This step includes compaction or condensation of DNA using polycations like chromatin structure, which is condensed by the electrostatic attractions between the lysine residues around the histone octamer and DNA. This compaction and condensation step is useful for the polyplex to become a nanoparticle that is not recognized by the immune system and not digested by nuclease.

The second step in this process is the internalization of DNA into the target cells. The positively charged DNA complexes are known to interact with anionic substances on the cell surface, such as proteoglycan and sialic acid, by electrostatic interaction. The internalization process of the polyplexes may be affected by their size and zeta potential. Administration of positively charged polyplexes into the blood gives aggregation with negatively charged proteins that are abundantly reserved in the blood and rapidly removed by reticulo-endothelial systems. On the other hand, the polyplexes in neutralized condition form secondary aggregates, which may cause the formation of large precipitates. The polyplex system consisting of cationic homopolymer and DNA is appropriate for examining the functional properties of complexation and condensation of DNA. However, it is not suitable for intravenous administration.

Cationic block copolymer, consisting of hydrophilic PEG and cationic polymer, produces DNA condensate through the formation of polyion complex (PIC) micelles that are surrounded by a hydrophilic layer (Katayose 1998) presenting a nano-sized structure, as shown in the electron microscopic image (Fig. 2) (Harada-Shiba 2002). The formation of micelles with cationic block copolymer and DNA gives colloidal stability in a protein-containing medium, a favorable property for *in vivo* use.

Receptor-mediated Gene Delivery

Wu et al. developed a system for delivering genes to hepatocytes via a unique receptor that can bind and internalize the galactose terminal of asialoglycoprotein (Wu et al. 1987). It was shown that asialoorosomucoid-poly-L lysine carriers delivered DNA specifically to hepatoma cells (Wu et al. 1988a, b). The cell-specific delivery received much attention, and ligands including lactose, folate, transferrin, oligopeptides with arginin-glycine-aspartic acid (RGD), low-density lipoprotein (LDL), antibody, and

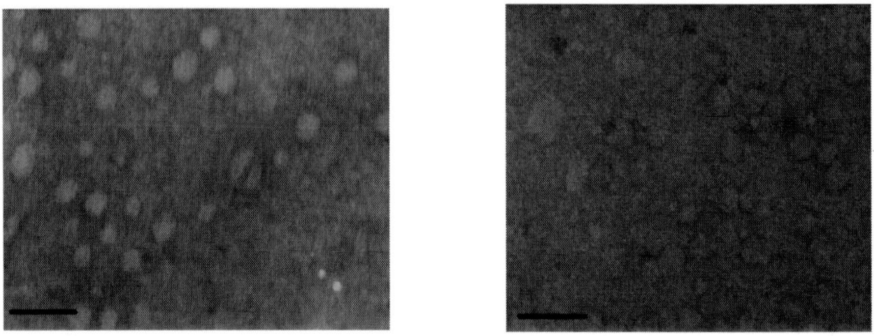

Fig. 2. Electron microscopic findings of negatively stained polyion complex (PIC) nanomicelles. The nanomicelles were prepared by mixing pGL3-control plasmid and PEG-poly-L-lysine block copolymer at the charge ratio of 1:2 (a) and 1:4 (b). Bar equals 100 nm. The data are taken from Harada-Shiba (2003).

so on, were tried targeting hepatocytes, hematopoietic stem cells, epithelial cells, and other cells (Table 1). The polyethyleneimine (PEI) derivatives conjugated to the integrin-binding peptide CYG-GRGDTP via a disulfide bridge led to transgene expression in integrin-expressing epithelial cells (Hela) and fibroblasts at expression levels 10 to 100 times those obtained by PEI alone. The advantage of receptor-mediated gene delivery is not only the specificity of the cells but also the controlled intracellular trafficking of the delivered complex (Erbacher et al. 1999).

Receptor-mediated gene delivery has also been developed using cationic liposomes, whose transfection efficiency is higher than that of cationic polymers. Effective hepatocyte targeting was reported by using galactosylated cationic liposomes. However, because the introduction of asialoglycoproteins to liposomes appeared to have some problems in reproducibility and immunogenicity, low-molecular-weight glycolipids were found to be more promising by virtue of their low immunogenicity and high reproducibility. Intravenously administered DNA/cationic liposome

Table 1. Receptors and ligands targeted for gene transfer.

Receptor	**Ligand**	**Target cells**
Asialoglycoprotein receptor	Asialoorosomucoid Glycosylated polylysine	Hepatocytes
LDL receptor	LDL	Hepatocytes
Folate receptor	Folate	Ubiquitous
Epidermal growth (EGF) receptor	Anti-EGF antibody	Ubiquitous
Integrins	RGD	Ubiquitous
Mannose receptor	Glycosylated polylysine	Macrophage
Transferrin receptor	Transferrin	Ubiquitous

complexes are reported to interact with erythrocytes, which suggests that DNA/cationic liposome complexes are aggregated by non-specific interaction with erythrocytes. The modification of cationic liposomes with PEG lipids was reported to reduce the aggregation (Eliyahu et al. 2002).

A System that Can Enable the Endosomal Escape

When DNA is incorporated into the target cell via a non-viral vector—whether by non-specific endocytosis, phagocytosis, pinocytosis, or receptor-mediated endocytosis—it is separated from the cytoplasm by the vesicle membrane, called endosome. The internal pH of endosomes containing DNA complexes gradually decreased to about 5.5 to become late endosomes, after which the endosomes fuse with lysosomes to become secondary lysosomes where the incorporated DNA complexes are usually hydrolyzed by lysosomal enzymes. This step is recognized as the biggest barrier to successful gene delivery and expression (Fig. 3). In order to let the transferred gene express, the gene must escape from the endosomal vesicles and reach the nucleus. When naked DNA is introduced into the cytosol of the cell by microinjection or osmotic shock, much higher expression of the gene is obtained than in the coculture method. This suggests that lysosomal digestion is the key to getting better transfection efficiency.

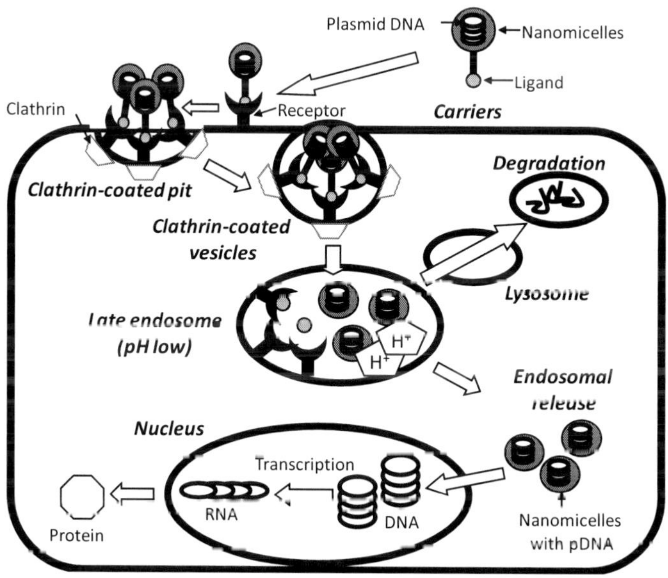

Fig. 3. The schema of intracellular trafficking of non-viral gene delivery systems. The route of nanomicelles from the binding to the cell surface, incorporation, endosomal release, nuclear targeting, transcription, and translation is shown.

Much attention has been paid to promote endosomal escape to get better transfection efficiency. Actually, viruses have functions to fuse and destabilize the endosomal or lysosomal membrane in their envelopes. For example, adenoviruses have capsid proteins that change their conformation under acidic conditions, such as in the late endosomes, to cause an interaction between the protein and vesicle membrane, disrupting the endosomes. These results suggest that endosomal disruption improves the expression efficiency. Wagner et al. showed that peptide sequences derived from the influenza hemagglutinin HA-2 bound to the transferrin-polylysine molecular conjugate markedly increased the level of transgene expression in cells (Wagner et al. 1992). There have been many attempts to construct synthetic peptides that imitate the endosomolytic functions of viral proteins.

Recently, polyplexes assembled from poly(aspartamide) derivatives bearing 1,2-diaminoethane side chains, [PAsp(DET)], were reported to display amplified *in vitro* and *in vivo* transfection activity (Miyata et al. 2008). PAsp(DET) revealed minimal membrane destabilization at physiological pH, yet there was a significant increase in membrane destabilization at acidic pH, mimicking the late endosomal compartment (pH approximately 5), suggesting the potential for successful transfection.

Nuclear Transport

The final goal of gene delivery is to deliver exogenous DNA to the nucleus, where it can proceed to transcription and translation to get functional peptides or proteins. The molecules are transported between the cytoplasm and the nucleus through the nuclear pore complex (NPC), which is a large protein complex of about 125 MDa composed of 50–100 different proteins. To freely diffuse through the NPC channel, a molecule must be less than 40 kDa, and molecules that are larger than 40 kDa need an active process in order to pass through. Nuclear localization signals (NLS), oligopeptides mainly composed of cationic residues, are used to enhance nuclear trafficking through the NPC. Several viral proteins are known to provide translocation activity of the viral genome to the nucleus.

A great deal of effort has been made to enhance transfection efficiencies with non-viral vectors by conjugating NLS peptides to DNA. Plasmid DNA coupled to the NLS derived from simian virus 40 (SV-40) large T antigen effectively targeted zebrafish embryo nuclei.

THERAPEUTIC GENE

For the successful therapeutic application of gene therapy, the delivery of several kinds of genes and nucleotides has been tested in animal models of genetic diseases as well as in patients to supply missing proteins in order to maintain cellular function, or to deliver proteins that induce proliferation or apoptosis of the target cells. In cardiovascular medicine, gene therapy has been applied to treat coronary artery disease, peripheral artery disease, restenosis after vascular interventions and graft failure, hyperlipidemia, thrombosis, and pulmonary arterial hypertension (PAH). Table 2 lists therapeutic targets and genes used for gene therapy.

Table 2. Therapeutic target diseases and genes used for gene therapy in cardiovascular disease.

Therapeutic target disease	Treatment genes
Therapeutic angiogenesis	VEGF-A, -B, -V, -D, -E, FGF-1, -2, 4, 5, angiopoetin-1, HGF, MCP-1, PDGF, eNOS, iNOS, adrenomedullin
Restenosis, vein-graft failure	VEGF-A, C, eNOS, iNOS, COX, Thymidine kinase, CNP Fas ligand, p16, p21, p27, p53, NFkB, and E2F decoys, cdk-2, cdc-2, c-myb, c-myc, ras, bcl-x, PCNA antisense oligonucleotide ribozimes blocking PDGF or TGF-β expression or their receptors
Atherosclerosis, hyperlipidaemia	LDL receptor, VLDL receptor, apoA-1, lipoprotein lipase Hepatic lipase, LCAT, apoB, lipid transfer proteins Lp(a) inhibition, soluble scavenger-receptor decoy, Soluble VCAM or ICAM, SOD, PAF-AH
Thrombosis	Hirudin, tPA, thrombomodulin, COX, TFPI
Pulmonary hypertension	Prepro-calcitonin gene-related peptide, ANP, eNOS prostacyclin synthase, VEGF-A, adrenomedullin
Vasospasm after SAH	Endothelial NOS, ECSOD, CuZnSOD, Antisense preproendothelin-1, Prepro-CGRP

VEGF, vascular endothelial growth factor; FGF, fibroblast growth factor; HGF, hepatocyte growth factor; eNOS, endothelial nitric oxide synthase; CGRP, calcitonin gene-related peptide.

Coronary Artery Disease and Peripheral Artery Disease

Atherosclerosis is the most prevalent process to affect adult coronary and peripheral arteries. Atherosclerotic lesions reduce the lumen of the arteries, leading to a reduction in the arterial blood supply to the myocardium or skeletal muscle. Stimulation of collateral vessel formation by the use of gene therapy will help to increase perfusion of the ischemic tissues. Gene-encoding growth factors, such as vascular endothelial growth factor, fibroblast growth factor, and hepatocyte growth factor, have been successfully tested in animal models and clinical trials for therapeutic angiogenesis.

Restenosis after Vascular Interventions and Vein Graft Failure

The occlusion of arteries after balloon angioplasty, stenting, or the failure of bypass vein graft is a major factor in the prognosis of peripheral and coronary artery disease. Smooth muscle cell proliferation, remodeling, matrix deposition, thrombosis, and platelet and leukocyte adhesion may play roles in the development of arterial restenosis in these settings. To decrease vascular cell proliferation, various gene therapy strategies have been employed. Anti-proliferative strategies designed for the treatment of experimental cardiovascular disease can be grouped into two main categories: (1) antisense approaches, ribozymes, transcription-factor decoys, and siRNA strategies to inactivate positive cell-cycle regulators; and (2) over-expression of negative regulators of cell growth.

Pulmonary Arterial Hypertension

PAH is characterized by a progressive increase in pulmonary vascular resistance, leading to right heart failure and death. Recent advances in therapeutic approaches to PAH show promising targeting pathways believed to play critical pathogenic or pathophysiologic roles. Despite these findings, PAH remains a challenging condition. The average survival from the time of diagnosis is 2.8 y (Nagaya 2004). To reduce pulmonary vascular resistance, the transfer of genes encoding endothelial nitric oxide synthase, calcitonin gene-related peptide, prostacyclin synthase, and adrenomedullin (AM) has been shown to be effective in animal models (Nagaya et al. 2000; Harada-Shiba et al. 2009).

SELECTION OF ROUTE DEPENDING ON THE TARGET TISSUE

Intravenous Administration

The administration of a DNA complex has been considered a promising way to deliver genes to organs, and many attempts have been reported. However, it is not an easy way to deliver genes successfully because there are many barriers for the gene to go through. For example, the intravenous administration of positively charged polyplex gives aggregation with negatively charged proteins abundantly reserved in the blood. Endothelial barriers are another hurdle on the way to the target. Many attempts have been made to deliver specific organs via specific receptors, as mentioned in the subsections on receptor-mediated gene delivery in the section on development of gene carriers.

Intratracheal Administration

Intratracheal administration can avoid the propensity of nuclease to degrade in the blood compartment. The lung has an enormous surface area that can serve as a therapeutic bioreactor for the delivery of therapeutic genes. Pulmonary administration is a promising therapeutic route, especially in the clinic, by virtue of its high patient compliance with the use of an inhaler or nebulizer.

Special features of pulmonary gene delivery via airways are that the lung has characteristics that critically influence transfection efficiency, such as the presence of surfactant, alveolar macrophages, and mucociliary clearance mechanisms. In the early 1990s, lipoplex was used by aerosol delivery or intratracheal instillation. However, cationic lipids were shown to have decreased transfection efficiency due to interaction with lung surfactants compared to a cationic polymer like PEI. To overcome the surfactant barriers, cationic emulsion was used and showed much higher transfection activity than lipoplexes such as Lipofectin, Lipofectamine, and DMRIE/c. However, even for cationic emulsion, the gene's expression efficiency was limited. Polyplexes and polyplex nanomicelles made from cationic polymer or block copolymer were reported to show higher transfection efficiency than cationic lipoplexes for pulmonary gene delivery via airways.

VECTORS USED IN GENE DELIVERY

Polymer

Synthetic polymer vectors possess cationic charges and form a PIC with negatively charged DNA based on electrostatic interactions. The transfection efficacy depends on their chemical structure, charge density, and molecular weight. The list of polymer vectors is shown in Table 3.

Polyethyleneimine (PEI)

PEI, a polycation with high ionic charge density, has been used as a gene delivery material. PEI binds to DNA by an electrostatic interaction to form small complexes that are internalized into cells and can localize to the nucleus. PEI's high transfection efficiency can be caused by its large buffering capacity, which may change endosomal osmolarity. Endosomal release may enhance gene delivery to the nucleus (Yamagata et al. 2007). Intravenous delivery of PEI is not suitable for targeting of the lung. PEI

Table 3. List of polymer vectors.

Polymers	Modified polymers
Block copolymer	PEG-poly(aspartate) (PEG-PAsp) PEG-poly(glutamate) (PEG-PGlu) PEG-poly(lysine) (PEG-PLys) PEG-chitosan PEG-b-P[Asp(DET)]
PLGA poly(lactic-coglycolic acid)	Chitosan-modified PLGA PEG-PLGA
Chitosan	Glycol chitosan PEI-graft-chitosan chitosan lactate
PEI poly(ethyleneimine)	Branched PEI Galactose-PEG-PEI Glucosylated PEI Cell-penetrating peptides-PEG-PEI
Poly-L-lysine	
Dendrimer	G2/G3 polyamidoamine (PAMAM) G9 PAMAM

or modified PEI has been shown to be one of the most effective agents for constructing gene delivery systems available today, with high levels of pulmonary gene transfer via the airways (Densmore 2006).

Chitosan

Chitosan is produced by the deacetylation of chitin, which is the structural component in the exoskeleton of crabs and shrimp. Chitosan is also a promising biopolymer as a carrier material because of its biocompatibility, low toxicity, and low degradation activity. This product is made at relatively low cost because natural sources of it are abundant. Though complexes of chitosan-DNA are stable against DNase, the transfection efficiency of chitosan-DNA is generally lower than those of other non-viral vectors. The degree of deacetylation, molecular weight, and N/P ratio (the ratio of the number of nitrogen molecules in chitosan against that of phosphate in a gene) are effective factors for controlling the gene delivery efficiency.

Dendrimers

Dendrimers are typically symmetric around the core and are composed of hyperbranched amino-acid units. Dendrimers are classified by generation, which refers to the number of repeated branching cycles that are performed during their synthesis. In general, dendrimers form a neutral complex with plasmid DNA or siRNA; they have properties by which free DNA is released from the complex more easily than from other polymers, thus providing high transfection efficiency.

Liposome

Cationic liposomes condense DNA to form particles (100 to 200 nm), called lipoplexes, based on electrostatic interactions and protect the DNA from degradation. In most cases, after lipoplexes are intravenously injected, the highest levels of gene expression are obtained in the lung because the lung capillaries are the first traps. Cationic liposomes have been clinically evaluated and are popular carriers for gene delivery to the lungs (Desmore 2006).

Liposomes conjugated with cell-penetrating peptides are used for intracellular delivery to the lung because they enhance the liposome uptake by cells. Pulmonary gene delivery via inhalation has an advantage in that the lung has critical features influencing transfection efficiency, such as the presence of alveolar surfactant, macrophage, and mucociliary clearance mechanisms. Lipoplexes were used with aerosol delivery or intratracheal instillation in the early 1990s. However, in gene delivery, cationic lipids were shown to have reduced transfection efficiency due to interaction with lung surfactants compared with a cationic polymer like PEI (Wiseman et al. 2003; Bragonzi et al. 1999).

Polyplex Nanomicelles

To overcome the drawbacks of positively charged DNA complexes that bind to proteins and erythrocytes and form aggregates that are cleared by the reticulo-endothelial system, we have developed a PIC with a cationic block copolymer possessing a hydrophilic segment, PEG (Fig. 1). The complexes of DNA and block copolymers form self-assembling particles with core-shell structures called PIC nanomicelles. PIC nanomicelles are water-soluble and nuclease-resistant nanoparticles with small zeta potential (Katayose et al. 1998). These nanomicelles show high colloidal stability under physiological conditions and substantial transfection activity against various cell types even after pre-incubation with serum proteins. Nanomicelles remained intact for 3 h after intravenous administration, suggesting high stability in circulating blood (Harada Shiba et al. 2002).

High transfection efficiency and low cytotoxicity with the use of PIC nanomicelles formed by PEG-block-poly(aspartamide) copolymers carrying the N-(2-aminoethyl)-2-aminoethyl group in the side chain (PEG-P[Asp(DET)]) were reported (Kanayama et al. 2006). These successful *in vivo* gene therapies have been explained by the specific structure of the side chain of P[Asp(DET)], in which the 1,2-ethanediamine moiety of N-(2-aminoethyl)-2-aminoethyl group exhibits distinct two-step protonation behavior, suggesting the potential proton sponge capacity of Asp(DET) units for efficient endosomal escape.

IN VIVO GENE DELIVERY BY INTRATRACHEAL ADMINISTRATION USING POLYPLEX NANOMICELLES FOR TREATMENT OF ANIMAL MODELS OF PAH

The large number of human diseases presenting poor prognoses and limited efficacy under current therapeutic regimens necessitates the development of alternative approaches. PAH is such a disease that lacks a highly efficacious therapeutic regimen. PAH patients are currently treated with a variety of drugs, including prostacyclin, prostacyclin analogs, calcium channel blockers, nitric oxide (NO) inhalation, angiotensin-converting enzyme inhibitors, endothelin receptor antagonists, and phosphodiesterase 5 inhibitors; in severe cases, lung transplantation and subsequent immunosuppression are necessary. Viral or viral-related vectors have been used for the delivery of therapeutic genes, and these gene carriers have the potential for immunogenicity and inflammatory response. In diseases where a single dose can cure or provide palliative care, viral vectors may be suitable. However, PAH therapy requires repeated administration for efficacy, hence the utility of viral or viral-based gene therapy is contraindicated.

P[Asp(DET)], a poly(aspartamide) derivative bearing an N-(2-aminoethyl)aminoethyl group as the side chain, showed improved transfection efficiency and biocompatibility compared to linear poly(ethyleneimine) (LPEI) (Kanayama et al. 2006). PEG-b-P[Asp(DET)] was applied to the *in vivo* delivery of therapeutic plasmids for a murine, skull bone defect model and a rabbit carotid artery with neointima model; its successful therapeutic efficacy in these mammalian studies provided the impetus for expanded application to gene therapy for intractable diseases (Itaka et al. 2007; Akagi et al. 2007).

We applied the gene delivery system to introduce genes into mice and found a 100-fold increase in transgene expression by the intratracheal administration of PEG-b-P[Asp(DET)] compared with LPEI (Fig. 4) (Harada-Shiba et al. 2009). The high levels of gene expression persisted for 14 d. PEG-*b*-P[Asp(DET)] polyplex nanomicelles loaded with pDNA bearing the yellow fluorescence protein (YFP) gene (N/P = 80) or LPEI/pYFP polyplexes (N/P = 6) were sprayed intratracheally into mice. The results showed that significantly higher fluorescence intensity was clearly seen in the lungs treated with the PEG-*b*-P[Asp(DET)] polyplex nanomicelle than in the LPEI polyplex controls. The lungs administered with LPEI/pLuc showed moderate infiltration of neutrophils at the terminal bronchiole and alveoli. However, in the lungs administered with PEG-*b*-P[Asp(DET)]/pLuc, neutrophilic infiltration was scattered and either minimal or absent, suggesting increased biocompatibility with the

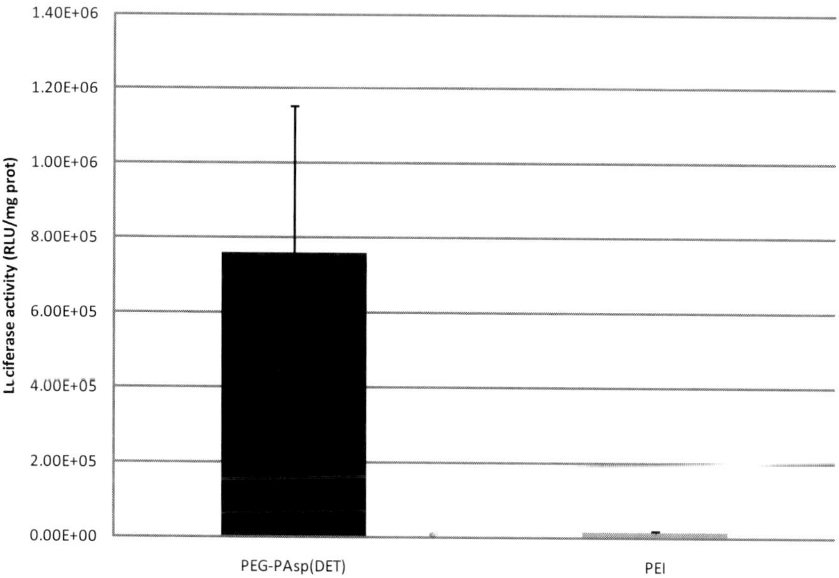

Fig. 4. Luciferase gene expression by intratracheal administration of LPEI polyplex (N/P = 6) or PEG-*b*-P[Asp(DET)] polyplex nanomicelle (N/P = 80). Samples of the polyplex and the polyplex nanomicelle were prepared before the administration and left for 1 d. The mice (five per group) were anesthetized and the polyplex or the polyplex nanomicelle was administered intratracheally. At 24 h post-administration, the lung tissues were harvested, homogenized, and measured for luciferase activity (mean + SEM, N = 5). With permission from Harada-Shiba et al. (2009).

PEG-*b*-P[Asp(DET)] polyplex micelle. Pro-inflammatory cytokine mRNA levels did not increase for intratracheally administered naked pLuc in saline or for the PEG-*b*-P[Asp(DET)] polyplex micelles. However, LPEI/pLuc polyplexes revealed a two-fold increase in tumor necrosis factor α, IL-6, 1L-10, and Cox-2 compared to the control, suggesting that polyplex nanomicelles have favorable properties for *in vivo* use.

Then, a PEG-*b*-P[Asp(DET)] polyplex nanomicelle system was tested to introduce the AM gene via intratracheal administration in PAH model rats. After 4 wk of monocrotaline injection, right ventricular pressure was increased to twice the normal value. Notably, right ventricular pressure was decreased significantly by an intratracheal spray of the PEG-*b*-P[Asp(DET)] polyplex nanomicelle loaded with the expression vector of AM (N/P = 40) (Fig. 5). On the other hand, right ventricular pressure did not change significantly after administration of naked plasmid encoding the AM gene in saline or the LPEI polyplex loaded with the AM gene, or with the administration of the polyplex nanomicelle loaded with the luciferase gene. The lungs transfected with the polyplex nanomicelle

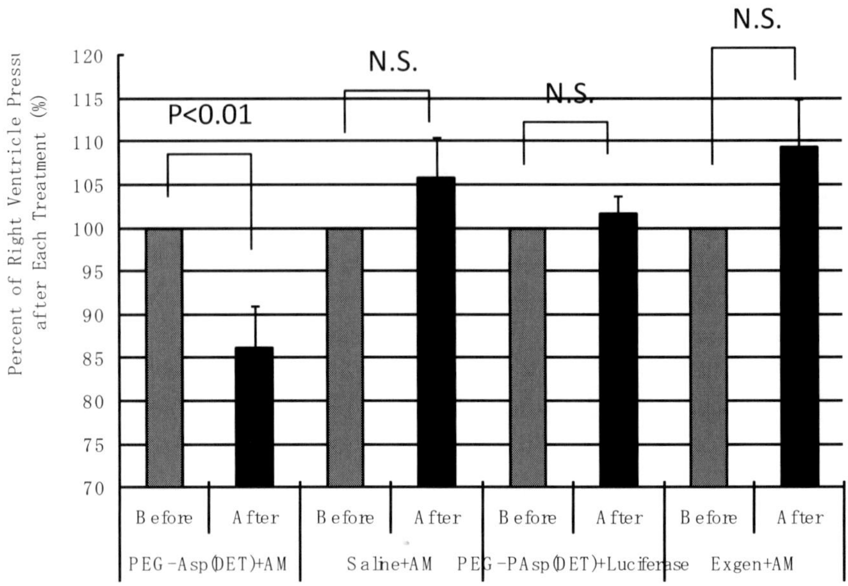

Fig. 5. Effect of gene transfer on right ventricle pressure in the PAH rat model. The PEG-*b*-P[Asp(DET)] polyplex nanomicelle loaded with AM expression vector; AM expression vector in saline; PEG-*b*-P[Asp(DET)] polyplex micelle loaded with luciferase gene; or the LPEI polyplex loaded with AM expression vector was sprayed intratracheally. Three days later, a hemodynamic study was performed again to measure the right ventricle pressure, with the results indicated as "After". With permission from Harada-Shiba et al. (2009).

loaded with the AM expression vector had high levels of AM mRNA. Alternatively, the levels were much lower in the lungs transfected with the LPEI polyplex loaded with the AM expression vector. These results suggest that successful delivery of the therapeutic gene was performed by intratracheal administration using PEG-*b*-P[Asp(DET)] polyplex nanomicelle.

APPLICATIONS TO OTHER AREAS OF HEALTH AND DISEASE

Intratracheal administration of therapeutic genes using PEG-*b*-P[Asp(DET)] polyplex nanomicelle can be applied to lung diseases such as cystic fibrosis, alpha 1-antitrypsin deficiency, asthma, or lung cancer. It can also be applied to other diseases for the lungs to use as a reservoir organ.

Summary Points

- To successfully deliver genes to target tissue, three primary factors must be addressed: the development of an appropriate gene carrier, the selection of a therapeutic gene, and the selection of an administration route.
- For successful gene delivery to target cells, carriers should provide properties that can break through four major barriers: (1) a system that can protect plasmids from the environment during the route to target cells, (2) a system that can be incorporated into the target cells, (3) a system that can enable endosomal escape, and (4) a system that can be transported to the nucleus.
- In cardiovascular medicine, gene therapy has been applied to treat coronary artery disease, peripheral artery disease, restenosis after vascular interventions and graft failure, hyperlipidemia, thrombosis, and pulmonary arterial hypertension.
- Development of non-viral vectors such as polymer, block copolymer, and liposome have been developed and used for the treatment of animal models as well as patients.
- A therapeutic gene was successfully delivered by intratracheal administration using a PEG-*b*-P[Asp(DET)] polyplex nanomicelle, which is water soluble and nano-sized, has low zeta potentials, and can act as a proton sponge for efficient endososomal escape. We succeeded in the treatment of PAH animal models by delivering the AM gene using the PEG-*b*-P[Asp(DET)] polyplex nanomicelle without adverse effect.

Key Facts

- Plasmid DNA needs to be compacted and condensed in order to remain intact during the route to the target cells.
- Cationic block copolymer, consisting of hydrophilic PEG and cationic polymer, produces DNA condensate through the formation of polyion complex (PIC) micelles.
- Gene delivery can be targeted by using receptor-mediated delivery.
- The endosomal escape step is recognized as the biggest barrier to successful gene delivery and expression.
- Nuclear localization signals, oligopeptides mainly composed of cationic residues, are used to enhance nuclear trafficking.

- In cardiovascular medicine, gene therapy has been applied to treat coronary artery disease, peripheral artery disease, restenosis after vascular interventions and graft failure, hyperlipidemia, thrombosis, and pulmonary arterial hypertension.
- Pulmonary administration of a gene is a promising therapeutic route of administration, especially in the clinic, because patients show high degrees of compliance with the use of inhalers or nebulizers.
- The complexes of DNA and block copolymer form self-assembling particles with core-shell structures, PIC nanomicelles.
- PIC nanomicelles have high transfection efficiency and low toxicity.
- Successful delivery of the adrenomedullin gene was performed by intratracheal administration using PEG-*b*-P[Asp(DET)] polyplex nanomicelle, ameliorating pulmonary arterial hypertension in rat models.

Definitions

AM: The most potent peptide that relaxes smooth muscle cells.

Endosomal escape: A property that non-viral vectors should have to get better expression of genes, which are the mechanisms that are not degraded in the endosome.

Lipoplex: A complex consisting of a liposome and a gene.

Nuclear pore complex: A complex of proteins on the surface of the nucleus that plays a role in the gate of the nucleus.

Polyplex nanomicelle: A complex consisting of a block copolymer and a gene.

Polyplex: A complex consisting of a polymer vector and a gene.

Transfection: Introduction of the genes to the cells.

Abbreviations

AM	:	adrenomedullin
DDS	:	drug delivery systems
IL	:	interleukin
LDL	:	low-density lipoprotein
LPEI	:	linear polyethyleneimine
N/P ratios	:	nitrogen/phosphate ratios
NLS	:	nuclear localization signal
NO	:	nitric oxide
NPC	:	nuclear pore complex
PAH	:	pulmonary arterial hypertension
PEG	:	polyethylene glycol

PEG-*b*-P[Asp(DET)]	:	PEG*b*-poly{*N*-[*N*-(2-aminoethyl)-2-aminoethyl]aspartamide}
PEI	:	polyethyleneimine
PIC	:	polyion complex
SV-40	:	simian virus-40
YFP	:	yellow fluorescence protein

References

Akagi, D., M. Oba, H. Koyama, N. Nishiyama, S. Fukushima, T. Miyata, H. Nagawa and K. Kataoka. 2007. Biocompatible micellar nanovectors achieve efficient gene transfer to vascular lesions without cytotoxicity and thrombus formation. Gene Ther. 14: 1029–1038.

Bae, Y. and K. Kataoka. 2009. Intelligent polymeric micelles from functional poly(ethylene glycol)-poly(amino acid) block copolymers. Adv. Drug Deliv. Rev. 61: 768–784.

Bragonzi, A., A. Boletta, A. Biffi, A. Muggia, G. Sersale, S.H. Cheng, C. Bordignon, B.M. Assael and M. Conese. 1999. Comparison between cationic polymers and lipids in mediating systemic gene delivery to the lungs. Gene Ther. 6: 1995–2004.

Densmore, C.L. 2006. Advances in noninvasive pulmonary gene therapy. Curr. Drug Deliv. 3: 55–63.

Eliyahu, H., N. Servel, A.J. Domb and Y. Barenholz. 2002. Lipoplex-induced hemagglutination: potential involvement in intravenous gene delivery. Gene Ther. 9: 850–858.

Erbacher, P., J.S. Remy and J.P. Behr. 1999. Gene transfer with synthetic virus-like particles via the integrin-mediated endocytosis pathway. Gene Ther. 6: 138–145.

Harada-Shiba, M., K. Yamauchi, A. Harada, I. Takamisawa, K. Shimokado and K. Kataoka. 2002. Polyion complex micelles as vectors in gene therapy—pharmacokinetics and in vivo gene transfer. Gene Ther. 9: 407–414.

Harada-Shiba, M., I. Takamisawa, K. Miyata, T. Ishii, N. Nishiyama, K. Itaka, K. Kangawa, F. Yoshihara, Y. Asada, K. Hatakeyama, N. Nagaya and K. Kataoka. 2009. Intratracheal gene transfer of adrenomedullin using polyplex nanomicelles attenuates monocrotaline-induced pulmonary hypertension in rats. Mol. Ther. 17: 1180–1186.

Itaka, K., S. Ohba, K. Miyata, H. Kawaguchi, K. Nakamura, T. Takato, U.I. Chung and K. Kataoka. 2007. Bone regeneration by regulated in vivo gene transfer using biocompatible polyplex nanomicelles. Mol. Ther. 15: 1655–1662.

Kanayama, N., S. Fukushima, N. Nishiyama, K. Itaka, W.D. Jang, K. Miyata, Y. Yamasaki, U.I. Chung and K. Kataoka. 2006. A PEG-based biocompatible block catiomer with high buffering capacity for the construction of polyplex micelles showing efficient gene transfer toward primary cells. ChemMedChem. 1: 439–444.

Katayose, S. and K. Kataoka. 1998. Remarkable increase in nuclease resistance of plasmid DNA through supramolecular assembly with poly(ethylene glycol)-poly(L-lysine) block copolymer. J. Pharm. Sci. 87: 160–163.

Marshall, E. 2000. FDA halts all gene therapy trials at Penn. Science. 287: 565–567.

Marshall, E. 2002. Clinical research. Gene therapy a suspect in leukemia-like disease. Science 298: 34–35.

Masago, K., K. Itaka, N. Nishiyama, U.I. Chung and K. Kataoka. 2007. Gene delivery with biocompatible cationic polymer: pharmacogenomic analysis on cell bioactivity. Biomaterials 28: 5169–5175.

Miyata, K., M. Oba, N. Nakanishi, S. Fukushima, Y. Yamasaki, H. Koyama, N. Nishiyama and K. Kataoka. 2008. Polyplexes from poly(aspartamide) bearing 1, 2-diaminoethane side chains induce pH-selective, endosomal membrane destabilization with amplified transfection and negligible cytotoxicity. J. Am. Chem. Soc. 130: 16287–16294.

Nagaya, N. 2004. Drug therapy of primary pulmonary hypertension. Am. J. Cardiovasc. Drugs 4: 75–85.

Nagaya, N., C. Yokoyama, S. Kyotani, M. Shimouchi, R. Morishita, M. Uematsu, T. Nishikimi, N. Nakanishi, T. Ogihara, M. Yamagishi, K. Miyatake, Y. Kaneda and T. Tanabe. 2000. Gene transfer of human prostacyclin synthase ameliorates monocrotaline-induced pulmonary hypertension in rats. Circulation 102: 2005–2010.

Thomas, C.E., A. Ehrhardt and M.A. Kay. 2003. Progress and problems with the use of viral vectors for gene therapy. Nat. Rev. Genet. 4: 346–358.

Wagner, E., C. Plank, K. Zatloukal, M. Cotten and M.L. Birnstiel. 1992. Influenza virus hemagglutinin HA-2 N-terminal fusogenic peptides augment gene transfer by transferrin-polylysine-DNA complexes: toward a synthetic virus-like gene-transfer vehicle. Proc. Natl. Acad. Sci. 89: 7934–7938.

Wakebayashi, D., N. Nishiyama, K. Itaka, K. Miyata, Y. Yamasaki, A. Harada, H. Koyama, Y. Nagasaki and K. Kataoka. 2004. Polyion complex micelles of pDNA with acetal-poly(ethylene glycol)-poly(2-(dimethylamino)ethylmethacrylate) block copolymer as the gene carrier system: physicochemical properties of micelles relevant to gene transfection efficacy. Biomacromolecules 5: 2128–2136.

Wiseman, J.W., C.A. Goddard, D. McLelland and W.H. Colledge. 2003. A comparison of linear and branched polyethylenimine (PEI) with DCChol/DOPE liposomes for gene delivery to epithelial cells *in vitro* and *in vivo*. Gene Ther. 10: 1654–1662.

Wu, G.Y. and C.H. Wu. 1987. Receptor-mediated in vitro gene transformation by a soluble DNA carrier system. J. Biol. Chem. 262: 4429–4432.

Wu, G.Y. and C.H. Wu. 1988a. Evidence for targeted gene delivery to Hep G2 hepatoma cells *in vitro*. Biochemistry 27: 887–892.

Wu, G.Y., and C.H. Wu. 1988b. Receptor mediated gene delivery and expression in vivo. J. Biol. Chem. 263: 14621–14624.

Yamagata, M., T. Kawano, K. Shiba, T. Mori, Y. Katayama and T. Niidome. 2007. Structural advantage of dendritic poly(L-lysine) for gene delivery into cells. Bioorg. Med. Chem. 15: 526–532.

16

Use of Microbubbles and Nanobubbles for Diagnostic Vascular Molecular Imaging and Therapeutic Applications

Chih-Hsien Lee,[1,a] *Cheng-An J. Lin,*[1,b] *Rajkumar Rajendram*[2] and *Walter H. Chang*[1,c,*]

ABSTRACT

Microbubbles are tiny bubbles (diameter 1–999 μm) that consist of a gas core surrounded by a shell. Nanobubbles are even smaller (diameter 1–999 nm). Many different types of microbubbles and nanobubbles are currently available and others are in advanced stages of development. The physical properties of these bubbles influence the biological effects that occur when bubbles are insonated *in vivo* and therefore determine the advantages, disadvantages

[1]Department of Biomedical Engineering, Chung Yuan Christian University, Chung-Li, Taiwan, R.O.C.

[a] E-mail: petertrue1@yahoo.com.tw

[b]E-mail: chengan_lin@cycu.edu.tw

[c]E-mail: whchang@cycu.edu.tw

[2]Departments of Anaesthesia and Intensive Care, Royal Free Hospital, Hampstead, London, England; E-mail: rajkumarrajendram@doctors.org.uk

*Corresponding author

List of abbreviations after the text.

and potential uses of each type of bubble. Microbubbles have an established role in medical diagnostics as ultrasound contrast agents. Recent advances in ultrasound and nanotechnology have led to the development of targeted bubbles. A ligand conjugated to a bubble can target a specific ligand-receptor interaction site. Targeted bubbles can be used to determine relative blood volume and assess individual organ perfusion *in vivo*. Ligand-targeted vascular imaging can focus on receptors at the cellular and even molecular levels to detect processes such as tumour angiogenesis, inflammation and thrombus formation. The development of microbubbles and nanobubbles as drug carriers is therefore very exciting. These bubbles could, for example, encapsulate anticancer drugs and genes for targeted tumour and gene therapy. These advances in bubble technology could be used to develop novel theranostic agents that combine and simplify diagnosis and treatment. However, many limitations must be overcome before this technology can revolutionize clinical practice.

INTRODUCTION

There are three important stages in the clinical management of disease: diagnosis, treatment and confirmation of successful treatment. Different techniques are used at each stage that can prolong and complicate the management of disease. The relatively novel concept of 'theranostics' aims to use a single technique to diagnose, treat and confirm successful treatment of a disease.

Microbubbles are tiny gas bubbles with diameters between 1 and 999 μm that consist of a gas core surrounded by a shell. Nanobubbles have similar structures but are even smaller (diameter 1–999 nm). Microbubbles are already in clinical use and recent developments in nanotechnology could revolutionize the diagnosis and treatment of disease in humans.

Ultrasound is an inaudible high frequency (> 20 kHz) mechanical vibration. It is widely used in clinical practice for diagnostic and therapeutic procedures. Ultrasound is generated by the conversion of electrical energy to acoustic energy by mechanical deformation of a piezoelectric crystal located within a transducer. Ultrasound waves are propagated by molecular collision and vibration. Acoustic energy is progressively lost (attenuation) while passing through tissues because of absorption, dispersion or scattering of the wave. The scattered ultrasound waves that are reflected back to the transducer are converted back to an electrical signal by the piezoelectric crystal and used to generate an image of the insonated tissue.

Imaging blood-filled cavities with standard ultrasound is difficult because the tiny amount of ultrasound scattered by blood at conventional imaging frequencies is below the displayed dynamic range. So blood appears black on standard ultrasound images. However, microbubbles greatly increase the strength of the ultrasound signal returning from within blood-filled cavities *in vivo* and so have a prominent role in medical diagnostics as intravascular ultrasound contrast agents.

When insonated with higher power, ultrasound bubbles can induce various biological effects *in vivo*. These biological effects and their potential clinical uses are discussed below. High power ultrasound can also disrupt bubbles. Microbubbles and more recently nanobubbles have been developed as potential drug carriers. Ultrasound targeted microbubble destruction could be harnessed to release drugs in specific areas. Ligand-conjugated carrier bubbles could be used to carry drugs or genes to the targeted area. These bubbles could then be visualized and destroyed by ultrasound emitted by standard ultrasound machines. The contents of the bubbles would then be deposited at the target site.

This ultrasound targeted microbubble destruction (UTMD) is highly organ specific and could improve drug delivery and gene transfer efficiency. UTMD could be applied in any well-perfused tissue accessible to ultrasound. Some of the most interesting cardiac applications for UTMD include the management of coronary artery disease, heart failure and cardiac remodeling (Jones and Koch 2005). In non-cardiac tissue, UTMD can be applied in blood vessels to treat diseases such as arteriosclerosis and thrombus formation. Tumour imaging and treatment with UTMD could also be better, simpler and more cost-effective.

The combination of advances in ultrasound and nanotechnology has many potential diagnostic and therapeutic applications. Microbubbles and more recently nanobubbles are among the most promising theranostic agents currently in development. However, fine control of the chemical and physical properties of the bubbles is required for biomedical use in humans *in vivo*. The thickness and mechanical properties of the bubble shell influence the acoustic response and destruction threshold of insonated bubbles. Advanced synthetic techniques are required to manipulate the size, biodegradability, composition, functions and safety profiles of bubbles.

This chapter reviews the physical and chemical properties of bubbles. The influence of these properties on the biological effects that occur when bubbles are insonated are also considered. The chapter subsequently focuses on the development of carrier bubbles for gene and drug delivery and their diagnostic and therapeutic applications. It concludes with discussion of the challenges that limit the diagnostic and therapeutic use of bubbles *in vivo*.

BUBBLE-ULTRASOUND INTERACTION

The acoustic properties of gas bubbles (unlike solid tissues) vary with the strength of the ultrasound applied. Initially bubbles oscillate in a linear fashion, but as the power of the ultrasound increases they oscillate non-linearly. Increasing the power ultrasound still further can destroy bubbles (scintillation).

Bubbles as Ultrasound Contrast Agents

The use of air bubbles as ultrasound contrast agents (UCAs) to enhance ultrasound imaging was first reported by Gramiak and Shah (1968). However, these large simple air bubbles rapidly decompose in circulating blood *in vivo* and are highly unstable when insonated. When injected intravenously they cannot cross the pulmonary capillaries and so can only be used to enhance ultrasound imaging of the right heart and systemic veins.

Several methods are used to improve the stability of bubbles within the blood circulation *in vivo*. For example, the use of low diffusivity core gases with high molecular weight molecules (e.g., perfluorocarbons or PFC; sulphur hexafluoride) improves the stability of bubbles.

The composition of the shell also affects the stability of bubbles *in vivo*. The UCAs can be classified by the composition of their shells into soft and hard shell bubbles. Soft shell bubbles are surrounded by monolayers, such as surfactants, lipids, and phospholipids. Hard shell bubbles are encapsulated by harder materials, such as proteins, polymers (e.g., polycyanoacrylate) or inorganic materials (e.g., silica). Soft shell bubbles are less stable but more echogenic than hard shell bubbles, because soft shells are more sensitive to pressure changes (i.e., flexible) and resonate better when insonated.

Changes to the core gas and shells of bubbles have increased their stability within the blood circulation *in vivo* up to at least a few minutes. After intravenous injection these bubbles can pass through pulmonary capillaries. These UCAs can be used to enhance imaging of the left heart and systemic arteries.

Imaging with ultrasound and bubble contrast has several advantages over the alternative imaging modalities: computer-assisted tomography (CT), nuclear medicine and magnetic resonance imaging (MRI). Besides the minimal cost of contrast-enhanced ultrasound in comparison to CT and MRI, ultrasound machines are portable and can be used at the bedside. Furthermore, ultrasound imaging does not use ionizing radiation (unlike CT and nuclear medicine) and can therefore be repeated frequently for serial follow-up.

Historical Classification of Contrast Agents

The first intravenous UCAs capable of left heart contrast (first generation agents) were air bubbles stabilized by encapsulation (Albunex) by a fatty acid, lipid, or protein shell or adherence to microparticles (Levovist). However, the lifespan of the first-generation bubbles after intravenous injections *in vivo* was still very short because the air core rapidly dissolved in blood.

Second-generation microbubbles (e.g., Optison, SonoVue) replaced the air core with gases that are much less soluble in blood (e.g., high molecular weight PFC or sulphur hexafluoride). These gases greatly increased the stability and circulation time of bubbles within the body *in vivo*. Commercially available bubbles are classified by shell composition and core gas, as summarized in Table 1. These UCAs are commonly used to enhance the ultrasound imaging of the right and left ventricles, systemic arteries and perfused tissues.

Table 1. Overview of ultrasound contrast agents (UCAs) that are commercially available and used in clinical practice.

Name	Manufacture	Core gas	Shell material	Notes
First-Generation UCAs				
Echovist	Schering (Germany)	Air	Galactose matrix	Only available in Europe
Levovist	Schering (Germany)	Air	Palmitic acid	Only available in Europe
Albunex	Molecular Biosystems (CA)	Air	Albumin	First FDA-approved UCA
Second-Generation UCAs				
Optison	GE Healthcare (NJ)	Octafluoropropane	Albumin	First FDA-approved second-generation UCA
Definity	BMS (MA)	Octafluoropropane	Phospholipid	Activated by agitation before use
Imagnet	IMCOR (CA)	Perfluorohexane	Phospholipid	Similar to Definity and Optison
SonoVue	Bracco (Nertherland)	Sulphur hexafluoride	Phospholipid	FDA-approved yet Approved, used in Europe. Use in non-cardiac imaging, extra caution should be exercised in patients (e.g., hypotension, bradycardia, cardiac arrest and myocardial infarction).

Third-generation agents are in development. These will incorporate a low solubility gas surrounded by a polymer shell. This should result in UCAs with much more reproducible acoustic properties. Liposomes are another promising material from which to develop UCAs for theranostic use. These shells have advantages over albumin or nonionic shells. Liposome-based bubbles induce less host response and are completely biodegradable (Verma 1997) but are still sufficiently stable *in vivo* to reach their target (Bloch et al. 2004).

Non-linear Oscillation and Harmonic Imaging

When bubbles oscillate non-linearly the alternate expansion and contraction of the bubble is not equal. Non-linear oscillation results in the generation of ultrasound harmonics, i.e., the frequency of the ultrasound produced by non-linear oscillation of bubbles is a multiple of the insonating frequency. The harmonic response is dependent upon the physical properties of the bubbles (both size and mechanical properties), the incident pressure of the ultrasound field, and the frequency.

Harmonic imaging was initially developed to optimize contrast-enhanced ultrasound. Tuning the receiver to receive ultrasound at double the transmitted frequency generates an image from the first harmonic signal. This greatly improves the signal-to-noise ratio. As tissue also generates harmonics, harmonic imaging can also enhance conventional grey scale imaging of tissues. Harmonic imaging is a standard feature on most ultrasound machines available today. Thus, contrast imaging can be optimized by choosing the most appropriate agent and settings for the ultrasound equipment in use.

Tissue perfusion can potentially be identified by imaging contrast agent and not tissue. Harmonic imaging greatly improves sensitivity to contrast and enables excellent left ventricular cavity opacification, but resolution is not sufficient for tissue perfusion imaging. Pulse inversion and power modulation imaging have been developed to improve the differentiation of contrast from tissue still further.

Cavitation

When any acoustic pressure wave travels through a medium, the medium expands (rarefaction phases; Fig. 1a) and contracts (compression phases; Fig. 1a). Figure 1b illustrates the passage of an ultrasound wave through a medium in which molecules can be displaced. The dark dots represent the individual molecules within the medium. The molecules spread apart in low pressure areas and are compressed together in high pressure areas. Similarly, insonated bubbles cavitate (i.e., contract then expand) when exposed to high pressure and low pressure, respectively (Fig. 1c and d).

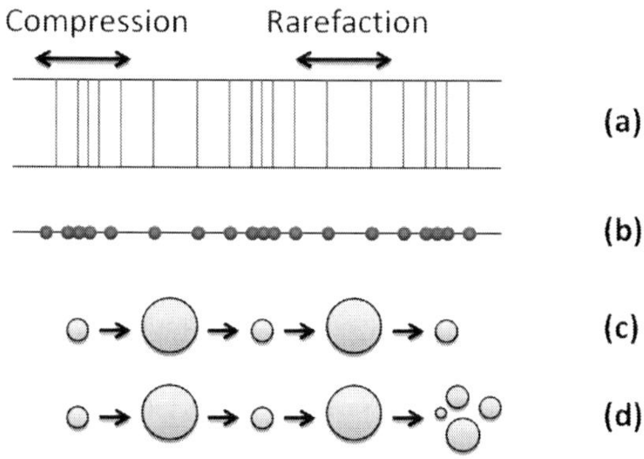

Fig. 1. Ultrasound cavitation. (a) Schematic representation of an acoustic pressure wave, (b) the displacement of molecules in a pressure wave, (c) stable cavitation of bubbles, and (d) inertial cavitation of bubbles.

There are two types of cavitation: inertial cavitation and non-inertial (stable) cavitation. Bubbles exhibit stable cavitation when insonated by low amplitude ultrasound (Fig. 1c). At higher ultrasound intensities, bubbles expand rapidly during the rarefaction phase, until they collapse with the inertia of inrushing fluid. This can cause bubbles to fragment into many smaller bubbles (inertial cavitation; Fig. 1d). The characteristics of the ultrasound beam that affect cavitation include the pulse phase, centre frequency, and acoustic pressure. These parameters alter the expansion that occurs during cavitation and the potential for fragmentation. Inversion and reduced frequency of ultrasound transmission increase the probability of bubble fragmentation. Increasing the acoustic pressure intensity can also cause bubble fragmentation.

Cavitation can be captured by high-speed cameras. Various studies have investigated bubble oscillation and fragmentation, and established models for different bubble shell materials. The stability and dynamic behaviour of bubbles can be simulated and predicted. Some important properties of bubbles can be defined and determined. For example, the maximum radius (R_{max}) during the rarefaction phase determines the cavitation and fragmentation thresholds. These properties are important for the application of bubble technology.

Radiation Force

An acoustic wave exerts a force on any interface across which ultrasonic intensity decreases in the direction of wave propagation. Over the past decade, various research groups have published theoretical analyses of the effects of radiation force on spheres. These studies suggest that a primary radiation force translates the bubble along the axis of the beam and a secondary radiation force attracts or repels bubbles (Bloch et al. 2004). Bloch et al. in 2004 used ultrasound radiation force to greatly enhance a specific molecular interaction. They showed ultrasound images of microbubbles bound to blood vessels undergoing angiogenesis stimulated by fibroblast growth factors in a Matrigel plug. Prior to insonation, microbubbles conjugated to echistatin-targeted ligands were uniformly distributed within the blood vessel. Targeting was enhanced by insonation that deflected bubbles to the vessel wall.

Figure 2 is a diagram demonstrating the application of radiation force in vascular targeting. Some studies have suggested that the effects of radiation force are closely related to bubble size, pulse frequency, and pressure amplitude. A bubble is deflected tiny distances (of the order of a fraction of micrometres) by a pulse cycle if the pulse frequency is near the bubble's resonant frequency. Increasing pulse pressure amplitude can increase the distance the bubble is deflected, but increasing pressure increases the risk of fragmentation. So long pulses with low pressure amplitude and frequencies near the bubbles' resonant frequencies are used. Taking into account the diversity of bubbles, high-time-frequency bandwidth pulses have also been considered to maximize the volume of deflected bubbles.

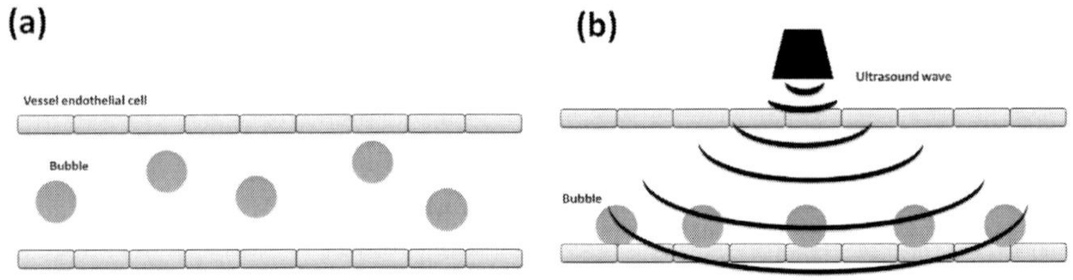

Fig. 2. Schematic diagram of the use of radiation force in vascular targeting: (a) Bubbles are uniformly distributed within the blood vessel before the application of ultrasound radiation force. (b) After the application of radiation force, bubbles are deflected to the vessel wall. This effect can be used to enhance targeting.

THE BIOEFFECTS OF BUBBLE-ULTRASOUND INTERACTION

The bioeffects of the bubble-ultrasound interaction have been studied extensively. However, the detailed mechanisms of these effects are still unclear. The study of the bioeffects of bubble-ultrasound interaction is a multidisciplinary field, including physicists, engineers and biologists. The common acoustic parameters and bubble properties known to change cell membranes and vascular permeability are discussed below. The effects of the physical properties of bubbles and various ultrasound parameters on the oscillation of bubbles are considered.

Changing Cell Membrane and Vascular Permeability

The echogenicity and oscillation of bubbles used to enhance ultrasound are discussed above. Bubbles are highly echogenic because of their large differences in acoustic impedance compared with surrounding tissue or blood. As well as scattering ultrasound waves when insonated, bubbles also cavitate in the ultrasound field.

Power Output from Ultrasound Machines (Mechanical Index)

Ultrasound machine settings can be varied to optimize the imaging modality in use. This is usually achieved by changing the power output from the ultrasound machine. In clinical practice, the ultrasound power is expressed as the mechanical index (MI). This is an estimate of the peak negative pressure (PNP) within insonated tissue. The MI is calculated by division of the PNP within the insonated tissue by the square root of the ultrasound frequency (Eq. 1).

$$MI = \frac{P}{\sqrt{f}} \tag{1}$$

where P is the peak rarefactional acoustic pressure normalized by 1 MPa and f is the centre frequency of the ultrasound normalized by 1 MHz.

The MI depends on characteristics of the ultrasound beam and the insonated tissue. However, the characteristics of insonated tissues are assumed to be similar in all patients.

The MI can be used as a safety feature to estimate the effects of ultrasound on tissues. It estimates the probability of a mechanical biological effect on an object exposed to diagnostic ultrasound. A higher MI indicates that it is more likely that a bioeffect will occur.

As the MI also reflects the mechanical effect of ultrasound on bubbles, it has been used to optimize ultrasound machine settings for bubble contrast ultrasound. It is imprecise and not directly comparable between machines, but is still the most important parameter to set correctly for an ultrasound contrast study.

When insonated with ultrasound with a low MI or low power ultrasound (generally MI < 0.2), the cavitation of bubbles is stable (inertial). Stable (inertial) cavitation creates strong currents in liquids flowing around the bubbles (microstreaming). Microstreaming applies a shear stress to cell membranes that may cause transient opening of the cell membranes (Fig. 3).

At a higher acoustic energy amplitude (generally MI > 0.4), cavitation is non-inertial. Some non-linear oscillation occurs and bubbles emit sound in harmonics of fundamental frequencies. Thus, the bubble is not simply

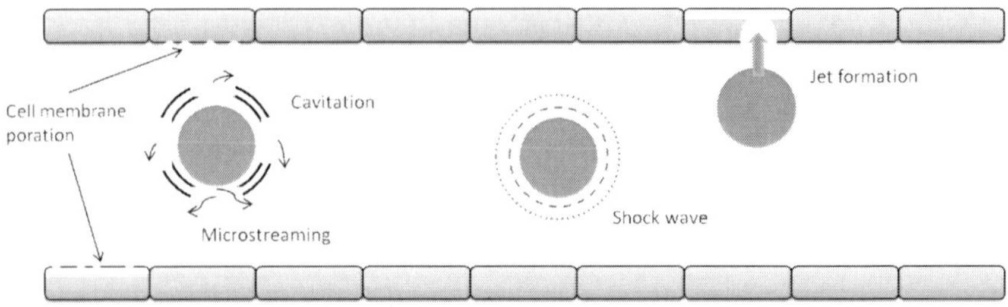

Fig. 3. Schematic representation of the biological effects of cavitation and the implosion of bubbles.

a passive reflector of ultrasound, but also a source of sound to generate harmonic frequencies that biological tissues do not produce at the same MI. Thus, the contrast generated can be used to image the bubbles and improve the signal-to-noise ratio.

Under these conditions, the bubbles first expand and then implode violently. The imploding bubbles can generate high temperatures and highly reactive free radicals. At higher power intensities, a bubble may collapse and form many small bubbles if the kinetic energy exceeds the surface tension of the collapsed bubble. The bubble's collapse generates shock waves in the fluid surrounding the bubble (Fig. 3). If the collapsing bubble is close to the surface of the cell membrane or vascular wall, collapse may be asymmetric. Asymmetric collapse results in the formation of a jet of liquid towards the surface (Fig. 3). The shock wave and liquid jet can generate sufficient force to perforate cell membranes and increase the permeability of blood vessels (Marmottant and Hilgenfeldt 2003). The changes in cell membranes and vascular permeability caused by bubble cavitation can be used for drug and gene delivery.

Standard clinical ultrasound imaging uses an MI around 1.0, but lower powers (0.5), are usually sufficient for left ventricular opacification. Myocardial perfusion imaging uses extremes of power output. High power (MI > 1.2) is used to destroy bubbles in power Doppler imaging. Low power (MI < 0.1) is used to induce linear oscillation of microbubbles for real time myocardial perfusion imaging.

Important Parameters and Their Impacts on Bioeffects

The upper MI limit was chosen to be consistent with the acoustic output produced by ultrasound systems prior to 1976. Thus, American guidelines recommend limiting the MI (<1.9) when diagnostic ultrasound is used in the absence of a bubble contrast agent. However, the MI was initially described before bubbles were in common use. Therefore the equations for calculation of the MI do not consider whether bubbles reduce the MI threshold or take into account the biological effects of insonated bubbles *in vivo*. The physical models of gas body behaviour indicate that inertial cavitation may occur when the MI value is greater than approximately 0.4. In theory this represents boundary between insonation regimens designed to induce either inertial or non-inertial cavitation activity (Miller et al. 2008).

The relationship between acoustic pressure and centre frequency that predicts the biological effects of UCAs *in vivo* is still under investigation. In *in vitro* studies, two different cell models, suspensions and monolayers, are used to try to quantify the bioeffects induced when UCAs are insonated. Diagnostic ultrasound with modest acoustic pressure amplitude can kill attached monolayers of cells when in contact with contrast agent gas bodies. The cell injury can be caused by microstreaming shear stress that occurs below the inertial cavitation threshold.

In *in vivo* studies of bioeffects induced by contrast-enhanced diagnostic ultrasound, biologically significant microscale effects, such as petechial haemorrhage, have been demonstrated with ultrasound exposures corresponding to an MI above 0.4. This is consistent with the theoretical threshold for inertial cavitation in blood. With an MI above 0.4, ultrasound-induced bioeffects increase rapidly. These bioeffects include sonoporation, microvascular leakage, capillary rupture, microlesions with inflammation, and premature ventricular contractions. Therefore, an MI of 0.4 is recognized as an important threshold by the American Institute of Ultrasound in Medicine in the Bioeffect of Diagnostic Ultrasound with Gas Body Contrast Agents.

Many other parameters besides the MI determine the thresholds for the occurrence of biological effects. These parameters include pulse duration/duty cycle, time between destructive pulses, UCA size distribution and shell parameters, and dose of UCA. Bubbles can repeatedly fragment and re-combine during a pulse and may remain as discrete fragments between or after pulses. Therefore, pulse duration and duty cycle are important parameters for drug delivery sequences and the induction of biological effects. However, these parameters cannot be adjusted in most clinical ultrasound diagnostic systems.

Some studies suggest that longer pulses or duty cycles improve the efficiency of drug and gene delivery (Mcdannold et al. 2008; Caskey et al. 2009). Moreover, increasing pulse length can decrease the acoustic pressure amplitude thresholds for the disruption of the blood-brain barrier and gel boundary (Mcdannold et al. 2008; Caskey et al. 2009). Therefore, longer pulses are advantageous because lower acoustic pressures generate therapeutic effects while avoiding complications associated with high acoustic pressures, such as unintended cavitation.

The physical properties of bubbles can also cause biological effects and, so, must be considered. Generally, larger bubbles interact with endothelial cells more than small bubbles so larger bubbles cause more biological effects (Caskey et al. 2007). Pislaru et al. (2003) compared two commercial bubbles, PESDA (8.3 ± 2.9 μm) and Optison (6.7 ± 0.7 μm), with the same shell material (albumin) in the size range of 6–8 μm. They reported that PESDA increased drug and gene delivery *in vitro* and *in vivo* more than Optison (Pislaru et al. 2003).

Increasing the dose of bubbles should also decrease the threshold for biological effects as the number of bubbles near boundaries increases with increasing concentration (Ward et al. 2000).

DRUG AND GENE DELIVERY FOR BUBBLE SYNTHESIS AND DESIGN

Bubbles loaded with genes and drugs are fascinating and promising tools for drug and gene delivery and theranostic applications. There are many potential benefits of loading drugs or genes onto bubbles. First, drug-carrying bubbles can be targeted to a specific site and then release their contents locally when insonated at that site. This could increase the concentration of drug at sites affected by disease and reduce the delivery of drugs to normal tissues. It could greatly reduce the toxicity and side effects of drugs and could revolutionize chemotherapy for the treatment of malignancy, which currently causes significant systemic side effects.

Bubbles and ultrasound can induce biological effects to enhance drug uptake efficiency. For example, cavitation can enhance the vascular permeability and improve drug transport through the endothelium. If a drug is close to the bubble surface, the drugs could be taken into cells by microstreams or shockwaves.

Some biological agents, like nucleic acids and peptides, rapidly degrade after intravenous injection. Protecting these agents by binding with bubbles may protect them from degradation *in vivo*. Finally, drug-loaded bubbles could allow confirmation that drugs have been delivered. For example, an ultrasound image of the target could be obtained with a low acoustic pressure beam. Bubbles seen at the target site could then be ruptured by a high acoustic pressure ultrasound beam. Continuous imaging of the target site could then confirm drug delivery. Therefore, drug-loaded bubbles could reduce drug doses, the cost of treatment and toxic side effects.

Several methods can be used to load drugs or genes onto bubbles (Fig. 4). These include allowing drugs to dissolve in the inner layer of the oil phase of the bubble, incorporation into the bubble shell, attaching

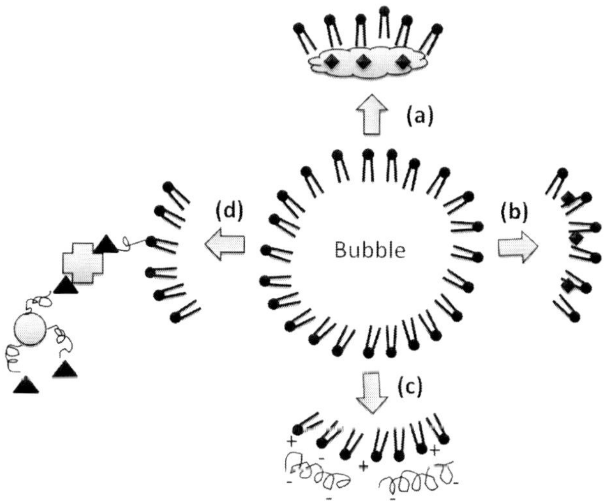

Fig. 4. Schematic representation of sites where drugs may be loaded onto bubbles. Drugs can be (a) dissolved in an oil layer between the perfluorocarbon core and bubble shell, (b) incorporated in the bubble shell, (c) loaded outside the shell by electrostatic interaction, or (d) loaded outside the shell by avidin–biotin linkage. Albumin or polymer shells can be used instead of a lipid shell.

them to the bubble surface by covalent bonds, electrostatic interaction and affinity interaction. If drugs are incorporated inside the bubble or shell the shell can prevent release of drugs before bubbles are insonated at the target site. In 1998, Unger et al. introduced a lipid microbubble in which the perfluorocarbon core was separated from the lipid shell by a paclitaxel containing oil (Unger 1998). The structure of this bubble is similar to that shown in Fig. 4a. However, it remains difficult to achieve efficient ultrasound-mediated drug release from this bubble. Another important drawback is that these bubbles can only incorporate hydrophobic drugs.

Various strategies have been used to incorporate hydrophilic drugs in the inner shells of bubbles. However, these methods still have several limitations, including lower loading efficiency or lower bubble yields.

Agents that can be incorporated into shells include deoxyribonucleic acid (DNA), peptides, or anticancer drugs such as doxorubicin. These drugs can be incorporated in the shell through non-specific binding or hydrophobic interaction between the drug and shell materials.

Loading drugs on the surface of a bubble through electrostatic interaction is generally used for delivery of DNA. However, the loading efficiency of electrostatic interaction is low, because the surface of the bubble has very few cationic charges. Developing shells from polymers rich in positive charges enhances nucleic acid absorption. Layer-by-layer techniques can be used to attach several layers of nucleic acids. Positive polyelectrolytes can greatly increase the amount of nucleic acid carried by a bubble and the polymer shell also reduces gas diffusion out of the bubble, stabilizing the bubble. However, one of the biggest drawbacks is that multi-layered bubbles aggregate into clumps too large to enter cells, then implode and fragment. Some small aggregated vesicles containing DNA can cross cell membranes to enter the cytoplasm. However, transcription does not occur if the DNA remains bound to the shell fragment and cannot enter the nucleus.

Linking drug-loaded nanoparticles to a microbubble shell is another promising design for a drug delivery system (Fig. 4d). Drugs are first loaded into the nanoparticle (such as a liposome or polystyrene nanoparticle). The loaded nanoparticles are then attached to the surfaces of bubbles. This concept is promising because it greatly increases the drug-carrying capacity of the bubble. Futhermore, liposome nanoparticles can be used to load both hydrophilic and hydrophobic drugs onto the surface of the bubble.

Attachment of nanoparticles to bubbles via a biotin-avidin binding system may induce an immune response *in vivo*. However, nanoparticles can be bound to bubbles with covalent bonds. Some studies suggest that modifying nanoparticles with polyethylene glycol (PEG) can reduce premature capture by the reticulo-endothelial system after *in vivo* injection. However, PEGylating nanoparticles reduces endocytic uptake and endosomal escape. This reduces transfection efficiency. However, Lentacker et al. (2007) observed that coupling of PEGylated lipoplexes onto microbubbles significantly improves transfection efficiency when bubbles are insonated.

BUBBLE APPLICATIONS FOR ULTRASOUND MOLECULAR IMAGING AND THERAPY

Vascular Perfusion Imaging

The most widespread clinical application of bubbles is echocardiography. Bubbles are used to improve detection of the endocardial border and myocardial contractility (Cosgrove et al. 2009). The assessment of myocardial perfusion is crucial for diagnosing coronary artery disease and acute myocardial infarction. It allows assessment of the amount of myocardium at risk of necrosis, measurement of tissue flow velocities, and assessment of the success of thrombolysis after myocardial infarction. Myocardial perfusion imaging initially uses high MI ultrasound pulses to destroy the bubbles, which then disperse into the vessels. Then, low MI ultrasound pulses are used to detect bubbles within the blood perfusing the vessels. The territory affected by the myocardial infarction is less well perfused and so is darker (contains fewer bubbles) than the healthy myocardium. This perfusion technique can be used in other organs and tissues, to assess hepatic and renal perfusion, for example.

Abnormalities of arterial walls such as atherosclerotic plaques can be detected with the assistance of UCAs. Insonating an artery filled with bubbles can generate an ultrasound image that accurately delineates the internal surface of the artery. This helps to demonstrate the presence of plaques.

Molecular Ultrasound Imaging

Bubbles are being developed as a tool for molecular imaging. Molecular imaging offers high resolution insights into biological processes at the molecular and cellular levels. Molecular imaging could play a key role in the early detection of cancer. Bubbles prepared for targeted molecular imaging are conjugated to specific ligands or antibodies. Thus, targeted bubbles can be selectively retained within diseased tissue. As targeted bubbles are confined to the intravascular space, only molecular targets within the blood or on the vascular endothelial cell surface can be imaged. The main vascular molecular imaging targets include inflammation, angiogenesis and thrombus formation. These processes produce characteristic intravascular changes that can be traced by targeted bubbles.

Specific sites can be targeted by modifying the chemical properties of the surface of the bubble or conjugating targeting ligands associated with specific diseases to the bubble's surface. Engineering the chemical properties of the bubble's surface to passively target specific tissues or organs is "passive targeting". An example of passive targeting is the use of a bubble with an albumin shell to bind a leukocyte because albumin interacts with the integrins on the surface of the leukocytes. Active targeting describes the conjugation of specific targeting ligands to bubble surfaces to target specific cells or tissues through ligand-receptor interaction.

Targeting efficiency is affected by several factors, including the affinity and density of targeting ligands and receptors, blood flow, and shear rate. Leukocytes, macrophages and Kupffer cells tend to capture charged bubbles. Therefore, to avoid unwanted systemic effects, the electrical charge of bubble surfaces should be finely controlled. Modifying bubbles with PEG (PEGylation) can reduce complement-mediated events. Polyethylene glycol can be used as a spacer to protrude a ligand from the bubble surface, thereby increasing targeted bubble affinity and decreasing steric hindrance. Unfortunately, "tethering" targeting bubbles to a solid surface damps the oscillation of insonated bubbles, reducing echogenicity, which is a disadvantage.

Tumours develop in two major phases: an initial avascular phase and a subsequent angiogenic phase. In the initial avascular phase, malignant cells obtain sufficient nutrients and oxygen by passive diffusion so that a direct blood supply is not required. However, tumours larger than 1–2 mm cannot survive on diffusion alone. The resulting hypoxia induces the expression of pro-angiogenic factors by malignant cells. These pro-angiogenic factors degrade the extracellular matrix, destabilize the surrounding vasculature and stimulate angiogenesis (mediated by the vascular endothelial growth factor; VEGF). The newly formed vessels enable further invasive growth and tumour metastasis to other organs. The endothelial cells of the activated tumour express specific marker molecules. These markers can be used as targets for angiogenic imaging of the tumour.

One of the most important angiogenic markers, VEGF type 2 (VEGF2), has been extensively studied as a target for molecular ultrasound imaging. Rychak et al. (2007) reported that bubbles conjugated to an antibody against VEGF2 accumulated within tumour tissue significantly more than control bubbles. $\alpha_v\beta_3$ integrin has also been extensively studied as an alternative molecular target for imaging angiogenesis (Rychak et al. 2007). $\alpha_v\beta_3$ integrins are highly expressed on activated endothelial cells but almost absent on the remaining vessels. Thus, bubbles can conjugate specific antibodies or cyclic RGD (arginine-glycine-aspartate) peptides to target to $\alpha_v\beta_3$ integrins. Other markers associated with angiogenesis include CD105 (endoglin), P/E-selectin, and VCAM-1. Van Wamel et al. compared bubbles conjugated to various different angiogenic target ligands. They suggested that a combination of different markers may be useful in investigating the development of tumours.

Inflammation is a common physiological process that occurs in a vast array of diseases. Inflammation occurs when leukocytes within the blood pool are activated and transmigrate to the extravascular compartment. The recruitment and transmigration of leukocytes are medicated by adhesion molecules on the surface of leukocytes and endothelial cells. The initial recruitment of leukocytes is mediated by expression of E- and P-selectin on the inner vessel wall. Rolling leukocytes are stopped by another group of adhesion molecules: intercellular adhesion molecule 1 (ICAM-1) and vascular adhesion molecule 1 (VCAM-1). Hence, the inflammatory markers, E- and P-selectin, as well as ICAM-1 and VCAM-1, have been used as the targets for molecular ultrasound imaging of inflammation. Other inflammatory applications associated with ultrasound molecular imaging were recently summarized by Kiessling et al. (2009).

APPLICATIONS TO AREAS OF HEALTH AND DISEASE

Nanotechnology, such as liposomes, polymeric micelles, hollow particles, and emulsion droplets, could allow controlled targeted drug delivery and reduce systemic accumulation of unnecessary drugs. These drug carriers can intelligently release drugs at targeted sites using stimuli-responsive systems that release the drug payload only in response to specific physical stimuli, such as pH, temperature, light or ultrasound. Ultrasound can focus energy non-invasively (through the skin) or minimally invasively through intraluminal, laparoscopic or percutaneous procedures. Although the lifespan of bubbles within circulating blood *in vivo* is still short, several research groups have developed bubble-based drug delivery systems (Unger et al. 2004; Ferrara 2008). A variety of drugs and therapeutic molecules (oligonucleotides, proteins, peptides, nucleic acids) can be incorporated into bubbles and released at the site of interest when insonated (Fig. 5).

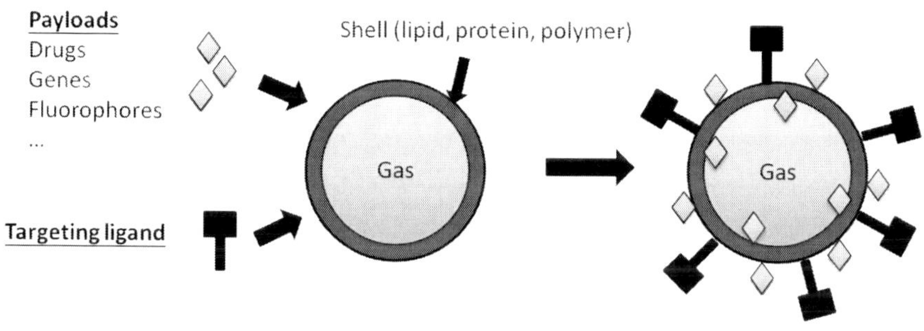

Fig. 5. Schematic representations of drugs carried by targeted bubbles. Drugs can be embedded in the shell of the bubble or dissolved in an oil layer between the gas core and the shell.

FUTURE PERSPECTIVES

Bubbles have been used in clinical practice as UCAs for many years. Recently, more novel applications for bubbles have been conceived from research into bubble synthesis and the physical properties of bubbles. Developments in nanotechnology and biotechnology have brought these concepts closer to clinical practice.

When using bubbles for these novel applications the problems of the instability of bubbles within the circulation *in vivo* must be considered and overcome. The lifespan of bubbles has been increased by using PFC instead of air. The use of polymeric shells could extend the lifespan of bubbles even further.

The size of bubbles is also relevant. Currently all commercially available bubbles are over a micrometre in size (1–10 μm). Although the vessels supplying tumours are particularly permeable and often have large endothelial gaps, microbubbles are generally still too large to traverse the gaps in these vessels. Thus, nanobubbles are generally required for drug delivery into tissue parenchyma.

To improve drug delivery, the development of bubbles with higher loading and transfection efficiencies is an important focus for translational research. Safety is also critically important. Although cavitation may be harmful, the intensity of ultrasound required to cause significant tissue damage *in vivo* is quite high. The properties of carrier bubbles should be engineered to promote cavitation activity that results in effective drug release at an ultrasound stimulation threshold well below that which causes tissue damage.

Key Facts

- Bubbles used in current clinical practice consist of a core of gas surrounded by a shell.
- Bubbles are echogenic (reflect ultrasound) and are currently used clinically as contrast agents for ultrasound imaging.
- Bubbles can be conjugated to specific targeting ligands, such as antibodies, peptides, or DNA. These bubbles can be used for assessment of organ perfusion, drug delivery and vascular molecular imaging.
- A drug encapsulated in a bubble can be released into the systemic circulation but only deposited in a specific area targeted by ligands conjugated to the bubbles and ultrasound stimulation. The bioeffects of ultrasound can also enhance the uptake of drug.
- Vascular molecular imaging with bubbles is mainly used to study molecules related to tumour angiogenesis (such as vascular endothelial growth factor receptor and $\alpha_v\beta_3$ integrin) and inflammation.
- Nanobubbles can easily pass through vessel walls in malignant tissue by passive diffusion and enhanced permeability and retention effect.

Definitions

Cavitation: The behaviour of gas bubbles within a flowing liquid in a region where the pressure of the liquid falls below its vapour pressure. Two different types of cavitation may occur: transient and stable.

Mechanical index (MI): An ultrasound index that can be used to estimate the degree of bioeffects that a given set of ultrasound parameters may induce. A higher mechanical index generally induces a larger bioeffect.

Molecular imaging: The visual representation, characterization, and quantification of biological processes at the cellular and subcellular levels within intact living organisms. The images obtained reflect cellular and molecular pathways and *in vivo* mechanisms of disease present within the natural physiological environment. (Definition as published in Genes & Dev, 2003, 17: 545–580.)

Steric hindrance: Steric hindrance occurs when the size of a molecule prevents chemical reactions that can occur with related molecules that are smaller.

Ultrasound contrast agent (UCA): Microbubbles that consist of a gas core encapsulated by a shell that are highly echogenic (reflect ultrasound very well) and are used to enhance ultrasound imaging of blood-filled cavities.

Ultrasound imaging: A non-invasive imaging technique that uses very high frequency sound waves.

Summary Points

- Bubbles, which have a shell (lipid, albumin, or polymer) and a gas core, are echogenic (they reflect ultrasound waves). Bubbles are used clinically to improve conventional ultrasound imaging of the heart and blood vessels.
- Bubbles rapidly decompose within the blood circulation *in vivo*. The use of high molecular perfluorocarbon or sulphur hexafluoride gases as substitutes for the air core greatly increases the stability of bubbles.
- Ultrasound stimulation (insonification) of bubbles can cause bioeffects such as changing cell membrane and vascular permeability because of cavitation. Factors that affect the occurrence and magnitude of these bioeffects include the mechanical index (MI), ultrasound pulse length, bubble size, and bubble concentration.
- There are several methods that can be used to load drugs or genes into bubbles (Fig. 4). Drugs may be dissolved in an oil layer between the gas core and the shell, incorporated in the bubble shell, or loaded on the outside of the bubble shell by electrostatic forces or avidin-biotin linkage.
- Bubbles are now being developed as a tool for molecular imaging.
- Bubbles are promising tools for many potential applications. However, many challenges and limitations still need to be overcome, such as the bubble stability, lifespan within the circulation, and drug loading efficiency.

Abbreviations

CT	:	computer-assisted tomography
DNA	:	deoxyribonucleic acid
ICAM-1	:	intercellular adhesion molecule 1
MI	:	mechanical index
MRI	:	magnetic resonance imaging
PEG	:	polyethylene glycol
PFC	:	perfluorocarbon
PNP	:	peak negative pressure
VCAM-1	:	vascular adhesion molecule 1
VEGF	:	vascular endothelial growth factor
VEGF2	:	vascular endothelial growth factor type 2
UCA(s)	:	ultrasound contrast agent(s)
UTMD	:	ultrasound targeted microbubble destruction

References

Bloch, S.H., M. Wan, P.A. Dayton and K.W. Ferrara.. 2004. Optical observation of lipid- and polymer-shelled ultrasound microbubble contrast agents. Appl. Phys. Lett. 84: 631–633.

Caskey, C.F., S.M. Stieger, S. Qin, P.A. Dayton and K.W. Ferrara. 2007. Direct observations of ultrasound microbubble contrast agent interaction with the microvessel wall. J. Acoustical Soc. Am. 122: 1191–1200.

Caskey, C.F., S.P. Qin, P.A. Dayton and F.W. Ferrara. 2009. Microbubble tunneling in gel phantoms. J. Acoustical Soc. Am. 125: El183–El189.

Cosgrove, D. and C. Harvey. 2009. Clinical uses of microbubbles in diagnosis and treatment. Med. Biol. Eng. Comput. 47: 813–826.

Ferrara, K.W. 2008. Driving delivery vehicles with ultrasound. Adv. Drug Delivery Rev. 60: 1097–1102.

Gramiak, R. and P.M. Shah. 1968. Echocardiography of the aortic root. Investigative Radiol. 3: 66.

Jones, J.M. and W.J. Koch. 2005. Gene therapy approaches to cardiovascular disease. Methods Mol. Med. 112: 15–35.

Kiessling, F., J. Huppert and M. Palmowski. 2009. Functional and molecular ultrasound imaging: concepts and contrast agents. Curr. Medicinal Chem. 16: 627–642.

Klibanov, A.L., P.T. Rasche, M.S. Hughes, J.K. Wojdyla, K.P. Galen, J.H. Wible and G.H. Brandenburger. 2002. Detection of individual microbubbles of an ultrasound contrast agent: Fundamental and pulse inversion imaging. Acad. Radiol. 9: S279–S281.

Lentacker, I., S. De Smedt, J. Demeester, V. Van Marck, M. Bracke and N. Sanders. 2007. Lipoplex-loaded microbubbles for gene delivery: A Trojan horse controlled by ultrasound. Human Gene Therapy 18: 1046–1046.

Marmottant, P., and S. Hilgenfeldt. 2003. Controlled vesicle deformation and lysis by single oscillating bubbles. Nature 423: 153–156.

Mcdannold, N., N. Vykhodtseva and K. Hynynen. 2008. Effects of acoustic parameters and ultrasound contrast agent dose on focused-ultrasound induced blood-brain barrier disruption. Ultrasound Medicine Biol. 34: 930–937.

Miller, D.L., M.A. Averkiou, A.A. Brayman, E.C. Everbach, C.K. Holland, J.H. Wible and J.R. Wu. 2008. Bioeffects considerations for diagnostic ultrasound contrast agents. J. Ultrasound Med. 27: 611–632.

Pislaru, S.V., C. Pislaru, R.R. Kinnick, R. Singh, R. Gulati, J.F. Greenleaf and R.D. Simari. 2003 Optimization of ultrasound-mediated gene transfer: comparison of contrast agents and ultrasound modalities Eur. Heart J. 24: 1690–1698.
Rychak, J.J., J. Graba, A.M.Y. Cheung, B.S. Mystry, J.R. Lindner, R.S. Kerbel and F.S. Foster. 2007. Microultrasound molecular imaging of vascular endothelial growth factor receptor 2 in a mouse model of tumor angiogenesis. Mol. Imaging 6: 289–296.
Schutt, E.G., D.H. Klein, R.M. Mattrey and J.G. Riess. 2003. Injectable microbubbles as contrast agents for diagnostic ultrasound imaging: The key role of perfluorochemicals. Angewandte Chemie-Intl. Ed. 42: 3218–3235.
Unger, E.C. 1998. Acoustically active lipospheres containing paclitaxel—A new therapeutic ultrasound contrast agent. Investigative Radiol. 33: 886.
Unger, E.C., T. Porter, W. Culp, R. Labell, T. Matsunaga and R. Zutshi. 2004. Therapeutic applications of lipid-coated microbubbles. Adv. Drug Delivery Rev. 56: 1291–1314.
Verma, I.M. 1997. Gene therapy—promises, problems and prospects. Nature 389: 239.
Ward, M., J.R. Wu and J.F. Chiu. 2000. Experimental study of the effects of Optison (R) concentration on sonoporation in vitro. Ultrasound Med. Biol. 26: 1169–1175.

17

Multifunctional Nanoagents for the Detection and Treatment of Thromboses

S. Sibel Erdem[1,a] and *Jason R. McCarthy*[1,b,*]

ABSTRACT

Pathological thromboses occur as a consequence of the dysregulation of the normal hemostatic system, which is responsible for the minimization of blood loss due to vessel injury. Their formation is the cause of a number of thrombosis-related syndromes, including myocardial infarction, stroke, pulmonary embolism, and deep vein thrombosis. Clinically, thromboses may be detected by a number of imaging modalities, including ultrasound, magnetic resonance, and X-ray computed tomographic imaging. Whereas these methodologies may give the location and severity of the disease, they cannot determine biological hallmarks of the thrombus, including age and constitution, both of which may assist in determining the optimal therapeutic intervention. In this regard, molecular imaging has become an enabling

[1]Center for Systems Biology, Harvard Medical School and Massachusetts General Hospital, 149 13th St, 6th floor, Charlestown, MA 02129, USA.
[a]E-mail: SERDEM@PARTNERS.ORG
[b]E-mail: jason_mccarthy@hms.harvard.edu
*Corresponding author

List of abbreviations after the text.

technology, as blood clots, which are constituted of platelets, fibrin, and coagulation factors, provide a number of potential targets for the development of imaging agents. The advent of nanotechnology has furthered the creation of targeted agents for the detection of thromboses, allowing for the creation of multifunctional agents bearing a number of different ligands, including those used for targeting, imaging, and therapy. This chapter focuses on the detection and characterization of intravascular thrombi using multifunctional nanoparticles. The utility of these nanoagents in the treatment of thrombosis is also discussed.

INTRODUCTION

The formation of a blood clot within a vessel is normally a beneficial process, as it prevents blood loss caused by trauma. Dysregulation of this process can lead to bleeding or clotting disorders, including hemophilia and thrombophilia. There are three causal factors of clot formation: the disturbance of blood flow, as can occur when an individual is seated for a prolonged length of time; increased blood coagulability, as caused by genetic or autoimmune disorders; and vessel injury. The last cause can be a result of external factors, such as an injury, or pathological conditions, including atherosclerotic vascular disease.

Thrombi, or blood clots, are formed on a regular basis and do not usually cause pathological conditions. When the vessel heals, the clot is broken down enzymatically and cleared by the body. When the clotting process occurs relatively unchecked, pathological thrombi form, occluding the vessel and potentially causing serious damage, including myocardial infarctions and stroke, each of which is associated with increased morbidity and mortality. For many of the patients who present with these disorders, the re-establishment of blood flow is of paramount importance, and may be accomplished via the administration of thrombolytic drugs, although such interventions are not guaranteed to work. In fact, 20% of infarct-related arteries fail to open, while 15% of reperfused arteries re-close. If reocclusion does occur, most of the benefits of thrombolysis will be lost, since sustained patency is associated with the best prognosis (Anderson and Willerson 1993).

The course of therapeutic interventions is currently determined on the basis of a combination of blood and other tests, including ultrasound, magnetic resonance (MR), and X-ray computed tomographical (CT) imaging. While this allows for the detection of thrombi, it does not allow for the characterization of the clot. What are its main constituents? How old is it? Whereas knowing the location of the clot is important, this additional information may help in determining the optimum therapeutic strategy.

Molecular imaging allows for the examination of disease via the targeted detection of distinct signatures, such as specific cell types, receptors, or enzymatic activity (Gambhir 2010). With regard to thrombosis, this may include activated platelets, fibrin, or coagulation factors. Whereas a number of small molecule-based agents have been described, the advent of nanomedicine has allowed for the creation of multifunctional agents capable of simultaneously targeting, imaging, and treating disease. This is due, in part, to the large surface area-to-volume ratio of nanoparticles, which allows for the incorporation of numerous ligands on the particle surface, including antibodies and peptides for targeting, fluorophores and radionuclides for imaging, and chemotherapeutics for therapy.

This chapter focuses on the detection and characterization of thromboses using multifunctional nanoagents. While this review is not designed to be all-encompassing, it highlights some of the most poignant recent examples of the applicability of nanomedicine to the detection and treatment of intravascular thombi. More in-depth reviews of the application of nanotechnology in cardiovascular disease can be found elsewhere (Cormode et al. 2009; McCarthy 2010a, b).

THROMBOSIS

Hemostasis, Thrombosis, and Thrombolysis

Hemostasis is the process by which the body presents a barrier to blood loss (Furie and Furie 2008). This process occurs in two phases. In the initial phase, endothelial cell injury causes the interaction of coagulation factors and platelets with the subendothelium, including collagen and von Willebrand Factor. This interaction promotes the adhesion of the platelets, initiating primary hemostasis, although this initial clot is fragile and can be easily dislodged. In order to strengthen the hemostatic plug, secondary hemostasis facilitates the deposition of insoluble fibrin. Once the vessel has healed, and the clot is no longer needed, it is broken down via fibrinolysis.

The deposition of fibrin can occur via two different pathways and is a direct result of the coagulation cascade (Fig. 1A). Upon tissue damage, a number of coagulation factors are activated, which in turn activate thrombin, one of the main mediators of thrombogenesis. Thrombin is formed by the cleavage of circulating prothrombin by activated factor X. In healthy vasculature, thrombin formation is regulated by antithrombin activity, including thrombomodulin and active protein C. In the absence of this feedback mechanism, thrombin is formed, which readily cleaves fibrinogen to form fibrin. While fibrin is a major constituent of thrombi, biologically acute thrombi are further stabilized by the activity of activated

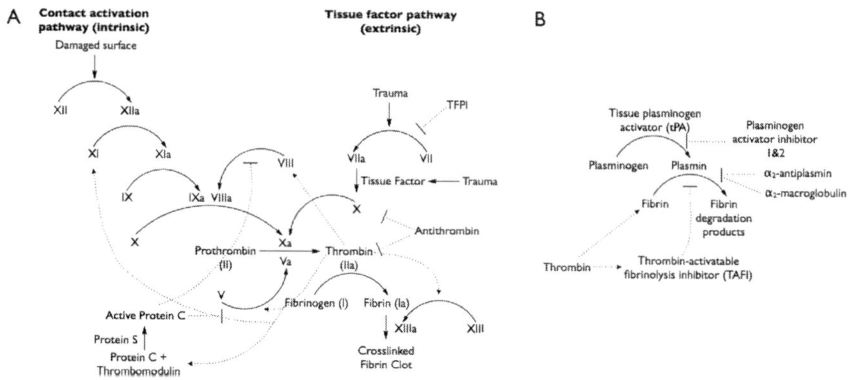

Fig. 1. Coagulation and thrombolysis. (A) The coagulation cascade. Coagulation can result from trauma (extrinsic pathway) or damage to the endothelial surface (intrinsic pathway) and leads to the activation of tissue factors and ultimately the deposition and cross-linking of fibrin. (B) Fibrinolysis is affected by the conversion of plasminogen to plasmin, which can efficiently degrade fibrin. In (A) and (B), dotted lines terminating in arrowheads represent species responsible for the cooperative activation, whereas those that terminate in bars represent inhibitory actions. Figure reproduced under the Creative Commons Attribution-Share Alike 3.0 license.

factor XIII (FXIIIa). FXIII is itself activated by thrombin and covalently cross-links fibrin monomers, enhancing thrombus stability. Additionally, FXIIIa cross-links α_2-antiplasmin into the clot, thereby increasing fibrinolytic resistance.

In the case of pathological thrombosis, the hemostatic system runs amuck, generating thrombi that partially or completely block blood flow distal to the obstruction. When over 75% of the surface area of the lumen is obstructed, this results in hypoxia, whereas blockages over 90% can result in anoxia and infarction. In these cases, it is of the utmost importance to restore blood flow in order to prevent end-organ tissue death. Pharmacologically, this is attempted via the administration of thrombolytic agents that activate the fibrinolytic system in order to effect the dissolution of the blood clot (Fig. 1B). These plasminogen activators convert the proenzyme plasminogen into an active enzyme, plasmin, allowing for the digestion of fibrin into its soluble degradation products. Plasminogen is released from the liver and is normally found in circulation. During thrombogenesis, fibrin presents binding sites for plasminogen, promoting its localization within thrombi.

Clinical Diagnosis of Thrombosis

A number of different methodologies are used for the diagnosis of pathological thrombosis syndromes. When a symptomatic patient arrives at the emergency room or clinic and the suspicion of thrombosis is low, initial blood tests, such as D-dimer levels, can assist in the exclusion of a clot. D-dimer is a cross-linked fibrin degradation product resulting from plasmin-mediated thrombolysis. This test has a high negative predictive value.

When thrombosis is suspected, the resulting battery of tests is based upon the location of the clot. For deep vein thrombosis, compression ultrasound imaging is used most commonly and is capable of giving both the architecture of the vasculature and the flow analysis. Less commonly, pelvic vein MRI and CT imaging are used. For the diagnosis of pulmonary embolism, which is caused by fragments of clots traveling through the vasculature to the lungs, contrast-enhanced CT pulmonary angiography is the gold standard. In this imaging modality, iodinated contrast agents cause the blood to appear bright in the CT image, while emboli appear as darkened/grey areas. The diagnosis of stroke, which can be caused by embolization or the rupture of atherosclerotic lesions in the carotid artery, uses a combination of CT and MR imaging.

While the above-mentioned methodologies are able to give functional information as to the presence of a thrombus and the relative blood flow in its vicinity, they are not able to examine the biological hallmarks of thrombi. Pathologically, the constituent components and age of a thrombus may be deemed as important as its presence, especially with regard to the determination of the optimum therapeutic intervention. A number of molecular imaging strategies have been developed to answer these questions, some of which are further elucidated below.

MULTIFUNCTIONAL NANOAGENTS

Molecular Imaging

Molecular imaging is a non-invasive technique using biomarkers to visualize the biological processes of living organisms at the cellular and subcellular level. The power of monitoring and understanding these interactions provides numerous advantages for diagnosis, treatment, and follow-up of various life-threatening conditions. Molecular imaging not only can provide for the early detection of disease but also greatly contributes to development of personalized treatments as well as monitoring the effectiveness of these therapies. Thus, the field of molecular imaging has gained significant interest from researchers in a multitude of areas, including oncology and cardiology.

As was stated above, molecular imaging may play an increasing role in the investigation of thrombosis. To date, several imaging modalities, including MRI, positron emission tomography and optical imaging, have been used to study the anatomy, physiology, and molecular level interactions of thrombosis. This section details relevant examples of thrombus-targeted imaging agents.

Agents for the Investigation of Thromboses

A number of strategies have been developed to image the components of thrombi, including platelets, fibrin, and clotting factors. For the most part, these strategies have relied upon antibodies or peptides as affinity ligands, which have been either fluorescently or radiochemically labeled. For platelets, the most common protocols involve their isolation from whole blood, labeling with a fluorescent dye, and infusion into an animal undergoing thrombosis (Flaumenhaft et al. 2007; Massberg et al. 1999; Sim et al. 2004). Alternatively, activated platelets can be labeled *in vivo*, as described by Tung et al. using a branched peptide-based probe (Tung et al. 2005), or with fluorescently labeled antibodies, such as CD41, a platelet-specific antigen (Falati et al. 2002).

Aside from platelets, fibrin is the largest component of thrombi. It is also constitutively present during thrombogenesis, as it is polymerized to form and stabilize the clot, and thus serves as a good target for the detection of intravascular thrombi regardless of thrombus age. A number of targeting strategies have been used, including fluorescently or radiolabeled antibodies to fibrin (Falati et al. 2002; Stratton et al. 1994) or its D-dimer degradation product (Morris et al. 2004). One other interesting methodology involved the targeting of paramagnetic nanoparticles using an anti-fibrin antibody for MRI (Yu et al. 2000). Radiolabeled peptides have also been used to image fibrin. One sequence in particular, GPRPP, which is related to the N-terminus of the fibrin α-chain, has been used to identify vascular thrombosis and thromboemboli in rabbit and swine models (Aruva et al. 2006; Kawasaki et al. 1993; Thakur et al. 2000).

As was discussed above, activated FXIII plays a significant role in the promotion of thrombus stability via the cross-linking of fibrin. Thus, this coagulation factor may serve as an adequate target for the detection of acute thrombosis (Tung et al. 2003). To this end, Jaffer and co-workers have developed a near-infrared fluorescent (NIRF) probe derived from the N-terminus of α_2-antiplasmin to image FXIIIa activity *in vivo* (Jaffer et al. 2004). The FXIIIa-specific peptide sequence, GNQEQVSPLTLLKWC, was synthesized and conjugated to a near-infrared (NIR) fluorophore, Alexa Fluor 680 (A15). As a control imaging agent (C15), the authors designed a peptide differing from the imaging agent by a single amino acid residue, replacing one of the glutamine residues with an alanine, which prevents the cross-linking of the agent to the clot. Both agents were tested *in vitro* in plasma clots. Following the incubation of clots with the agents, they were examined by NIRF imaging, revealing a 347-fold NIRF signal increase for the targeted agent over saline incubated clots (Fig. 2A). Further confirmation of the specificity of A15 for FXIIIa was obtained by incubation of the clots with A15 in the presence of iodoacetamide, a FXIIIa inhibitor, which demonstrated suppression of the fluorescence signal by 93%.

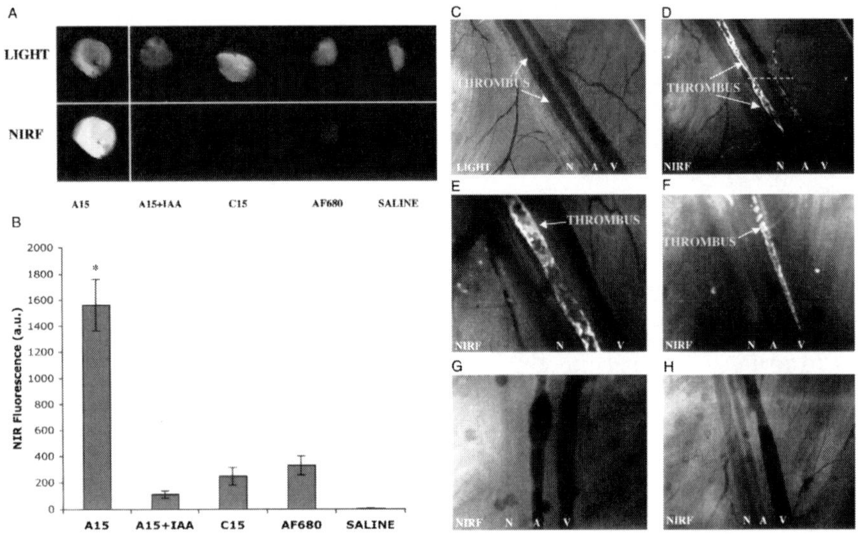

Fig. 2. Imaging of FXIIIa activity. (A) FFP clots incubated with A15 show strong NIRF signal. (B) While A15 incubated clots demonstrated five- to six-fold increase in NIRF signal over other control groups, they showed 347-fold NIRF signal enhancement over saline group. (C) Light image of intravascular thrombi after application of ferric chloride. (D) NIRF image of arterial and venous thrombi (arrows) obtained 82 min after A15 injection. (E) High magnification of A15 enhanced thrombus acquired 62 min after the agent injection. (F) NIRF enhancement of venous thrombus imaged 83 min after A15 injection. (G) 28 h after thrombus formation, A15 injection showed minimal or no NIRF signal enhancement due to declined FXIIIa activity. (H) NIRF image of control thrombi obtained 72 min after C15 injection. Reproduced from Jaffer et al. (2004), with permission.

The A15 imaging agent was next examined *in vivo*. Using a murine model of arterial thrombosis in which an $FeCl_3$-soaked piece of filter paper was applied to the exposed artery to elicit thrombus formation, six mice were injected with either A15 or C15 at 0, 30, or 120 min following injury. For subacute thrombus studies, three mice received the imaging agent 28 h after $FeCl_3$ application. Intravital fluorescence microscopy (IVFM) was then used to detect agent accumulation. The acquired images demonstrated that A15 possessed a $15.0 \pm 6.7\%$ contrast-to-noise ratio 90 min after agent injection, whereas minimal signal enhancement was detected with C15 probe. As expected for a covalently bound probe, the signal increased significantly over time (Fig. 2C-H). In subacute thrombus groups, minor enhancement of the thrombosis was observed *only* in small branched vessels. This is consistent with the expected FXIIIa activity, which is present in acute thrombi and diminishes over time. The authors thus concluded that the A15-based imaging agent covalently binds to fibrin and is able to detect active thrombi by amplification of thrombus signal.

Another example has been published by Fayad and co-workers (Klink et al. 2010), who aimed to monitor the activity of platelets in thrombosis by targeting the glycoprotein $\alpha_{IIb}\beta_3$, which is expressed when the platelets are activated. Because of its abundance, specific conformational change upon activation, and binding of adhesion proteins via the arginine-glycine-aspartic acid (RGD) amino acid motif, the authors have selected $\alpha_{IIb}\beta_3$ as an ideal targeting molecule. To accomplish this, a paramagnetic contrast agent (P975) was synthesized by the coupling of RGD sequence bearing cyclic peptide (P977) to Gd-DOTA at a 1:1 molar ratio. Binding affinity of P975, P977 and trofiban, a non-peptide inhibitor used as a positive control, to the activated platelets was examined *in vitro* by measuring the degree of inhibition of FITC-fibrinogen binding to thrombin receptor-activating peptide stimulated platelets. While Gd-DOTA showed no specific binding, the contrast agent and the native peptide showed high binding affinity with inhibitory values as low as 1.5 μmol/L and 1.1 μmol/L, respectively.

For *in vivo* studies, an arachidonic acid (AA) model was employed to initiate inflammatory thrombogenesis via platelet and coagulation cascade activation. The critical role of platelets in thrombosis was imaged in real time by IVFM after the injection of fluorophore-labeled platelets in the mouse with AA-induced thrombosis. MR images were obtained before and after the injection of P975 and Gd-DOTA and clearly showed that after the administration of P975, the signal intensity of the lumen of the thrombosed carotid artery increased within the first 30 min and was maintained for up to 120 min post-injection with excellent contrast properties (Fig. 3). Unlike P975, Gd-DOTA signal enhancement was observed at 30 min post-injection, yet the signal intensity dramatically decreased after 30 min because of the clarification of the agent from the circulation. The authors were thus able to image thrombogenesis by employing a contrast agent, P975, that specifically binds to activated platelets.

Multifunctional Nanomaterials for the Investigation of Thromboses

Nanomedicine has enabled the synthesis of targeted agents with high selectivity and increased binding affinity, as compared to their non-nanoparticulate counterparts. These two properties can be attributed to the multivalent nature in which nanomaterials present their targeting ligands. These agents are also capable of bearing a number of different

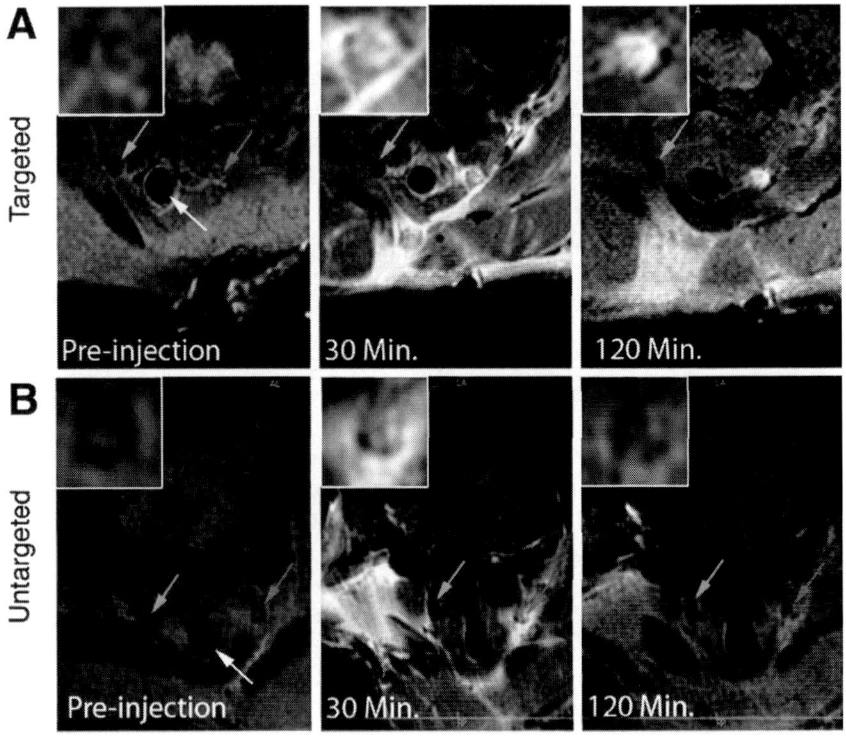

Fig. 3. *In vivo* MRI images of AA-induced arterial thrombosis. The trachea (white), the non-injured left carotid (green) and the injured right carotid (red) are indicated with arrows. Inserts on the upper left corner of each image are the magnified image of the injured carotid. (A) 30 min after the injection of P975, T1 signal enhancement was observed and the signal intensity was persistent throughout 120 min. (B) Following the Gd-DOTA (control) injection, signal intensity increased for the first 30 min, yet it rapidly decreased in 120 min due to the clearance of the control agent from the circulation. Reproduced from Klink et al. (2010), with permission.

Color image of this figure appears in the color plate section at the end of the book.

imaging modalities, allowing for the extraction of the maximum amount of information from one preparation. This section focuses on the detection of intravascular thrombi with multifunctional nanomaterials. For additional information on nanoparticulate preparations used in cardiovascular molecular imaging, please see Chapter 1 of this book.

As was described above, fibrin and FXIIIa are excellent targets for the development of thrombus imaging agents, and multifunctional magnetofluorescent nanoagents targeted to these sites have therefore been created (McCarthy et al. 2009). In order to enable multimodal imaging of thrombosis, fluorescently labeled cross-linked iron oxide (CLIO) nanoparticles were decorated with either fibrin (GPRPPGGSKGC) or

FXIIIa (described above) targeting peptides to yield approximately ten peptides per particle. To compare the targeting ability of the peptides, corresponding control peptides were also prepared by a single amino acid residue substitution of the original peptides. The binding affinity of the fibrin-targeted (CLIO-GPRPP) and FXIIIa-targeted magnetofluorescent nanoparticles (CLIO-FXIII) were tested *in vitro* in fresh frozen plasma (FFP) clots. Following incubation with the imaging agents, the clots were imaged by fluorescence reflectance imaging (FRI) using the appropriate channel based on the spectral properties of the conjugated fluorophore (Cy7 or VivoTag 680). These studies revealed that targeted imaging agents displayed significantly stronger fluorescence signal than the respective control agents (Fig. 4C-D). In order to demonstrate the multimodal nature of the nanoagents, T_2*-weighted MR imaging was performed, with both targeted preparations exhibiting strong binding to clots, with 50.4 and 30.6 thrombus-to-agar contrast-to-noise ratios, respectively (Fig. 4B)

In vivo evaluation of the magnetofluorescent imaging agents were performed in wild-type mice after thrombosis induction with $FeCl_3$. Thirty minutes after vessel injury, the nanoagents were injected and allowed to circulate for an additional 30 min, at which time the thrombosed and contralateral control jugular veins were resected and imaged by FRI. As is shown in Fig. 4E-F, the fibrin and FXIIIa targeted magnetofluorescent nanoparticles selectively accumulated within thrombi, possessing 1.6 and 2.7 signal-to-background ratios, respectively.

Another recent example of thrombosis imaging has been published by Gupta and co-workers (Srinivasan et al. 2010). The authors developed platelet-targeted RGD-modified liposomes and have tested the particles *in vitro* and *in vivo*. The authors chose liposomes for their ability to encapsulate both hydrophobic and hydrophilic moieties, their biocompatibility, and their ability to fuse with a cell membrane, thereby delivering therapeutic agents and/or imaging probes. The used cyclic RGD peptide strongly binds to GPIIb-IIIa integrin receptors on the activated platelets and was immobilized on the nitrobenzofurazan-labeled liposomes encapsulating gold nanoparticles. As a control agent, RGE cyclic peptide bearing liposomes were also prepared. Both agents were initially tested on coverslip-adhered adenosine diphosphate activated human platelets. *In vitro* fluorescence and phase contrast microscopy images indicated that RGD-liposomes showed considerably higher binding to activated platelets than did the RGE-liposomes.

For *ex vivo* studies, a rat carotid injury model was employed in which an acute vascular injury was created on the luminal wall of the left common carotid artery. After balloon catheter induction, the right common carotid and the descending aorta were ligated to ensure blood flow through only the injured left carotid vessel. Fluorescence microscopy and scanning

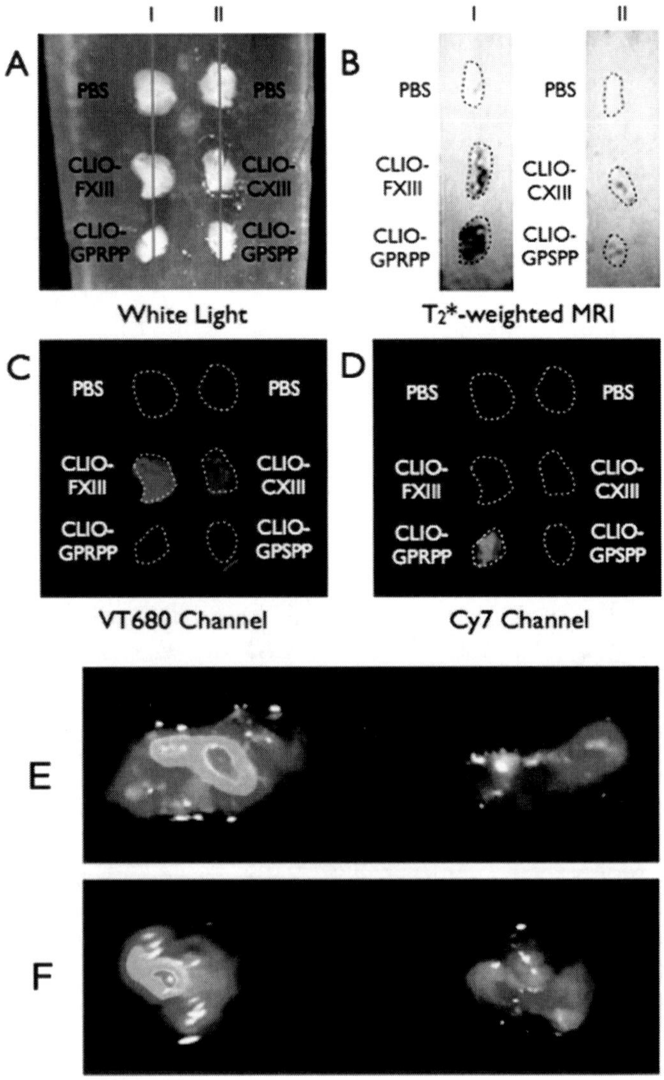

Fig. 4. *In vitro* and *in vivo* imaging of thrombi. (A) White light image of FFP clots. Lines indicated the sagittal view of the clots acquired in MRI. (B) T_2-weighted MRI image of CLIO-FXIII and CLIO-GPRPP showed T_2 signal decrease as a result of strong binding of agents to the clots. (C) FRI image of the CLIO-FXIII incubated clots in VT680 channel. (D) FRI image of the CLIO-GPRPP incubated clots in Cy7 channel. (E) *Ex vivo* FRI image of fibrin-targeted nanoagents in Cy7 channel indicated strong binding of the agent to the clots. (F) Increased FRI signal in VT680 channel confirmed selective binding of FXIII targeted nanoagents to the clots. Reproduced from McCarthy et al. (2009), with permission.

Color image of this figure appears in the color plate section at the end of the book.

electron microscope (SEM) images of the carotid wall were acquired before and after the injection of the nitrobenzofurazan-labeled RGD- and RGE-modified liposomes. SEM images confirmed that, while healthy tissue had a striated appearance, acute disruption of the luminal wall was readily visible by its non-uniform appearance. Platelet targeting ability of the nanogold-labeled RGD-liposomes was confirmed by secondary electron mode images. While the images verified the presence of dense fibrin mesh and adhered platelets for both RGD- and RGE-liposomes exposed injured carotid wall, backscattered electron images showed significantly bright contrast only for the RGD-liposome exposed injury site, confirming binding of the gold-labeled RGD-liposomes to the activated platelets (Fig. 5). The authors are currently seeking further *in vivo* evidence for specific binding of RGD-liposomes to the platelets in an acute vascular injury environment.

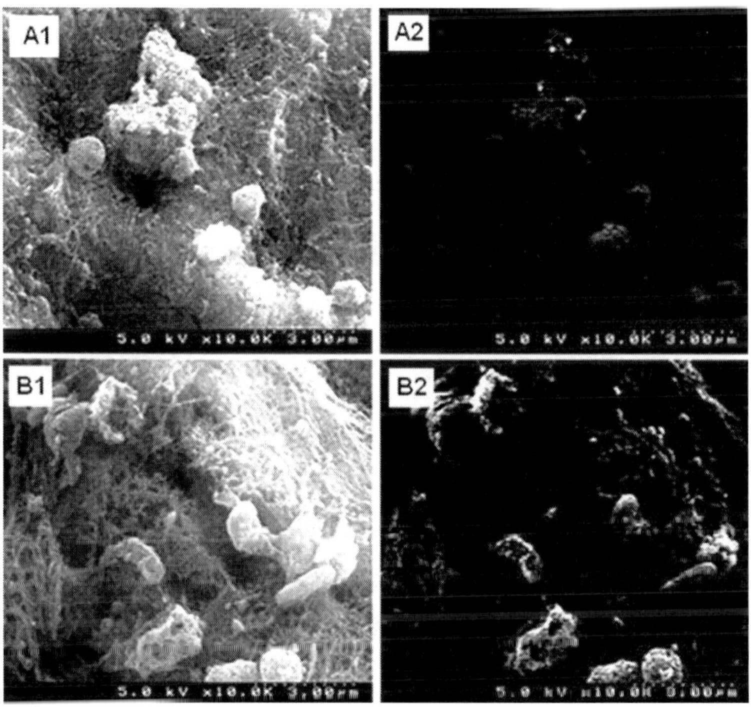

Fig. 5. *Ex vivo* SEM imaging of activated platelets. (A1) SEM image of RGE-liposome incubated platelets. (B1) SEM image of cRGD-incubated platelets. Both images (A1 and B1) showed similar acute thrombotic environment with platelets, which were densely surrounded by fibrin mesh. (A2) Backscatter image of RGE-liposome incubated platelets demonstrated minimal binding of the agent to the target. (B2) Backscatter image of cRGD-liposome incubated sample showed enhanced contrast clearly indicating strong binding of cRGD-liposomes to platelets. Reproduced from Srinivasan et al. (2010), with permission.

Lanza and co-workers also developed a fibrin-targeted contrast agent, a lipid-encapsulated perfluorocarbon nanoparticle incorporating tens of thousands of Gd-DPTA moieties (Yu et al. 2000). These particles were modified with biotinylated anti-fibrin antibodies in order to affect thrombus localization. In their study, the authors investigated the extent of enhancement of T_1 and T_2 weighted MR images of fibrin-rich clots by targeted nanoparticles. For *in vitro* studies, fibrin clots were exposed to the targeted-paramagnetic nanoparticles and SEM images were acquired before and after treatment (Fig. 6A-C). SEM images verified the binding of the paramagnetic nanoparticles to the dense networked fibrin strands. As seen in Fig. 6C, the density of the nanoparticles was dramatically decreased when the clots were pretreated with anti-fibrin antibody.

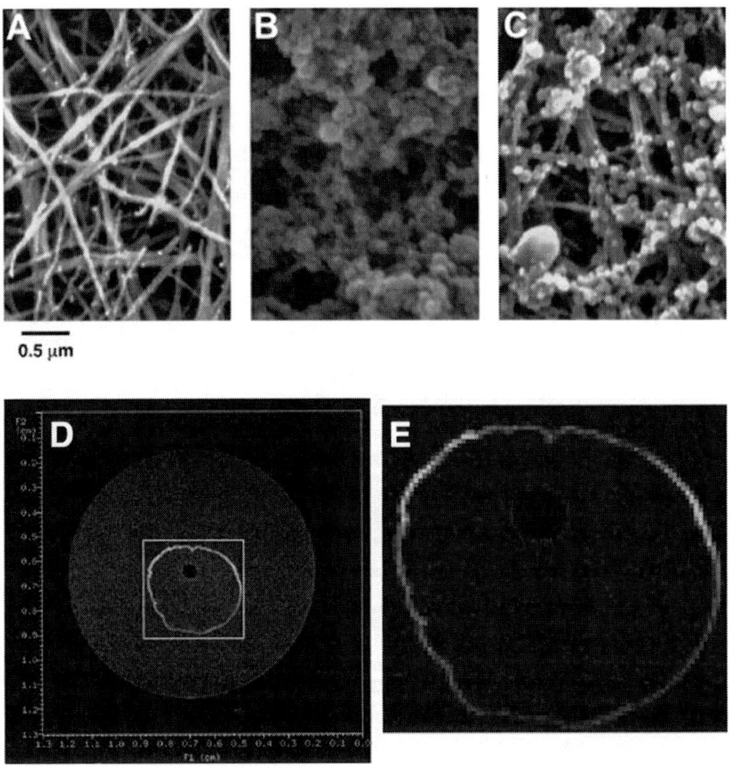

Fig. 6. SEM and high resolution MRI images of clots. (A) SEM image of dense fibrin strands in untreated clots. (B) SEM image of nanoparticle incubated clots showed strong binding of the agents to the fibrins. (C) SEM image of anti-fibrin antibody pre-treated clots confirmed covalent binding of the nanoagents to the clots. (D) High resolution ^{1}H image of treated clot demonstrated deposition of the contrast agent in a thin layer of less than 100 µm. (E) Magnified image of highlighted area of the clot in the image D. Reproduced from Yu et al. (2000), with permission.

The high resolution MR image of the clots showed that the targeted paramagnetic nanoparticles were deposited on the clot surface as a thin layer less than 100 μm thick (Fig. 6D-E). Further experiments were carried out to investigate the ability of the nanoparticles to detect microthrombi within the microfissures of atherosclerotic plaques. As is shown in Fig. 7, the targeted agent was capable of detecting clots as small as 0.5 mm in diameter, while untreated clots showed no contrast. Interestingly, the probe was so sensitive that it allowed for the delineation of clot geometry. As a result of this study, the authors concluded that the abundant payload of the contrast agent on the surface of the nanoparticles plays a crucial role in the detection of the fibrin clots.

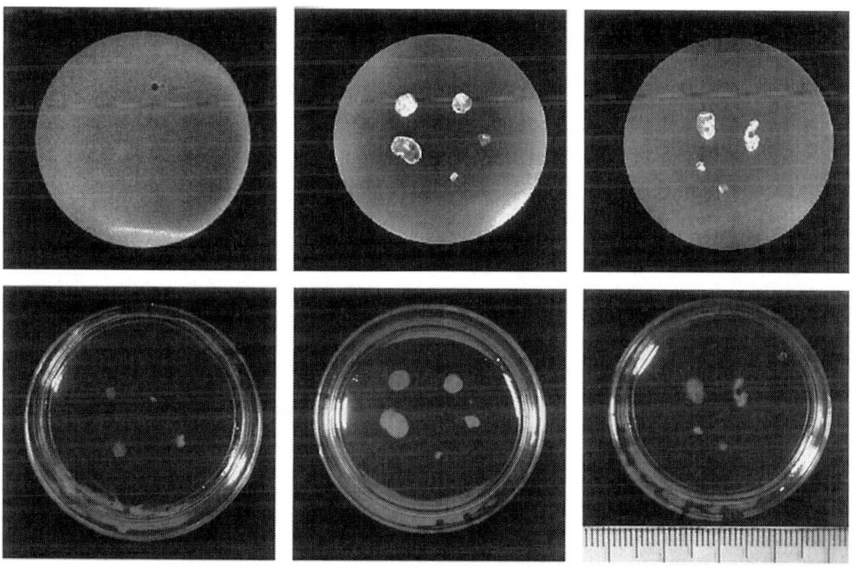

Fig. 7. T_1-weighted spin echo images of clots embedded in agarose gel on a petri dish. While untreated clots (left) showed no contrast, treated (middle) and partially blocked clots (right) showed significant contrast. Optical images (lower panels) of the clots confirmed that MR images accurately map the morphology of the clots. Reproduced from Yu et al. (2000), with permission.

THERANOSTIC NANOAGENTS

Given the multifunctional nature of the nanoparticle-based imaging agents described above, it is only logical that these targeted particles be further modified to concomitantly deliver therapeutic moieties. These theranostic nanoparticles, or those bearing both diagnostic and therapeutic compounds, are expected to allow for feedback mechanisms to determine the localization, release, and therapeutic efficacy of treatments.

One recent example of the use of theranostic magnetic nanoparticles for targeted thrombolysis was published by Bi et al. (2009). Urokinase (UK) was covalently conjugated to an amine-modified dextran-coated magnetic nanoparticle and its efficacy and safety were investigated *in vivo*. The authors used arteriovenous shunt thrombosis model in Sprague-Dawley rats to test the developed magnetic nanoparticles. In order to induce thrombus formation, the right carotid artery and left jugular vein of the rat were connected via a polyethylene tube containing silk thread. Following thrombus formation, the rats were injected with saline (control), UK, the magnetic nanoparticles, or the magnetic nanoparticles under external magnetic field in order to affect magnetically targeted localization. Following 30 min circulation time, the silk thread was removed to determine the weight of residual thrombus. As compared to saline-injected animals, the other agents demonstrated a decrease in the size of the thrombus, with the best result being obtained with UK-bearing nanoparticles under external magnetic field, resulting in 5-fold efficacy over UK and 2.6-fold efficacy over non-targeted nanoparticles. Moreover, examination of circulation of the residual fibrinogen and bleeding times confirmed that under the external magnetic field, the UK-conjugated magnetic nanoparticles could be employed as specific thrombolytic agents with decreased fibrinogen depletion and minimum prolongation of bleeding times.

A similar approach was also used by Ma et al., who investigated feasibility and efficacy of target thrombolysis with magnetic iron oxide nanoparticles (Ma et al. 2009). Following the synthesis of polyacryclic acid–coated magnetic nanoparticles, recombinant tissue plasminogen activator (rtPA), a serine protease that plays a role in the conversion of zymogen plasminogen to the active fibrin-degrading enzyme plasmin, was covalently conjugated to the nanoparticles. rtPA-conjugated nanoparticles and free rtPA, which served as a control, were tested *in vitro* and *in vivo* in a rat hind limb ischemia model. While free rtPA showed equivalent efficacy and amidolytic activity *in vitro*, rtPA-conjugated nanoparticles exhibited 5-fold thrombolytic efficacy *in vivo* when administered under a targeting external magnetic field.

Recently, nanoparticles were studied by Kempe et al. for implant-assisted magnetic drug targeting in thrombolytic therapy (Kempe and Kempe 2010). The magnetic nanoparticles were silanized and coated with triethylene glycol and/or polyethylene glycol to enhance the biocompatibility of the agents. The drug applied in the study, rtPA, was covalently immobilized on the ethylene glycol–coated nanoparticles, which were subsequently tested *in vitro* and *in vivo* using a Kanthal D ferromagnetic wire as a stent mimic. The idea behind using the stent was to increase the magnetic gradient and consequently to enhance the magnetic force and capture efficiency of the particles. To mimic the blood circulatory system, the wire was looped to a coil which was inserted to a glass capillary tube. Nanoparticles were circulated in the closed loop at various flow rates. While more than half of the nanoparticles were captured during the first 10 min circulation time at 25 mL per min flow rate, in the absence of the stent, significant retention in the capillary was obtained only at 1 mL per min flow rate with 90 min circulation time.

For *in vivo* studies in a porcine model, the stent was inserted into the brachial artery of the right forelimb and magnets were placed at both ends of the forelimb to generate a magnetic field of 0.1T. Following thrombus formation, tPA-bearing nanoparticles (5.4 mg tPA-nanoparticle conjugate in 40 mL of water) were injected into the stented vessel. The administered amount of tPA-nanoparticle conjugate corresponded to 0.38 mg of free tPA, which is significantly lower than the dose required for humans (100 mg). *In vivo* experiments confirmed lysing of the thrombus and re-opening of the vessel suggesting low dose of tPA may be adequate to lyse thrombus when the drug was conjugated to magnetic nanoparticles (Fig. 8).

APPLICATIONS TO OTHER AREAS OF HEALTH AND DISEASE

Whereas thrombosis is mostly associated with cardiovascular disease, there is evidence that clotting factors are also relevant in the detection of other diseases, including cancers. Using *in vivo* phage screening, Simberg et al. identified a peptide that homed to tumor vasculature (Simberg et al. 2007). After further investigation, the authors discovered that this peptide, CREKA (Cys-Arg Glu-Lys-Ala), actually bound to clotted plasma proteins. When conjugated to a fluorescently labeled iron oxide nanoscaffold, the particles readily detected breast cancer xenografts via *in vivo* FRI. Interestingly, the peptide also brought about further localized clotting, which resulted in biomimetic amplification of the nanoparticle homing, which may also provide for a therapeutic effect via the cessation of tumoral blood supply via embolization.

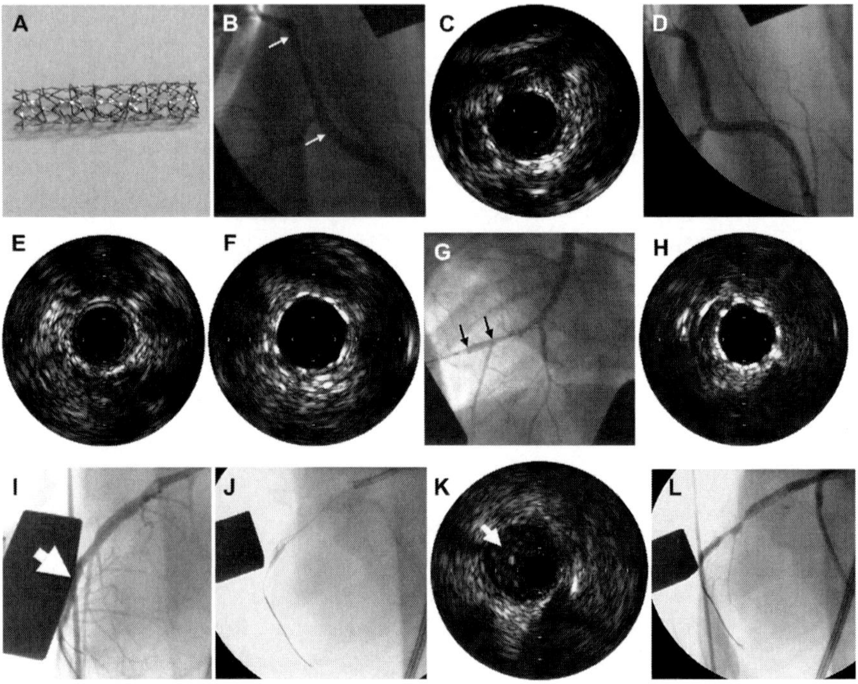

Fig. 8. *In vivo* pig experiments on the brachial artery (A-F) and left anterior descending artery (G-L). (A) Coronory stent with an interwoven Kanthal D wire in a coiled configuration. (B) After the stent was inserted in the brachial artery, the angiogram was obtained. Stented segment in the artery was indicated by arrows. (C) Intravascular ultrasound (IVUS) image of brachial artery was obtained directly after the stent insertion. (D) Angiogram was obtained directly after the injection of nanoparticles. (E) IVUS image was obtained directly after the injection of nanoparticles. (F) IVUS image was acquired 30 min after injection of nanoparticles. (G) Angiogram indicating the position of the stent in the left anterior descending artery distal to the first diagonal branch. (H) IVUS image of brachial artery obtained directly after the stent insertion showed a patent lumen. (I) Angiogram obtained after injection of nanoparticles indicated that the artery remained patent after the injection. Arrow indicates the position of the stent and the permanent magnet is the dark object on the left. (J) Angiogram obtained after thrombus formation indicating the blood flow ceased in the left anterior descending artery. (K) IVUS image showed formation of thrombus as the dark area indicated by the arrow. (L) Following the injection of tPA-nanoparticles, lysis of the thrombus was clearly confirmed by the angiogram. Reproduced from Kempe and Kempe (2010), with permission.

CONCLUSIONS

Current imaging techniques are capable of providing precise information about the presence of life-threatening thrombus and analysis of the blood flow within the vessel. However, significant biological details such as the age and anatomy of the thrombus are omitted with the present methodologies. Thus, new avenues are being opened to develop thrombus-targeted probes using nanotechnology. As recent studies have demonstrated, this field has shown great potential in the imaging and therapy of the thrombus.

Nanotechnology allows for the creation of agents bearing multiple functionalities, including targeting, imaging, and therapeutic ligands. These agents possess a number of advantages, such as allowing for the simultaneous detection and treatment of disease. Interestingly, cardiovascular disease and its sequelae have not been the main focus of nanomedicine. As compared to cancers, vascular targets are much more readily accessible, without the many barriers encountered with tumors. As is exemplified by this chapter, the prevalence of nanotechnology in the analysis of thrombotic events is increasing. Although, at this time, none of these materials have transitioned to clinical use, the generation of multifunctional nanoagents has the potential to revolutionize the detection and treatment of disease.

Summary Points

- Current imaging techniques provide information about the presence of thrombus, yet lack the ability to investigate thrombus age or other biological characteristics.
- Nano- or non–nano-based agents have been developed and examined *in vitro* and *in vivo* for imaging thromboses over the last two decades.
- Key components of thrombus such as coagulation factors, active platelets, and fibrin have been examined as target molecules.
- With the recent advances in nanotechnology, theranostic nanoagents, particularly those based on magnetic nanoparticles, have been investigated for the simultaneous detection and treatment of the disease.
- Targeted nanoagents have used a variety of imaging modalities, including MR, positron emission tomography, and optical imaging.

Definitions

Hemophilia: A group of bleeding disorders in which the body's ability to form a blood clot is impaired.

Hemostasis: The process by which bleeding from an injured vessel is stopped. This involves the activation of platelets and a number of coagulation factors.

Intravascular: Within a vessel.

Multifunctional: Bearing or possessing more than one functionality. For nanoparticles this can be targeting, imaging, or therapeutic ligands.

Multimodal: Bearing complementary imaging moieties capable of being imaged by more than one modality. This may be a combination of MR, nuclear, optical, ultrasound, or CT imaging agents.

Thrombogenesis: The formation of a thrombus.

Thrombophilia: A coagulation abnormality in which there is a propensity for blood clotting.

Thrombosis: The formation or presence of a blood clot within a blood vessel. Pathological thrombosis can result in significant morbidity and mortality.

Thrombolysis: The process by which a clot is broken down pharmacologically.

Zymogen: An inactive enzyme precursor.

Key Facts

- Hemostasis is the process by which the body presents a barrier to blood loss via the formation of a clot.
- Pathological thrombosis is caused by a dysregulation of the normal hemostatic system, causing partial or total blockage of a blood vessel. When thromboses occupy 75% of the lumen, they create hypoxic conditions, whereas blockages greater than 90% result in infarction.
- Thromboses are responsible for a myriad disorders, including myocardial infarctions (heart attacks), pulmonary emboli, stroke, and deep vein thromboses.
- Thrombolysis is the process by which a pharmaceutical is administered in order to dissolve the clot, restoring blood flow. Thrombolytic drugs include urokinase, streptokinase, and recombinant tissue plasminogen activators.
- For patients diagnosed with stroke, thrombolytic therapies with rtPA result in a 6.4% rate of intracerebral hemorrhage. For myocardial infarction, this rate was 0.93%, with 55% of those resulting in death. Of those who survived, 55% had a residual neurological deficit (Armstrong and Collen 2001).

Abbreviations

AA	:	arachidonic acid
CLIO	:	cross-linked iron oxide
CT	:	X-ray computed tomography
FFP	:	fresh frozen plasma
FRI	:	fluorescence reflectance imaging
FXIIIa	:	activated factor XIII
Gd-DPTA	:	gadolinium-chelating diethylenetriamine pentaacetic acid

Gd-DOTA	:	gadolinium-chelating 1, 4, 7, 10-tetraazacyclododecane-1, 4, 7, 10-tetraacetic acid
IVFM	:	intravital fluorescence microscopy
IVUS	:	intravascular ultrasound
MRI	:	magnetic resonance imaging
NIRF	:	near infrared fluorescence
rtPA	:	recombinant tissue plasminogen activator
SEM	:	scanning electron microscope
UK	:	urokinase

References

Anderson, H.V. and J.T. Willerson. 1993. Thrombolysis in acute myocardial infarction. N. Engl. J. Med. 329: 703–709.

Armstrong, P.W. and D. Collen. 2001. Fibrinolysis for acute myocardial infarction: current status and new horizons for pharmacological reperfusion, part 1. Circulation 103: 2862–2866.

Aruva, M.R., J. Daviau, S.S. Sharma and M.L. Thakur. 2006. Imaging thromboembolism with fibrin-avid ^{99m}Tc-peptide: evaluation in swine. J. Nucl. Med. 47: 155–162.

Bi, F., J. Zhang, Y. Su, Y.C. Tang and J.N. Liu. 2009. Chemical conjugation of urokinase to magnetic nanoparticles for targeted thrombolysis. Biomaterials 30: 5125–5130.

Cormode, D.P., T. Skajaa, Z.A. Fayad and W.J. Mulder. 2009. Nanotechnology in medical imaging: probe design and applications. Arterioscler. Thromb. Vasc. Biol. 29: 992–1000.

Falati, S., P. Gross, G. Merrill-Skoloff, B.C. Furie and B. Furie. 2002. Real-time in vivo imaging of platelets, tissue factor and fibrin during arterial thrombus formation in the mouse. Nat. Med. 8: 1175–1181.

Flaumenhaft, R., E. Tanaka, G.J. Graham, A.M. De Grand, R.G. Laurence, K. Hoshino, R.J. Hajjar and J.V. Frangioni. 2007. Localization and quantification of platelet-rich thrombi in large blood vessels with near-infrared fluorescence imaging. Circulation 115: 84–93.

Furie, B. and B.C. Furie. 2008. Mechanisms of thrombus formation. N. Engl. J. Med. 359: 938–949.

Gambhir, S.S. 2010. General principles of molecular imaging. In: R. Weissleder, S.S. Gambhir, B.D. Ross and A. Rehemtulla (eds.). Molecular Imaging: Principles and Practice. People's Medical Publishing House, Shelton, CT. 1–9.

Jaffer, F.A., C.H. Tung, J.J. Wykrzykowska, N.H. Ho, A.K. Houng, G.L. Reed and R. Weissleder. 2004. Molecular imaging of factor XIIIa activity in thrombosis using a novel, near-infrared fluorescent contrast agent that covalently links to thrombi. Circulation 110: 170–176.

Kawasaki, K., M. Miyano, K. Hirase and M. Iwamoto. 1993. Amino acids and peptides. XVIII. Synthetic peptides related to N-terminal portion of fibrin alpha-chain and their inhibitory effect on fibrinogen/thrombin clotting. Chem. Pharm. Bull. (Tokyo) 41: 975–977.

Kempe, H. and M. Kempe. 2010. The use of magnetite nanoparticles for implant-assisted magnetic drug targeting in thrombolytic therapy. Biomaterials doi:10.1016/j.biomaterials.2010.07.107.

Klink, A., E. Lancelot, S. Ballet, E. Vucic, J.E. Fabre, W. Gonzalez, C. Medina, C. Corot, W.J. Mulder, Z. Mallat and Z.A. Fayad. 2010. Magnetic resonance molecular imaging of thrombosis in an arachidonic acid mouse model using an activated platelet targeted probe. Arterioscler. Thromb. Vasc. Biol. 30: 403–410.

Ma, Y.H., S.Y. Wu, T. Wu, Y.J. Chang, M.Y. Hua and J.P. Chen. 2009. Magnetically targeted thrombolysis with recombinant tissue plasminogen activator bound to polyacrylic acid-coated nanoparticles. Biomaterials 30: 3343–3351.

Massberg, S., M. Sausbier, P. Klatt, M. Bauer, A. Pfeifer, W. Siess, R. Fassler, P. Ruth, F. Krombach and F. Hofmann. 1999. Increased adhesion and aggregation of platelets lacking cyclic guanosine 3', 5'-monophosphate kinase I. J. Exp. Med. 189: 1255–1264.

McCarthy, J.R. 2010a. Nanomedicine and cardiovascular disease. Curr. Cardiovasc. Imaging Rep. 3: 42–49.

McCarthy, J.R. 2010b. Multifunctional agents for concurrent imaging and therapy in cardiovascular disease. Adv. Drug Deliv. Rev. doi:10.1016/j.addr.2010.07.004.

McCarthy, J.R., P. Patel, I. Botnaru, P. Haghayeghi, R. Weissleder and F.A. Jaffer. 2009. Multimodal nanoagents for the detection of intravascular thrombi. Bioconjug. Chem. 20: 1251–1255.

Morris, T.A., J.J. Marsh, P.G. Chiles, R.G. Konopka, C.A. Pedersen, P.F. Schmidt and M. Gerometta. 2004. Single photon emission computed tomography of pulmonary emboli and venous thrombi using anti-D-dimer. Am. J. Respir. Crit. Care Med. 169: 987–993.

Sim, D.S., G. Merrill-Skoloff, B.C. Furie, B. Furie and R. Flaumenhaft. 2004. Initial accumulation of platelets during arterial thrombus formation *in vivo* is inhibited by elevation of basal cAMP levels. Blood 103: 2127–2134.

Simberg, D., T. Duza, J.H. Park, M. Essler, J. Pilch, L. Zhang, A.M. Derfus, M. Yang, R.M. Hoffman, S. Bhatia, M.J. Sailor and E. Ruoslahti. 2007. Biomimetic amplification of nanoparticle homing to tumors. Proc. Natl. Acad. Sci. USA 104: 932–936.

Srinivasan, R., R.E. Marchant and A.S. Gupta. 2010. In vitro and in vivo platelet targeting by cyclic RGD-modified liposomes. J. Biomed. Mater. Res. A 93: 1004–1015.

Stratton, J.R., M.D. Cerqueira, T.A. Dewhurst and T.R. Kohler. 1994. Imaging arterial thrombosis: comparison of technetium-99m-labeled monoclonal antifibrin antibodies and indium-111-platelets. J. Nucl. Med. 35: 1731–1737.

Thakur, M.L., V.R. Pallela, P.M. Consigny, P.S. Rao, D. Vessileva-Belnikolovska and R. Shi. 2000. Imaging vascular thrombosis with 99mTc-labeled fibrin alpha-chain peptide. J. Nucl. Med. 41: 161–168.

Tung, C.H., N.H. Ho, Q. Zeng, Y. Tang, F.A. Jaffer, G.L. Reed and R. Weissleder. 2003. Novel factor XIII probes for blood coagulation imaging. ChemBioChem 4: 897–899.

Tung, C.H., L. Quinti, F.A. Jaffer and R. Weissleder. 2005. A branched fluorescent peptide probe for imaging of activated platelets. Mol. Pharm. 2: 92–95.

Yu, X., S.K. Song, J. Chen, M.J. Scott, R.J. Fuhrhop, C.S. Hall, P.J. Gaffney, S.A. Wickline and G.M. Lanza. 2000. High-resolution MRI characterization of human thrombus using a novel fibrin-targeted paramagnetic nanoparticle contrast agent. Magn. Reson. Med. 44: 867–872.

18

Biodegradable Nanofibers in Cardiovascular Medicine: Drug Delivery Application

Masato Mutsuga,[1,a] *Aika Yamawaki-Ogata,*[1,b] *Makoto Satake,*[2,a] *Hiroaki Kaneko,*[2,b] *Yuichi Ueda*[1,c] and *Yuji Narita,*[1,d,*]

ABSTRACT

Nanotechnology will play a great role in the field of cardiovascular disease in the near future. It will be a very useful tool in drug delivery approaches to treat cardiovascular disease. Nanomedicine (medical treatment using nanotechnology) can be considered a refinement of molecular medicine and integrates advances in genomics and proteomics to facilitate the development of personalized medicine. Nanomedicine for cardiovascular disease

[1]Department of Cardiac Surgery, Nagoya University, Graduate School of Medicine, 65, Tsurumai-cho, Showa-ku, Nagoya, Aichi 466-8550, Japan.
[a]E-mail: mutsuga@med.nagoya-u.ac.jp
[b]E-mail: aika@med.nagoya-u.ac.jp
[c]E-mail: yueda@med.nagoya-u.ac.jp
[d]E-mail: ynarita@med.nagoya-u.ac.jp
[2]Department of Tissue Engineering Development, Technology Innovation Center, Teijin Limited, 3-2, Asahigaoka 4-chome, Hino, Tokyo 191-8512, Japan.
[a]E-mail: ma.satake@teijin.co.jp
[b]E-mail: hi.kaneko@teijin.co.jp
*Corresponding author

List of abbreviations after the text.

has focused on materials at super-molecular levels to deliver increased therapeutic efficacy and diminish adverse effects, which have ultimately resulted in their clinical application. All the aforementioned developments have revolutionized the application of nanotechnologies and have led to improved understanding of the possibilities for drug delivery to diseased tissues and organs.

In this chapter we introduce a novel drug delivery system (DDS) using biodegradable nano-scaled fiber for use in cardiovascular medicine. Occasionally, systemic administration of the drug for cardiovascular diseases not only fails to work sufficiently, but also causes adverse side effects due to the limited effective threshold of the drug. Therefore, these medicines may need to be released locally and gradually. Our newly developed DDS applied biodegradable polymer mixed with a drug used for surgical treatment of cardiovascular diseases, to achieve sustained release. The mechanisms of sustained release of a drug are quite simple; the drug is released with diffusion, hydrolysis of the biodegradable polymer and/or degradation of polymer by enzymatic or phagocytotic effects *in vivo*. An electrospinning procedure is used to fabricate the nanofiber configurations for these materials. In this chapter, we review the DDS scaffold using electrospinning nanofiber technology, and also highlight our recent results using our new DDS with biodegradable nano-scaled fiber for cardiovascular diseases to prevent anastomotic stricture and treat aortic aneurysm.

INTRODUCTION

Despite significant clinical advancements, cardiovascular diseases continue to rank among the greatest health problems worldwide. Nanotechnology will have an increasing impact on cardiovascular diseases in coming years. This chapter is focused on development of a drug delivery system (DDS) for treatment of cardiovascular disease by use of biodegradable nanofiber fabricated by an electrospinning procedure. Electrospinning is a process that uses electrostatic forces to produce synthetic fiber. This technique has been known for over 100 years, and it can produce polymer fibers of nanometric diameter (Sill and von Recum 2008). Recently, because of its ease of fabrication and resemblance to the natural extracellular matrix (ECM) of tissues, electrospinning has drawn great attention worldwide as a modeling technique to create scaffold for tissue engineering, a technology well known in the field of regenerative medicine (Vasita and Katti 2006). Electrospun nanofiber can also be used as carriers for drug delivery by applying biodegradable polymer for controlled release of the drug. Controlled and tailored release of drugs from polymers will improve control of pharmacokinetics and bioavailability. This technology will increase delivery of the effective concentration of the drug to the site of action and decrease systemic loading and adverse or toxic side effects of the drug. Thus, the new technology has enormous potential for improving the treatment of cardiovascular diseases and may be applicable to tissue-engineered scaffolds for cardiovascular regeneration.

Various types of materials (such as hydrogels, micelles, liposomes and biodegradable nanofibers/particles) to establish controlled and sustained release of the drug using nano-technology have been reported to date (Goldberg et al. 2007). In addition, numerous studies have demonstrated use of biodegradable polymers as valuable carriers for drug delivery (Venugopal et al. 2008). In the area of cardiovascular disease, for example, nanoparticles of polymers are used to prevent late thrombosis of drug-eluting stents (DES) for percutaneous coronary intervention (Lanza et al. 2006). Electrospun nanofiber is one type of carrier available for drug delivery. Although various agents and proteins (including growth factor, plasmid DNA, antibiotics, anti-inflammatory agents, anti-neoplastic drug) with biodegradable polymers (polyglycolic acid, PGA; polylactic acid, PLA; poly caprolactone, PCL; and various co-polymers) have been used for establishment of a DDS by using electrospinning nanofibers, few investigations have focused on the area of cardiovascular diseases for application of electrospinning nanofibers. Therefore, it is important here to describe the therapeutic strategies using DDS with biodegradable nanofiber for treatment of cardiovascular diseases.

The advantages of this system are that almost any number of drugs can be released site-specifically from the fiber scaffold into the body. The electrospinning procedure does not require a large and/or complex system for fabrication; it is relatively easy to spin nanofibers. Moreover, the desired scaffold configuration can be readily created. Thus, products can be customized for various situations. On the other hand, the disadvantage of this system is that surgery is necessary to implant the materials. Conversely, these products may be a useful tool in surgical intervention.

FABRICATION OF ELECTROSPUN NANOFIBER FOR CONTROLLED RELEASE OF DRUGS

Electrospinning is a highly sophisticated technology with a myriad variable applications that can fabricate non-woven fibrous architecture with fiber diameters ranging from micro to nano-levels. This technology is basically a solution spinning method, and the process uses electrostatic forces. The processes of electrospinning may be briefly described as follows. The polymer solution is charged with high-positive voltage (typically around 10 kV) jets and forms fibrous structures in the atmosphere. The fibers are then deposited on a cathode collector. Because the cathode collector is able to form any desired shape in this technique, the fibrous structures fabricated on the collector also take a desired shape. Since the system of electrospinning is very simple and intuitively understandable, many parameters affect the state of the spun fiber, including the type of equipment, polymer types and condition of solution (Sill and von Recum 2008). Figure 1 provides a schema of a typical mono-axial nozzle electrospinning setup. It consists of a high-voltage power supply connected to the needle to provide the charging potential. A polymer solution is dispensed through a needle using a syringe pump and the fiber jet is eventually ejected. Fibers produced using this process typically have diameters on the order of a few micrometers down to 10 or more nanometers. Application of this technology has attracted attention from a wide variety of industries. Particularly in the field of medicine, this technology is drawing attention to the use of tissue engineering and DDS because the diametric size of the material is similar to biological ECMs.

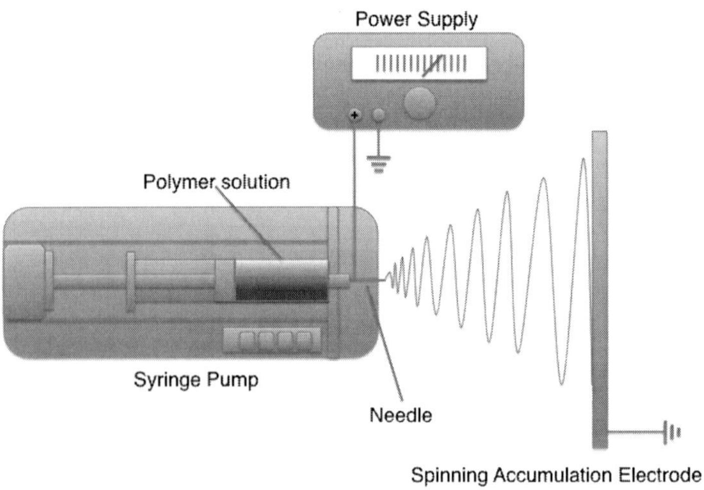

Fig. 1. Schema of setup for electrospinning procedure. Polymer solution is held at the tip of the needle attached to a syringe by virtue of its surface tension. When a high voltage is applied to the polymer solution, the electrical forces overcome the surface tension and cause the formation of a jet that leads to formation of nanofibers that are collected on a grounded metal substrate.

Electrospinning has been known for more than a century as a process of forming synthetic fiber by use of electrostatic forces and originated from electrospraying. Electrospraying was described in the literature as early as 1914, and the process was patented along with an apparatus that used electric charges to spin synthetic fibers in 1934. Since then, many technical improvements and further development of fabricating nano-scaled fiber using electrospinning have been made. Nowadays, complex structural electrospinning of nanofiber can be done using a co-axial nozzle. This technology has shown rapid growth in the past few years, and our understanding of it continues to improve thanks to the benefit of on-going research.

DDS using electrospinning nanofiber is in itself a final product. The electrospinning technique has flexibility in selecting materials for DDS applications. Either biodegradable or non-biodegradable materials can be used to control whether drug release occurs via diffusion alone or with fiber degradation. Moreover, the wide variety of electrospinning techniques provides manifold fabrication methods used for drug delivery, which includes coating, embedding, coaxial and emulsion. These technical factors can be used for the controlled kinetic release of drugs. The drug solubility and stability in the polymer solution might be especially important factors for fine and well-shaped fibers in those fabrications. Finally, a number of drugs might be deliverable to any kind of cells, tissues and organs using fibers. They include low-molecular-weight drugs, antibody drugs and oligonucleotide drugs (Liang et al. 2007). The various electrospun nanofibers for drug delivery are listed in Table 1. They include polymers, solvents, and drugs, and the clinical applications and references are also listed.

DDS USING BIODEGRADABLE NANOFIBER FOR CARDIOVASCULAR DISEASES (OUR RECENT RESULTS)

Few articles have described a DDS using electrospinning nanofiber for use in cardiovascular diseases. Jiang et al. reported that an ibuprofen-loaded biodegradable membrane was fabricated by an electrospinning procedure for treatment of post-operative atrial fibrillation. However, they did not demonstrate preventive effects against atrial fibrillation *in vivo* (Jiang et al. 2004). On the other hand, we have developed a new DDS using electrospun nano-scaled fiber for cardiovascular disease for prevention of post-operative anastomotic stricture and aortic aneurysms. We have demonstrated it *in vivo* experimentally and have recently obtained the following results.

Table 1. Electrospun nanofibers for drug delivery.

Polymer	Solvent	Drug	Application	*in vivo* demonstration	Reference
poly(L-lactic acid) and poly(ethylene-co-vinyl acetate)	Chloform	Tetracycline hydrochloride	Prevention of surgical site infection	–	Kenawy et al. 2002
poly(L-lactic acid)	Chloroform and acetone	Rifampin or paclitaxel	–	–	Zeng et al. 2003
hydroxylpropylmethylcellulose	Ethanol and methylene chloride	Itraconazole	Prevention of surgical site infection	–	Verreck et al. 2003
poly(L-lactic-co-glycolide) or PLGA / PEG-b-PLA diblock copolymer / PLA	N,N-dimethyl formamide	Mefoxin® (cefoxitin sodium)	Prevention of post-surgical adhesions and infections	–	Kim et al. 2004
poly(L-lactic-co-glycolide)	Tetrahydrofuran and N,N-dimethyl formamide	Cefazolin	Prevention of surgical site infection	–	Katti et al. 2004
poly(L-lactic-co-glycolide) and poly (ethylene glycol)-g-chitosan (PEG-g-CHN)	DMSO-d6 or CDCl3	Ibuprofen	Prevention of atrial fibrillation	–	Jiang et al. 2004
poly(L-lactic-co-glycolide)	Dichloromethane	Paclitaxel	Anti-neoplasm	–	Xie and Wang 2006
poly(ethylene glycol)– poly (L-lactic acid)	Chloform	BCNU	Anti-neoplasm	–	Xu et al. 2006
poly(ε-caprolactone)	Chloroform and N,N-dimethyl formamide	Biteral	Prevention of post-surgical adhesions	+	Bolgen et al. 2007
poly(L-lactic acid) and/or poly(ε-caprolactone)	Chloroform, tetrahydrofuran or toluene	Chlorotetracycline hydrochloride, Tetracycline hydrochloride, Amphotericin B	Prevention of surgical site infection	–	Buschle-Diller et al. 2007
poly(L-lactide-co-glycolide)	Dichloromethane and EtOH	Tacrolimus	Prevention of arterial or venous anastomotic stricture	+	Mutsuga et al. 2009
poly(L-lactide-acid)	Dichloromethane and EtOH	Doxycycline	Prevention of aortic aneurysm	+	Yamawaki-Ogata et al. 2010

PLGA, poly(L-lactic-co-glycolide); PEG, poly(ethylene glycol); PLA, poly(L-lactic acid); CHN, chitosan; DMSO, dimethyl sulfoxide; DCDI3, deuterated chloroform; EtOH, ethanol; BCNU, 1, 3-Bis(2-Chloroethyl)-1-Nitrosourea.

Prevention of Anastomotic Strictures

Anastomotic strictures have remained the main reason for arterial, venous, arteriovenous, and prosthetic graft failure (Kapadia et al. 2008). Moreover, peculiar obstructions have occurred in pulmonary veins after correction of a total anomalous pulmonary venous connection (Caldarone et al. 1998). These kinds of anastomotic stricture are largely affected by the severity of neo-intimal hyperplasia. Thus, preventing this neo-intimal hyperplasia improves not only patient morbidity but the prognosis as well.

Our newly developed device consists of biodegradable nano-scaled fiber based on the prevention of neo-intimal overgrowth by sustained release of a drug in line with the DES concept. Neo-intimal hyperplasia is composed of mesenchymal cells and their excessively produced ECM. Therefore, neo-intimal hyperplasia can be prevented by the cytostatic effects of a drug to inhibit mesenchymal cell proliferation and ECM production. Tacrolimus was chosen as the drug to prevent neo-intimal hyperplasia in our experimental study, because it has been carefully evaluated in previous studies as an agent to prevent neo-intimal hyperplasia with respect to release kinetics, effective dosage, safety in clinical practice, and benefit. Tacrolimus is mixed with biodegradable polymer (poly(l-lactide-co-glycolide), which consists of 50% of each polymer), and it is spun as a nano-scaled fiber using the electrospinning technique to achieve sustained release to create tacrolimus-eluting biodegradable nano-scaled fiber (TEBN). Dichloromethane and ethanol were used as a polymer solvent. The polymer solution was added to a 1 ml syringe with a right-angled metal capillary attached to it. The circular orifice of the capillary has an inner diameter of 1.2 mm. A round, flat counter electrode was located at about 35 cm from the capillary tip. Pressure was applied to the solution in a syringe by putting a certain weight on top of the piston to maintain a steady flow of the solution from the capillary outlet. The flow rate of the polymer solution was 0.3 ml/min. The applied voltage was in the range of 10 to 15 kV. The fibers released in the atmosphere were electrostatically removed and trapped with a rod-shaped collector located 20 cm from the capillary tip. The diameter of nanofiber was 100 to 800 nm. Our new device has a cotton wool–like configuration that is suitable for any type of anastomosis (Fig. 2).

Our data (Mutsuga et al. 2009a) showed that TEBN gradually released tacrolimus for at least 1 mon both *in vitro* and *in vivo* (Fig. 3). Approximately 30% of the tacrolimus was released into the solution from TEBN on the first day, but the release continued, showing 48% release 2 wk and 49% release 4 wk after incubating TEBN in phosphate-buffered saline. Early burst release was also observed *in vivo*. Approximately 70% of the tacrolimus was released from TEBN within the first day. After the

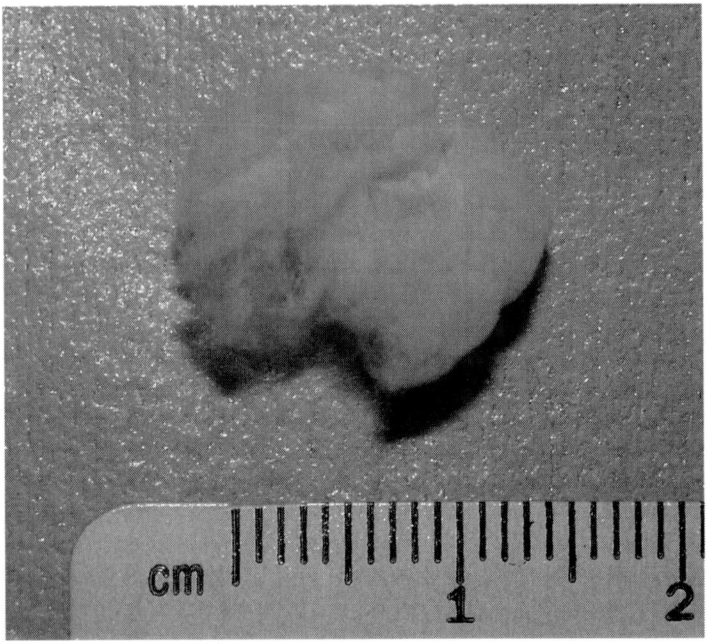

Fig. 2. Cotton wool–like appearance of tacrolimus-eluting biodegradable nano-scaled fiber.

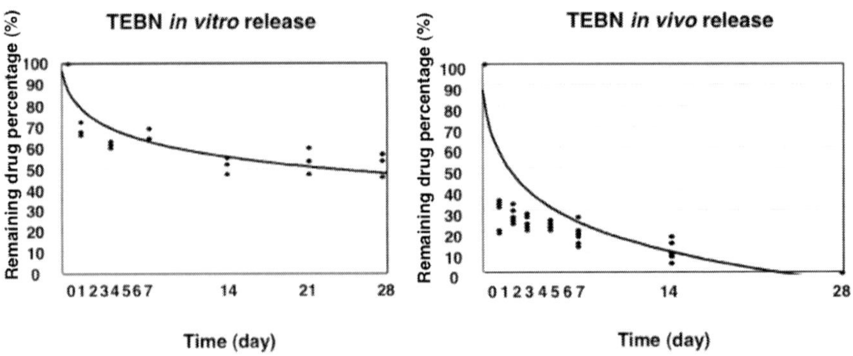

Fig. 3. *In vitro* and *in vivo* tacrolimus-releasing test with tacrolimus-eluting biodegradable nano-scaled fiber (TEBN). Abscissa indicates the remaining percentage of tacrolimus in the TEBN (%). The remaining tacrolimus curve seems to be logarithmic. Tacrolimus was released at a fast initial burst phase during the first few days followed by a slow sustained phase during the following 4 wk (Mutsuga et al. 2009a).

initial burst phase during the first few days, a slow-release phase followed for 4 wk. The *in vivo* experimental study confirmed the gradual sustained release of tacrolimus from TEBN.

Prevention of Arterial Anastomostic Stricture

An animal arterial anastomostic stricture model was used with Wistar rats using the end-to-end abdominal aortic anastomosis by a continuous running suture with 9-0 silk. Rats were randomly divided into two groups: (1) control without TEBN and (2) 5 mg of TEBN containing 1.0 wt% tacrolimus. Rats were sacrificed 2 wk after operation and evaluated with the following procedures.

For the histological analysis, the anastomotic sites were resected beyond the suture lines, fixed with formaldehyde, and embedded in paraffin. Short axial consecutive 5 µm serial sections were obtained just around the suture anastomosis. The sections were stained with hematoxylin-eosin (H&E) for morphometric analysis (Fig. 5). The area of the intima (including neo-intima) and media at the anastomosis site was measured on the sections stained with H&E. The ratios of the intimal area (i.e., intimal area divided by the intimal added medial area) in the control group and the TEBN group were 0.26 ± 0.09 and 0.08 ± 0.04, respectively (Fig. 4). TEBN with 1.0 wt% of tacrolimus significantly inhibited neo-intimal hyperplasia. Immunohistochemical staining was performed to identify the cells, which composed the intimal hyperplasia, and to evaluate re-endothelialization at the injured anastomotic site. To label mesenchymal cells (e.g., smooth muscle cell, SMC, or myofibroblast) and endothelial cells, alpha-smooth muscle actin (ASMA), smooth muscle myosin heavy chain embryo (SMemb), and factor VIII were used as

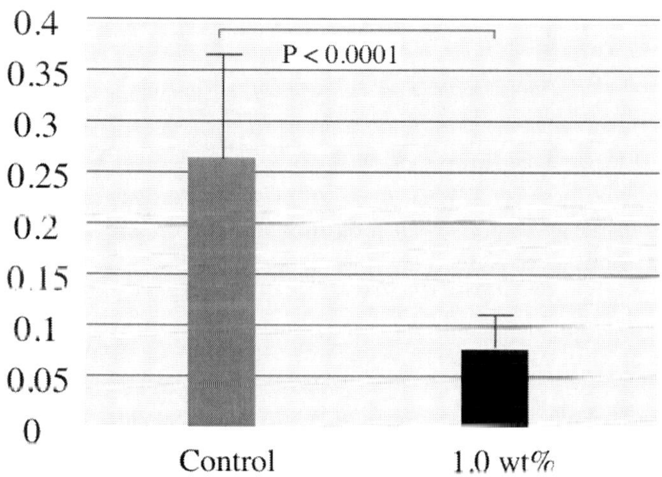

Fig. 4. Ratio of intimal area. There is a significant difference between control group and TEBN 1.0 wt% (Mutsuga et al. 2009a).

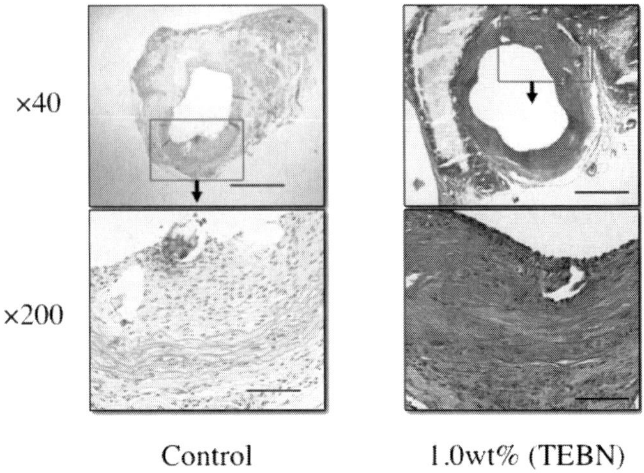

Fig. 5. Histological findings in the anastomotic site 2 wk after operation (arterial model). H&E staining of control group magnification ×40 (above) ×200 (below) and TEBN containing 1.0 wt% tacrolimus magnification ×40 (above) ×200 (below). Neo-intimal hyperplasia is significantly decreased in TEBN containing 1.0 wt% tacrolimus. Scale bar = 100 μm (above), scale bar = 500 μm (below) (Mutsuga et al. 2009a).

Color image of this figure appears in the color plate section at the end of the book.

primary antibodies. The cells constituting the intimal hyperplasia were ASMA- and SMemb-positive. Endothelial cells, which were stained with factor VIII immunohistochemically, covered the surface of the aortic lumen even in the 1.0 wt% TEBN group, in which neo-intimal hyperplasia was significantly inhibited.

Prevention of Venous Anastomostic Stricture

An animal venous anastomostic stricture model was made with Wistar rats using an inferior vena cava partial anastomosis by a continuous running suture (9-0 nylon) with the assumption that a pulmonary venous obstruction had developed following a total anomalous pulmonary venous return correction (Mutsuga et al. 2009b). Histological and immunohistological analysis is the same as with an arterial model. The ratios of intimal area in the control without TEBN and with 5 mg of TEBN (tacrolimus containing 1.0 wt%) were 0.43 ± 0.26 and 0.07 ± 0.04 at 1 wk, 0.39 ± 0.19 and 0.05 ± 0.05 at 2 wk, and 0.31 ± 0.15 and 0.09 ± 0.04 at 4 wk, respectively. TEBN inhibited neo-intimal hyperplasia significantly at each time point (1, 2 and 4 wk). Representative micrographs of anastomotic sections stained with H&E are shown in Fig. 6. Immunohistochemical staining for mesenchymal cells revealed that the cells constituting the intimal hyperplasia were positive for ASMA and SMemb. These findings meant that the cells that constituted neo-intimal hyperplasia consisted of dedifferentiated SMCs, myofibroblasts, and/or bone marrow–derived mesenchymal stem/progenitor cells. Endothelial cells, which were stained with factor VIII immunohistochemically, covered the surface of the venous lumen in both groups and at each time point. There was endothelialization even at the site of inhibited neo-intimal hyperplasia in the TEBN group.

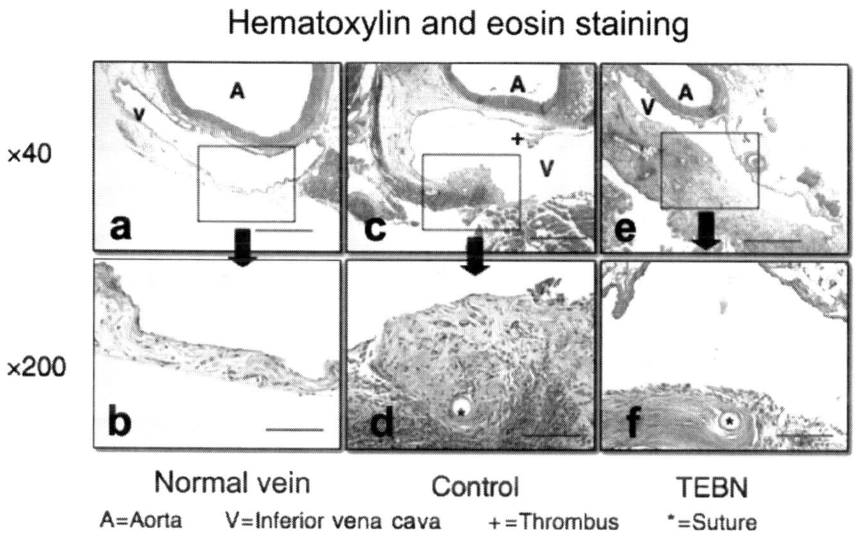

Fig. 6. Histological findings in the anastomotic site 1 wk after operation (venous model). HE staining of normal group (a) ×40 (b) ×200, control group (c) ×40 (d) ×200 and TEBN group (e) ×40 (f) ×200. Neo-intimal hyperplasia occurred around the surgical suture (d) and clearly decreased in TEBN group (f). Scale bar in (a), (c) and (e) 500 μm. In (b), (d) and (f): 100 μm (Mutsuga et al. 2009b).

Color image of this figure appears in the color plate section at the end of the book.

Prevention of Aortic Aneurysm

An aortic aneurysm is a silent but life-threatening condition associated with aging and atherosclerosis. Such aneurysms progress over time, resulting in a rupture of the diseased aorta, with a mortality rate of 80–90% (Ernst 1993). Surgical aortic replacement using artificial prosthesis is a powerful tool for prevention of aortic rupture. However, mortality and morbidity following surgical treatment (especially thoracic or thoracoabdominal aortic aneurysm) are still high. In addition, a patient of advanced age with complications, such as chronic obstructive lung disease and stroke, cannot undergo surgery because of high risk. Therefore, alternative and/or additional therapeutic strategies will be needed for particularly extended and complicated cases of aortic aneurysm.

The pathogenesis of an aortic aneurysm reflects a failure in the balance of the ECM degradation and synthesis in the aortic wall. Inflammation, which is a major pathological component of atherosclerosis, including aortic aneurysms, causes an ECM degradation of the aortic wall. Such inflammation leads to macrophages infiltrating the aortic wall and secreting matrix metalloproteinase (MMP)-2 and -9 that play a pivotal role in the degradation of the aortic ECM and aortic aneurysm development. The protein levels and activities of cathepsin K, L and S are more elevated in the tissue of an aortic aneurysm than in healthy arteries. In addition, the infiltrating inflammatory cells secrete various inflammatory cytokines and chemokines such as IL-1β, IL-6, tumor necrosis factor (TNF) α and monocyte chemoattractant protein 1 (MCP-1). These cytokines and chemokines function for the stimulation, activation and inflammation of various cells, such as regulation of MMP secretion and recruitment of monocytes to atherosclerotic lesions. Meanwhile, ECM synthesis also affects the progression of an aortic aneurysm. Elastin is a major fibrillar component of the aortic ECM, and elastin decreases in the tissue of an aortic aneurysm (Thompson et al. 2002). Elastin polypeptides are known to be synthesized by SMCs. The gene expression of elastin is modulated by growth factors and cytokines including transforming growth factor (TGF) β1, insulin-like growth factor (IGF) 1, interleukin (IL) 1β and lysyl oxidase (LOX). To elucidate the molecular mechanisms of the pathogenesis of aortic aneurysms, it is obvious that a deeper understanding of the aortic ECM metabolism, including its degradation and synthesis, will be needed to achieve a new and alternative therapeutic strategy for aortic aneurysms.

It is also well known that doxycycline (DOXY) (a tetracycline derivative antibiotic agent) inhibits MMPs and suppresses the progression of aortic aneurysms (Pyo et al. 2000). The systemic administration of DOXY has been shown to effectively suppress abdominal aortic aneurysm in an experimental rodent model and to reduce the aortic wall and plasma levels of MMP-9 in patients with abdominal aortic aneurysm in clinical testing (Baxter et al. 2002). However, the oral administration of DOXY is associated with dose-dependent side effects, such as dental discoloration and microbial substitution, making it difficult to use for the clinical treatment of aortic aneurysms.

In consideration of the above, local administration and sustained release of DOXY on the aortic wall might prove effective for the treatment of aortic aneurysm without side effects. We tried to develop a DDS composed of DOXY and a biodegradable polymer (poly-l-lactide) to achieve locally sustained release of DOXY without the need for complex devices. We fabricated a DOXY controlled-release biodegradable nano-scaled fiber (DCRBF) using a biodegradable polymer containing DOXY in an electrospinning procedure. The polymer solvent and fabricating procedure were the same as TEBN. The DCRBF showed a cotton wool–like formation and was about 2.25 ± 0.42 μm in diameter (Fig. 7). The maximum DOXY content in the DCRBF was 5.0 wt%. The *in vitro* and *in vivo* release of DOXY from the DCRBF was evaluated by measuring the DOXY concentration in the remaining DCRBF. In the *in vitro* release, DOXY was released gradually during the experimental process, and it was observed that 40% of DOXY was released at 2 wk and 45% at 4 wk, respectively (Fig. 8A). In *in vivo* release, after the initial burst during the first few days, a sustained-release phase was observed during the following 12 wk (Fig. 8B). A screening blood examination was performed to investigate the side effects caused by DCRBF. There was no difference between the DCRBF-treated mice and the normal mice in terms of complete blood count, serologic liver functional examination (γ-GTP, GOT, GPT), renal functional examination (BUN, Cre), and pancreatic enzymes (AMY).

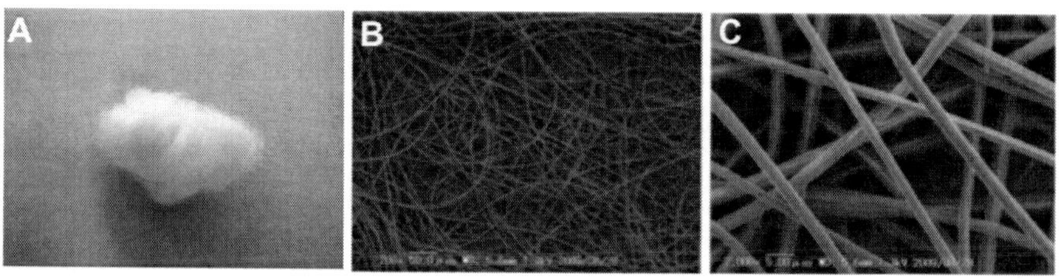

Fig. 7. Photograph and scanning electron microscopic (SEM) image of doxycycline controlled-release biodegradable fiber (DCRBF). (A) Overall appearance of DCRBF with cotton wool–like formation. (B) SEM (original magnification, ×200) of DCRBF resembles non-woven fabric. Scale bar: 100 μm. (C) SEM (original magnification, ×2000) of DCRBF shows mean fiber diameter of polymer to be 2.25 ± 0.42 μm. Scale bar: 10 μm (Yamawaki-Ogata et al. 2010).

We demonstrated the effects of DCRBF on the expression of various aortic aneurysms and on related markers in *in vitro* and *ex vivo* co-culture systems using cell culture inserts. The 2 mg DCRBF was co-cultured with SMCs, macrophages and aortic tissue that were harvested after 4 wk from C57BL/6 mice. We evaluated gene and protein expression of proteases. In the presence of DCRBF, gene expression of MMP-12 was significantly decreased, gene expression of TGF-β1 and LOX was significantly increased in SMC (Table 2), and gene expression of MMP-9 and -12 was significantly decreased in macrophages (Table 3). In addition, the DCRBF preserved elastin content (Fig. 9) and decreased protein expression of MMP-2 and -9 in aortic tissue (Fig. 10).

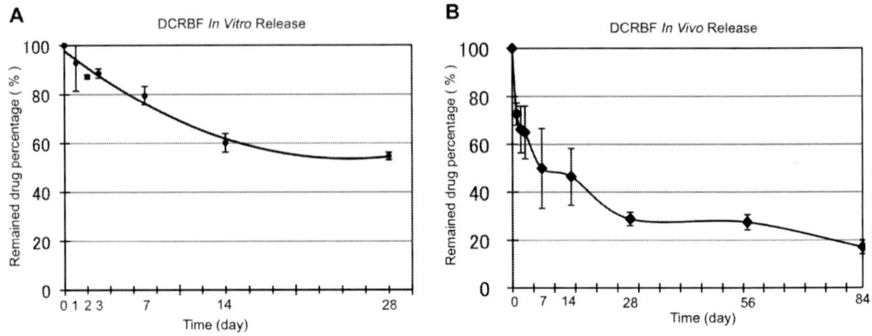

Fig. 8. (A) *In vitro* and (B) *in vivo* doxycycline (DOXY) release assay for doxycycline controlled-release biodegradable fiber (DCRBF). The releasing curve seems to be a logarithmic one. Half of DOXY in the DCRBF remained 1 mon after cultivation (A). DOXY was released in a fast initial-burst phase during the first few days followed by a slow sustained phase over the following 12 wk *in vivo* (B). (Yamawaki-Ogata et al. 2010.)

Table 2. Indirect effects of doxycyclin controlled-release biodegradable nano-scaled fiber (DCRBF) on mouse smooth muscle cell (mSMC) gene .

	24 h		48 h		96 h	
	mSMCs mono-cultured (n=3)	co-cultured with DCRBF (n=4)	mSMCs mono-cultured (n=4)	co-cultured with DCRBF (n=3)	mSMCs mono-cultured (n=4)	co-cultured with DCRBF (n=5)
MMP2	2.20 ± 1.10	3.05 ± 1.51	2.05 ± 0.41	2.45 ± 0.55	0.94 ± 0.30	0.97 ± 0.25
MMP3	0.82 ± 0.45	1.12 ± 0.24	0.77 ± 0.63	0.66 ± 0.51	0.92 ± 0.12	0.85 ± 0.26
MMP9	3.32 ± 2.14	2.48 ± 1.22	1.20 ± 0.81	1.14 ± 0.84	1.06 ± 0.19	1.12 ± 0.30
MMP12	3.08 ± 0.25	3.77 ± 1.77	0.99 ± 0.80	0.98 ± 0.79	1.47 ± 0.29	0.99 ± 0.19*
Cathepsin K	0.25 ± 0.04	0.16 ± 0.05	0.31 ± 0.24	0.14 ± 0.01	0.21 ± 0.16	0.28 ± 0.17
Cathepsin L	1.08 ± 0.57	0.73 ± 0.13	0.99 ± 0.47	0.66 ± 0.28	0.84 ± 0.45	1.04 ± 0.29
Elastin	3.19 ± 0.64	3.12 ± 0.32	1.64 ± 0.42	1.88 ± 0.74	1.08 ± 0.60	2.00 ± 0.99
TGF-β1	8.84 ± 2.18	7.35 ± 1.10	3.01 ± 1.02	3.08 ± 1.02	1.79 ± 0.51	5.34 ± 1.92**
Lox	1.27 ± 0.66	0.57 ± 0.37	0.93 ± 0.22	1.60 ± 0.30*	0.86 ± 0.16	1.08 ± 0.13

To investigate the effects of DCRBF on the gene expression of mSMCs, aortic aneurysm-related gene expression such as MMP-2, 3, 9, 12, 13, cathepsin K, L, elastin, TGF-β1 and LOX were evaluated by a qRT-PCR. qRT-PCR findings revealed that MMP-12 was significantly decreased over 96 h. In addition, gene expressions of LOX at 48 h and TGF-β1 at 96 h were significantly increased in co-cultured. *$p < 0.05$ compared with mono-cultured, **$p < 0.01$ compared with mono-cultured (Yamawaki-Ogata et al. 2010).

An *in vivo* aortic aneurysm model, angiotensin-II (Ang-II), was administered by osmotic pump implanted subcutaneously in apolipoprotein E-deficient (apoE-/-) mice at 7 to 11 mon of age. Concomitantly, the mice were implanted with 25 mg of DCRBF in the abdominal aorta (cranial side

Table 3. Indirect effects of controlled-release biodegradable nano-scaled fiber (DCRBF) on macrophage gene expression.

	96 h	
	macrophages mono-cultured (n=5)	co-cultured with DCRBF (n=5)
MMP2	1.01 ± 0.31	0.67 ± 0.35
MMP3	1.04 ± 0.44	0.99 ± 0.36
MMP9	0.79 ± 0.20	0.35 ± 0.11**
MMP12	0.89 ± 0.30	0.35 ± 0.15**
Cathepsin K	0.93 ± 0.32	0.87 ± 0.28
Cathepsin L	0.91 ± 0.38	0.52 ± 0.11
Cathepsin S	1.22 ± 0.33	0.80 ± 0.32
CC3	1.19 ± 0.14	0.90 ± 0.25
CC5	0.65 ± 0.31	0.72 ± 0.19
CXC10	0.84 ± 0.27	0.60 ± 0.17
TNF-α	0.71 ± 0.16	0.54 ± 0.12
Alox5ap	0.77 ± 0.22	0.60 ± 0.11

To investigate the effects of DCRBF on the gene expression of macrophages, and aortic aneurysm-related gene expression such as MMP-2, 3, 9, 12, 13, cathepsin K, L, S, CC3, 5, CXC10, TNF-α and Alox5ap were evaluated by a qRT-PCR. qRT-PCR findings revealed that MMP-9 and MMP-12 at 96 h were significantly decreased in co-cultured. $^{**}p < 0.01$ compared with mono-cultured (Yamawaki-Ogata et al. 2010).

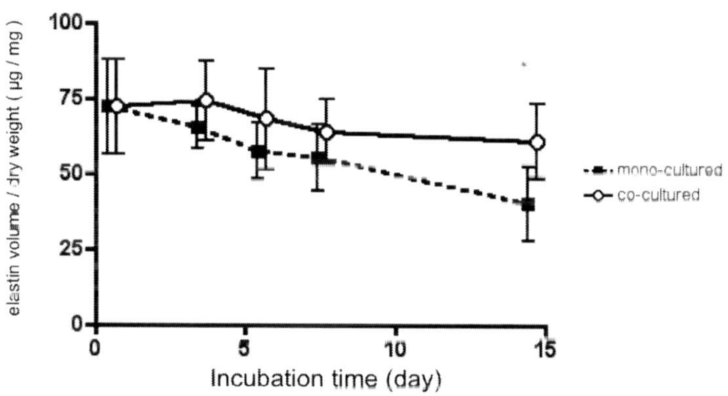

Fig. 9. Elastin volume of aortic tissues *ex vivo* co-cultured (open circle) or mono-cultured (closed square) with DCRBF throughout 14 d. Elastin volume in mono-cultured tissues decreased gradually over time. However, when co-cultured with DCRBF, the volume was maintained for 14 d. Elastin volume of tissues co-cultured with DCRBF at day 14 was significantly greater than that of mono-cultured. $^{*}p<0.05$ compared with mono-cultured (Yamawaki-Ogata et al. 2010).

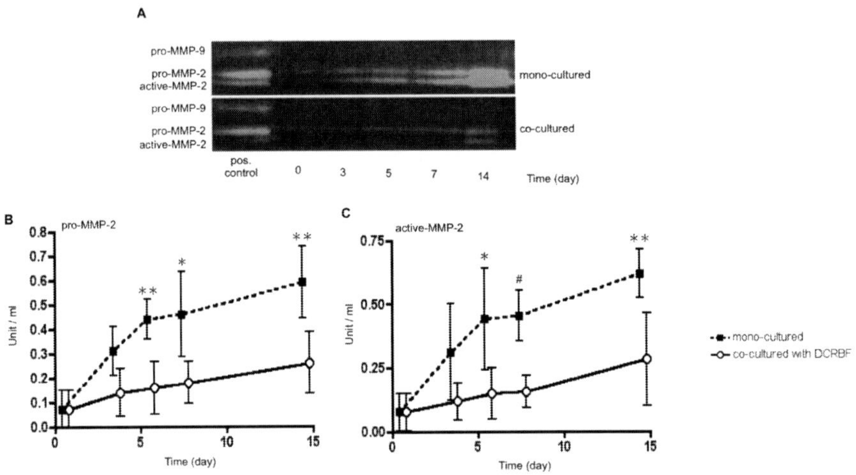

Fig. 10. Gelatin zymography of aortic tissues *ex vivo* co-cultured or mono-cultured with DCRBF. (A) MMP-9 (pro-MMP-9) was not detected in either group. In contrast, expression of MMP-2 (pro- and active-) increased over time in both groups. Concentration of MMP-2 (pro-MMP-2 (B) and active-MMP-2 (C)) co-cultured with DCRBF was significantly lower than in mono-cultured for 14 d. *$p<0.05$, **$p<0.01$, #$p<0.001$ compared with mono-cultured (Yamawaki-Ogata et al. 2010).

to the renal artery). Both treated and untreated mice were sacrificed and evaluated 4 wk after the operation. The overall aorta (from the ascending to the iliac bifurcation) was evaluated by elastin volume, MMP expression and other protein expression. In the presence of DCRBF, the elastin volume was preserved and equal to that of the normal group (Fig. 11). Band intensity of the expression levels of the MMPs, pro-, active-MMP-2, 9 of apoE-/- with the DCRBF group was significantly decreased in comparison with that of the apoE-/- group (Fig. 12). In addition, protein expression of IGF-1 and TIMP-1 was significantly increased, and protein expression of IL-6 and TNF-α was significantly decreased with DCRBF group (Fig. 13).

The mechanisms of DOXY-induced inhibitory effects on aortic aneurysms have been reported, such as inhibition of MMPs (Garcia et al. 2005), increase of TIMPs (Abdul-Hussien et al. 2009), and decrease of IL-6 and TNF-α (Cazalis et al. 2008). Similarly to MMPs, DOXY down-regulates inflammatory reactions that also serve to inhibit aortic aneurysm progression. Meanwhile, they are powerful inducers of elastin expression (Liu and Davidson 1988; Conn et al. 1996), though no study has indicated

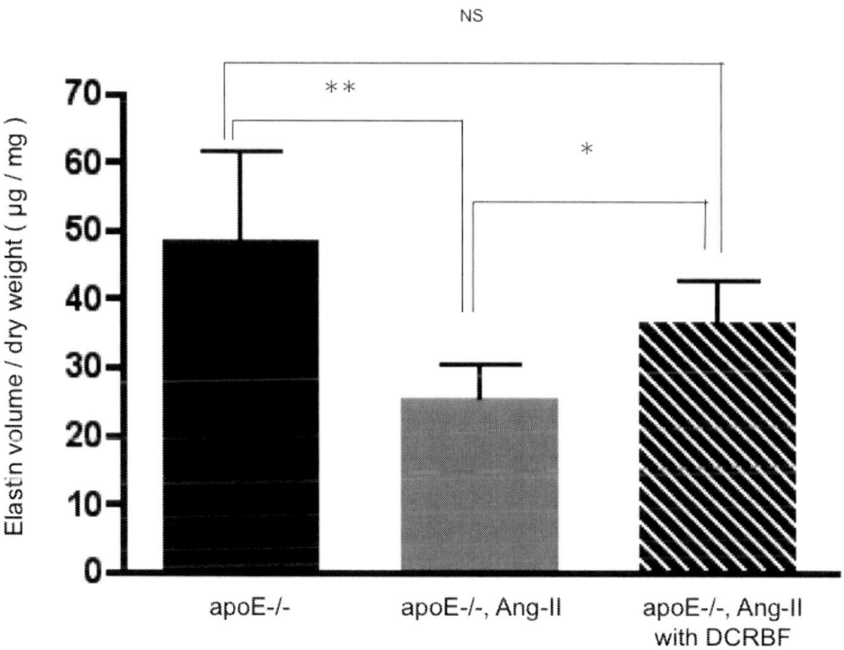

Fig. 11. Elastin volume of the aortic tissue, *in vivo* treated and not treated with DCRBF. Elastin volume of the aortic tissue of the (apoE-/-, Ang-II) group was significantly decreased compared with the other two groups. In contrast, in the group treated with DCRBF, the elastin volume was preserved, and no significant differences were observed in the apoE-/- group. *$p<0.05$, **$p<0.01$ compared with (apoE-/-, Ang-II) group (Yamawaki-Ogata et al. 2010).

that TGF-β1 and IGF-1 increased because of the presence of DOXY. In the present study, although the mechanisms of TGF-β1 and IGF-1 up-regulation remain unclear, it seems very likely that they are involved in the maintenance of elastin in the aortic wall. DOXY induced elastin synthesis in SMC and might induce IL-1β secretion. It is worth considering that the acceleration of ECM synthesis may indeed be one of the anti-aneurysmal effects of DOXY.

In conclusion, DCRBF improved the balance of ECM degradation/synthesis in the wall of aortic aneurysms and suppressed the progression of aneurysms without side effects. Our results suggest that DCRBF may become a new alternative therapeutic strategy for treating aortic aneurysms.

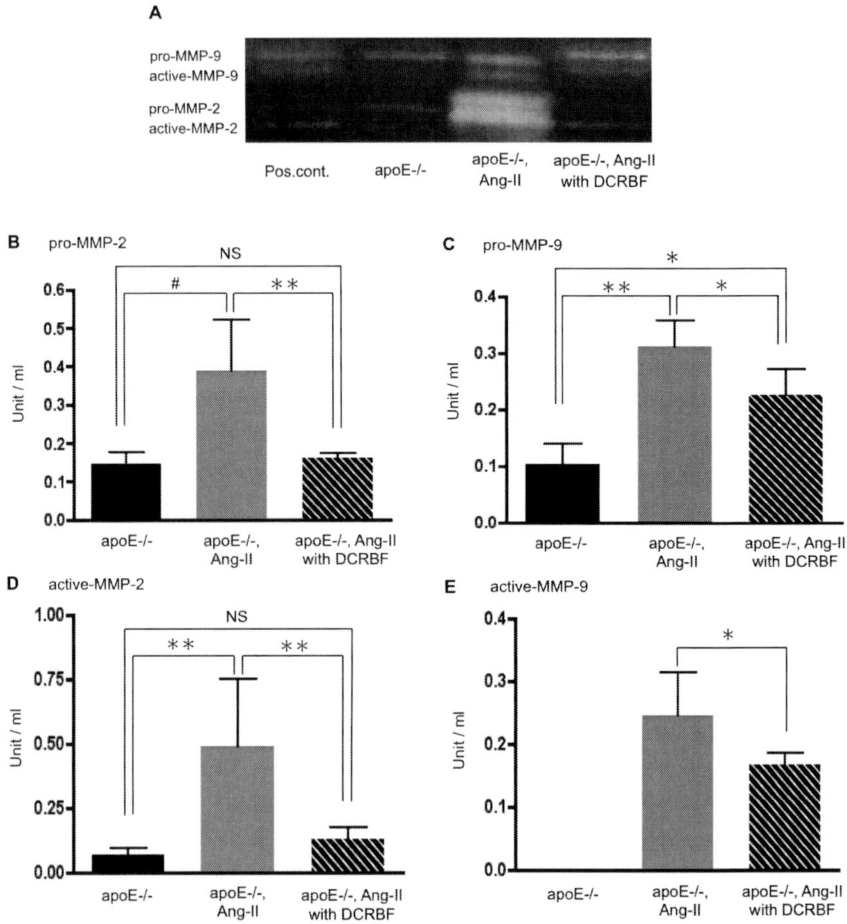

Fig. 12. Gelatin zymography for MMPs in the aortic tissue-treated or not treated with DCRBF *in vivo*. (A) Band intensity of MMP-2 and -9 (pro- and active-) were brighter in the apoE-/-, Ang-II group than in other groups. Concentration of MMPs (pro-MMP-2, -9 and active-MMP-2, -9) expression levels for the aortic tissues are indicated in (B) (C) (D) and (E). Concentration of MMPs in (apoE-/-, Ang-II) groups was significantly increased compared with the other two groups. $^*p<0.05$, $^{**}p<0.01$ compared with (apoE-/-, Ang-II) group (Yamawaki-Ogata et al. 2010).

Key Facts

- An electrospinning procedure can create nanofiber without large and/or complex systems, a technology that provides the basis for the tissue engineering scaffold and/or carrier of drug delivery.

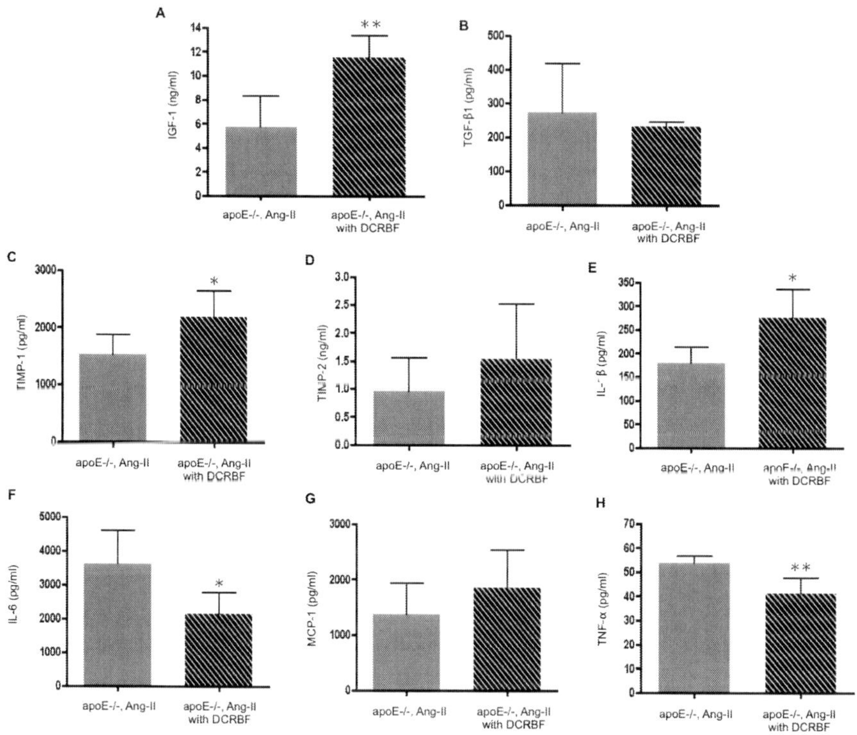

Fig. 13. Protein expression (growth factors, cytokines and chemokine) of the aortic tissues treated or not treated with DCRBF. Although there were no differences between (B) TGF-β1 and (C) TIMP-2 in each group, expressions of (A) IGF-1 and (E) IL-1β that promote elastin synthesis and expression of (C) TIMP-1 were significantly increased in the (apoE-/-, Ang-II with DCRBF) group compared with the (apoE-/-, Ang-II) group. Though there were no differences in (G) MCP-1 in either group, expressions of (F) IL-6 and (H) TNF-α that promote inflammatory reaction were significantly decreased in the (apoE-/-, Ang-II) with DCRBF group compared with the (apoE-/-, Ang-II) group. *$p<0.05$ and **$p<0.01$ compared with (apoE-/-, Ang-II) group (Yamawaki-Ogata et al. 2010).

- A drug delivery system using biodegradable nano scaled fiber can release the drug slowly and locally, to avoid adverse side effects and serve as a useful adjuvant treatment in association with surgical procedure.
- Sustained release of drugs can be achieved using electrospinning biodegradable nanofiber.
- An anastomotic stricture can be prevented by a drug that has cytostatic effects. Tacrolimus prevents anastmotic stricture by inhibition of mesenchymal cells.

- Growth of aortic aneurysms can be suppressed by inhibition of matrix metalloproteinase (MMP) secretion. Doxycycline is a candidate for treatment of aortic aneurysms by inhibiting MMP secretion.
- The sustained drug-releasing biodegradable nano-scaled fibers we developed have a cotton wool–like configuration suitable for any sites and situations.

Definitions

Angiotensin II: There is considerable evidence implicating angiotensin II (Ang II) in the atherogenic process. Ang II increases blood pressure.

Aortic aneurysm: A silent but life-threatening condition associated with aging and atherosclerosis. Abdominal aortic aneurysms affect 6–9% of men over 65 and progress over time, resulting in a rupture of diseased aorta, with a mortality rate of 80–90%. It is known that patients with a large aortic aneurysm (at least 5 cm in diameter in the abdominal aorta and 6 cm in diameter in the thoracic aorta) have an increased risk of rupture. The histological features of aortic aneurysm include chronic medial and adventitial inflammation with medial degeneration including smooth muscle cell apoptosis and excessive loss of extracellular matrix, especially extensive elastin fragmentation.

Apolipoprotein E knockout mouse: A mouse lacking the apolipoprotein E gene, which develops extensive atherosclerosis with lesions progressing from lipid-laden fatty streaks to advanced fibroproliferative lesions by age 20–30 wk.

C57BL6 mice: The inbred strain most widely used in research in cardiovascular biology, developmental biology, diabetes and obesity, genetics, immunology, neurobiology, and sensorineural research.

Cathepsins: Members of the cysteine protease family. Cathepsins are the most potent mammalian elastases and collagenases, widely expressed at basal levels in most tested tissues and cell types, and regulated by pro-inflammatory stimuli.

Chemokines: A superfamily of structurally related small chemotactic cytokines (most being 8–10 kDa) involved in leukocyte trafficking and activation. To date, nine receptors for CC chemokines (CCR1-9) and five for CXC chemokines (CXCR1-5) have been cloned and characterized.

Drug delivery system (DDS): A means of administering a pharmaceutical compound to achieve a therapeutic effect using the drug releasing profile, absorption, distribution and elimination.

Drug-eluting stent (DES): A coronary stent placed in diseased and narrowed coronary arteries. It slowly releases the drug to inhibit cell proliferation and thus prevent re-stenosis.

Doxycycline: A member of the tetracycline antibiotics group. Besides its use in the general indications for all members of the tetracycline antibiotics group, doxycycline is frequently employed to treat chronic prostatitis, sinusitis, syphilis, chlamydia, pelvic inflammatory disease, acne, rosacea, and Rickettsial infections.

Extracellular matrix (ECM): The extracellular part of animal tissue that usually provides structural support for scaffolding in the cell adhesion and plays the role of a retention growth factor. The ECM is composed of fibrous proteins (such as collagen, elastin, fibronectin and laminin) and proteoglycans (such as glycosaminoglycan, heparan sulfate, chondroitin sulfate and keratan sulfate).

Elastin: The most abundant extracellular protein in the aortic wall, which maintains the elasticity of the aortic wall. Elastin is secreted from vascular smooth muscle cells as tropoelastin protein molecules and synthesized insoluble elastin in extracellular space.

Inflammatory cytokines: Cytokines are a diverse group of soluble short-acting proteins, glycoproteins and peptides produced by various immune cells and vascular cells. Cytokines act in picomolar to nanomolar concentrations to activate specific receptors and modulate the functions of many cells and tissues. Inflammatory cytokines are produced predominantly by activated macrophages involved in the up-regulation of inflammation.

Intimal hyperplasia: The thickening of the intima of a blood vessel. It is induced by vascular injury after vascular anastomosis, intravascular treatment and endarterectomy. Intimal hyperplasia is the cause of bypass graft failure and re-stenosis of percutaneous coronary intervention.

Macrophages: White blood cells within tissues, produced by the differentiation of monocytes. Monocytes and macrophages are phagocytes, acting for non-specific defense (innate immunity) as well as to help initiate specific defense mechanisms (adaptive immunity) of vertebrate animals.

Matrix metalloproteinases (MMPs): A family of proteolytic enzymes that degrade various components of the ECM. Members of the MMP family include collagenases, gelatinases, stromelysins, matrilysins and membrane-type MMPs. Pro- and active-MMPs are cleaved into active forms that promote degradation of ECM proteins.

Pulmonary venous obstruction (PVO): The most fatal complication after TAPVR repair. Although the detailed mechanism of PVO is unknown, anastomotic stricture plays a pivotal role in its progression.

Smooth muscle cells: Cells that compose the tunica media of the wall of blood vessels. The smooth muscle layer acts as a functional unit to contract and relax as well as to support mechanical strength of the tissue.

Tacrolimus (FK506): An immunosuppressive drug that is the product of bacteria *Streptomyces tsukubaensis* discovered in Japan. It reduces interleukin-2 production by T-cells.

Tissue inhibitors of metalloproteinases (TIMPs): The matrix metalloproteinases are inhibited by specific endogenous TIMPs, which comprise a family of four protease inhibitors: TIMP1, TIMP2, TIMP3 and TIMP4.

Total anomalous pulmonary venous return (TAPVR): A cyanotic congenital heart defect. All four pulmonary veins do not return to the left atrium and have anomalous connections to systemic venous circulation. Most patients need surgical reconstruction within the first month of life; occasionally, an emergency operation is required because of obstruction of pulmonary venous drainage.

Summary Points

- This chapter introduces a novel DDS using biodegradable nanofiber in cardiovascular medicine.
- The electrospinning procedure can create nanofiber, which is used for the carrier in drug delivery, without the need for large and/or complex systems.
- Biodegradable nanofiber has great advantages to encapsulate various drug molecules and sustained release properties without systemic toxicity and hypersensitive reactions in the treatment of cardiovascular disease.
- TEBN reduces neo-intimal hyperplasia and preserves endothelialization in the animal model, suggesting that it is a promising material for vascular anastomosis.
- DCRBF improves the balance of ECM degradation/synthesis in the wall of aortic aneurysms and suppresses the progression of aneurysms without side effects in an animal model. It thus might become a new alternative therapeutic strategy for treating aortic aneurysms.

Abbreviations

Ang-II	:	angiotensin-II
apoE-/-	:	apolipoprotein E-deficient
ASMA	:	alpha-smooth muscle actin
DCRBF	:	doxycycline controlled-release biodegradable nano-scaled fiber
DDS	:	drug delivery system

DES : drug-eluting stent
DOXY : doxycycline
ECM : extracellular matrix
H&E : hematoxylin-eosin
IGF-1 : insulin-like growth factor-1
IL : interleukin
LOX : lysyl oxidase
MCP-1 : monocyte chemoattractant protein-1
MMP : matrix metalloproteinase
PCL : poly caprolactone
PGA : polyglycolic acid
PLA : polylactic acid
SMC : smooth muscle cell
SMemb : smooth muscle myosin heavy chain embryo
TEBN : tacrolimus-eluting biodegradable nano-scaled fiber
TGF-β1 : transforming growth factor-β1
TIMP-1 : tissue inhibitor matrix metalloproteinase-1
TNF-α : tumor necrosis factor-α

References

Abdul-Hussien, H., R. Hanemaaijer, J.H. Verheijen, J.H. van Bockel, R.H. Geelkerken and J.H. Lindeman. 2009. Doxycycline therapy for abdominal aneurysm: Improved proteolytic balance through reduced neutrophil content. J. Vasc. Surg. 49: 741–749.

Baxter, B.T., W.H. Pearce, E.A. Waltke, F.N. Littooy, J.W. Jr. Hallett, K.C. Kent, G.R. Jr. Upchurch, E.L. Chaikof, J.L. Mills, B. Fleckten, G.M. Longo, J.K. Lee and R.W. Thompson. 2002. Prolonged administration of doxycycline in patients with small asymptomatic abdominal aortic aneurysms: report of a prospective (Phase II) multicenter study. J. Vasc. Surg. 36: 1–12.

Bolgen, N., I. Vargel, P. Korkusuz, Y.Z. Menceloglu and E. Piskin. 2007. *In vivo* performance of antibiotic embedded electrospun PCL membranes for prevention of abdominal adhesions. J. Biomed. Mater. Res. B Appl. Biomater. 81: 530–543.

Buschle-Diller G., J, Cooper, Z. Xie, Y. Wu, J. Waldrup and X. Ren. 2007. Release of antibiotics from electrospun bicomponent fibers. Cellulose 14: 553–562.

Caldarone, C.A., H.K. Najm, M. Kadletz, J.F. Smallhorn, R.M. Freedom, W.G. Williams and J.G. Coles. 1998. Surgical management of total anomalous pulmonary venous drainage: impact of coexisting anomalies. Ann. Thorac. Surg. 66: 1521–1526.

Cazalis, J., C. Bodet, G. Gagnon and D. Grenier. 2008. Doxycycline reduces lipopolysaccharide-induced inflammatory mediator secretion in macrophage and *ex vivo* human whole blood models. J. Periodontol. 79: 1762–1768.

Conn, K.J., C.B. Rich, D.E. Jensen, M.R. Fontanilla, M.M. Bashir, J. Rosenbloom and J.A. Foster. 1996. Insulin-like growth factor-I regulates transcription of the elastin gene through a putative retinoblastoma control element. A role for Sp3 acting as a repressor of elastin gene transcription. J. Biol. Chem. 271: 28853–28860.

Ernst, C.B. 1993. Abdominal aortic aneurysm. N. Engl. J. Med. 328: 1167–1172.

Garcia, R.A., D.P. Pantazatos, C.R. Gessner, K.V. Go, V.L. Woods Jr. and F.J. Villarreal. 2005. Molecular interactions between matrilysin and the matrix metalloproteinase inhibitor doxycycline investigated by deuterium exchange mass spectrometry. Mol. Pharmacol. 67: 1128–1136.

Goldberg, M., R. Langer and X. Jia. 2007. Nanostructured materials for applications in drug delivery and tissue engineering. J. Biomater. Sci. Polym. Ed. 18: 241–268.

Jiang, H., D. Fang, B. Hsiao, B. Chu and W. Chen. 2004. Preparation and characterization of ibuprofen-loaded poly(lactide-co-glycolide)/poly(ethylene glycol)-g-chitosan electrospun membranes. J. Biomater. Sci. Polym. Ed. 15: 279–296.

Kapadia, M.R., D.A. Popowich, and M.R. Kibbe. 2008. Modified prosthetic vascular conduits. Circulation 117: 1873–188.

Katti, D.S., K.W. Robinson, F.K. Ko and C.T. Laurencin. 2004. Bioresorbable nanofiber-based systems for wound healing and drug delivery: optimization of fabrication parameters. J Biomed Mater Res B Appl Biomater. 2004. 70: 286–296.

Kenawy, R., G.L. Bowlin, K. Mansfield, J. Layman, D.G. Simpson, E.H. Sanders and G.E. Wnek. 2002. Release of tetracycline hydrochloride from electrospun poly(ethylene-co-vinylacetate), poly(lactic acid), and a blend. J. Control Release 81: 57–64.

Kim, K., Y.K. Luu, C. Chang, D. Fang, B.S. Hsiao, B. Chu and M. Hadjiargyrou. 2004. Incorporation and controlled release of a hydrophilic antibiotic using poly(lactide-co-glycolide)-based electrospun nanofibrous scaffolds. J. Control Release 98: 47–56.

Lanza, G., P. Winter., T. Cyrus, S. Caruthers, J. Marsh, M. Hughes and S. Wickline. 2006. Nanomedicine opportunities in cardiology. Ann. NY Acad. Sci. 1080: 451–465.

Liang, D., B.S. Hsiao and B. Chu. 2007. Functional electrospun nanofiber scaffolds for biomedical applications. Adv. Drug Deliv. Rev. 59: 1392–1412.

Liu, J.M., and J.M. Davidson. 1988. The elastogenic effect of recombinant transforming growth factor-beta on porcine aortic smooth muscle cells. Biochem. Biophys. Res. Commun. 154: 895–901.

Mutsuga, M., Y. Narita, A. Yamawaki, M. Satake, H. Kaneko, Y. Suematsu, A. Usui and Y. Ueda. 2009a. A new strategy for prevention of anastomotic stricture using tacrolimus-eluting biodegradable nanofiber. J. Thorac. Cardiovasc. Surg. 137: 703–709.

Mutsuga, M., Y. Narita, A. Yamawaki, M. Satake, H. Kaneko, A. Usui and Y. Ueda. 2009b. Development of novel drug-eluting biodegradable nano-fiber for prevention of post operative pulmonary venous obstruction. Interact. Cardiovasc. Thorac. Surg. 8: 402–407.

Pyo, R., J.K. Lee, J.M. Shipley, J.A. Curci, D. Mao, S.J. Ziporin, T.L. Ennis, S.D. Shapiro, R.M. Senior and R.W. Thompson. 2000. Targeted gene disruption of matrix metalloproteinase-9 (gelatinase B) suppresses development of experimental abdominal aortic aneurysms. J. Clin. Invest. 105: 1641–1649.

Sill, T.J. and H.A. von Recum. 2008. Electrospinning: Applications in drug delivery and tissue engineering. Biomaterials 29: 1989–2006.

Thompson, R.W., P.J. Geraghty and J.K. Lee. 2002. Abdominal aortic aneurysms: basic mechanisms and clinical implications. Curr. Prob. Surg. 39: 110–230.

Vasita, R. and D.S. Katti. 2006. Nanofibers and their applications in tissue engineering. Int. J. Nanomed. 1: 15–30.

Venugopal, J., M.P. Prabhakaran, S. Low, A.T. Choon, Y.Z. Zhang, G. Deepika and S. Ramakrishna. 2008. Nanotechnology for nanomedicine and delivery of drugs. Curr. Pharm. Des. 14: 2184–2200.

Verreck, G., I. Chun, J. Peeters, J. Rosenblatt and M.E. Brewster. 2003. Preparation and characterization of nanofibers containing amorphous drug dispersions generated by electrostatic spinning. Pharm. Res. 20: 810–817.

Xie, J. and C.H. Wang. 2006. Electrospun micro- and nanofibers for sustained delivery of paclitaxel to treat C6 glioma in vitro. Pharm. Res. 23: 1817–1826.

Xu, X., X. Chen, T. Lu, X. Wang, L. Yang and X. Jing. 2006. BCNU-loaded PEG-PLLA ultrafine fibers and their in vitro antitumor activity against Glioma C6 cells. J. Control Release 114: 307–316.

Yamawaki-Ogata, A., R. Hashizume, M. Satake, H. Kaneko, S. Mizutani, T. Moritan, Y. Ueda and Y. Narita. 2010. A doxycycline loaded, controlled-release, biodegradable fiber for the treatment of aortic aneurysm. Biomaterials 31: 9554–9564.
Zeng, J., X. Xu, X. Chen, Q. Liang, X. Bian, L. Yang and X. Jing. 2003. Biodegradable electrospun fibers for drug delivery. J. Control Release 92: 227–231.

19

Europium Hydroxide Nanorods and Angiogenic Processes

Chitta Ranjan Patra

ABSTRACT

Nanobiotechnology is the application of nanotechnology in biology and medicine that will hopefully lead to useful research tools, advanced drug delivery systems, and new ways to detect and treat disease or repair damaged tissues and cells. Among nanoparticles, lanthanide salts, chelates, nanoparticles and nanorods have been widely used for diagnostic and therapeutic applications in biomedical nanotechnology because of their fluorescence and pro-angiogenic properties to endothelial cells and *in vivo* model. Angiogenesis is the physiological process involving the growth of new blood vessels from existing vessels. In this chapter, we discuss the pro-angiogenic properties of europium (III) hydroxide [$Eu^{III}(OH)_3$] nanorods, synthesized by the microwave technique and characterized by several physicochemical techniques. These nanorods may be used as alternative pro-angiogenic agents of vascular endothelial growth factor A and basic fibroblast growth factor in *in vitro* and *in vivo* model. Finally, we have demonstrated the non-toxic behavior of $Eu^{III}(OH)_3$ nanorods by *in vitro* and *in vivo* model. We also discuss here the possible molecular

Chemical Biology Laboratory, Indian Institute of Chemical Technology, Uppal Road, Hyderabad - 500007, AP, INDIA; E-mail: CRPatra@iict.res.in

List of abbreviations after the text.

mechanism of inorganic nanorods–induced angiogenesis. Considering the overall impact of angiogenesis in our health care system, the discovery of pro-angiogenic properties of $Eu^{III}(OH)_3$ nanorods could be used to develop new treatment strategies for cardiovascular diseases. Additionally, fluorescent properties of lanthanide nanorods offer useful and alternative inorganic fluorescent probes for targeting various molecules in living cells because of their non-toxic behavior.

INTRODUCTION

Angiogenesis is a physiological process that involves the growth of new blood vessels from existing vasculature. It is a complex process involving both pro-angiogenic and anti-angiogenic factors (Folkman 1995; Risau 1997; Yancopoulos et al. 2000). Angiogenesis plays an important role in physiological processes such as the menstrual cycle or embryonic development, and in pathologic conditions (e.g., wound healing, proliferative diabetic retinopathy, atherosclerosis, and post-ischemic vascularization of the myocardium, tumor growth and metastasis, rheumatoid arthritis) (Folkman 1995; Risau 1997). However, one of the most essential roles of angiogenesis has also been used to enhance blood flow through collateral blood vessel formation in patients with cardiovascular diseases (CVD) including ischemic heart disease (IHD) and peripheral vascular diseases. Cardiovascular ischemia is a leading cause of morbidity and mortality in the United States and, in fact, all over the world (Huston et al. 1999).

In this context, several strategies including prevention through diet and exercise, lipid-lowering drugs, medical intervention via angioplasty, stent placement, surgery, and pro-angiogenic factors such as VEGF-A and bFGF (Folkman 1995; Risau 1997) have been used to enhance blood flow in ischemic tissues via formation of collateral blood vessels that have also been associated with pathological angiogenesis, thrombosis, fibrosis and/or the proliferation of tumor cells (Claffey et al. 1996). The latter observation warrants the further development of novel pro-angiogenic as well as anti-angiogenic molecules that are very important for ischemic diseases where nanotechnology can play a pivotal role (Claffey et al. 1996). This, therefore, necessitates identification of novel compounds that may specifically promote angiogenesis in ischemic tissues without affecting the process of angiogenesis in other pathological conditions. In this context, we have found the pro-angiogenic properties of europium (III) hydroxide [$Eu^{III}(OH)_3$] nanorods using several *in vitro* analyses and also by chick chorioallantoic membrane (CAM) assays. These inorganic nanorods are pro-angiogenic like

VEGF-A and bFGF. The most important observation is that these nanorods are non-toxic to *in vitro* (endothelial cells) and *in vivo* model. The idea of $Eu(OH)_3$ nanorod-induced angiogenesis was first reported by Patra et al. (2008b). Therefore, this technique can be described as state-of-the-art. We strongly believe that the pro-angiogenic properties of $[Eu^{III}(OH)_3]$ nanorods can be used to develop new treatment strategies for CVD including IHD, limb ischemic disease and peripheral vascular diseases. Additionally, these inorganic nanorods have fluorescent properties that can be used as inorganic fluorescent label in live cells as it is non-toxic.

U.S. MARKET VALUE FOR TREATMENT OF PATIENTS WITH CARDIOVASCULAR DISEASE

The development of drugs that stimulate revascularization of ischemic tissues remains an exciting but unrealized goal in cardiovascular therapeutics. The unmet medical need for such a therapy is considerable. The global cardiovascular therapeutics market was valued at $76 billion in 2008 and expected to fall to $67 billion with a compound annual growth rate of –1.7% by 2015 (Market Research Report: Global Information, Inc. http://www.the-infoshop.com/report/gbi108042-cardiovascular.html). Approximately 127 million patients received pharmacological treatment for cardiovascular diseases.

VALUE OF NANOTECHNOLOGY IN BIOLOGY AND MEDICINE

The dimension of nanoscale devices is similar to cellular components, such as DNA (width of a DNA ~ 2.5 nm) and proteins (dimensions of most proteins ~ 1–20 nm) (Bertozzi and National Institutes of Health (U.S.) 2005); tools developed through nanotechnology may be used to detect or monitor several diseases at the molecular level (El-Sayed et al. 2005). Because of their small size compared to human cells (100 to 10,000 times as small as human cells), nanoparticles can enter the cells and readily interact with bio-molecules/proteins on both the cell surface and inside.

$[Eu^{III}(OH)_3]$ NANORODS AND THEIR APPLICATION IN ANGIOGENESIS

Importance of $Eu^{III}(OH)_3$ Nanorods in Angiogenesis

Earlier reports suggest that transition elements/nanoparticles (Va, Mg, Fe, Zn, Cu, Mn, Mo) and lanthanide elements/nanoparticles (Eu, Tb, Gd) play important roles in biology and medicine (Patra et al. 2005, 2006, 2007,

2008a, b, c, 2009; Thompson and Orvig 2006). In this context, lanthanide-based inorganic fluorescents, especially Eu- and Tb-phosphate/oxide/hydroxide nanoparticles/nanorods, have attracted a great deal of attention in cell biology and medicine for their fluorescent, angiogenic properties and non-toxic effect (Patra et al. 2005, 2006, 2007, 2008a, b, c, 2009). For the first time, we have shown that inorganic lanthanide phosphate fluorescent nanorods can be used as fluorescent labels in cell biology (Patra et al. 2006, 2007). Toxicity study of nanomaterials (nanospheres or nanorods) is presently a topic of considerable importance, so we have carried out the toxicity study of $Eu^{III}(OH)_3$ nanorods *in vitro* as well as in *in vivo* model and demonstrated for the first time the pro-angiogenic properties (Risau 1997) of $[Eu^{III}(OH)_3]$ nanorods (Patra et al. 2008b). Considering the critical role of angiogenesis and its impact on CVD, we strongly believe that $Eu^{III}(OH)_3$ nanorod-induced angiogenesis could be used to develop new treatment strategies for cardiovascular ischemic and peripheral vascular diseases in the near future. In this chapter, we discuss the synthesis of $Eu^{III}(OH)_3$ nanorods, their pro-angiogenic properties, molecular mechanism and toxicity study *in vivo* model and their future perspectives.

Synthesis of $Eu^{III}(OH)_3$ Nanorods in Domestic Microwave Oven

$Eu^{III}(OH)_3$ nanorods were fabricated in a domestic microwave oven by the interaction of an aqueous solution containing europium (III) nitrate and aqueous ammonium hydroxide (NH_4OH) (Patra et al. 2005, 2006) according to the following equation.

$$Eu(NO_3)_3 + aq.NH_4OH + \text{MW-Heating} \longrightarrow Eu(OH)_3\,[\text{Nanorods}] + NH_3\uparrow$$

Characterization

The crystal structure, morphologies, and chemical nature of microwave-assisted as-synthesized materials were characterized by different physicochemical techniques such as transmission electron microscopy (TEM), X-ray diffraction (XRD), thermo-gravimetric analysis (TGA), and differential scanning calorimetric (DSC) analysis, respectively (Patra et al. 2005, 2006). The results from XRD, TEM, TGA, and DSC indicate that the microwave-assisted as-synthesized $[Eu^{III}(OH)_3]$ products are hexagonal (Fig. 1), with nanorods having diameter 35–50 nm and length 200–300 nm.

Advantages of Microwave Heating

Microwave heating has several advantages over conventional heating. It does not need high temperatures (though local high temperatures are obtained), high pressures, catalyst, templates, surfactants, vacuum

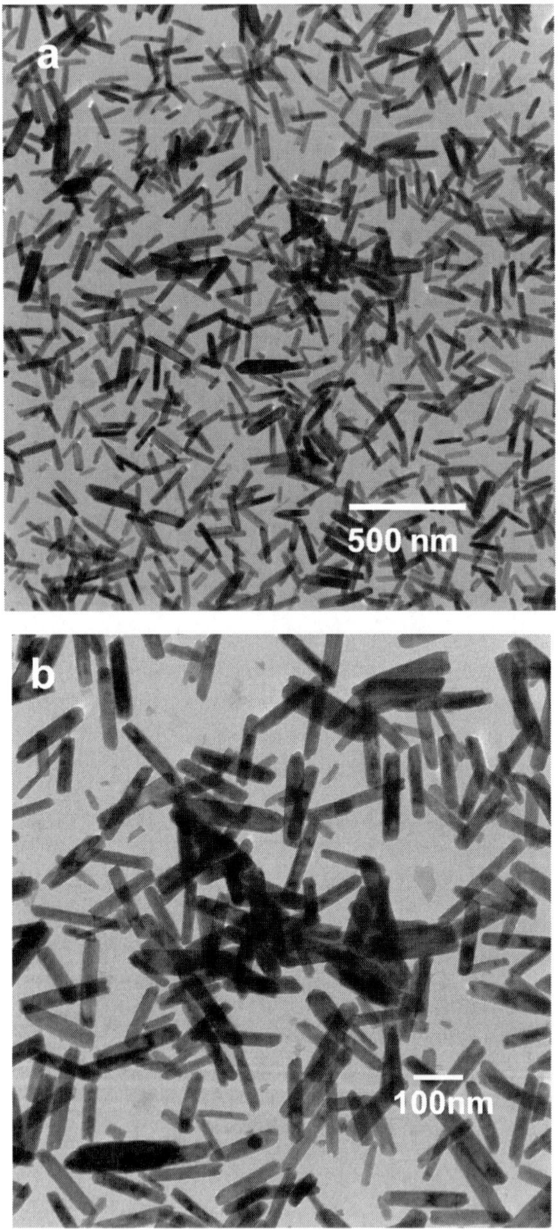

Fig. 1. TEM image of microwave-assisted as-synthesized $Eu^{III}(OH)_3$ nanorods. The pictures show that as-synthesized materials consist entirely of nanorods with diameter 35–50 nm and length 200–300 nm. Reprinted with permission from Patra et al. (2009). Copyright ©2009, Elsevier Limited.

conditions or preprocessing. The microwave-assisted products obtained are pure with high yields and are structurally uniform and well crystallized (Patra et al. 2005, 2006, 2009). Finally, it is simple, fast, clean, efficient, economical, and non-toxic (Patra et al. 2005, 2006).

Cell Proliferation Assay

The angiogenesis process contains several steps such as the stimulation of endothelial cells by growth factors, degradation of the extracellular matrix by proteolytic enzymes, migration and endothelial cell proliferation, and finally, capillary tube formation. Among these steps, endothelial cell proliferation is one of the most critical steps in angiogenesis (Risau 1997). Growth factors such as VEGF play a major role in the regulation of angiogenesis as well as tumorigenesis (Risau 1997). Interestingly, we have investigated the pro-angiogenic properties of $Eu^{III}(OH)_3$ nanorods such as VEGF-A and bFGF nanorods by several *in vitro* and *in vivo* assays including cell proliferation assay.

To evaluate the proliferation of human umbilical vein endothelial cells (HUVEC), we performed a [^{3}H]-thymidine incorporation assay in presence and absence of $Eu^{III}(OH)_3$ nanorods, using VEGF as a positive control experiment (Patra et al. 2008b). These crystalline nanorods promote a dose-dependent increase in HUVEC proliferation as compared to non-treated cells, whereas amorphous $Eu^{III}(OH)_3$, other lanthanide hydroxide, and oxide materials did not induce significant proliferation (data not shown). As expected, HUVECs treated with VEGF (10 ng/ml) also resulted in increased proliferation (Fig. 2a). Hence, our results clearly show that $Eu^{III}(OH)_3$ nanorods, although inorganic, function like other pro-angiogenic cytokines such as VEGF.

Cell Cycle Assay

The cell cycle in eukaryotes is broadly classified into four phases: G1, preparation of the chromosomes for replication; S, synthesis of DNA and centrosomes; G2, preparation for mitosis; and M, mitosis (Patra et al. 2008b). Cell cycle analysis was performed in order to determine $Eu^{III}(OH)_3$-induced EC proliferation (Fig. 2a). Cell cycle analysis using PI staining in HUVECs revealed a significant increase in the percentage of cells in the S-phase when treated with 50 μg/ml of $Eu^{III}(OH)_3$ nanorods as compared with that of untreated control cells (Fig. 2d). These data further support the results obtained from the cell proliferation assays (Fig. 2a).

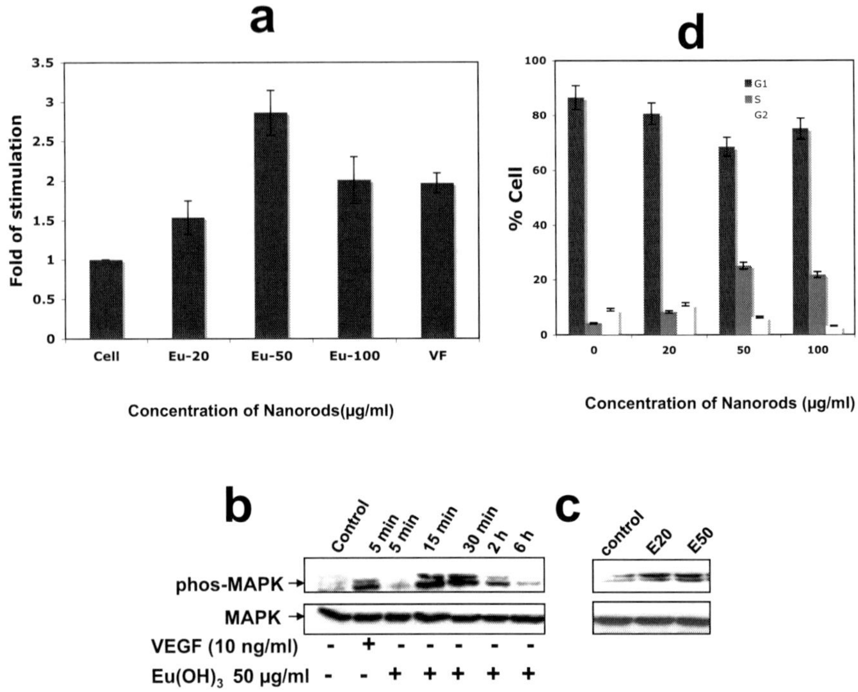

Fig. 2. Effect of $Eu^{III}(OH)_3$ nanorods on cell proliferation assay, cell cycle assay and MAPK activation. (a) Effect of $Eu^{III}(OH)_3$ nanorods [Eu-20, -50, -100 indicates 20, 50 and 100 µg/ml and VF = VEGF (10 ng/ml)] on HUVECs, observed by cell proliferation assay, represented as fold stimulation. (b) Western blot analysis for phospho-mapkinase and total mapkinase from lysate of (i) untreated control HUVECs, HUVECs treated with (ii) $Eu^{III}(OH)_3$ nanorods of 50 µg/ml between 5 min and 6 h, and (iii) VEGF (10 ng/ml) for 5 min. (c) HUVECs treated with $Eu^{III}(OH)_3$ nanorods at different concentrations for 24 h. (d) Cell cycle analysis of HUVECs treated with Eu^{III} $(OH)_3$ nanorods (0–100 µg/ml). The data are statistically significant where $p \leq 0.05$ (mean ± SD of three separate experiments performed in triplicates). Reprinted with permission from Patra et al. (2008b). Copyright ©2008, Wiley-VCH Verlag GmbH & Co. KGaA.

Internalization of $Eu(OH)_3$ Nanorods

A number of methods, such as differential interference contrast microscopy, confocal microscopy and TEM, were used to determine cellular trajectories of nanorods (Patra et al. 2006). The direct proof of internalization of nanorods is the TEM images of endothelial cells treated with $Eu(OH)_3$ nanorods. The TEM images of HUVEC (after cross-section) treated with $Eu(OH)_3$ nanorods (50 µg/mL) at different magnifications indicated cytoplasmic internalization of the nanorods inside the cells (Patra et al. 2008b). We expect that internalization occurred at double-membrane layered early endosome-like compartments away from the nucleus or late endosome-like structures close to the nucleus. However, the nanorods appear to be amorphous after internalization. This might be due to low pH (~3.5) inside ER where these nanorods get internalized (Patra et al. 2008b).

MAPK Activation (Signaling Pathway)

According to published literature, mitogen-activated protein kinase (MAPK) activation is one of the key signaling pathways for angiogenesis (Patra et al. 2008b; Zeng et al. 2001). In order to confirm that, western blot analysis for phospho-MAPK of HUVEC treated with and without $Eu^{III}(OH)_3$ nanorods was performed in a dose-dependent (20–50 µg/mL) (Fig. 2b) and time-dependent manner (5 min to 6 h) (Fig. 2c). As a positive control, HUVEC cells were treated with VEGF at a concentration of 10 ng/ml for 5 min (Patra et al. 2008b). Western blot analysis for phospho-MAPK of HUVEC treated with and without $Eu^{III}(OH)_3$ nanorods between 15 and 30 min is comparable to VEGF-treated 5 min sample. Again, MAPK phosphorylation was increased and reached maximum at 50 µg/ml in a dose-dependent concentration of $Eu^{III}(OH)_3$ nanorods (20–50 µg/ml). These results support the cell proliferation analysis and cell cycle assay at 50 µg/ml concentration.

CAM Assay (*In Vivo* Assay)

According to *in vitro* observation of angiogenesis, we hypothesize that the inorganic nanorods should work in *in vivo* model. To determine the *in vivo* relevance of our findings, we performed chick CAM assays to measure nanoparticle-induced *in vivo* angiogenesis and vascular sprouting (Patra et al. 2008b). Treatment with Eu^{III}(OH)3 nanorods (Fig. 3c and d) in different concentrations (1 µg and 10 µg) promotes significant vascular sprouting (at higher magnification) as compared to that of CAMs treated with TE buffer (negative control experiment) (Fig. 3a) and VEGF (positive control experiment, Fig. 3b). Angiogenesis was quantified by counting branch points arising from tertiary vessels (Fig. 3e), which indicates that nanorods have unique angiogenic properties not only *in vitro* but also *in vivo*.

ROS-dependent Angiogenesis (Molecular Mechanism)

Recent reports suggest that reactive oxygen species (ROS) play an important role in angiogenesis (Maulik 2002; Ushio-Fukai 2006). In endothelial cells NADPH oxidase is the major source of intracellular ROS such as super oxide anion (O_2•–), hydroxyl radical (OH), and hydrogen peroxide (H_2O_2) that are produced primarily in cells as a by-product of normal cellular metabolism during the conversion of molecular oxygen (O_2) to

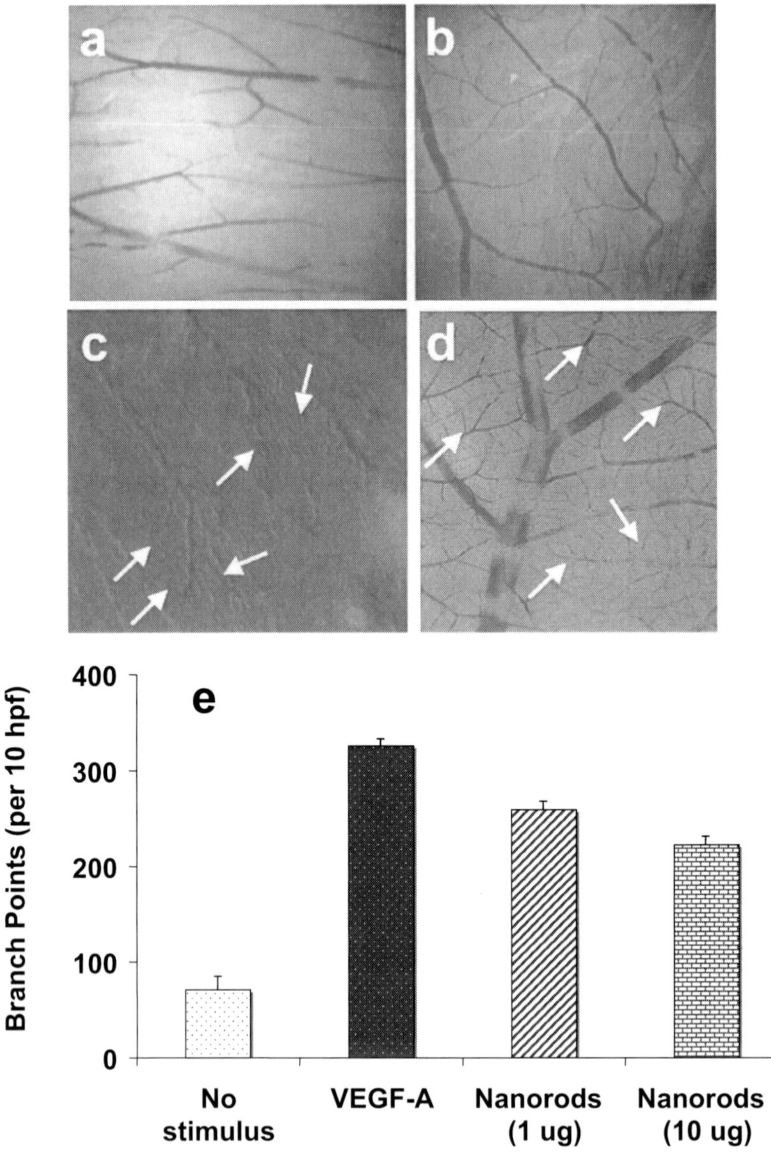

Fig. 3. Inorganic nanorods induced *in vivo* angiogenesis observed by CAM assay. CAMs treated with (a) TE (tris-EDTA) buffer, (b) VEGF (50 ng), (c, d) 1 µg and 10 µg of nanorods in TE buffer. (e) Angiogenesis was quantified by counting branch points arising from tertiary vessels from a minimum of 10 specimens from three separate experiments. Reprinted with permission from Patra et al. (2008b). Copyright ©2008, Wiley-VCH Verlag GmbH & Co. KGaA.

Color image of this figure appears in the color plate section at the end of the book.

water molecule (H_2O) (Katusic 1996). In endothelial cells super oxide (O_2•−) and hydrogen peroxide (H_2O_2) are two of the most biologically important ROS that stimulate the cell proliferation and migration, basic and key steps for angiogenesis in the cardiovascular system (Maulik 2002). On the other hand, hydroxyl radical is toxic, and it does not help angiogenesis. It should be noted that ROS at low concentration is pro-angiogenic and at high concentration it is toxic (Fig. 4). This information may be useful to understand the new mechanism of inorganic nanorod-induced angiogenesis and to develop new alternate therapeutic strategies for cardiovascular disease by targeting ROS signaling in the future (Fig. 4).

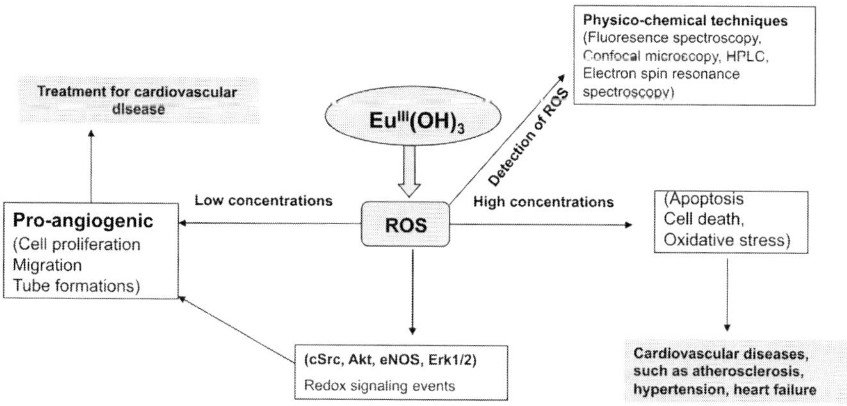

Fig. 4. Generation of ROS and their role in angiogenesis and in signaling pathways. Schematic diagram for the generation of ROS from $Eu^{III}(OH)_3$ nanorods, their role as pro-angiogenic and anti-angiogenic depending on the concentration, their detection by several physicochemical methods, and their role in redox-sensitive signaling pathways.

ROS Generation in the Biological System

ROS is a collective term that includes both oxygen radicals and certain non-radicals that are oxidizing agents and/or are easily converted into radicals (O_2•−, H_2O_2, OH•). Molecular oxygen is a biradical reduced successively by four one-electron transfer reactions via the super oxide anion radical (O_2•−) and hydrogen peroxide (H_2O_2) to the hydroxyl radical (OH•) and finally to water (H_2O) (Simon and Brömmeb 2005).

$O_2 + e^- \rightarrow O_2$•− (super oxide radical anion)

O_2•− $+ e^- \rightarrow O_2^{2-} + 2H^+ = H_2O_2$ (hydrogen peroxide)

$H_2O_2 + e^- \rightarrow$ OH• $+ OH^-$ (hydroxyl radical and hydroxide anion)

OH• $+ e^- + H^+ \rightarrow H_2O$

Apart from that, ROS are produced in vascular cells by a number of oxidases, including the NADPH oxidases (Nox) and xanthine oxidase (Maulik 2002; Ushio-Fukai 2006).

Metal Ion-induced ROS Formations

There is one possibility for the generation of (OH•) by transition metal ions (Simm and Brömmeb 2005; Stohs and Bagchi 1995)as follows:

$Fe^{2+} + H_2O_2 \rightarrow OH\bullet + OH^- + Fe^{3+}$ (Fenton chemistry) (Simm and Brömmeb 2005)

$Fe^{3+} + OH\bullet + OH^- \rightarrow Fe^{2+} + H_2O_2$ (Stohs and Bagchi 1995)

$Cu^+ + H_2O_2 \rightarrow Cu^{2+} + OH\bullet + OH$ (Stohs and Bagchi 1995)

$2O_2\bullet- +2\,H^+ \rightarrow O_2 + H_2O_2$ (Stohs and Bagchi 1995)

Our previous results demonstrated that $Eu^{III}(OH)_3$ nanorods show significant pro-angiogenic activity to endothelial cells as observed by the cell proliferation assay using ^{3}H-thymidine. We therefore hypothesized that $Eu^{III}(OH)_3$ nanorods may be the source of ROS, especially super oxide ($O_2\bullet-$) and hydrogen peroxide (H_2O_2), which is responsible for nanorod-induced angiogenesis (Fig. 4). Since hydroxyl radical is toxic and it does not help cell proliferation or migration, we assume that in our case hydroxyl radical is not involved in nanorod-induced angiogenesis. However, the mechanism for the formation of ROS from $Eu^{III}(OH)_3$ nanorods is yet unknown. We speculate the following metal ion–induced chemical reaction for the generation of H_2O_2 from $Eu^{III}(OH)_3$ nanorods using Fenton chemistry:

$Eu^{3+} + OH\bullet + OH^- \rightarrow Eu^{2+} + H_2O_2$ (Fenton chemistry)

In light of reports (Maulik 2002; Ushio-Fukai 2006) regarding ROS-mediated angiogenesis, we investigated the formation of ROS in HUVECs in presence and absence of nanorods. Control untreated HUVECs show much weaker green fluorescence due to formation of endogenous ROS (Fig. 5a-c). The green fluorescence as seen in Fig. 5d-f indicates that the production of ROS inside HUVECs induced with 100 μM of tert-butyl hydro peroxide (TBHP) for 1h served as a positive inducer. Figure 4g-i and j-l indicates the generation of ROS in the presence of $Eu^{III}(OH)_3$ nanorods at the concentrations of 20 and 50 μg/ml, respectively, suggesting nanorods induce angiogenesis most likely because of increased production of ROS. Most recent results observed by our group indicate that formation of super oxide ($O_2\bullet-$) and hydrogen peroxide (H_2O_2), are responsible for nanorod-induced angiogenesis in HUVECs and *in vivo* model (unpublished data).

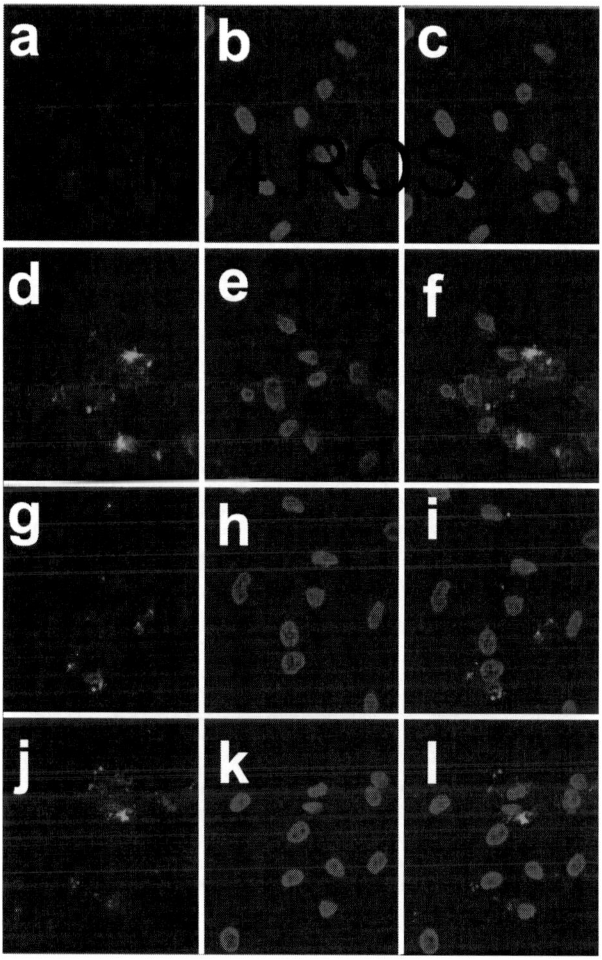

Fig. 5. Generation of ROS in HUVECs in presence of $Eu^{III}(OH)_3$ nanorods. ROS are produced during the endothelial cell proliferation in presence of $Eu^{III}(OH)_3$ nanorods at different concentrations. (a-c) control untreated HUVECs, (d-f) HUVECs treated with 100 μM/ml of tert-butyl hydro peroxide (TBHP) as positive ROS inducer. (g-l) HUVECs treated with 20 μg/ml (g-i) and 50 μg/ml (j-l) of $Eu^{III}(OH)_3$ nanorods. Reprinted with permission from Patra et al. (2008b) Copyright ©2008, Wiley VCH Verlag GmbH & Co. KGaA.

Color image of this figure appears in the color plate section at the end of the book

Toxicity Study of $Eu^{III}(OH)_3$ Nanorods (*In Vitro*)

Applications of any nanoparticles and nanorods in biology and medicine require that the nanoparticles/nanorods should be completely non-toxic *in vivo* (Hirano and Suzuki 1996; Ogawa et al. 1995). In this context, we demonstrated that $Eu^{III}(OH)_3$ nanorods are non-toxic to endothelial cells, observed by *in vitro* apoptosis assay (Ogawa et al. 1995; Patra et al. 2005, 2006, 2007, 2008a, b, c, 2009).

Apoptosis Assay (dUTP Nick End Labeling-TUNEL Assay)

We performed the TUNEL assay on HUVECs in the presence and absence of $Eu^{III}(OH)_3$ nanorods in order to investigate the *in vitro* toxicity. Results from the TUNEL assay clearly demonstrated that at concentrations up to 50 µg/ml, there was no induction of apoptosis in HUVECs (Patra et al. 2008b). Cell viability test also indicates that these nanorods do not affect cell viability (Patra et al. 2008b, 2009).

Toxicity Study in Mouse Model

In vivo toxicity study of $Eu^{III}(OH)_3$ is one of the important steps before applying these nanomaterials for *in vivo* use (Patra et al. 2009) as nanoparticles may interact with biological systems in unforeseen ways. We reported an *in vivo* toxicity study using $Eu^{III}(OH)_3$ nanorods in a systematic way (Patra et al. 2008a, 2009). $Eu^{III}(OH)_3$ nanorods suspension in TE buffer with different concentrations (1.25 $mgkg^{-1}day^{-1}$, 12.5 $mgkg^{-1}day^{-1}$ and 125 $mgkg^{-1}day^{-1}$) was administered to each mouse through intraperitoneal (IP) injection every day with the help of a tuberculin syringe over a period of 7 d. We examined *in vivo* toxicity of $Eu^{III}(OH)_3$ nanorods and their bio-availabilities in mouse models over the short term (mice sacrificed on day 8 after 7 d of consecutive IP injections of nanorods) (acute) and long term (mice sacrificed on day 60) (chronic). IP injection of $Eu^{III}(OH)_3$ nanorods in mice at different doses (1.25 to 125 $mgkg^{-1}day^{-1}$) has not shown biochemical or hematological toxicities, with the exception of a slight elevation of liver enzymes (Tables 1 and 2). Blood tests done on day eight (short term) of the acute toxicity experiments do not suggest significant systemic toxicity at any of the dose levels (Tables 1 and 2), except the slight elevation of liver enzymes, especially AST, for only the short-term toxicity study. In particular, even at the highest dose levels (dose 125 $mgkg^{-1}day^{-1}$), inorganic nanorods do not cause anemia, neutropenia, lymphopenia, or thrombocytopenia. That suggests there is no hematological toxicity (Tables 1 and 2). Again, for the long-term toxicity study also we did not find any significant difference of ALT and AST values between control untreated mice and nanorod-treated mice even at the high dose of 125 $mgkg^{-1}day^{-1}$ (Patra et al. 2009).

We collected the various vital organs (liver, spleen, lungs, kidney) and performed histopathological examination for short term using H&E staining to compare the acute and chronic toxicity of control tissue vs. nanorod-treated tissue and they showed normal histology of liver, kidney,

Table 1. Short term (S-T) toxicity study in mouse model: Blood hematology of mice after intra-peritonial injection with 0 (-ve control, 0.1ml of TE buffer), 1.25 mg $Kg^{-1}day^{-1}$ (0.1 ml), 12.5 mg $Kg^{-1}day^{-1}$ (0.1ml) and 125 mg $Kg^{-1}day^{-1}$ (0.1ml) of europium hydroxide nanorods suspended in TE buffer in the blood, sampled at 8 days. 10 animals were used per measurements, and all values were within normal rage.

	Control (TE buffer)	**1.25 mg $Kg^{-1}day^{-1}$**	**12.5 mg $Kg^{-1}day^{-1}$**	**125 mg $Kg^{-1}day^{-1}$**
CBC without differential hemoglobin (g/dL)	12.1 (± 0.6)	11.9 (± 0.4)	11.9 (± 0.7)	12.4 (± 0.5)
Hematocrit (%)	34.4 (± 1.1)	34.1 (± 1.0)	34.5 (± 1.5)	35.3 (± 1.5)
Erythrocytes [x10(12)/L]	7.5 (± 0.3)	7.4 (± 0.2)	7.6 (± 0.4)	7.9 (± 0.3)
MCV (fL)	45.8 (± 0.4)	45.9 (± 0.4)	45.4 (± 0.4)	44.9 (± 0.4)
RBC Distrib Width (%)	37.5 (± 1.7)	37.0 (± 2.3)	36.4 (± 0.2)	39.7 (± 0.9)
Leukocytes [x10(9)/L]	1.5 (± 0.3)	0.8 (± 0.1)	1.7 (± 0.6)	1.7 (± 0.2)
Platelet count [x10(9)/L]	841.7 (± 269)	804.5 (± 73.7)	807(± 301)	831 (± 175.6)

CBC: Complete blood, count; MCV: mean cell volume; RBC: Red blood cells. Numerical values (±) in the parenthesis are considered as 'Standard Deviation (SD)'. P values were calculated using 'Student-Newman-Keuls Multiple Comparisons Test ' (ANOVA) by comparing between different groups (control group vs treatment groups) and we found that these values are considered not significant ($P > 0.05$). Reprinted with permission from Patra et al.(2009). In Vivo Toxicity Studies of Europium Hydroxide Nanorods in Mice. Toxicol Appl Pharmacol *240*: 88-98. Copyright ©2009, Elsevier Limited.

Table 2. Short term (S-T) toxicity study in mouse model: Serum clinical chemistry of mice after intra-peritonial injection with 0 (-ve control, 0.1ml of TE buffer), 1.25 mg $Kg^{-1}day^{-1}$ (0.1 ml), 12.5 mg $Kg^{-1}day^{-1}$ (0.1ml) and 125 mg $Kg^{-1}day^{-1}$ (0.1ml) of europium hydroxide nanorods suspended in TE buffer in the blood, sampled at 8 days. 10 animals were used per measurements, and all values were within normal rage.

	Control	**1.25 mg $Kg^{-1}day^{-1}$**	**12.5 mg $Kg^{-1}day^{-1}$**	**125 mg$Kg^{-1}day^{-1}$**
ALP (u/L)	123.0 (± 8.6)	122.4 (± 7.3)	132.8 (± 10.1)	95.2 (± 16.7)
AST (u/L)	158.0 (± 77.7)	250.8 (±60.9)	236.6 (± 71.5)	225.8 (± 86)
ALT (u/L)	49.4 (± 15.5)	105 (± 60.5)	63.8 (± 37.0)	74.0 (± 50.1)
Phosphorous (mg/dL)	11.0 (± 0.6)	11.7 (± 0.5)	11.2 (± 2.1)	11.2 (± 3)
Calcium (mg/dL)	9.4 (± 0.1)	9.4 (± 0.2)	9.3 (± 0.1)	9.6 (± 0.2)
Albumin (g/dL)	3.2 (± 0.1)	3.3 (± 0.1)	3.0 (± 0.2)	2.9 (± 0.1)
Creatinine, P/S (mg/dL)	0.1 (± 0.0)	0.1 (± 0.0)	0.1 (± 0.0)	0.1 (± 0.0)
Glucose (mg/dL)	209.2 (± 43.9)	222.4 (± 42.9)	216.6 (± 63)	191.8 (± 55.9)
Protein-total (g/dL)	4.6 (± 0.2)	4.6 (± 0.1)	4.4 (± 0.2)	4.4 (± 0.1)
Bilirubin total, S (mg/dL)	0.1 (± 0.0)	0.1 (± 0.0)	0.1 (± 0.0)	0.1 (± 0.0)
BUN) (mg/dL)	22.8 (± 2)	23.8 (± 2)	22.0 (± 2)	20.4 (± 2.6)

ALP: Alkaline phosphates, AST: Aspartate Aminotransferase, ALT: Alanine Aminotransferase, BUN: Blood Urea nitrogen. Numerical values (±) in the parenthesis are considered as 'Standard Deviation (SD)'. P values were calculated using 'Student-Newman-Keuls Multiple Comparisons Test ' (ANOVA) by comparing between different groups (control group vs treatment groups) and we found that these values are considered not significant ($P > 0.05$). Reprinted with permission from Patra et al.(2009). In Vivo Toxicity Studies of Europium Hydroxide Nanorods in Mice. Toxicol Appl Pharmacol *240*: 88-98. Copyright ©2009, Elsevier Limited.

spleen and lung (Fig. 6a-p). In lung histopathology, with a dose of 12.5 $mgkg^{-1}day^{-1}$, mild thickening of the alveolar membrane and localized para bronchiolar lipophagocytic changes were detected with both 12.5 $mgkg^{-1}day^{-1}$ and 125 $mgkg^{-1}day^{-1}$ doses, indicating mild lung toxicity (Fig. 6c, d). Similarly, in liver (Fig. 6e-h), kidney (Fig. 6i-l) and spleen histopathology (Fig. 6m-p), we observed mild or no toxicity (Patra et al. 2009).

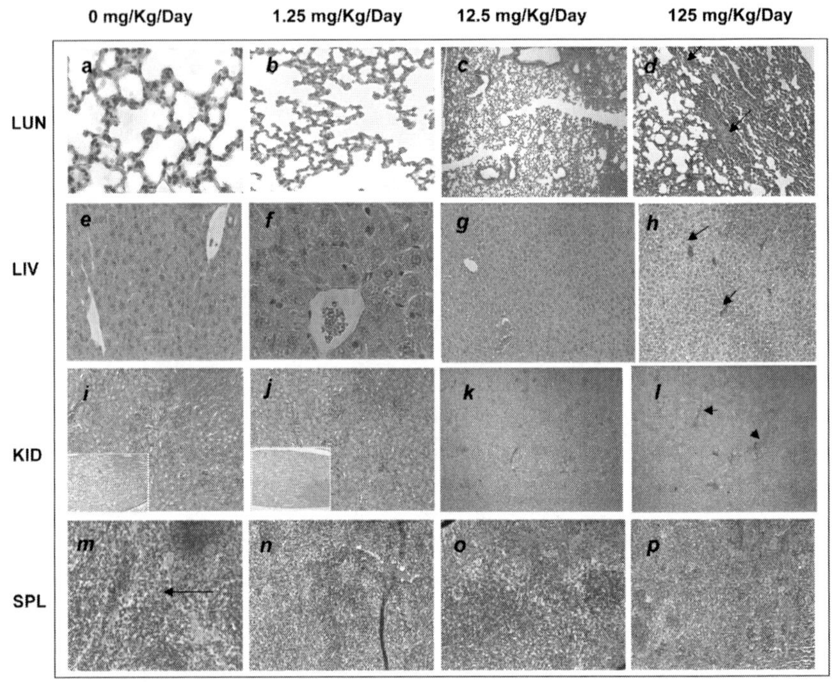

Fig. 6. Histological specimens of mice tissues (lung, liver, kidney and spleen) collected from mice sacrificed on day 60 stained with hematoxylin and eosin (H&E). Long-term survival of mice was also observed even at highest does of $Eu^{III}(OH)_3$ nanorods (125 $mgkg^{-1}day^{-1}$). (a) Control animal lung section showing normal alveolar geometry and normal-looking alveolar septum. (b) Normal alveolar geometry and normal-looking alveolar septum with dose of 125 $mgkg^{-1}day^{-1}$. (c) Normal lung parenchymal blood vessels with dose of 125 $mgkg^{-1}day^{-1}$. (d) Control animal liver section showing normal hepatic architecture. (e) Normal portal tract, sinusoids and hepatocytes with dose of 125 $mgkg^{-1}day^{-1}$. (f) Normal central vein with dose of 125 $mgkg^{-1}day^{-1}$. (g) Kidney sections from control animals show normal renal cortex with normal-looking glomerular tufts and tubules. (h) Normal-looking glomerular tufts with dose of 125 $mgkg^{-1}day^{-1}$. (i) Normal-looking renal tubules and blood vessels. (j) Splenic sections from control animals show normal splenic lymphoid follicles. (k) Normal splenic lymphoid follicles after a dose of 1.25 $mgkg^{-1}day^{-1}$. (l) Normal splenic sinuses are still seen after a dose of 1.25 $mgkg^{-1}day^{-1}$. The histological pictures were taken at following magnifications: a at > x10, b and c at > x100, d at > x40, e and f at > x100, g at > x40, h and i at > x100, j at > x10, k at > x 40, i at > x100. Reprinted with permission from Patra et al. (2009). Copyright ©2009, Elsevier Limited.

Color image of this figure appears in the color plate section at the end of the book.

Bioavailabilities of Europium

We also determined bio-distribution of the europium element in different organs using inductively coupled plasma mass spectrometry (Patra et al. 2009). The europium element was detected in diverse organs such as liver, kidney, spleen and lungs and we observed that the nanorods were distributed in all organs (data not shown). However, the distribution of europium in liver, kidney and lungs was almost negligible at low doses.

Our ultimate goal is to use the pro-angiogenic properties and non-toxic behavior of $Eu^{III}(OH)_3$ nanorods for future therapeutic treatment strategies of severe IHD, peripheral ischemic disease, limb ischemic disease, and other conditions. Our therapeutic dose of $Eu^{III}(OH)_3$ nanorods would be administered locally to the ischemic part of the disease at a microgram level, which would be a far lower dose than the dose of our toxicity experiments (milligram level) (Patra et al. 2009). As the therapeutic dose of $Eu^{III}(OH)_3$ nanorods is very low compared to the toxicity dose, we believe that these nanorods can be used safely for the treatment of several ischemic diseases.

Additionally, the remarkable findings of pro-angiogenic and fluorescence properties and non-toxic behavior of $Eu^{III}(OH)_3$ nanorods suggests that these nanorods could be used for (1) the development of inorganic fluorescent nanorods in cell biology and (2) future therapeutic alternative treatment strategies for severe IHD, peripheral ischemic disease, limb ischemic disease, and wound healing (Patra et al. 2007, 2008b, 2009).

APPLICATIONS TO AREAS OF HEALTH AND DISEASE

Cardiovascular diseases are globally the leading cause of morbidity and mortality. They accounted for 16.7 million or 29.2% of total global deaths in 2003 (according to a World Health Organization report). Despite their high incidence rate, conventional treatment strategies for CVD have many limitations and urgently require more effective alternative therapeutic approaches. Conventionally, VEGF-A and bFGF have been used to enhance blood flow in ischemic tissues via formation of collateral blood vessels that have also been associated with pathological angiogenesis, thrombosis, fibrosis, and/or the proliferation of tumor cells. This latter observation warrants that the development of novel pro-angiogenic molecules be a major focus. In this context, we demonstrated that $[Eu^{III}(OH)_3]$ nanorods (inorganic nanorods) exhibit significant pro-angiogenic activities and fluorescence properties. Considering the critical role of angiogenesis and its impact in CVD, we strongly believe that the pro-angiogenic properties of $Eu^{III}(OH)_3$ nanorods can be used to develop novel treatment strategies for CVD, ischemic diseases, and wound healing. Additionally, since these non-toxic nanorods have fluorescence properties, they may be used for imaging techniques, pharmacotherapeutics and diagnostics in live cell and animal. The discovery of angiogenic properties of $Eu^{III}(OH)_3$ nanorods opens up an exciting new area in the field of CVD and imaging techniques.

Key Facts

- Angiogenesis is a pathophysiological process of formation of new capillaries from existing blood vessels. Angiogenesis has also been used to enhance blood flow through collateral blood vessel formation in patients with CVD, a leading cause of morbidity and mortality in the United States and throughout the world (Huston et al. 1999).
- Conventional method for the treatment of cardiovascular ischemia includes diet and exercise, lipid-lowering drugs, and medical intervention via angioplasty, stent placement, and surgery.
- VEGF-A and bFGF (Folkman 1995; Risau 1997) as pro-angiogenic factors have been used to enhance blood flow in ischemic tissues via formation of collateral blood vessels.
- Application of VEGF-A and bFGF has been associated with pathological angiogenesis, thrombosis, fibrosis and/or the proliferation of tumor cells (Claffey et al. 1996), major disadvantages for the treatment of CVD using these growth factors.
- Development of novel pro-angiogenic as well as anti-angiogenic molecules is very important in combating CVD.
- Nanotechnology, a multidisciplinary field, can play a pivotal role (Claffey et al. 1996) in treating CVD.
- Investigation by several assays shows that $[Eu^{III}(OH)_3]$ nanorods, synthesized in domestic microwave oven, are pro-angiogenic.
- The pro-angiogenic properties of $[Eu^{III}(OH)_3]$ nanorods can be used to develop new treatment strategies for CVD.

Definitions

Angiogenesis: The complex process, involving both pro-angiogenic and anti-angiogenic factors, of formation of new capillaries from existing blood vessels (Folkman 1995). Normal angiogenesis occurs, for example, during fetal development to create a baby's circulatory system, menstrual cycle in uterus, or wound healing.

Cardiovascular disease (CVD): The class of diseases that involve the heart or blood vessels (arteries and veins). CVDs are the number one cause of death in the United States.

Cell cycle assay: The cell cycle in eukaryotes is broadly classified into four phases: G1, preparation of the chromosomes for replication; S, synthesis of DNA and centrosomes; G2, preparation for mitosis; and M, mitosis (Patra et al. 2008b).

Cell proliferation: One of the critical steps in angiogenesis (Risau 1997), which is controlled by (a) growth factors that bind to receptors, (b) receptors on the cell surface that connect to signaling molecules, and (c) signaling molecules (signal transduction pathway) that convey message from receptor to the nucleus where transcription factors bind to DNA, turning on or off the production of proteins that cause cells to continue dividing.

Chicken chorioallantoic membrane (CAM) assay: A quick, technically simple, and inexpensive assay that determines the biological activity of angiogenic factors (Patra et al. 2008b). This assay has been used for studying neovascularization.

Microwave chemistry: The science of involving microwave irradiation in chemical reactions (de la Hoz et al. 2005). Microwave heating has several advantages over conventional heating (described before) (Patra et al. 2005, 2006).

Mitogen-activated protein kinase (MAPK) activation: One of the key signaling pathways for angiogenesis that regulate various cellular activities, such as gene expression, mitosis, differentiation, proliferation, and cell survival/ apoptosis (Pearson et al. 2001; Patra et al. 2008b; Zeng et al. 2001).

Nanomedicine: The bio-medical application of nanotechnology that will lead to useful research tools, advanced drug delivery systems, and new ways to detect and treat disease or repair damaged tissues and cells.

Nanorods: Nanoparticles shaped like long sticks with a diameter in nanoscale but having great length.

Nanotechnology: A cutting edge technology devoted to understanding, creating, designing and using material structures, devices and systems at the atomic, molecular, or macromolecular range (~1–100 nm) with fundamentally new properties and functions (Patra et al. 2008a).

Reactive oxygen species (ROS): A collective term that includes super oxide anion ($O_2\bullet-$), hydroxyl radical (OH), and hydrogen peroxide (H_2O_2). ROS in certain conditions has a significant role in angiogenesis in the cardiovascular system (Maulik 2002; Pearlman et al. 1995; Katusic 1996).

VEGF: Angiogenic cytokines (Folkman 1995, Yancopoulos et al. 2000) also known as vascular permeability factor (VPF) or vasculotropin, a homodimeric 34–42 kDa, heparin-binding glycoprotein with potent angiogenic, mitogenic and vascular permeability-enhancing activities specific for endothelial cells. It is an important signaling protein involved in both vasculogenesis and angiogenesis.

Summary Points

- The synthesis of $Eu^{III}(OH)_3$ nanorods has been carried out in our laboratory by microwave irradiation.
- Microwave method/chemistry is an economical and eco-friendly technique.
- The nanorods were characterized by several physicochemical techniques.
- We have demonstrated the pro-angiogenic properties of $Eu^{III}(OH)_3$ nanorods, observed in several *in vitro* and *in vivo* assays.
- $Eu^{III}(OH)_3$ nanorods might be used for diagnostic and therapeutic purposes in biomedical nanotechnology because of their fluorescence and pro-angiogenic properties, respectively.
- The non-toxic behavior of $Eu^{III}(OH)_3$ nanorods has been demonstrated in *in vitro* and *in vivo* models.
- $Eu^{III}(OH)_3$ nanorods could be used for diagnostic purpose in live cell and animal because of to its non-toxic behavior.
- $Eu^{III}(OH)_3$ nanorods could be used to develop new treatment strategies for CVD.
- CVD is the most common cause of death in the United States and in several other countries.

Acknowledgement

This work was supported by GAP 0305\DST\CP (Ramanujan Fellowship, DST, Govt. of India, New Delhi) to CRP. Dr. Patra, Department of Chemical Biology, Indian Institute of Chemical Technology (IICT), Hyderabad, is also thankful to DST, Govt. of India, New Delhi, for the award of Ramanujan Fellowship (SR/S2/RJN-04/2010).

Abbreviations

ALP	:	alkaline phosphates
ALT	:	alanine aminotransferase
AST	:	aspartate aminotransferase
bFGF	:	basic fibroblast growth factor
BUN	:	blood urea nitrogen
CAGR	:	compound annual growth rate
CAM	:	chick chorioallantoic membrane
CRN	:	creatinine
CVD	:	cardiovascular diseases
DSC	:	differential scanning calorimetric
Eu	:	Europium
$Eu^{III}(OH)_3$	:	Europium (III) hydroxide

H_2O_2	:	hydrogen peroxide
HUVEC	:	human umbilical vein endothelial cells
IP injection	:	intraperitoneal injection
IR Spectra	:	infrared spectroscopy
MAPK	:	mitogen-activated protein kinase
MCV	:	mean cell volume
PCI	:	percutaneous coronary intervention (PCI)
PDGF	:	platelet-derived endothelial cell growth factor
TEM	:	transmission electron microscopy
ROS	:	reactive oxygen species
TBHP	:	tert-butyl hydro peroxide
TBLi	:	total bilirubin
TE	:	Tris-EDTA
TGA	:	thermo gravimetric analysis
TGF	:	transforming growth factors
TP	:	total protein
VEGF-A	:	vascular endothelial growth factor A
XPS	:	X-ray photoelectron spectroscopy
XRD	:	X-ray diffraction

References

Bertozzi, C.R. and National Institutes of Health (U.S.). 2005. In: Chemistry in Living Systems New Tools for Probing the Glycome. Medical Arts and Photography Branch. National Institutes of Health, Bethesda, MD.

Claffey, K.P., L.F. Brown, L.F. delAguila, K. Tognazzi, K.T. Yeo, E.J. Manseau and H.F. Dvorak. 1996. Expression of vascular permeability factor vascular endothelial growth factor by melanoma cells increases tumor growth, angiogenesis, and experimental metastasis. Cancer Res. 56: 172–181.

de la Hoz, A., A. Díaz-Ortiz and A. Moreno. 2005. Microwaves in organic synthesis. Thermal and non-thermal microwave effects. Chem. Soc. Rev. 34: 164–178.

El-Sayed, I.H., X.H. Huang and M.A. El-Sayed. 2005. Surface plasmon resonance scattering and absorption of anti-EGFR antibody conjugated gold nanoparticles in cancer diagnostics: Applications in oral cancer. Nano Lett. 5: 829–834

Folkman, J. 1995. Angiogenesis in cancer, vascular, rheumatoid and other disease. Nat. Med. 1: 27–31.

Hirano, S. and K.T. Suzuki. 1996. Exposure, metabolism, and toxicity of rare earths and related compounds. Environ. Health Persp. Suppl. 104: 85–95.

Huston, S.L., E.J. Lengerich, E. Conlisk and K. Passaro. 1999. Trends in ischemic heart disease death rates for blacks and whites United States, 1981–1995 (Reprinted from MMWR, 47: 945–949, 1998). JAMA 281: 28–29.

Katusic, Z.S. 1996. Superoxide anion and endothelial regulation of arterial tone. Free Radic. Biol. Med. 20: 443–448.

Maulik, N. 2002. Redox signaling of angiogenesis. Antioxid. Redox Signaling 4: 805–815.

Ogawa, Y., S. Suzuki, K. Naito, M. Saito, E. Kamata, A. Hirose, A. Ono, T. Kaneko, M. Chiba, Y. Inaba et al. 1995. Toxicity study of europium chloride in rats. J. Environ. Pathol. Toxicol. Oncol. 14: 1–19.

Patra, C.R., S.S. Abdel Moneim, E.F. Wang, S. Dutta, S. Patra, M. Eshed, P. Mukherjee, A. Gedanken, V. Shah and D. Mukhopadhyay. 2009. *In vivo* toxicity studies of Europium hydroxide nanorods in mice. Toxicol. Appl. Pharmacol. 240: 88–98.

Patra, C.R., G. Alexandra, S. Patra, D.S. Jacob, A. Gedanken, A. Landau and Y. Gofer. 2005. Microwave approach for the synthesis of rhabdophane-type lanthanide orthophosphate (Ln = La, Ce, Nd, Sm, Eu, Gd and Tb) nanorods under solvothermal conditions. New J. Chem. 29: 733–739.

Patra, C.R., R. Bhattacharya, D. Mukhopadhyay and P. Mukherjee. 2008a. Application of gold nanoparticles for targeted therapy in cancer. J. Biomed. Nanotechnol. 4: 99–132.

Patra, C.R., R. Bhattacharya, S. Patra, S. Basu, P. Mukherjee and D. Mukhopadhyay. 2006. Inorganic phosphate nanorods are a novel fluorescent label in cell biology. J. Nanobiotechnol. 4: 11.

Patra, C.R., R. Bhattacharya, S. Patra, S. Basu, P. Mukherjee and D. Mukhopadhyay, 2007. Lanthanide phosphate nanorods as inorganic fluorescent labels in cell biology research. Clin. Chem. 53: 2029–2031.

Patra, C.R., R. Bhattacharya, S. Patra, N.E. Vlahakis, A. Gabashvili, Y. Koltypin, A. Gedanken, P. Mukherjee and D. Mukhopadhyay. 2008b. Pro-angiogenic properties of europium(III) hydroxide nanorods. Adv. Mater. 20: 753–756.

Patra, C.R., P. Mukherjee and D. Mukhopadhyay. 2008c. Application of Lanthanide (Ln = Eu & Tb) Nanoparticles in Biology and Medicine. Nova Science Publishers, Inc., New York, pp. 99–141.

Pearlman, J.D., M.G. Hibberd, M.L. Chuang, K. Harada, J.J. Lopez, S.R. Gladstone, M. Friedman, F.W. Sellke and M. Simons. 1995. Magnetic-resonance mapping demonstrates benefits of Vegf induced myocardial angiogenesis. Nat. Med. 1: 1085–1089.

Pearson, G., F. Robinson, T.B. Gibson, B. Xu, M. Karandikar, K. Berman, and M.H. Cobb. 2001. Mitogen-Activated Protein (MAP) Kinase pathways: regulation and physiological functions. Endocr. Rev. 22: 153–183.

Risau, W. 1997. Mechanisms of angiogenesis. Nature 386: 671–674.

Simm, A. and H.J. Brömmeb. 2005. Reactive oxygen species (ROS) and aging: Do we need them, can we measure them should we block them? Signal Transduction 3: 115–125.

Stohs, S.J. and D. Bagchi. 1995. Oxidative mechanisms in the toxicity of metal-ions. Free Radic. Biol. Med. 18: 321–336.

Thompson, K.H., and C. Orvig. 2006. Editorial: Lanthanide compounds for therapeutic and diagnostic applications. Chem. Soc. Rev. 35: 499–499.

Ushio-Fukai, M. 2006. Redox signaling in angiogenesis: Role of NADPH oxidase. Cardiovasc. Res. 71: 226–235.

Yancopoulos, G.D.S., N.W. Davis, N.W. Gale, J.S. Rudge, S.J. Wiegand and J. Holash. 2000. Vascular-specific growth factors and blood vessel formation. Nature 407: 242–248.

Zeng, H.Y., S. Sanyal and D. Mukhopadhyay. 2001. Tyrosine residues 951 and 1059 of vascular endothelial growth factor receptor-2 (KDR) are essential for vascular permeability factor/vascular endothelial growth factor-induced endothelium migration and proliferation, respectively. J. Biol. Chem. 276: 32714–32719.

20

Iron Oxide Nanoparticles and Cardiac Stem Cells

K.W. Au,[1] *April M. Chow,*[2] *Ed X. Wu*[3] and *H.F. Tse*[4,*]

ABSTRACT

Nanotechnology offers promising perspectives in biomedical research and clinics. Stem cell therapy is an emerging therapeutic approach for cardiovascular diseases. The combination of the two allows the tracking of the migration and fate of transplanted cells. Indeed, superparamagnetic iron oxide (SPIO) nanoparticles have been employed for cell labeling and non-invasive magnetic resonance imaging monitoring of cells in the living organism. In this chapter, the effect of nanoparticles on cells, labeling methods and the potential application of SPIO nanoparticles are reviewed.

[1]Division of Cardiology, Department of Medicine, L801, Laboratory Block, LKS Faculty of Medicine Building, 21 Sassoon Road, Pokfulam, The University of Hong Kong; E-mail: aukawing@hku.hk

[2]Department of Electrical and Electronic Engineering, Laboratory of Biomedical Imaging and Signal Processing, L1003, Laboratory Block, LKS Faculty of Medicine Building, 21 Sassoon Road, Pokfulam, The University of Hong Kong; E-mail: aprilcmk@eee.hku.hk

[3]Laboratory of Biomedical Imaging and Signal Processing, Department of Electrical and Electronic Engineering, L1004, Laboratory Block, LKS Faculty of Medicine Building, 21 Sassoon Road, Pokfulam, The University of Hong Kong; E-mail: ewu@eee.hku.hk

[4]Division of Cardiology, Department of Medicine, Rm 1928, 19/F, Block K, Queen Mary Hospital, 102 Pokfulam Road, The University of Hong Kong; E-mail: hftse@hku.hk

*Corresponding author

List of abbreviations after the text.

INTRODUCTION

Despite recent advances in pharmacological and interventional therapies, cardiovascular diseases remain the leading causes of morbidity and mortality worldwide. Stem cells are undifferentiated cells characterized by their functional capacity to self-renew and to generate a large number of differentiated progeny cells. Recently, stem cell transplantation has been investigated as a potential novel therapeutic approach in patients with myocardial infarction and heart failure (Siu et al. 2010). One of the major challenges in cardiac stem cell therapy is cell engraftment and survival after transplantation. *In vivo* experimental studies have demonstrated the feasibility of monitoring cell migration and cell fate after transplantation using either reporter genes or chimeric animals (Cahill et al. 2004). However, these methods are cumbersome and cannot be translated to human use. Besides, this approach lacks the temporal analysis of the donor cells and limits its practical use. Thus, methods for monitoring implanted stem cell non-invasively *in vivo* will greatly facilitate the clinical realization and optimization of cell-based therapies. Cellular imaging can permit non-invasive visualization of cellular processes, including cell migration, and sequential fate and efficacy in tissue regeneration and maintenance, in living subjects. Unfortunately, there is a very limited method to allow serial monitoring of stem cells in clinic after transplantation (Ransohoff and Wu 2010). The recent development of superparamagnetic iron oxide (SPIO) nanoparticles labeling and cardiac magnetic resonance imaging (MRI) may provide a powerful tool to allow non-invasive tracking of cell migration and survival after transplantation. Nevertheless, the feasibility, efficiency and safety issues of this method should be addressed before clinical applications.

MAGNETIC RESONANCE IMAGING

Non-invasive imaging modalities provide *in vivo* information on anatomical, physiological and functional aspects of living organisms in normal, developing, or pathological states. X-ray computed tomography, positron emission tomography, single proton emission computed tomography, magnetic resonance imaging, ultrasound imaging, radionuclide imaging and optical imaging have been employed clinically for diagnosis as well as assessment of treatment efficacy and prognosis. This chapter focuses on current MR techniques in monitoring stem cell therapies *in vivo*.

MRI is non-invasive in nature and allows excellent soft-tissue characterization and good spatial and temporal resolution. In medical application, MRI has become one of the important techniques for diagnosis of pathological abnormalities through imaging various organs in any plane without exposing patients to ionizing radiation. In addition to anatomical imaging, MRI possesses advanced capabilities of perfusion imaging, diffusion imaging, functional imaging, molecular imaging and chemical shift imaging, hence providing information on associations of tissue functions, hemodynamics, molecular pathways and metabolisms in normal and disease conditions. Regarding the heart, since MRI requires minimum movement during the period of image acquisition, with the help of ECG triggering, cardiac MRI can be achieved. Cardiac MRI offers promise in measuring regional and global ventricular function and allows a wide range of tissue characterization including the quantitative assessment of non-viable, edematous, inflamed, infiltrated or hypoperfused myocardium (Kedziorek and Kraitchman 2010).

MRI provides excellent soft tissue contrast when compared with other imaging modalities. The image contrast depends on tissue intrinsic, magnetic susceptibility and pulse sequence parameters. Because of tissue differences, the MR contrast is determined by several intrinsic parameters, including proton density, longitudinal relaxation time (T_1) and transverse relaxation time (T_2). Proton density refers to the density of mobile hydrogen nuclei. T_1, also known as spin-lattice relaxation time, is the time constant for recovering of the equilibrium state of the longitudinal component of net magnetization after being disturbed by a radiofrequency pulse, through the exchange of energy between the spin system and surrounding thermal reservoir known as lattice. T_2, also known as spin-spin relaxation time, is the time constant for the transverse component of net magnetization to return to its equilibrium state, through the loss of phase coherence among spins due to slight differences in Larmor frequencies. When the magnetic field in homogeneity is also taken into account, the time constant for transverse magnetization to return to the equilibrium is called effective transverse relaxation time (T_2^*). Image contrast in MRI can be flexibly controlled by varying pulse sequence and its parameters, for example, pulse repetition time, echo time and flip angle. By choosing appropriate pulse sequence and parameters, proton density–weighted, T_1-weighted, T_2-weighted, or T_2^*-weighted images can be obtained. Intrinsic parameters can also be measured directly by specific pulse sequences for better tissue characterization. MR contrast enhancement agents can also be used to modify the image contrast by manipulating tissue T_1, T_2 or T_2^*.

Cellular MRI

Cellular imaging allow the visualization of targeted cells in living organisms longitudinally and non-invasively (Modo et al. 2005). Cellular T_1, T_2 or T_2^* can be manipulated through labeling with MR contrast agents *in vivo* or pre-labeling *in vitro,* and hence provide different signal intensity of the targeted cells from surrounding tissues. The different signal intensity, thus, allows the detection of the targeted cells and their migration using MRI. With the ability to achieve near-cellular resolution and whole body imaging, MRI can be used to locate labeled cells after transplantation or transfusion. In stem cell therapy, MRI allows tracking of cells; after local grafting or systemic injection, the cells may migrate from the site of transplantation or injection to relevant foci of disease. In addition, complementary morphological, functional and metabolic information can be obtained within the same imaging session for efficacy assessment of stem cell therapy serially.

For *in vivo* cellular labeling, polymeric paramagnetic gadolinium and manganese chelates have mainly been used to inject intravenously for hepatocyte imaging in detecting hepatic lesions and staging of hepatocellular carcinomas. Although SPIO nanoparticles could also serve the purpose of hepatocyte imaging, imaging the monocyte/macrophage activity has been the key application of SPIO-based cellular imaging in various regions such as lymph nodes, bone marrow, brain, kidney and pancreas. Recently, injection of SPIO or micron-sized iron oxide particles (MPIO) into ventricles near the subventricular zone has been developed to label the neural stem/progenitor cells for monitoring neurogenesis (Shapiro et al. 2006a). Because non-phagocytic cells do not readily take up MR contrast agents for cellular imaging, pre-labeling of these cells *in vitro* is required for monitoring the distribution of labeled cells *in vivo*, after transplantation or transfusion in various disease models.

As MRI is able to image anatomical structure with high spatial resolution non-invasively, it has been applied in abundant biological applications, including monitoring of cell therapy in ventricular myocardial infarction with cardiac MRI (Fig. 1). However, several drawbacks of MRI cannot be disregarded. For example, dilution of labels may result from cell division, causing loss in contrast, the effluence of labels from cells may cause confusion, and metabolic inactive or dead cells carrying SPIO leading to signal voids may lead to misinterpretation.

Contrast Agents in MRI

MR contrast agents are usually some paramagnetic ions used to modify the image contrast by manipulating tissue T_1 or T_2. Their effects are usually described by r_1 and r_1 relaxivities in the unit of $mM^{-1}s^{-1}$. Contrast agents

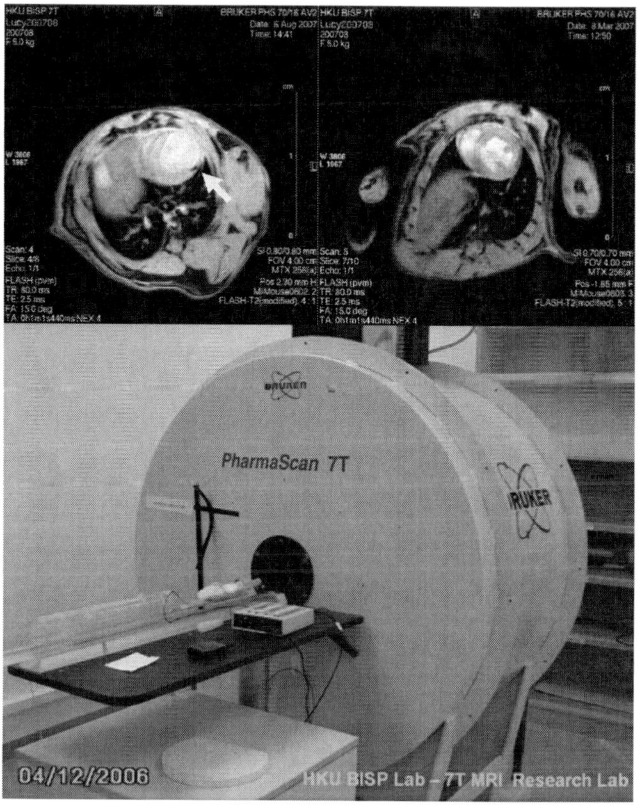

Fig. 1. (a) MION-PLL labeled mouse embryonic stem cells were observed as hypointensities at the injection site (yellow arrow) in myocardial infarcted mouse with T_2^*-weighted MR image, while (b) no apparent hypointensities were observed in myocardial infarcted mouse without cell transplantation using (c) 7 Tesla MRI scanner.

Color image of this figure appears in the color plate section at the end of the book.

thus naturally fall into two categories, T_1 agent or T_2 agent, based on their relaxivity ratio (r_2/r_1). Positive contrast agents, known as T_1 agents, include gadolinium chelates and manganese chelates. In general, they have lower relaxivity ratios of 1 to 2. They increase MR image intensity by shortening T_1. In contrast, negative contrast agents, or T_2 agents, such as SPIO nanoparticles have relaxivity ratios as high as 10 or more. They predominantly induce signal loss through spin dephasing. The characteristics of an ideal agent for non-invasive tracking of therapeutic cells are listed in Table 1.

Table 1. Characteristics of an ideal MR contrast agent for cell tracking.

• Low toxicity, without altering cell viability, growth, differentiation, or its functions
• Durable but ideally having some elimination pathway
• Permitting non-invasive serial detection
• Detectable with sufficient contrast-to-noise ratio, spatial and temporal resolution at practical doses
• Allowing quantitative assessment of the migration and fate of labeled cells

Superparamagnetic Iron Oxide Nanoparticles

For a cell of interest to be distinguished from complex background signal, it must contain a label capable of producing significant positive or negative contrast when imaged with the appropriate pulse sequence. SPIO nanoparticles are negative MR contrast agents that reduce the signal intensity in T_2-weighted or T_2*-weighted MR images due to magnetic susceptibility effect on the transverse relaxation. Because of its biocompatibility and relatively effective cellular internalization, SPIO is commonly used as contrast agent in cellular MRI. SPIO is made of small magnetites (1–10 nm) with the general form of $Fe_2^{3+}O_3Fe^{2+}O$ and surrounded by a stabilizing coating. Even below the Curie temperature, the thermal energy is sufficient to overcome the force between magnetites and the magnetites behave like some randomly oriented magnetic domains. In the absence of an applied magnetic field, the magnetic domains are randomly oriented and the net magnetization is averaged to zero. However, under the influence of an external magnetic field, these magnetic domains reorient and align with the external magnetic field, forming a net magnetization that is stronger than sum of magnetic dipoles from the unpaired electrons. Currently, there are different types of SPIO that can be classified by their hydrodynamic diameters: ultrasmall superparamagnetic iron oxide (USPIO, below 50 nm); superparamagnetic iron oxide (SPIO, 80–150 nm); and micron-sized iron oxide particles (MPIO, ~ 1 μm) (Fig. 2).

Endorem® (Guerbet) and Feridex® (Berlex Inc.) are SPIO coated with dextran, while Resovist® (Bayer Schering Pharma AG) is coated with carboxydextran. They are administered intravenously for liver/spleen applications. These SPIO would be taken up readily by macrophages in liver/spleen and provide passive targeting for reticuloendothelial system in liver and spleen. Sinerem® (Guerbet), also know as Combidex® (Advanced Magnetics Inc.), is dextran-coated USPIO. This contrast

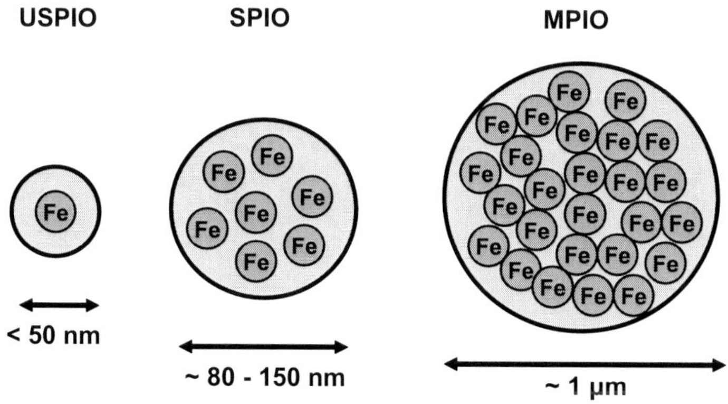

Fig. 2. Different types of SPIO nanoparticles, classified by their hydrodynamic diameters. USPIO, ultrasmall superparamagnetic iron oxide; SPIO, superparamagnetic iron oxide; MPIO, micron-sized iron oxide particles.

agent has a long plasma half-life of about 2–24 h and can function as an intravascular contrast agent. It is mainly used in lymphography, like SPIO, and also in MR angiography and cardiac perfusion owing to its low iron dose, T_1 shortening effect and long half-life. Monocrystalline iron oxide nanoparticle (MION) is another type of SPIO that has a molecular size of about 2–4 nm with a hydrodynamic diameter of 20 nm. It has a simple monocrystalline structure and small size that can easily pass through capillary endothelium and distribute in tissue. MION has a very strong r_2 relaxivity and can be used to target the reticuloendothelial system especially in lymph nodes, like USPIO. However, it has slow uptake in liver/spleen. Very small superparamagnetic iron oxide (VSOP) consists of SPIO nanoparticles with a citrate coating and the overall particle size of 4–8 nm. Because of their highly anionic surface, they possess the shortest blood half-life in humans, approximately 1 h. All compounds are developed for intravenous administration targeting tumors in the liver and spleen. Besides their clinical roles in diagnostic imaging, SPIO, USPIO, MION and VSOP have also been used to label stem cells for cell tracking. Recently, MPIO, which is originally developed for magnetic cell separation in laboratory settings, has been shown to image single cell because of the increased susceptibility effect resulting from its large size (Shapiro et al. 2006b). Table 2 summarizes experimental studies on the use of various commercial SPIO agents to monitor stem cells with cardiac MRI after transplantations.

Table 2. Commercial SPIO agents for cell tracking of stem cell therapy in cardiac applications.

Class	Common names	Mean particle size	Stem cell type, species	Disease model
USPIO	Sinerem	< 50 nm	CPC, Mouse (Kustermann et al. 2005)	Myocardial infarction
	Combidex			
SPIO	Feridex	80–150 nm	MSC, Porcine (Kraitchman et al. 2003)	Myocardial infarction
	Endorem		MSC, Canine (Kraitchman et al. 2005)	Myocardial infarction
	Resovist		Macrophage, Rat (Leor et al. 2006)	Myocardial infarction
			Myoblast, Rat (Cahill et al. 2004)	—
			MSC, Canine (Bulte et al. 2005)	Myocardial infarction
			ESC, Mouse (Tallheden et al. 2006)	—
			ESC, Mouse (Arai et al. 2006)	Myocardial infarction
			MSC, Swine (He et al. 2007)	Myocardial infarction
			MSC, Rat (Amsalem et al. 2007)	Myocardial infarction
			ESC, Rat (Qiao et al. 2009)	Myocardial infarction
			EPC, Rat (Yao et al. 2010)	Myocardial infarction
MPIO		~ 1 μm	MSC, Swine (Hill et al. 2003)	Myocardial infarction
			MSC, Swine (Dick et al. 2003)	Myocardial infarction
			MSC, Rat (Khan et al. 2009)	Myocardial infarction
			BMPC, Swine (Graham et al. 2010)	Myocardial infarction

BMPC, bone marrow progenitor cell; CPC, cardiac progenitor cell; EPC, endothelial progenitor cell; ESC, embryonic stem cell; MSC, mesenchymal stem cell.

Methods of Magnetic Labeling

For cells of interest to be detectable by MRI, they must be labeled with relatively large amount of contrast agents. However, SPIO nanoparticles cannot label stem cells or other mammalian cells efficiently in their unmodified form (Arbab et al. 2004). To facilitate labeling of cells that lack phagocytic capacity, optimizations and modifications of SPIO have been developed in penetrating the cell membrane.

The use of peptide- or antibody-coated nanoparticles is one way to label cells effectively. When SPIO is conjugated with human immunodeficiency virus transactivator (tat) peptide, it can be efficiently transported into cells (Lewin et al. 2000). The use of internalization monoclonal antibodies promotes intracellular uptake of SPIO via receptor-mediated endocytosis (Bulte et al. 1999). However, monoclonal antibodies are species-specific; development of new monoclonal antibodies is required in different animals. Non-specific endocytosis can also be induced via membrane bending by coating SPIO with dendrimers, called magnetodendrimers (Bulte et al. 2001), in various cells of mouse, rat and human.

Besides, transfection agents such as poly-amines and lipids are often employed to increase the labeling efficiency (Frank et al. 2002). The transfection agents coat SPIO nanoparticles via electrostatic interaction and chaperone these MR contrast agents into cells. They are highly charged macromolecules that form a complex stimulating endocytosis when combined with opposite surface-charged SPIO, thus facilitating the incorporation of the contrast agents into endosomes. Prussian blue staining of cultured mouse embryonic stem cells (ESC) labeled with MION using poly-L-lysine (PLL) for 24 h showed intracytoplasmic iron inclusions as dense blue-stained vesicles (Fig. 3a). In comparison, no iron was detected in the unlabeled ESCs (Fig. 3b). Note that the major drawback in using transfection agents is potential endosomal capture. It is also well known that transfection agents are toxic to cells, and their toxic effect is proportional to their concentration. However, when optimal ratio of SPIO is added to create complexes through the electrostatic interaction, the transfection agents can become an effective, stable and non-toxic complex to internalize the contrast agents into cells.

Alternatively, magnetoelectroporation technique can also be used to facilitate and expedite the cell labeling (Walczak et al. 2005). Electrical pulses are used to temporarily "shock" cells and hence induce electromechanical permeability changes in the cell membrane, enabling the introduction of SPIO nanoparticles in cells. Recently, magnetosonoporation has been introduced to label cells magnetically (Qiu et al. 2010). Ultrasound waves are used to temporarily permeabilize cells, facilitating intracellular uptake of contrast agents. With magnetoelectroporation or magnetosonoporation, rapid labeling can be achieved and no transfection agents are required. Cell culture procedures can also be eliminated and hence freshly isolated cells can be transfused back without significant delay or contamination. Figure 4 shows three major approaches of magnetic labeling of non-phagocytic cells.

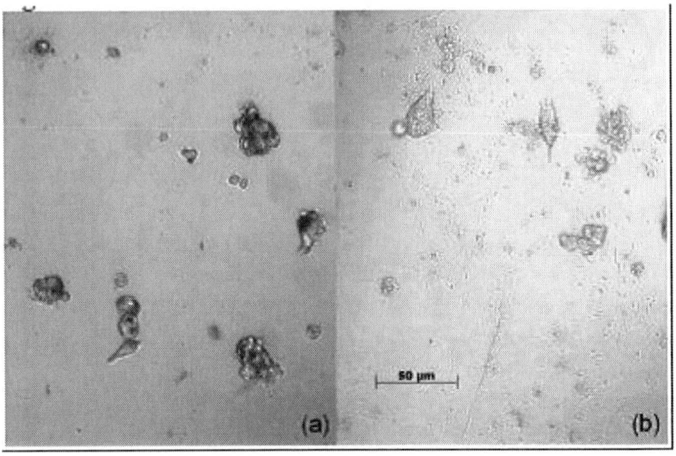

Fig. 3. Micrographs show (a) clumps of mouse ESCs stained with Prussian blue to demonstrate the uptake of MION particles. MION are visible as blue iron deposits; (b) mouse ESCs without labeling of MION-PLL showing no apparent MION uptake.

Color image of this figure appears in the color plate section at the end of the book.

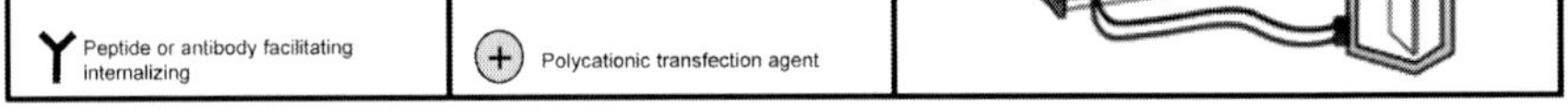

Fig. 4. Three major approaches of magnetic labeling of non-phagocytic cells: (a) use of SPIO conjugated with peptide or antibody, which facilitates internalizing; (b) use of highly charged transfection agents to form complexes with SPIO nanoparticles via electrostatic interaction and chaperone these MR contrast agents into cells; and (c) mechanical methods including magnetoelectroporation and magnetosonoporation.

On the other hand, without using transfection agents or modification of the surface, cells readily endocytose MPIO from 0.96 μm in diameter to as large as 5.80 μm (Shapiro et al. 2005). Labeling can be accomplished simply by incubating adherent cells with the particles for a time suitable for good labeling efficiency.

Detection Limit for SPIO-labeled Cells

The detection range of magnetically labeled cells depends on various factors, including magnetic field strength, pulse sequence and its parameters, relaxivity of the contrast agent, cell number, as well as the iron loading per cell. In general, about 10^7 to 10^8 SPIO-labeled cells could be readily detected by MRI at 1.5 T (Kraitchman et al. 2003). Using MPIO, single cell can be detected at 7 T *in vivo* following cell labeling and transplantation (Shapiro et al. 2006b). Upon cell transplantation, dilution of labels may result from cell division. Longitudinal monitoring of SPIO-labeled mesenchymal stem cells (MSCs) has been shown to be detectable up to 3 wk in porcine myocardium after myocardial infarction at 1.5 T (Kraitchman et al. 2003).

Biological Effects of Nanoparticles Labeling on Stem Cells

Other than the detection limit, the effects of contrast agents on cell, with respect to toxicity and stability, should also be taken into consideration before clinical use. Long-term viability, growth rate, and apoptotic indexes of the labeled human MSCs were found to be unaffected by the endosomal incorporation of SPIO using the PLL as transfection agent (Arbab et al. 2003). However, labeled human MSCs did not differentiate to other lineage. Similar inhibition of chondrogenesis was also observed in human MSCs after SPIO labeling with PLL (Kostura et al. 2004). In comparison, magnetic labeling using protamine sulfate as transfection agent did not inhibit the differentiation capacity of human MSCs (Arbab et al. 2005). The effects of MION-PLL complex on cardiac differentiation of mouse embryonic stem cell were investigated (Au et al. 2009). There was no significant alteration in term of cardiogenic capacity and effect on calcium-handling property. These findings demonstrated the feasibility of *in vivo* cell tracking without affecting the cardiac differentiation potential and functional properties of cells. For magnetoelectroporation or magnetosonoporation, cell viability and functional differentiation were not altered after SPIO labeling (Qiu et al. 2010; Walczak et al. 2005).

Taking the benefit that MRI allows non-invasive cellular tracking; and minimal effect of using SPIO nanoparticles as tracking agent, there is an increasing trend of using MRI to monitor the cell migration and cell fate. In recent advances in stem cell research, researchers attempt to use labeled stem cells in understanding the migration and homing of cells after injection in different disease model.

LIMITATIONS

Despite the advantage of monitoring cell migration and cell fate by means of nanoparticles, there are several limitations on the ion labeling methods. Apparently, labeled stem cells cannot be used to monitor cell survival and cell differentiation after injection. Toxicity of cell labeling is another issue that needs to be taken into consideration. Although no obvious toxicity has been reported for the incorporation of these nanoparticles, the determination of optimal dosage is important for safe application in clinical use.

APPLICATIONS TO AREAS OF HEALTH AND DISEASE

With the advances in the use of SPIO as tracking agent, several researchers attempt the use of these nanoparticles to label stem cell in order to understand the migration and homing of cells after injection. Recently, MPIO was used to label multipotent stromal cell to monitor the migration of cell, immediately after spinal cord injuries in animals. Labeling dendritic cells with MPIO also allows detection and quantification of the migration of cell. Regarding the heart, one of the main hurdles for cellular cardiomyoplasty is the low, variable retention of transplanted cells. Many injected cells are lost because of the washing out of cell in tissue blood flow, and squeezing out of cells in cardiac contraction. Regardless of the difficulties, one group successfully demonstrated the use of magnetic targeting to ensure homing of iron-labeled cardiosphere-derived cells, which improve retention of injected cells, enhance engraftment and improve heart function. This transient magnetic attraction after cell transplantation can attenuate cell loss during injection (Cheng et al. 2010). Cardiac stem cell therapy is now in a transition from bench to bedside. In preclinical experiments, most evidence suggests that treatment using various type of stem cells improves heart function (Siu et al. 2010). Nevertheless, detailed assessment is needed on timing of cell delivery, number of cells to administer, and route of cell delivery. In the future, MRI may provide information to ensure the homing of cells to the site of injury, and it will also be essential to label stem cells with a durable marker. Although stem cell therapy has been carried out in some clinical trials, it is important to highlight that the field is at an early stage, compared to drug development. There is clearly a long and difficult path ahead that is not helped by the lack of tools that we have for assessing mechanistic questions posed by the use of cell therapy in human.

Summary Points

- Stem cells have self-renewal properties and can differentiate into desired cell types. This property allows for an unlimited cell source and supply for stem cell therapy.
- Methods for monitoring implanted stem cell non-invasively *in vivo* will greatly facilitate the clinical realization and optimization of the cell-based therapies.

- Cells labeled with SPIO can be visualized with T_2-weighted or T_2^*-weighted sequences using MRI.
- No apparent bioeffects were found on SPIO-labeled cells.
- Safety is the most important concern for stem cell therapy.
- Tracking of cells using MRI is a useful tool for monitoring cell migration and fate after injection.

Key Facts

- Endorem®, Feridex® and Resovist® are agents approved by the U.S. Food and Drug Administration for MR liver imaging, Sinerem® and Combidex® are under phase III trial as intravascular contrast agent or in lymph node imaging, while other SPIO nanoparticles are currently only approved for research purposes.
- The first clinical trial in human demonstrated MR tracking of Feridex-labeled dendritic cells upon injection in lymph nodes at 3 T (de Vries et al. 2005).

Definitions

Magnetic resonance imaging (MRI): A non-invasive medical imaging technique used in radiology to visualize detailed internal structure and limited function of the body. MRI provides great contrast between the different soft tissues of the body. This new field of non-invasive imaging brings insights and breakthroughs towards therapeutic development and medical diagnosis. It is especially useful in neurological, musculoskeletal, cardiovascular, and cancer imaging.

Stem cell therapy: Therapy designed to "repair" damaged tissue and organs by augmentation with healthy cells provided by stem cell transplants. Stem cells can be differentiated into desired cell types and overcome the limitation of cell supply in therapeutic use.

Abbreviations

BMPC	:	bone marrow progenitor cell
CPC	:	cardiac progenitor cell
EPC	:	endothelial progenitor cell
ESCs	:	embryonic stem cells
MPIO	:	micron-sized iron oxide particles
MRI	:	magnetic resonance imaging
MSCs	:	mesenchymal stem cells
PLL	:	poly-L-lysine
SPIO	:	superparamagnetic iron oxide
USPIO	:	ultrasmall superparamagnetic iron oxide
VSOP	:	very small superparamagnetic iron oxide

References

Amsalem, Y., Y. Mardor, M.S. Feinberg, N. Landa, L. Miller, D. Daniels, A. Ocherashvilli, R. Holbova, O. Yosef, I.M. Barbash and J. Leor. 2007. Iron-oxide labeling and outcome of transplanted mesenchymal stem cells in the infarcted myocardium. Circulation 116: I38–45.

Arai, T., T. Kofidis, J.W. Bulte, J. de Bruin, R.D. Venook, G.J. Berry, M.V. McConnell, T. Quertermous, R.C. Robbins and P.C. Yang. 2006. Dual *in vivo* magnetic resonance evaluation of magnetically labeled mouse embryonic stem cells and cardiac function at 1.5 t. Magn. Reson. Med. 55: 203–209.

Arbab, A.S., L.A. Bashaw, B.R. Miller, E.K. Jordan, B.K. Lewis, H. Kalish and J.A. Frank. 2003. Characterization of biophysical and metabolic properties of cells labeled with superparamagnetic iron oxide nanoparticles and transfection agent for cellular MR imaging. Radiology 229: 838–846.

Arbab, A.S., G.T. Yocum, A.M. Rad, A.Y. Khakoo, V. Fellowes, E.J. Read and J.A. Frank. 2005. Labeling of cells with ferumoxides-protamine sulfate complexes does not inhibit function or differentiation capacity of hematopoietic or mesenchymal stem cells. NMR Biomed. 18: 553–559.

Arbab, A.S., G.T. Yocum, L.B. Wilson, A. Parwana, E.K. Jordan, H. Kalish and J.A. Frank. 2004. Comparison of transfection agents in forming complexes with ferumoxides, cell labeling efficiency, and cellular viability. Mol. Imaging 3: 24–32.

Au, K.W., S.Y. Liao, Y.K. Lee, W.H. Lai, K.M. Ng, Y.C. Chan, M.C. Yip, C.Y. Ho, E.X. Wu, R.A. Li, C.W. Siu and H.F. Tse. 2009. Effects of iron oxide nanoparticles on cardiac differentiation of embryonic stem cells. Biochem. Biophys. Res. Commun. 379: 898–903.

Bulte, J.W., T. Douglas, B. Witwer, S.C. Zhang, E. Strable, B.K. Lewis, H. Zywicke, B. Miller, P. van Gelderen, B.M. Moskowitz, I.D. Duncan and J.A. Frank. 2001. Magnetodendrimers allow endosomal magnetic labeling and *in vivo* tracking of stem cells. Nat. Biotechnol. 19: 1141–1147.

Bulte, J.W., L. Kostura, A. Mackay, P.V. Karmarkar, I. Izbudak, E. Atalar, D. Fritzges, E.R. Rodriguez, R.G. Young, M. Marcelino, M.F. Pittenger and D.L. Kraitchman. 2005. Feridex-labeled mesenchymal stem cells: cellular differentiation and MR assessment in a canine myocardial infarction model. Acad. Radiol. 12 Suppl. 1: S2–6.

Bulte, J.W., S. Zhang, P. van Gelderen, V. Herynek, E.K. Jordan, I.D. Duncan and J.A. Frank. 1999. Neurotransplantation of magnetically labeled oligodendrocyte progenitors: magnetic resonance tracking of cell migration and myelination. Proc. Natl. Acad. Sci. USA 96: 15256–15261.

Cahill, K.S., S. Germain, B.J. Byrne and G.A. Walter. 2004. Non-invasive analysis of myoblast transplants in rodent cardiac muscle. Int. J. Cardiovasc. Imaging 20: 593–598.

Cheng, K., T.S. Li, K. Malliaras, D.R. Davis, Y. Zhang and E. Marban. 2010. Magnetic targeting enhances engraftment and functional benefit of iron-labeled cardiosphere-derived cells in myocardial infarction. Circ. Res. 106: 1570–1581.

de Vries, I.J., W.J. Lesterhuis, J.O. Barentsz, P. Verdijk, J.H. van Krieken, O.C. Boerman, W.J. Oyen, J.J. Bonenkamp, J.B. Boezeman, G.J. Adema, J.W. Bulte, T.W. Scheenen, C.J. Punt, A. Heerschap and C.G. Figdor. 2005. Magnetic resonance tracking of dendritic cells in melanoma patients for monitoring of cellular therapy. Nat. Biotechnol. 23: 1407–1413.

Dick, A.J., M.A. Guttman, V.K. Raman, D.C. Peters, B.S. Pessanha, J.M. Hill, S. Smith, G. Scott, E.R. McVeigh and R.J. Lederman. 2003. Magnetic resonance fluoroscopy allows targeted delivery of mesenchymal stem cells to infarct borders in Swine. Circulation 108: 2899–2904.

Frank, J.A., H. Zywicke, E.K. Jordan, J. Mitchell, B.K. Lewis, B. Miller, L.H. Bryant Jr. and J.W. Bulte. 2002. Magnetic intracellular labeling of mammalian cells by combining (FDA-approved) superparamagnetic iron oxide MR contrast agents and commonly used transfection agents. Acad. Radiol. 9 Suppl. 2: S484–487.

Graham, J.J., W.D. Foltz, A.K. Vaags, M.R. Ward, Y. Yang, K.A. Connelly, R. Vijayaraghavan, J.S. Detsky, M.R. Hough, D.J. Stewart, G.A. Wright and A.J. Dick. 2010. Long-term tracking of bone marrow progenitor cells following intracoronary injection post-myocardial infarction in swine using MRI. Am. J. Physiol. Heart Circ. Physiol. 299: H125–133.

He, G., H. Zhang, H. Wei, Y. Wang, X. Zhang, Y. Tang, Y. Wei and S. Hu. 2007. *In vivo* imaging of bone marrow mesenchymal stem cells transplanted into myocardium using magnetic resonance imaging: a novel method to trace the transplanted cells. Int. J. Cardiol. 114: 4–10.

Hill, J.M., A.J. Dick, V.K. Raman, R.B. Thompson, Z.X. Yu, K.A. Hinds, B.S. Pessanha, M.A. Guttman, T.R. Varney, B.J. Martin, C.E. Dunbar, E.R. McVeigh and R.J. Lederman. 2003. Serial cardiac magnetic resonance imaging of injected mesenchymal stem cells. Circulation 108: 1009–1014.

Kedziorek, D.A. and D.L. Kraitchman. 2010. Superparamagnetic iron oxide labeling of stem cells for MRI tracking and delivery in cardiovascular disease. Methods Mol. Biol. 660: 171–183.

Khan, M., S. Meduru, I.K. Mohan, M.L. Kuppusamy, S. Wisel, A. Kulkarni, B.K. Rivera, R.L. Hamlin and P. Kuppusamy. 2009. Hyperbaric oxygenation enhances transplanted cell graft and functional recovery in the infarct heart. J. Mol. Cell. Cardiol. 47: 275–287.

Kostura, L., D.L. Kraitchman, A.M. Mackay, M.F. Pittenger and J.W. Bulte. 2004. Feridex labeling of mesenchymal stem cells inhibits chondrogenesis but not adipogenesis or osteogenesis. NMR Biomed. 17: 513–517.

Kraitchman, D.L., A.W. Heldman, E. Atalar, L.C. Amado, B.J. Martin, M.F. Pittenger, J.M. Hare and J.W. Bulte. 2003. *In vivo* magnetic resonance imaging of mesenchymal stem cells in myocardial infarction. Circulation 107: 2290–2293.

Kraitchman, D.L., M. Tatsumi, W.D. Gilson, T. Ishimori, D. Kedziorek, P. Walczak, W.P. Segars, H.H. Chen, D. Fritzges, I. Izbudak, R.G. Young, M. Marcelino, M.F. Pittenger, M. Solaiyappan, R.C. Boston, B.M. Tsui, R.L. Wahl and J.W. Bulte. 2005. Dynamic imaging of allogeneic mesenchymal stem cells trafficking to myocardial infarction. Circulation 112: 1451–1461.

Kustermann, E., W. Roell, M. Breitbach, S. Wecker, D. Wiedermann, C. Buehrle, A. Welz, J. Hescheler, B.K. Fleischmann and M. Hoehn. 2005. Stem cell implantation in ischemic mouse heart: a high-resolution magnetic resonance imaging investigation. NMR Biomed. 18: 362–370.

Leor, J., L. Rozen, A. Zuloff-Shani, M.S. Feinberg, Y. Amsalem, I.M. Barbash, E. Kachel, R. Holbova, Y. Mardor, D. Daniels, A. Ocherashvilli, A. Orenstein and D. Danon. 2006. *Ex vivo* activated human macrophages improve healing, remodeling, and function of the infarcted heart. Circulation 114: I94–100.

Lewin, M., N. Carlesso, C.H. Tung, X.W. Tang, D. Cory, D.T. Scadden and R. Weissleder. 2000. Tat peptide-derivatized magnetic nanoparticles allow *in vivo* tracking and recovery of progenitor cells. Nat. Biotechnol. 18: 410–414.

Modo, M., M. Hoehn, J.W. Bulte. 2005. Cellular MR imaging. Mol Imaging 4: 143–164.

Qiao, H., H. Zhang, Y. Zheng, D.E. Ponde, D. Shen, F. Gao, A.B. Bakken, A. Schmitz, H.F. Kung, V.A. Ferrari and R. Zhou. 2009. Embryonic stem cell grafting in normal and infarcted myocardium: serial assessment with MR imaging and PET dual detection. Radiology 250: 821–829.

Qiu, B., D. Xie, P. Walczak, X. Li, J. Ruiz-Cabello, S. Minoshima, J.W. Bulte and X. Yang. 2010. Magnetosonoporation: instant magnetic labeling of stem cells. Magn. Reson. Med. 63: 1437–1441.

Ransohoff, K.J. and J.C. Wu. 2010. Advances in cardiovascular molecular imaging for tracking stem cell therapy. Thromb. Haemost. 104: 13–22.

Shapiro, E.M., O. Gonzalez-Perez, J. Manuel Garcia-Verdugo, A. Alvarez-Buylla and A.P. Koretsky. 2006a. Magnetic resonance imaging of the migration of neuronal precursors generated in the adult rodent brain. Neuroimage 32: 1150–1157.

Shapiro, E.M., K. Sharer, S. Skrtic and A.P. Koretsky. 2006b. *In vivo* detection of single cells by MRI. Magn. Reson. Med. 55: 242–249.

Shapiro, E.M., S. Skrtic and A.P. Koretsky. 2005. Sizing it up: cellular MRI using micron-sized iron oxide particles. Magn. Reson. Med. 53: 329–338.

Siu, C.W., S.Y. Liao, Y. Liu, Q. Lian and H.F. Tse. 2010. Stem cells for myocardial repair. Thromb. Haemost. 104: 6–12.

Tallheden, T., U. Nannmark, M. Lorentzon, O. Rakotonirainy, B. Soussi, F. Waagstein, A. Jeppsson, E. Sjogren-Jansson, A. Lindahl and E. Omerovic. 2006. *In vivo* MR imaging of magnetically labeled human embryonic stem cells. Life Sci. 79: 999–1006.

Walczak, P., D.A. Kedziorek, A.A. Gilad, S. Lin and J.W. Bulte. 2005. Instant MR labeling of stem cells using magnetoelectroporation. Magn. Reson. Med. 54: 769–774.

Yao, Y., Y. Li, G. Ma, N. Liu, S. Ju, J. Jin, Z. Chen, C. Shen and G. Teng. 2010. *In vivo* magnetic resonance imaging of injected endothelial progenitor cells after myocardial infarction in rats. Mol. Imaging Biol. 48: 2094–2106.

21

Nitric Oxide from Nanoparticles and Applications to Cardiovascular Health

Pedro Cabrales,[1,*] *Adam J. Friedman*[2,a,3]
and *Joel M. Friedman*[2,b]

ABSTRACT

The current limitations of nitric oxide (NO) delivery systems has stimulated an extraordinary interest in the development of platforms that generate and release NO in a controlled and sustained manner, with a heavy emphasis on the treatment of cardiovascular disease states. This chapter describes the positive achievements of NO-releasing nanoparticles (NO-np) prepared using a novel platform based on hydrogel/glass hybrid nanoparticles capable of storing significant amounts of NO generated through thermal reduction of nitrite. The simplicity of the control release rate of

[1] Department of Bioengineering, University of California, San Diego, La Jolla, CA;
E-mail: pcabrales@ucsd.edu

[2] Departments of Physiology and Biophysics, Albert Einstein College of Medicine, Bronx, NY.
[a]E-mail: ajfriedm@aecom.yu.edu
[b]E-mail: jfriedma@aecom.yu.edu

[3] Division of Dermatology, Department of Medicine, Albert Einstein College of Medicine, Bronx, NY.

*Corresponding author

List of abbreviations after the text.

NO from NO-np is regulated by its hydration when exposed to moisture; therefore, it allows the nanoparticles to slowly release therapeutic levels of NO. In a dose-dependent manner, infusion of NO-np has significant direct cardiovascular effect on physiological conditions, in terms of decreasing blood pressure, increasing exhaled NO, inducing vasodilation and increasing microvascular perfusion, during several hours post-infusion. Furthermore, the NO released by the NO-np prevents the inflammatory response regenerated by immune response to macroscopic external agents in the circulation. In pathological conditions related to an increase in acellular Hb in circulation, NO-np are able to prevent negative cardiovascular reactions related to NO scavenging by the acellular Hb. Furthermore, the NO-np are capable of preventing the cardiovascular collapse related to decrease in NO generation after severe hemorrhage without fluid resuscitation. These implications suggest that NO release from NO-np is advantageous over other NO-releasing compounds, because it does not depend on chemical decomposition or enzymatic catalysis. On the basis of the observed physiological properties, NO-np has clear potential as a therapeutic agent and as a research tool to increase our understanding of NO signaling mechanisms within the cardiovascular system.

INTRODUCTION

The current limitations of nitric oxide (NO) delivery systems have stimulated an extraordinary interest in the development of compounds that generate NO in a controlled and sustained manner with a heavy emphasis on the treatment of cardiovascular disease states. Alternatives for intravascular NO therapy include formulations based on compounds containing either NO or an NO precursor in a stable form, which typically lack the capacity for controlled and sustained delivery. Thus, despite the considerable therapeutic potential of NO, systemic deployment of NO to the bedside has proven very difficult. This chapter describes the positive physiological response to the infusion of NO-releasing nanoparticles (NO-np) prepared using a new platform based on hydrogel/glass hybrid nanoparticles.

NITRIC OXIDE

NO is a heterodiatomic free radical that can participate in a wide range of biochemically relevant reactions to evoke innumerable biological responses. NO is surprisingly stable and soluble in aqueous solutions when compared to other radical species (Zacharia and Deen 2005). NO can diffuse over distances as great as several microns in aqueous solvent before engaging in collision-dependent reactions. Superoxide anion radicals and hydroxyl radicals, in comparison, are much more reactive, diffusing over much shorter reaction distances. The chemistry of NO is quite varied in physiological environments, which accounts for the rich and complex array of reactions, and serves as the basis for its broad range of biological effects.

In mammals, NO is produced enzymatically by three distinct NO oxide synthases (NOS) (Moncada et al. 1991). Each isoform differs considerably in expression and regulation, reflecting the site-specific functions for the generated NO. NOS isoforms share a similar overall catalytic scheme, in which an enzyme catalyzes the formation of NO plus L-citrulline by oxidizing the amino acid L-arginine (Busse and Fleming 1998; Moncada et al. 1991). However, the dependence of endothelial NOS (eNOS) isoform on Ca^{2+} constitutes the major difference in the regulatory mechanisms of inducible NOS (iNOS). The gene encoding for eNOS is located on human chromosome 7 and contains 26 exons spanning a region of 22 kbases. Immunohistochemical studies have located eNOS in various types of venous and arterial endothelial cells, and significant endogenous expression was also reported in myocytes, in neuronal cells, in platelets, and in various other tissues (Moncada et al. 1991). With respect to vascular function, shear stress exerted on the luminal (endothelial) surface stimulates vascular eNOS, regulating numerous vascular functions, principally smooth muscle tension (Busse and Fleming 1998). NO regulates vascular tissue homeostasis through a broad range of signaling pathways including the phospholipase, calcium signaling systems, tyrosine kinase signaling pathways, G-protein linked receptors, ion channels, and cyclic guanosine monophosphate– and cyclic adenosine monophosphate–dependent pathways.

STRATEGIES FOR INTRAVASCULAR NO SUPPLEMENTATION

NO supplementation has many complications, as NO is a short-lived molecule because it is quickly scavenged by hemoglobin (Hb) within erythrocytes (Lancaster 1997). NO is a crucial messenger in biological systems. This simple molecule functions as an important protective, regulatory, and signaling agent involved in homeostatic regulation of blood pressure and platelet aggregation, neurotransmission, and immune response.

Organic Nitrates

The term "nitrovasodilator" was used for chemical compounds that contain at least one biologically active NO moiety, which is potentially released as NO, and whose pharmacological effects include dilation of vascular smooth muscle (Feelisch and Kelm 1991). Despite the rapid acceptance of the use of organic nitrates, it has been realized that vasodilator response eventually loses clinical efficacy, a loss referred to as nitrate tolerance (Axelsson and Andersson 1983; Gori and Parker 2002). Thus, overall, it is clear that multiple enzyme systems are involved in nitrate biotransformation, with regional and efficiency variability on the resultant NO supplementation (Axelsson and Andersson 1983; Feelisch and Kelm 1991).

NONOates

Amine NO complexes are normally made by exposing nucleophilic amines in ether, or other suitable solvent, to NO in the absence of air (Homer and Wanstall 1998). One equivalent of amine reacts with two equivalents of NO and a second amine molecule extracts a proton from the initially formed stable complex in its anionic form (Homer and Wanstall 2000). These materials can be stored under nitrogen at refrigerator temperatures, for extended periods of time. These white powders are usually stable indefinitely, as solids in closed containers, but they release NO rapidly under acidic conditions and at slower rates in neutral solutions. However, when infused the NO and a nucleophile residue are released, and the nucleophile may have pharmacological or toxicological activity (Homer and Wanstall 1998). *In vitro*, their half-lives for NO release range from 2 sec (PROLI-NONOate) to 20 h (DETA-NONOate), but *in vivo* the biological effects are observed for shorter periods of time. They are advantageous research tools, e.g., for probing the influence of exposure rate on the course of NO-dependent physiological phenomena (Homer and Wanstall 1998).

Inhaled Nitric Oxide

Currently, the clinical therapeutic potential of NO has been exploited only via inhaled NO gas from pressurized tanks (Zapol 1996). The nature of NO makes it an appealing therapeutic target to treat or prevent disorders characterized by NO deficiency such as vasospasm and thrombosis. Inhaled NO gas produces selective pulmonary vasodilation and thus can be used to treat disorders, such as adult respiratory distress syndrome (Ichinose et al. 2004). Administration of exogenous NO by inhalation can be a successful way of reducing pulmonary vascular resistance in children with persistent pulmonary hypertension (Yu et al. 2009). At lung temperature and pressure, NO is a gas and behaves as such; therefore, inhalation therapy is viable. However, because the lifetime of this compound is extremely short *in vivo*, a continuous delivery of the gas is necessary (Zapol 1996). When administered by inhalation NO diffuses into the pulmonary vasculature of ventilated lung regions and reduces the pulmonary vasoconstrictor response to a variety of stimuli, including hypoxia. However, rebound hypoxemia may occur if treatment is stopped; this poses some serious limitations in inhalation therapy. While this approach is inconvenient and costly (Ichinose et al. 2004), inhaled NO gas is still the preferred and only approved NO treatment for acute pulmonary hypertension (Zapol 1996).

NANOENGINEERING INTRAVASCULAR RELEASING TECHNOLOGY

Ongoing research has revealed the therapeutic potential for NO in treating infection and in modulating vasoactivity, angiogenesis, and wound healing. Harnessing this potential poses a challenge, especially the development of a convenient or cost-effective approach. The objective, then, is to develop therapeutic strategies that sequester the NO functionality into a chemically stable carrier that can be directed to the deficient site to carry out its functions without systemic consequences (Singel and Stamler 2005).

Early approaches to delivering NO using nanotechnology included gold nanoparticles (1–5 nm) covered with amino groups reacted with NO. Gold nanoparticles were synthesized and functionalized with bromo-terminated alkanethiols and then reacted with amine groups. Lastly, the amine-functionalized gold nanoparticles were pressurized with NO and stirred to facilitate the synthesis of NO donors and prevent aggregation (Friedman and Friedman 2009). Although gold nanoparticles are a flexible technology to deliver NO, the total amount transported is limited.

HYBRID SOL-GEL NO-RELEASING NANOPARTICLES

A novel redox chemistry in solid matrices can be used as the basis for new platforms to generate powder formulations that not only are capable of delivering NO in a sustained fashion, but also are straightforward to store and apply. The general platform for preparing the NO-releasing materials is based on sugar-derived glassy matrices that support redox reactions, initiated using thermally generated electrons from glucose (Navati and Friedman 2006). This matrix supports the reduction of nitrite to NO. Additionally, it is a composite material having both sol-gel and glassy properties. Therefore, the NO generated remains trapped and stable within the dry matrix until the matrix is exposed to moisture

(Friedman and Friedman 2009). They also include a more robust material, tetramethoxysilane or tetraethoxysilane, as a backbone to create a nanoporous substrate suitable to trap all the molecules including sugar, water, nitrite, and NO (Ray et al. 2002). Additionally, both polyethylene glycol (PEG) and chitosan can be introduced into the porous matrix to control the hydration rate and NO release rate (Friedman et al. 2008). The resulting material is conveniently converted into nanoparticles capable of delivering therapeutic amounts of NO, over extended time periods (Fig. 1).

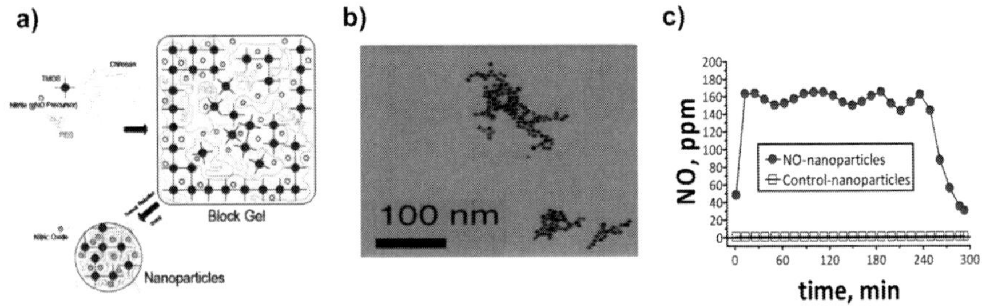

Fig. 1. NO-releasing nanoparticle synthesis and morphology. (a) Schematic of NO-releasing nanoparticle synthesis. The various components are prepared and combined to form a block gel consisting of a silica hydrogel matrix encompassing the other ingredients. The block gel is dried and the nitrite precursor converted to NO via thermal reduction, resulting in the formation of the nanoparticles. (b) NO-releasing nanoparticles. The scale bars represent 100 nm, the bottom one representing the lower left and right panes and the upper one the upper left image. (c) NO gas levels were measured using a chemiluminescent NO analyzer at 7.4 pH, 1 mg of NO-np. Control nanoparticles are prepared identically, but without nitrite.

Color image of this figure appears in the color plate section at the end of the book.

The release pattern of NO from the matrices is dependent on the preparation protocol, in all cases the release of NO initiated through exposure to moisture. The NO-containing dry powder remains stable for months. The pattern of NO release can be dramatically tuned by varying the composition of the constituents, especially PEG. The inclusion of large PEG molecules results in rapid release, whereas the inclusion of smaller PEGs results in slow, sustained release. All the materials comprising these particles are being used in drug delivery formulations, with no reported toxicities. The likely predominant product arising from the reduction of nitrite by glucose is gluconic acid, which is non-toxic and biocompatible. The potential of this NO-np to deliver pure NO is further enhanced by the nano-scale size of the particles. Despite the potential for intravascular systemic delivery, these powders appear ideal for immediate development and deployment for applications requiring cutaneous delivery of exogenous NO.

TOPICAL APPLICATION OF NO-RELEASING NANOPARTICLES

An obvious and immediate application for NO-np is the treatment of cutaneous infections and wounds. Staphylococcus aureus (SA) is a leading cause of both superficial and invasive infections in community and hospital settings, frequently resulting in chronic refractory disease. It is imperative that innovative therapeutics, to which the bacteria are unlikely to evolve resistance, be developed to curtail associated morbidity and mortality and ultimately improve the capacity to treat these infections. There is growing and compelling evidence that endogenously generated NO plays a significant role in the chemistry and physiology of antimicrobial action and wound healing. NO-releasing nanoparticles were applied to the treatment of methicillin-resistant SA (MRSA) wound infections (Fig. 2). The results showed that the nanoparticles exert antimicrobial activity against MRSA in a murine wound model (Han et al. 2009a; Martinez

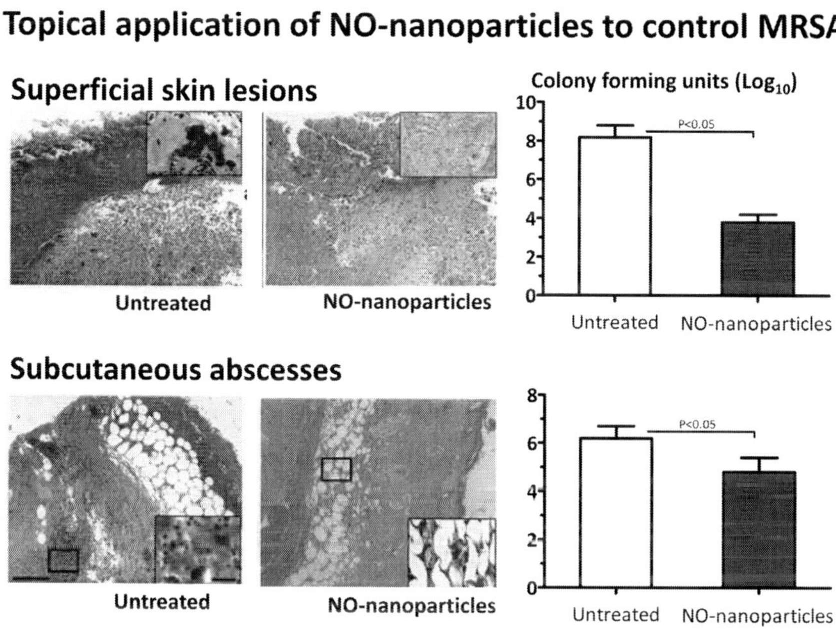

Fig. 2. Topical application of NO-releasing nanoparticles (NO np) to treat methicillin-resistant Staphylococcus aureus (MRSA), superficial and subcutaneous abscesses in mice, showed that the NO-np exerted antimicrobial activity against MRSA abscesses. Topical or intradermal NO-np treatment of abscesses reduced the area and bacterial load while improving skin architecture. Notably, inflammatory cytokines involved in immunomodulation and wound healing revealed that NO-np led to a reduction in angiogenesis, preventing bacterial dissemination. Data previously published by the authors (Han et al. 2009a, b).

Color image of this figure appears in the color plate section at the end of the book.

et al. 2009). Acceleration of infected wound closure in NO-treated groups was shown, compared with controls. The histology of wounds revealed that NO-np treatment decreased suppurative inflammation, reduced bacterial burden, and decreased collagen degradation (Han et al. 2009a; Martinez et al. 2009). The nanoparticulate powder could be used in the treatment of numerous types of wounds. Furthermore, wound healing, complicated by common comorbidities, such as obesity, diabetes, atopic dermatitis, and peripheral vascular disease, may benefit from an NO-based therapeutic. The beneficial effects of NO-np treatment on wound closure are likely attributable to antimicrobial activity and the promotion of collagen deposition (Han et al. 2009a; Martinez et al. 2009). The stability of the compound, combined with the simplicity of treatment, makes NO-np ideal for combat or disaster situations; soldiers or emergency personnel could carry packets with NO-np and topically apply it to traumatic wounds in the field to combat microbial infections before medical care becomes available.

CARDIOVASCULAR APPLICATIONS

The currently available NO delivery systems have significant limitations and drawbacks especially with respect to post-administration and efficacy. The use of gaseous NO is not practical given both the requirements for gas cylinders, flow modulation techniques, and the lack of suitability for non-hospital settings. While NO-releasing materials suffer limitation to varying degrees, the NO-np platform has already far exceeded the performance of previous NO-releasing materials (Cabrales et al. 2010b). The fact that the NO-np platform uses a spontaneous water-mediated release of NO, the actual molecule that is continuously bio-generated for physiological function, suggests that habituation to the NO-np platform is highly unlikely. The NO-np have been shown to penetrate skin effectively, in topical treatment of both abscesses and erectile dysfunction (Han et al. 2009a, b). In a rat erectile dysfunction model, topical application of NO-np has been shown to be quick and effective in restoring erectile activity, indicating both skin penetration of the particles and sustained release of the NO (Han et al. 2010).

INTRAVASCULAR INFUSION OF NO-RELEASING NANOPARTICLES

The largest family of NO donors currently in use is NONOates, also known as diazeniumdiolates. They are complexes of NO with nucleophiles, for which the decomposition rate depends on pH, temperature, and the nature of the nucleophile (Maragos et al. 1991, 1993). To compare the action of the NO-np platform to established NO donors, two particular NONOates were selected and examined on the basis of their disparate, but physiologically relevant, decomposition rates (DETA NONOate, half-life of 20 h at 37°C, pH 7.4; and DPTA NONOate, half-life of 3 h 37°C, pH 7.4) (Homer and Wanstall 1998, 2000). These results provide *in vivo* experimental evidence that NO-np intravascular exogenous supplementation of NO is more effective than molecular NO donors (Fig. 3). A NONOate theoretically releases 2 mol of NO per mole of parent compound, although experimentally at 37°C and pH 7.4, DETA NONOate liberates 10.7 μmol of NO/mg and DPTA NONOate liberates 9.4 μmol of NO/mg. In contrast to the NONOates, the NO-np half-life *in vitro* was found to be approximately 4 h at 37°C and pH 7.4, and 1 mg of powder was capable of releasing 0.32 μmol of NO (Fig. 1). Importantly, *in vitro* 1 mg of NONOates released more than 30 times the NO that NO-np did. Blood pressure decreased after infusion of NO-np in a dose-dependent fashion (Fig. 3). Maximal decrease in blood pressure after treatment with NO-np was measured at 90 min after infusion. Infusion of control particles (without NO) did not change blood pressure (Cabrales et al. 2010b). DETA NONOate and DPTA NONOate produced similar decreases in blood pressure (Fig. 3). The physiological impact of higher doses of NONOate was manifested by a significant increase on metHb formation with a subsequent decrease in oxygen-carrying capacity. Exhaled NO increased after infusion of NO-np, maximal exhaled NO levels were measured 1 h after infusion. Increased exhaled NO levels were maintained over a period of 2 h; after that, exhaled NO decreased. Additionally, infusion of fluorescently labeled NO-np allowed nanoparticles to be observed in circulation up to 6 h after infusion. However, control-np (without NO) were not observed in the circulation 4 h after infusion (Cabrales et al. 2010b). The physiological response to the infusion of NO-np in the circulatory system is consistent with a sustained release of physiologically relevant concentrations of NO. The efficacy and potential potency of the exogenous NO released by the NO-np is evident from the systemic hypotension and increase in exhaled NO (Cabrales et al. 2010b).

Circulating NO-np produced microvascular vasodilation and increased blood flow compared to baseline and control-np (Fig. 4). Infusion of control np did not affect arteriolar diameter or blood flow. Functional capillary density (FCD) was not affected in animals treated with NO-np, but control-np reduced the FCD (Fig. 4). This sustained release is reflected in both the decrease in blood pressure and the increase in vascular relaxation, compared to the absence of any vascular response following the infusion of the control-np (without NO). The number of immobilized leukocytes increased after infusion of control particles at 1 h and 2 h,

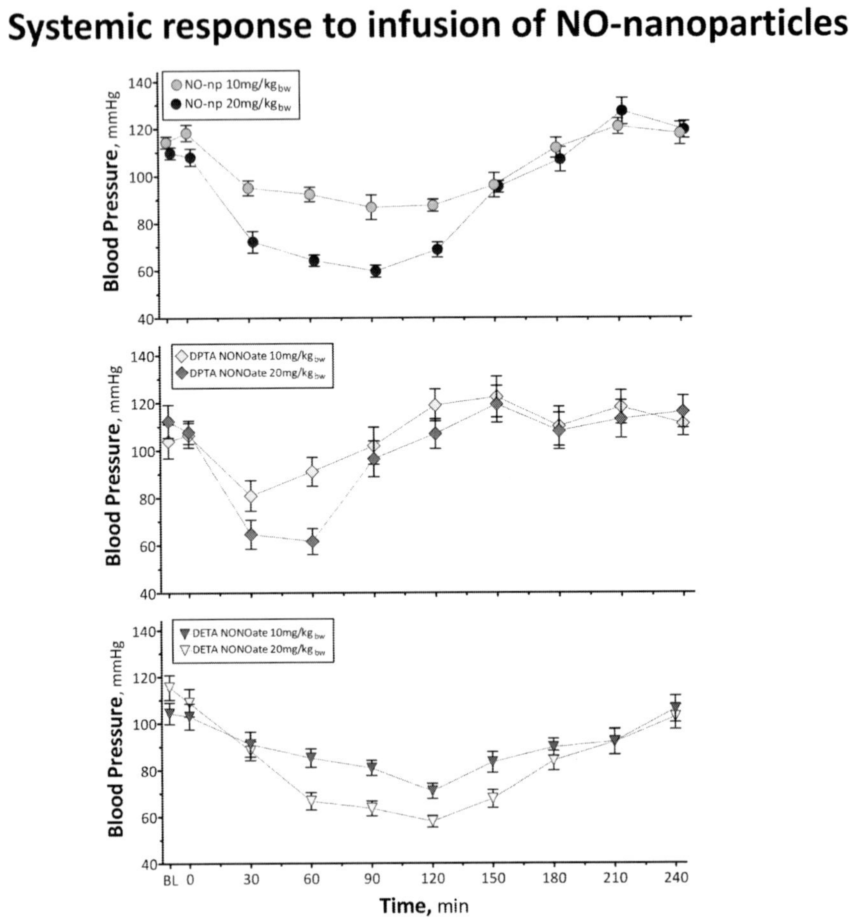

Fig. 3. Effects in blood pressure after infusion of NO-releasing nanoparticles (NO-np) and NONOates. The controlled and sustained release of NO observed from NO-np is regulated by its hydration. In a dose-dependent manner, circulating NO-np both decreased mean arterial blood pressure and increased exhaled NO concentrations over several hours. (a) Infusion of 10 mg/kg (n = 5) NO-np and 20 mg/kg (n = 5) NO-np. Maximal decreased pressure was measured 90 min after infusion. MetHb levels 4 h after infusion of NO-np were 9 ± 2% for 10 mg/kg and 14 ± 3% for 20 mg/kg, respectively. (b) Infusion of 10 mg/kg (n = 5) DPTA NONOate and 20 mg/kg (n = 5) DPTA NONOate. (c) Infusion of 10 mg/kg (n = 5) DETA NONOate and 20 mg/kg (n = 5) DETA NONOate. Investigations were performed using the hamster chamber window model, widely used for microvascular studies in the unanesthetized state, and the complete surgical technique is described in detail elsewhere (Endrich et al. 1980). Animals received NO-np suspended in deoxygenated saline (50 μl) via the jugular vein. NONOates (Cayman Chemicals, Ann Arbor, MI) were used: DETA-NONOate (Diethylenetriamine NONOate, CAS 146724-94-9) and DPTA-NONOate (Dipropylenetriamine NONOate, CAS 146724-95-0). Data previously published by the authors (Cabrales et al. 2010b).

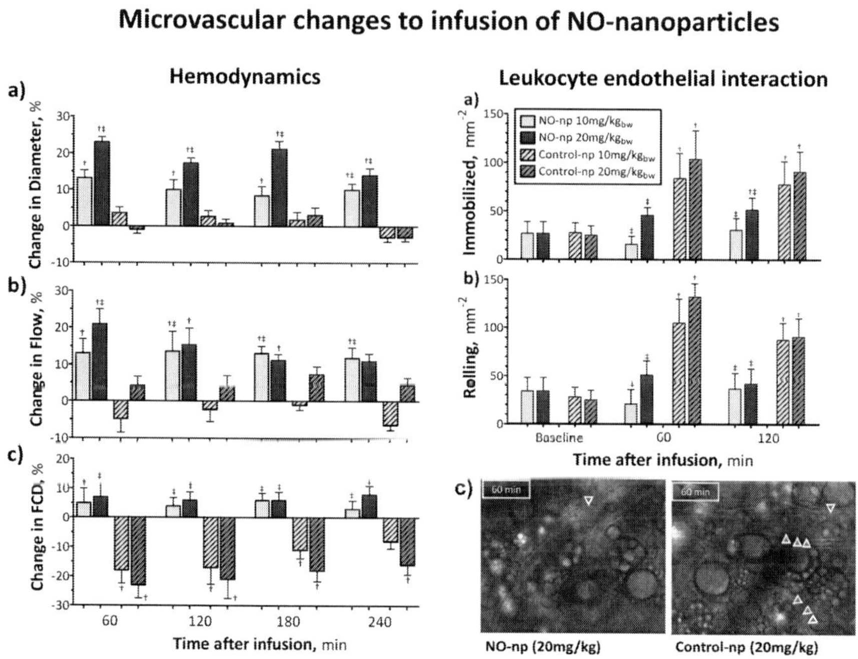

Fig. 4. Left panels, microvascular hemodynamic changes after infusion of NO-releasing nanoparticles (NO-np) and control nanoparticles (Control-np, without NO). Circulating NO-np induced vasodilatation and increased microvascular perfusion during their circulation lifetime of several hours. (a) Changes in arteriolar diameter after infusion of 10 and 20 mg/kg of NO-np (no pattern) and Control-np (pattern upward diagonal). (b) Changes in arteriolar blood flow after infusion of NO-np and Control-np. (c) Changes in functional capillary density (FCD) after infusion of NO-np and Control-np. †$P < 0.05$ to baseline; ‡$P < 0.05$ to Control-np at same concentration. Right panels, rolling and immobilized leukocytes after infusion of NO-np. (a) Immobilized leukocytes after infusion of 10 and 20 mg/kg of NO-np (no pattern) and Control-np (pattern upward diagonal). (b) Rolling leukocytes after infusion of NO-np and Control-np. Rolling leukocytes were quantified in the same animals and locations as Immobilized. †$P < 0.05$ to baseline; ‡$P < 0.05$ to Control-np at same concentration. (c) Immobilized and rolling leukocytes after infusion of 20 mg/kg of NO-np and Control-np. Red arrows point to immobilized leukocytes on the images, other leukocytes in the images are rolling or flowing. Investigations were performed using the hamster chamber window model, widely used for microvascular studies in the unanesthetized state, and the complete surgical technique is described in detail elsewhere (Endrich et al. 1980). Data previously published by the authors (Cabrales et al. 2010b).

Color image of this figure appears in the color plate section at the end of the book.

respectively (Fig. 4). NO-np did not increase immobilized leukocytes compared to baseline. The number of rolling leukocytes increased from baseline after infusion of control-np at all time points. NO-np did not increase the number of rolling leukocytes compared to baseline, and no difference was measured between doses of NO-np (Cabrales et al. 2010b).

There are strong indications that activated macrophages can recognize nanoparticles in the blood (Moghimi and Gray 1997). The presence of exogenous NO from the nanoparticles prevented inflammatory reaction to external agents, especially the number of immobilized and rolling leukocytes. As not all macrophages and monocytes are identical (Rutherford et al. 1993), macrophages found within microcirculatory regions are more active in cytokine production and have stronger phagocytic receptors (Ito et al. 1999). Therefore, NO-np technology could prevent long-circulating nanoparticles from being trapped by such macrophage populations. The NO-np technology seems to be optimal for macrophage evasion, providing a long-circulating drug reservoir from which the drug can be released into the vascular compartment in a continuous and controlled manner (Cabrales et al. 2010b).

PREVENTING ADVERSE RESPONSES OF ACELLULAR HEMOGLOBIN

Several formulations of Hb-based oxygen carrier plasma expander, including removal of red cell stroma to minimize renal toxicity, and Hb polymerization and cross-linking to resolve extravasation and extended circulating half-life, have been evaluated (Alayash 2004). All of these cross-linked and polymerized Hbs produced elevation of blood pressure and vasoconstriction, predominantly from scavenging of NO by plasma ferrous heme. Thus, NO bioavailability decreases upon hemolysis and as a result, endothelial vasomotor function, blood cell adhesion, and hemostasis are all adversely affected. The enhanced ability of acellular Hb to scavenge NO has been widely believed to be the cause of the hypertension, increased systemic and pulmonary vascular resistance, and morbidity and mortality associated with administration of Hb-based oxygen carrier plasma expander. This process shares many mechanistic similarities to intravascular hemolysis, as both situations eliminate the erythrocyte compartmentalization of the Hb, which removes the limit for NO scavenging by decreasing the reaction between NO and molecular oxygen (Lancaster 1994). NO-np can prevent/reverse the negative effects of acellular Hb, and therefore should function to produce vasodilation during acute or chronic hemolysis (Fig. 5). The levels of acellular Hb after infusion of Hb-based oxygen carrier plasma expanders are comparable with those seen in hemolysis resulting from sickle cell vaso-occlusive crisis (Reiter et al. 2002) and during other clinically relevant human hemolytic conditions (Gramaglia et al. 2006) (e.g., cardiopulmonary bypass, hemoglobinuria, alloimmune hemolysis).

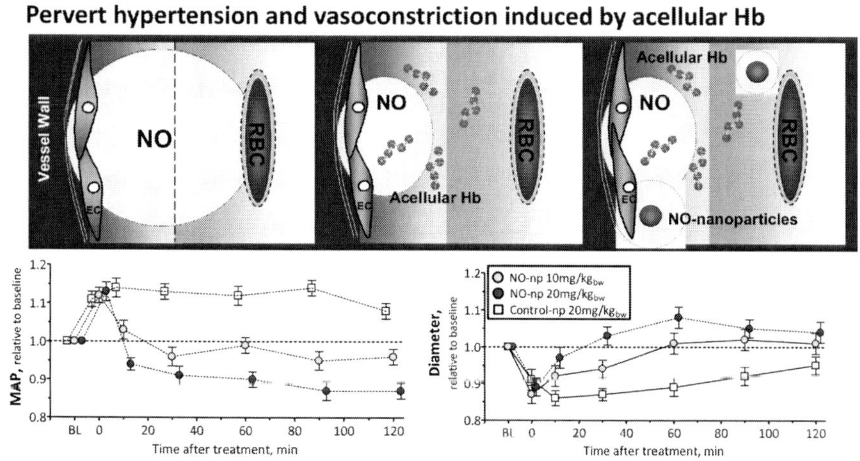

Fig. 5. Acellular Hb induces vasoconstriction due to NO scavenging, limiting the NO available for vascular smooth muscle. The central objective of this study was to restore NO bioavailability using NO-releasing nanoparticles (NO-np), to reverse the vasoconstriction and hypertension induced by polymerized bovine Hb (PBH) NO scavenging. PBH (13 g/dl) was infused in a volume equal to 10% of the animal blood volume. NO-np were provided at 10 or 20 mg by kg body weight. The study was performed using the hamster window chamber model to concurrently access systemic and microvascular hemodynamics (Endrich et al. 1980). Infusion of PBH increased blood pressure and induced vasoconstriction. Treatment with 10 and 20 mg/kg of NO-np reduced the blood pressure and vasoconstriction induced by PBH. Moreover, the higher dose of NO-np decreased blood pressure and induced vasodilation compared to baseline, respectively. Treatment with NO-np to decrease PBH-induced vasoconstriction increased metHb. NO-np has potential as a therapeutic agent to replenish NO in situations where NO production is impaired, insufficient or consumed, thereby preventing vascular complications. Data previously published by the author (Cabrales et al. 2010a).

Color image of this figure appears in the color plate section at the end of the book.

The systemic blood pressure and peripheral vasoconstriction associated with infusion of polymerized bovine Hb (PBH) can be reversed through administration of NO-np (Fig. 5). The results are consistent with a mechanism that involves restoring intravascular NO concentrations. If indeed the reversal is due to replenishing the NO scavenged by PBH, then the results imply that infusion of NO-np in situations during which vascular NO bioavailability is decreased, such as acellular Hb transfusions or hemolysis, can prevent or simply counteract NO scavenging–induced hypertension and vasoconstriction complications by replacing the NO consumed by acellular Hb (Cabrales et al. 2010a). The sustained release of NO from the NO-np is reflected by the continued reversal of the hypertension and vasoconstriction, as compared to changes produced by the control-np (without NO). Dose-dependent concentrations of exogenous NO released by the NO-np were evidenced by the significant decrease in blood pressure, vasodilation and acellular metHb formation. Furthermore, these data support the generally accepted concept that PBH-induced hypertension and vasoconstriction are due to the NO scavenging (Cabrales et al. 2010a).

Other strategies to prevent acellular Hb negative effects included inhaled NO, which prevented pulmonary hypertension, as the observed effects were on pulmonary hemodynamics (Yu et al. 2009). Any effects beyond the lung vasculature during NO inhalation can only be attributed NO-related metabolites (such as nitrite). On the other hand, nitrite can be a unique vasodilator, in part because of NO production through the allosterically regulated nitrite reductase activity of Hb (Minneci et al. 2008). Additionally, attempts to attenuate NO scavenging by acellular Hbs include sodium nitrite as a means for compensating NO scavenging of acellular Hb (Minneci et al. 2008), and during resuscitation with Hb oxygen carriers (Rodriguez et al. 2009). Another approach contemplated to overcome the NO scavenging effects is to co-infuse SNO-Hb along with unmodified Hb oxygen carrier; this approach seems to be effective in acute hypoxic, but not normoxic, conditions (Irwin et al. 2010). Katz et al. used nitroglycerin to reduce vasoconstriction in a swine model and were able to obtain a transient decrease in constriction (Katz et al. 2010). Nitroglycerine is an organic nitrate that releases NO from a three-electron reduction process, and the reducing agents involved lose effectiveness by continuous use (Gori and Parker 2002). Lastly, a combination of Sildenafil with NO-np, a phosphodiesterase type 5 (PDE5) inhibitor with NO-np, to prevent hypertension and vasoconstriction induced by acellular Hb was found to increase the efficacy of a given dose of NO-np. Therefore, the potential therapeutic collaboration between Sildenafil (or other PDE5 inhibitor) and NO-np will decrease the amount of NO-np needed in clinical scenarios where PDE5 inhibition may be applicable.

These findings demonstrate that NO-np can prevent/reverse the negative effects of acellular Hb, and therefore should function to produce vasodilation during acute or chronic hemolysis. NO-np may have a therapeutic role in minimizing the vascular toxicities of more severe episodes of intravascular hemolysis, such as erythrocyte parasitic diseases (Gramaglia et al. 2006). In these scenarios, the ability of NO to attenuate the pathophysiological effects of acellular Hb has been amply demonstrated, but a practical mechanism to administer NO remains to be defined. NO-np are a unique vasodilator under normal physiological conditions (Cabrales et al. 2010b), and in the presence of acellular Hb, they minimized the expected vascular adverse events.

RESTORATION OF NO HOMEOSTASIS AFTER HEMORRHAGIC SHOCK

Despite significant advances in the clinical setting for treating trauma, for hemorrhagic trauma there remains a tremendous need for efficacious volume interventions for first responders to rapidly stabilize single or mass casualties. Therefore, the development of a novel, low-weight and easy-to-store therapy that can greatly reduce logistical burdens is expected to provide life-saving treatments that would be broadly applicable. Hemorrhagic shock-induced cardiovascular collapse is related to disruption of NO generation as a consequence of (1) drastic reduction on endothelial shear stress (due to low blood pressure, cardiac output, circulating volume and blood viscosity), (2) free radical formation/accumulation (uncoupling eNOS) and (3) hypoxia (oxygen is a co-substrate for NO production). Because of the many roles of NO, restoration of bioavailability of the NO lost upon hemorrhagic shock is likely to reduce many shock complications including cardiovascular, inflammatory and thrombotic effects. Since NO synthesized by endothelial cells is the major source of NO to the vasculature, NO-np appears to be an optimal intraluminal source of NO during hemorrhagic shock.

Treatment of hemorrhagic shock with NO-np prevented cardiovascular collapse and allowed the animals to maintain superior systemic and microvascular hemodynamic conditions as compared to control-np (without NO). The importance of providing sustained exogenous NO after hemorrhage was evidenced in the form of maintenance of heart rate, microvascular function, FCD and vascular resistance. These findings relate to scenarios where an increase in NO bioavailability could potentially alter the course of treatment and ultimately increase survival, until volume resuscitation is available. By changing vascular reactivity during hemorrhagic shock through NO supplementation, the development of multiple organ dysfunction can be prevented. This proposed therapy to prevent hypovolemic cardiovascular collapse is capable of maintaining systemic hemodynamics and microvascular function in such a fashion that whole organism survivability and local tissue viability are ensured, to reduce hemorrhagic shock sequelae.

NO-RELEASING NANOPARTICLE EFFECTS ON HEART PERFORMANCE

The role of NO in the regulation of cardiac function and the circulatory system has been studied in both normal and pathological hearts. Some studies reported that NO had a bimodal effect on cardiac inotropy: positive inotropic effect at low amounts of NO exposure and a negative effect at higher amounts. A miniaturized pressure-volume conductance catheter to simultaneously measure ventricular pressure and volume allows the study of cardiac function in small animals (Chatpun and Cabrales 2010), and using this setup NO-nps were particularly effective in enhancing cardiac contractility (Fig. 6). The experimental results indicate that NO-np decreased pressure and increased stroke volume and heart rate. These cardiac effects include increased cardiac output, lowered arteriolar after-load and cardiac filling, and increased preload (venous return), because of NO vasodilatory properties and cardiac chronotropic effects (Fatehi-Hassanabad and Fatehi 2004). In conclusion, these findings demonstrate the beneficial effects of supplemented NO on cardiac function by increasing cardiac output, reducing vascular resistance, and lowering stroke work per ejected volume (Fig. 6).

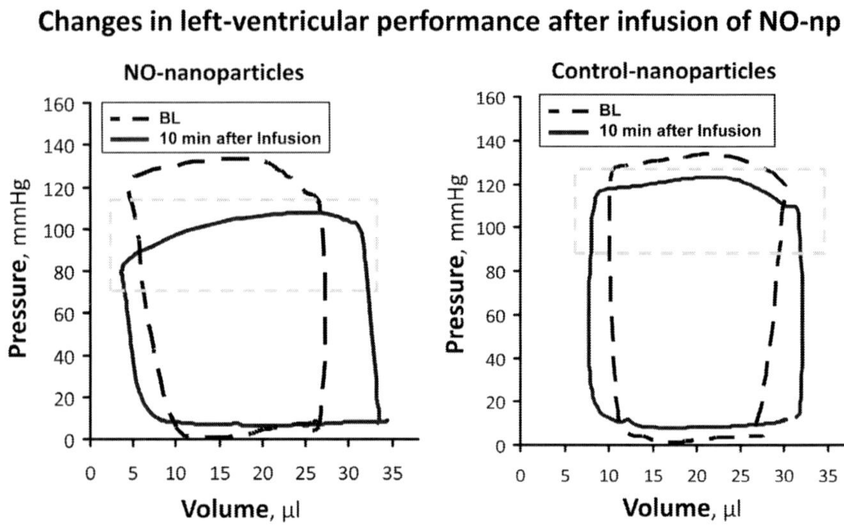

Fig. 6. This study investigated the left ventricular pressure volume (LV) changes after an acute increase in NO availability by administration of 5 mg by kg of body weight of NO-releasing nanoparticles (NO-nps). NO is a critical endothelial-derived relaxing factor, which plays a primordial role in the maintenance of vascular homeostasis. NO also affects organ function, including the heart, specifically the myocardium. NO-np showed effects on systemic and cardiac function parameters. Cardiac output increased by 20–22%. Stroke work was unchanged compared to baseline, although stroke work per stroke volume was significantly lower in the NO-np than in control-np ($p < 0.05$). The maximum rate of pressure change (dP/dt_{max}) was not different between groups but the minimum rate of pressure change (dP/dt_{min}) was significantly lower in the NO-np. Systemic vascular resistance in the NO-np was reduced to 55–62% of the baseline ($p < 0.05$). These findings demonstrated the beneficial effects on cardiac function of a small dose of NO-np. The relevance, stability, and efficacy of exogenous NO therapy in the form of NO-np will potentially facilitate the intended use in battlefield and trauma situations. Unpublished results by the author.

APPLICATIONS TO AREAS OF HEALTH AND DISEASE

NO-releasing nanoparticle (NO-np) platform has been developed to be injected into the circulation and provide sustained release of NO, or locally injected in a specific location to release NO. When in circulation, the NO-np increases the exhaled NO levels, decreases blood pressure, dilates blood vessels, decreases inflammation, extends the circulation time compared to NO-free nanoparticles, increases the formation of nitroso compounds and enhances phophodiesterase inhibitor effects, all over several hours. Infusions of the particles counteract vasoconstriction due to the presence of vasoconstrictive acellular Hb and hemorrhagic shock. While nanoparticles without NO are rapidly cleared from the bloodstream by the macrophages of the reticulo-endothelial system, NO-np circulates for long periods and can include other compounds serving as a platform to transport drugs, magnetic or radiopaque materials for diagnostic imaging, or bioactive molecules. The bioactive levels of NO released from the nanoparticles exert major biological effects including neurotransmission, regulation, macrophage-derived host-defense molecule, inhibition of platelet aggregation and endothelium adhesion molecules expression, antioxidant, cardiac chronotropic control, and potent vasodilation.

Key Facts

- NO-releasing nanoparticles are a cost-effective, easily deployed platform for sustained topical and systemic delivery of therapeutic levels of NO.
- NO has tremendous therapeutic potential for a broad range of clinical conditions.
- NO delivery requirements extend from highly localized and/or topical, to infusions for systemic applications.
- NO-np are an NO delivery platform that can allow for the elucidation of NO's role in physiological and pathophysiological processes.
- NO-np can be a general platform to transport and deliver therapeutic amounts of other small molecules such as peptides, drugs, chemotherapeutics, and nucleic acids.
- NO-np are a biocompatible platform capable of a targeted (topical, IV, IP), sustained release of NO with release rates that can be easily modulated.

Definition

NO homeostasis: The continuous production and release of endothelium-derived NO plays an important role in vascular and cardiac function. The signaling actions of NO in the vasculature are short-lived as a result of its rapid reaction with Hb, and the immense concentration of the latter, in blood.

Summary Points

- NO-releasing nanoparticles (NO-np) are a cost-effective, easily deployed platform for sustained topical and intravascular delivery of therapeutic levels of NO.
- Stable, biocompatible NO-np will have many applications.
- NO-np have the potential to be a routine drug or additive, as well as a long-lived circulation platform for drug delivery.
- NO-np can prevent and/or revert endothelial dysfunction by an NO supplementation strategy.
- NO-np has potential as a therapeutic agent to replenish NO in situations where NO production is impaired, insufficient, or consumed, thereby preventing vascular complications.

Acknowledgements

This work was supported by Program project P01 HL071064 to FJC. The authors thank George Han, Camille Roche, Parimala Nacharaju, Froilan P. Barra and Cynthia Walser for their contributions.

Abbreviations

eNOS	:	endothelial NOS
FCD	:	functional capillary density
Hb	:	hemoglobin
iNOS	:	inducible NOS
MetHb	:	methemoglobin
MRSA	:	methicillin-resistant Staphylococcus aureus
NO	:	nitric oxide
NO-np	:	NO-releasing nanoparticles
NOS	:	NO oxide synthase
PBH	:	polymerized bovine Hb
PDE5	:	phosphodiesterase type 5
PEG	:	polyethylene glycol
SA	:	Staphylococcus aureus

References

Alayash, A.I. 2004. Oxygen therapeutics: Can we tame haemoglobin? Nature Rev. Drug Discovery 3: 152–159.

Axelsson, K.L. and R.G. Andersson. 1983. Tolerance towards nitroglycerin, induced *in vivo*, is correlated to a reduced cGMP response and an alteration in cGMP turnover. Eur. J. Pharmacol. 881: 71–79.

Busse, R. and I. Fleming. 1998. Pulsatile stretch and shear stress: physical stimuli determining the production of endothelium-derived relaxing factors. J. Vasc. Res. 352: 73–84.

Cabrales, P., G. Han, P. Nacharaju, A.J. Friedman and J.M. Friedman. 2010a. Reversal of hemoglobin-induced vasoconstriction with sustained release of nitric oxide. Am. J. Physiol. Heart Circ. Physiol.

Cabrales, P., G. Han, C. Roche, P. Nacharaju, A.J. Friedman and J. M. Friedman. 2010b. Sustained release nitric oxide from long-lived circulating nanoparticles. Free Radic. Biol. Med. 494: 530–538.

Chatpun, S. and P. Cabrales. 2010. Effects of plasma viscosity modulation on cardiac function during moderate hemodilution. Asian J. Transfus. Sci. 42: 102–108.

Endrich, B., K. Asaishi, A. Götz and K. Messmer. 1980. Technical report: A new chamber technique for microvascular studies in unanaesthetized hamsters. Res. Exp. Med. 177: 125–134.

Fatehi-Hassanabad, Z. and M. Fatehi. 2004. Characterisation of some pharmacological effects of the venom from Vipera lebetina. Toxicon 434: 385–391.

Feelisch, M. and M. Kelm 1991. Biotransformation of organic nitrates to nitric oxide by vascular smooth muscle and endothelial cells. Biochem. Biophys. Res. Commun. 1801: 286–293.

Friedman, A. and J. Friedman. 2009. New biomaterials for the sustained release of nitric oxide: past, present and future. Expert Opin. Drug Deliv. 610: 1113–1122.

Friedman, A.J., G. Han, M.S. Navati, M. Chacko, L. Gunther, A. Alfieri and J.M. Friedman. 2008. Sustained release nitric oxide releasing nanoparticles: characterization of a novel delivery platform based on nitrite containing hydrogel/glass composites. Nitric Oxide 191: 12–20.

Gori, T. and J.D. Parker. 2002. The puzzle of nitrate tolerance: pieces smaller than we thought? Circulation 10618: 2404–2408.

Gramaglia, I., P. Sobolewski, D. Meays, R. Contreras, J.P. Nolan, J.A. Frangos, M. Intaglietta and H.C. van der Heyde. 2006. Low nitric oxide bioavailability contributes to the genesis of experimental cerebral malaria. Nat. Med. 1212: 1417–1422.

Han, G., L.R. Martinez, M.R. Mihu, A.J. Friedman, J.M. Friedman and J.D. Nosanchuk. 2009a. Nitric oxide releasing nanoparticles are therapeutic for Staphylococcus aureus abscesses in a murine model of infection. PLoS One 411: e7804.

Han, G., M. Tar, D.S. Kuppam, A. Friedman, A. Melman, J. Friedman and K.P. Davies. 2010. Nanoparticles as a novel delivery vehicle for therapeutics targeting erectile dysfunction. J. Sex. Med. 71 Pt 1: 224–233.

Han, G., J.H. Zippin and A. Friedman. 2009b. From bench to bedside: the therapeutic potential of nitric oxide in dermatology. J. Drugs Dermatol. 86: 586–594.

Homer, K. and J. Wanstall 1998. In vitro comparison of two NONOates (novel nitric oxide donors) on rat pulmonary arteries. Eur J Pharmacol. 3561: 49–57.

Homer, K.L. and J.C. Wanstall. 2000. Cyclic GMP-independent relaxation of rat pulmonary artery by spermine NONOate, a diazeniumdiolate nitric oxide donor. Br. J. Pharmacol. 1314: 673–682.

Ichinose, F., J.D. Roberts, Jr. and W.M. Zapol. 2004. Inhaled nitric oxide: a selective pulmonary vasodilator: current uses and therapeutic potential. Circulation 10925: 3106–3111.

Irwin, D., P.W. Buehler, A.I. Alayash, Y. Jia, J. Bonventura, B. Foreman, M. White, K.L. Hamilton and R.W. Gotshall 2010. Mixed S-nitrosylated polymerized bovine hemoglobin species moderate hemodynamic effects in acutely hypoxic rats. Am. J. Respir. Cell. Mol. Biol. 422: 200–209.

Ito, S., M. Naito, Y. Kobayashi, H. Takatsuka and S. Jiang, H. Usuda, L.D. Shultz, O. Elomaa and K. Tryggvason. 1999. Roles of a macrophage receptor with collagenous structure (MARCO) in host defense and heterogeneity of splenic marginal zone macrophages. Arch. Histol. Cytol. 621: 83–95.

Katz, L.M., J.E. Manning, S. McCurdy, C. Sproule, G. McGwin, Jr., P. Moon-Massat, C.B. Cairns and D. Freilich 2010. Nitroglycerin attenuates vasoconstriction of HBOC-201 during hemorrhagic shock resuscitation. Resuscitation 814: 481–487.

Lancaster, J.R. 1994. Simulation of the diffusion and reaction of endogenously produced nitric oxide. Proc. Natl. Acad. Sci. USA 91: 8137–8141.

Lancaster, J.R., Jr. 1997. A tutorial on the diffusibility and reactivity of free nitric oxide. Nitric Oxide 11: 18–30.

Maragos, C.M., D. Morley, D.A. Wink, T.M. Dunams, J.E. Saavedra, A. Hoffman, J.A. Hrabie and L.K. Keefer 1991. Complexes of NO with nucleophiles as agents for the controlled biological release of nitric oxide. Vasorelaxant effects. J. Med. Chem. 3411: 3242–3247.

Maragos, C.M., J.M. Wang, J.A. Hrabie, J.J. Oppenheim and L.K. Keefer 1993. Nitric oxide/nucleophile complexes inhibit the *in vitro* proliferation of A375 melanoma cells via nitric oxide release. Cancer Res. 533: 564–568.

Martinez, L.R., G. Han, M. Chacko, M.R. Mihu, M. Jacobson, P. Gialanella, A.J. Friedman, J.D. Nosanchuk and J.M. Friedman 2009. Antimicrobial and healing efficacy of sustained release nitric oxide nanoparticles against Staphylococcus aureus skin infection. J. Invest. Dermatol. 12910: 2463–2469.

Minneci, P.C., K. J. Deans, S. Shiva, H. Zhi, S.M. Banks, S. Kern, C. Natanson, S.B. Solomon and M.T. Gladwin 2008. Nitrite reductase activity of hemoglobin as a systemic nitric oxide generator mechanism to detoxify plasma hemoglobin produced during hemolysis. Am. J. Physiol. Heart Circ. Physiol. 2952: H743–754.

Moghimi, S.M., and T. Gray 1997. A single dose of intravenously injected poloxamine-coated long-circulating particles triggers macrophage clearance of subsequent doses in rats. Clin. Sci. (Lond.) 934: 371–379.

Moncada, S., and J.D. Erusalimsky 2002. Does nitric oxide modulate mitochondrial energy generation and apoptosis? Nat. Rev. Mol. Cell. Biol. 33: 214–220.

Moncada, S., M.J. Palmer and E.A. Higgs 1991. Nitric oxide: Physiology, pathophysiology, and pharmacology. Pharmacol. Rev. 43: 109–134.

Navati, M.S., and J.M. Friedman 2006. Sugar-derived glasses support thermal and photo-initiated electron transfer processes over macroscopic distances. J. Biol. Chem. 28147: 36021–36028.

Ray, A., B.A. Friedman and J.M. Friedman 2002. Trehalose glass-facilitated thermal reduction of metmyoglobin and methemoglobin. J. Am. Chem. Soc. 12425: 7270–7271.

Reiter, C.D., X. Wang, J.E. Tanus-Santos, N. Hogg, R.O. Cannon, A.N. Schechter and M.T. Gladwin 2002. Cell-free hemoglobin limits nitric oxide bioavailability in sickle-cell disease. Nat. Med. 812: 1383–1389.

Rodriguez, C., D.A. Vitturi, J. He, M. Vandromme, A. Brandon, A. Hutchings, J.D. Kerby and R.P. Patel 2009. Sodium nitrite therapy attenuates the hypertensive effects of HBOC-201 via nitrite reduction. Biochem. J. 4223: 423–432.

Rutherford, M.S., A. Witsell and L.B. Schook 1993. Mechanisms generating functionally heterogeneous macrophages: chaos revisited. J. Leukoc. Biol. 535: 602–618.

Singel, D.J., and J.S. Stamler 2005. Chemical physiology of blood flow regulation by red blood cells: the role of nitric oxide and S-nitrosohemoglobin. Annu. Rev. Physiol. 67: 99–145.

Yu, B., M.J. Raher, G.P. Volpato, K.D. Bloch, F. Ichinose and W.M. Zapol 2008. Inhaled nitric oxide enables artificial blood transfusion without hypertension. Circulation 11715: 1982–1990.

Yu, B., G.P. Volpato, K. Chang, K.D. Bloch and W.M. Zapol 2009. Prevention of the pulmonary vasoconstrictor effects of HBOC-201 in awake lambs by continuously breathing nitric oxide. Anesthesiology 1101: 113–122.

Zacharia, I.G., and W.M. Deen 2005. Diffusivity and solubility of nitric oxide in water and saline. Ann. Biomed. Eng. 332: 214–222.

Zapol, W.M. 1996. Inhaled nitric oxide. Acta Anaesthesiol. Scand. Suppl. 109: 81–83.

Index

About the Editors

Ross J. Hunter MD BSc MRCP trained in medical sciences at King's College London (Times University ranking 11th in UK). He spent a further year at Imperial College London (Times University ranking 3rd in UK) and was awarded his BSc in Cardiovascular medicine in 1998. Since returning to his medical training at King's College School of Medicine, he has remained an honorary research fellow at The Department of Nutritional Sciences, researching the effect of different nutritional states and alcoholism on the cardiovascular system. He was awarded his Bachelor of Medicine and Surgery (MBBS) with distinction in 2001. He trained in general medicine in London and Brighton and was made a member of the Royal College of Physicians (UK) in 2005. He trained as a Registrar in the London Deanery from 2005-2008. Since 2008 he has been a research fellow at the Department of Cardiology & Electrophysiology at St Bartholomew's Hospital London, conducting clinical research and clinical trials in cardiology and electrophysiology. He has published over 60 scientific articles of various kinds.

Victor R. Preedy BSc, PhD, DSc, FSB, FIBiol, FRCPath, FRSPH is Professor of Nutritional Biochemistry, King's College London, Professor of Clinical Biochemistry, Kings College Hospital (Hon) and Director of the Genomics Centre, King's College London. Presently he is a member of the King's College London School of Medicine. Professor Preedy graduated in 1974 with an Honours Degree in Biology and Physiology with Pharmacology. He gained his University of London PhD in 1981 when he was based at the Hospital for Tropical Disease and The London School of Hygiene and Tropical Medicine. In 1992, he received his Membership of the Royal College of Pathologists and in 1993 he gained his second doctoral degree, i.e. DSc, for his outstanding contribution to protein metabolism in health and disease. Professor Preedy was elected as a Fellow to the Institute of Biology in 1995 and to the Royal College of Pathologists in 2000. Since then he has been elected as a Fellow to the Royal Society for the Promotion of Health (2004) and The Royal Institute of Public Health (2004). In 2009, Professor Preedy became a Fellow of the Royal Society for Public Health. In his career Professor Preedy has carried out research at the National

Heart Hospital (part of Imperial College London) and the MRC Centre at Northwick Park Hospital. He is visiting lecturer at University College London (UCL) and has collaborated with research groups in Finland, Japan, Australia, USA and Germany. He is a leading expert on the pathology of disease and has lectured nationally and internationally. He has published over 570 articles, which include over 165 peer-reviewed manuscripts based on original research, 90 reviews and numerous books.

Color Plate Section

Chapter 1

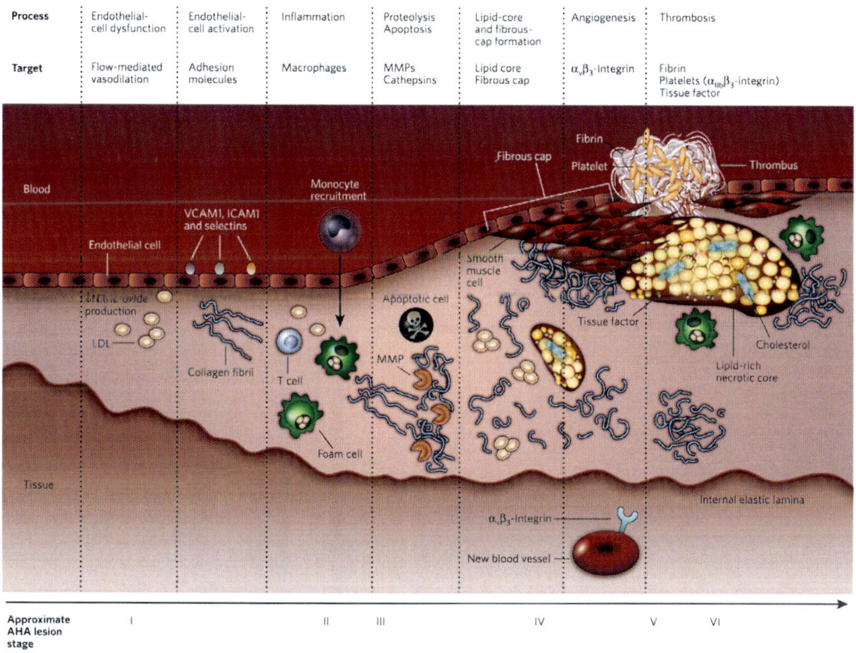

Fig. 1. A simplified depiction of the progression over time of an atherosclerotic lesion, indicating the different processes that occur and receptors expressed at different stages of lesion development. Reproduced with permission from Sanz and Fayad (2008).

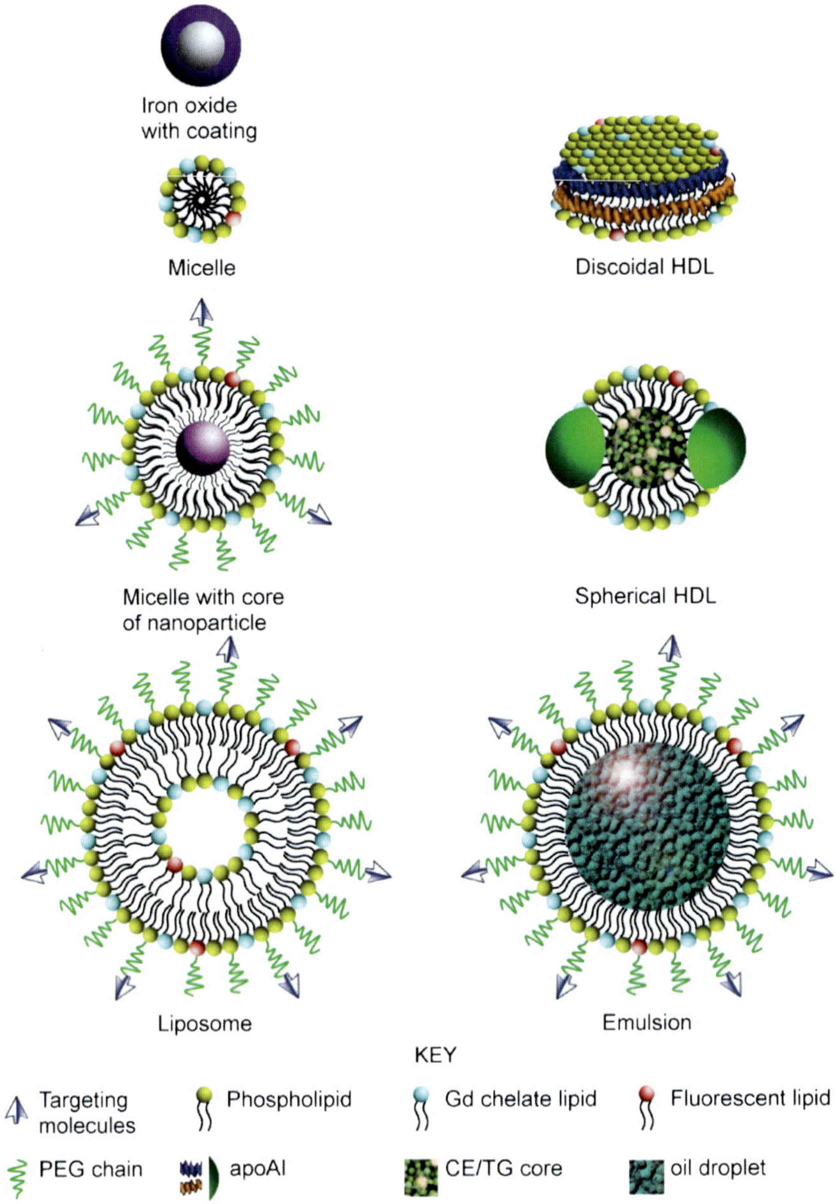

Fig. 3. Schematic depiction of many of the different nanoparticles discussed in this chapter for use in cardiovascular imaging. Reproduced with permission from Chen et al. (in press).

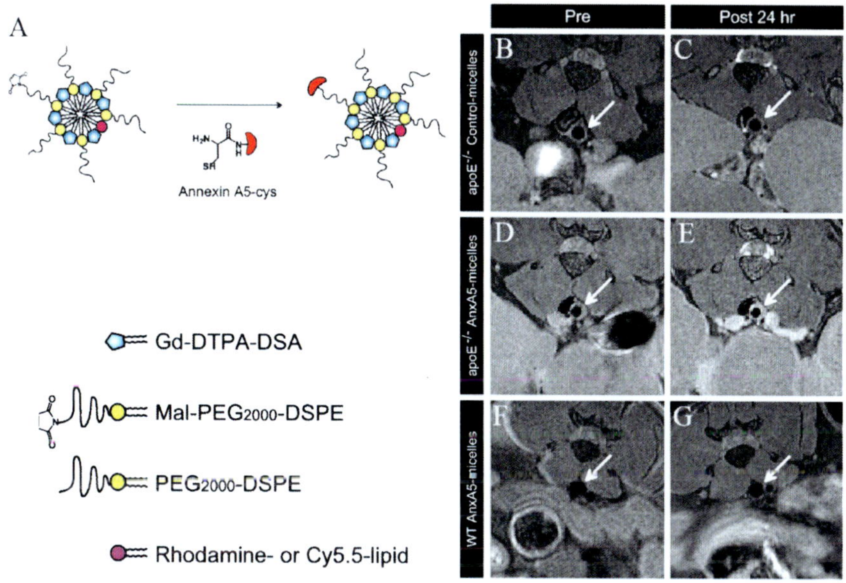

Fig. 8. Detection of apoptosis *in vivo*. (A) Schematic depiction of conjugation of AnxA5 to paramagnetic and fluorescent micelles. (B, C) Typical MR images obtained in atherosclerotic mice before and after injection of control micelles. No enhancement was noticed in that case, whereas injection of AnxA5-micelles produced a bright enhancement in the plaques of atherosclerotic mice (C, D). AnxA5-micelles did not produce any MR signal enhancement in healthy animals (E, F). Reproduced with permission from van Tilborg et al. (2010).

Chapter 2

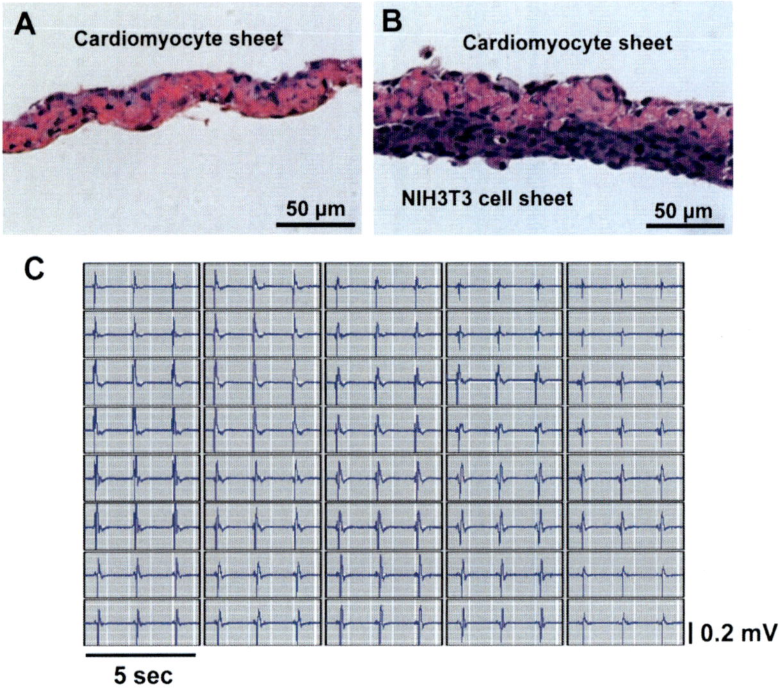

Fig. 2. Cross-sectional views of cell sheets and the electrograms of a cardiomyocyte sheet. A: The cross-sectional view of a single cardiomyocyte sheet. Hematoxylin and eosin staining of the cell sheet shows a cell-dense structure. B: The cross-sectional view of a double-layered cell sheet, a cardiomyocyte sheet (the upper layer) and an NIH3T3 cell sheet (the lower layer). The tight binding of the two cell sheets is detected by hematoxylin and eosin staining. C: Electrical potentials of a single cardiomyocyte sheet just after detachment. Spontaneous electrical potentials are recorded by 40 individual electrodes beneath the cardiomyocyte sheet, and all electrical potentials are completely synchronized. (Unpublished material of the authors.)

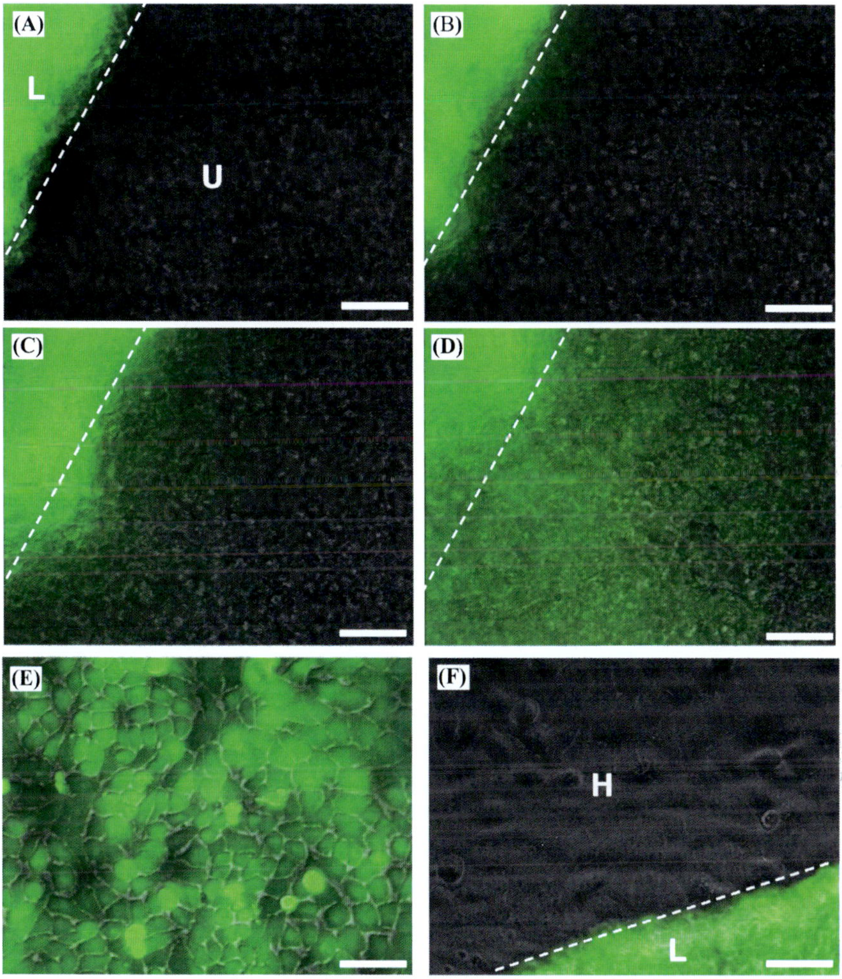

Fig. 4. Fluorescent dye transfer between a double-layered cardiomyocyte sheet. A calcein AM loaded cardiomyocyte sheet was layered on top of an unloaded cardiomyocyte sheet. Fluorescent pictures at the border region of the layered cardiomyocyte sheet were taken at 0 min (A), 30 min (B), 60 min (C), and 120 min (D) after layering. Calcein transfer to the unloaded sheet is observed at 30 min and expanded time-dependently. Though HeLa cells, which are GJ-negative cells, have the ability to take in calcein AM directly (E), no dye transfer from a calcein-loaded cardiomyocyte sheet (L) to the confluent HeLa cells (H) is observed even after 120 min (F). Broken lines show the border regions between the calcein-loaded cardiomyocyte sheet and unloaded cardiomyocytes or HeLa cells. L: calcein AM-loaded cardiomyocyte sheet; U: unloaded cardiomyocyte sheet; H: confluent HeLa cells. Scale bars show 50 μm. (Reprinted from Haraguchi et al. 2006 with permission from Elsevier.)

Chapter 3

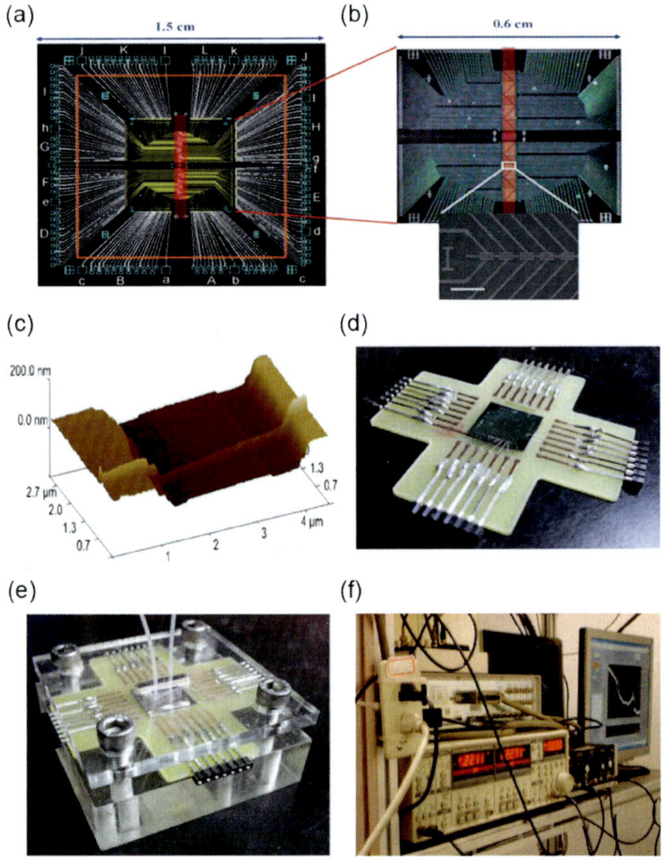

Fig. 1. Device fabrication and electrical measurement of SiNW-FETs in biosensory applications. (a) Mask design for the photolithographic fabrication of SiNW-FET device arrays. (b) Device arrays on a magnified scale. Top: image of the circuits in the area of the yellow square in (a); bottom: SEM image of a SiNW-FET array with a source-drain separation of 2 μm. The scale bar is 50 μm. In total, there are 180 pairs of source-drain electrodes on the silicon chip. (c) The topograph of a SiNW-FET scanned by AFM. A SiNW of ~50 nm in diameter is connected by two Ni/Al (70 nm/100 nm thick) electrodes of ~2 μm in separation. (d) The SiNW-FET device array on a silicon chip ($1.5 \times 1.5\ mm^2$) is connected to a plastic circuit board with aluminum wires (~30 μm in diameter). (e) A sample solution is delivered onto the SiNW-FET arrays through a polydimethylsiloxane microfluidic channel ($6.25 \times 0.5 \times 0.05\ mm^3$), which is designed to couple with the SiNW-FET arrays. (f) The variation of electrical signals is monitored by a detection system that combined a lock-in amplifier and a current preamplifier. (Wang et al. 2007. Reproduced with permission.)

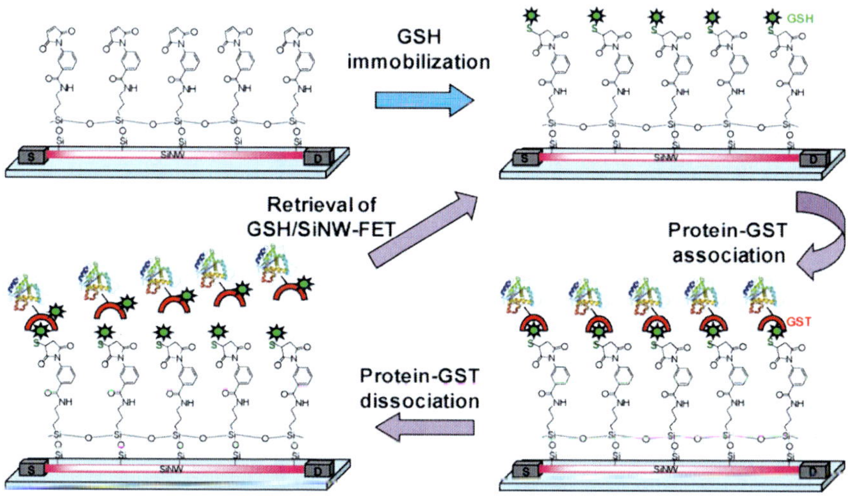

Fig. 2. Reversible high-throughput screening design of a reusable SiNW-FET biosensor. Upper left: The SiNW-FET is first modified with linkers prior to the immobilization of glutathione (GSH). Upper right and lower right: A protein fused with glutathione S-transferase (GST) is associated with the GSH/SiNW-FET to immobilize the protein on the SiNW-FET surface. Lower left: In the end of each measurement, the used protein-GSTs are removed with ≥1 mM GSH washing solution, making the GSH/SiNW-FET return to its original state. (Lin et al. 2009. Reproduced with permission.)

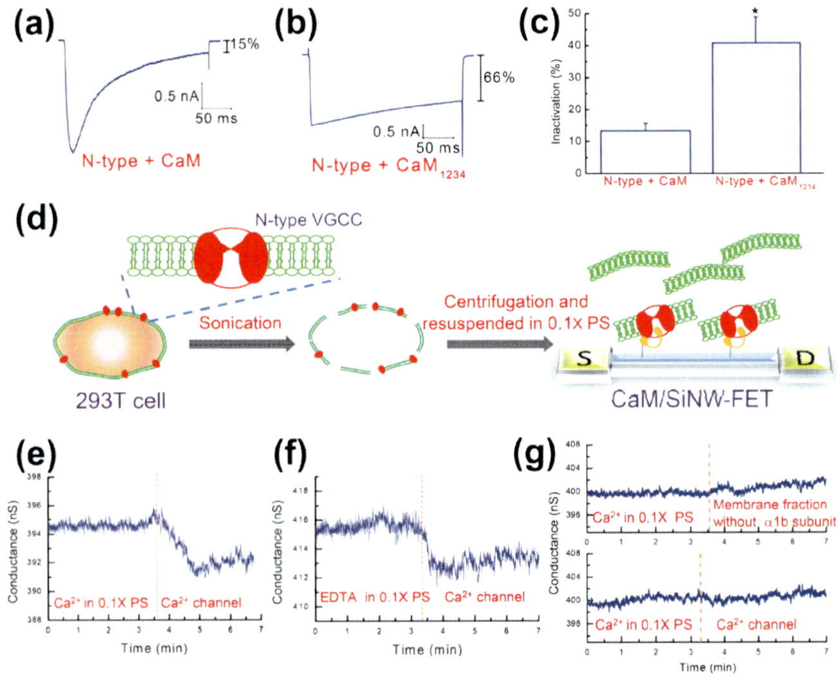

Fig. 3. Detection of N-type Ca^{2+} channels with a CaM/SiNW-FET. (a-c) Ca^{2+} currents in transfected 293T cells were recorded using a whole-cell patch configuration. The cells were voltage-clamped at –70 mV and depolarized to +20 mV for 250 ms to activate N-type VGCCs. Representative current traces from cells co-expressing N-type VGCCs with (a) CaM and (b) CaM_{1234} (a Ca^{2+} binding deficient mutant) are shown. Inactivation was obtained by normalizing the average currents between the 225 and 250 ms to the peak current during depolarization. (c) The bar graph depicts the average normalized inactivation values for cells expressing the different CaM constructs. The data represent the mean ± SEM. The sample numbers for N-type + CaM and N-type + CaM_{1234} are 7 and 5, respectively. *p value of < 0.05, as obtained by Student's t test. (d) Scheme for the detection of membrane fractions containing N-type VGCCs by CaM/SiNW-FET. (e-f) The electrical measurements represent the real-time detection of the binding of an N-type Ca^{2+} channel to the CaM/SiNW-FET in 0.1× PS (i.e., 0.1× phosphate solution, consisting of 0.76 mM Na_2HPO_4 and 0.24 mM NaH_2PO_4 at pH 7.4) supplemented with (e) 10^{-4} M Ca^{2+} or (f) 0.5 mM EDTA. (g) Two control experiments were conducted separately. Top graph: Sensing of the membrane fraction without the α1b subunit by the CaM/SiNW-FET in 0.1× PS supplemented with 10^{-4} M Ca^{2+}. Bottom graph: Sensing of a Ca^{2+} channel with a GST/SiNW-FET in 0.1× PS supplemented with 10^{-4} M Ca^{2+}. The protein concentration for the sensing experiments was 1.7 μg/μL (Lin et al. 2010. Reproduced with permission.)

Chapter 5

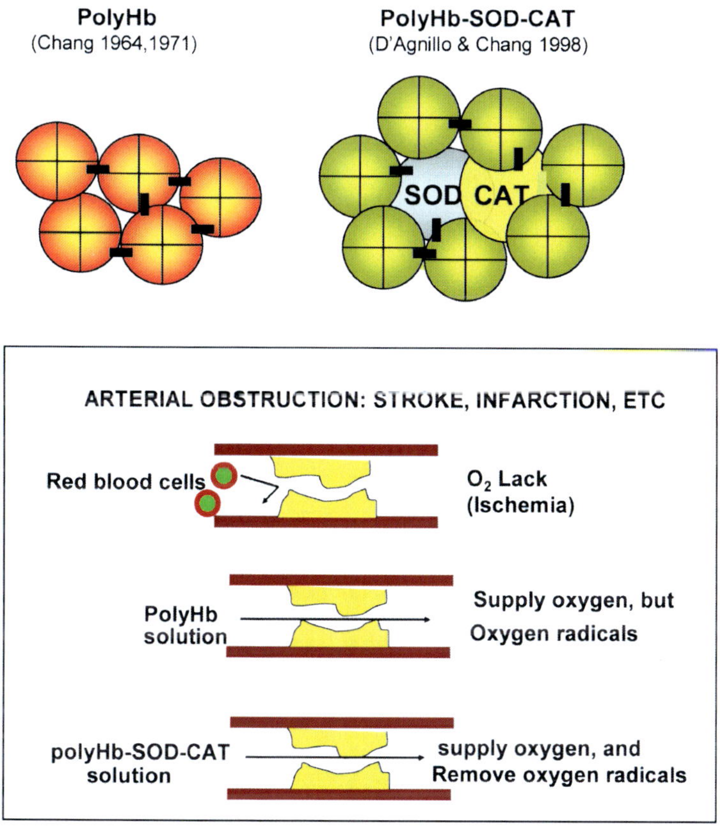

Fig. 3. Red blood cell and different types of blood substitutes in microcirculation. Arteriosclerosis causes narrowing and obstruction of blood vessels. Red blood cells, being 8 micron in diameter, have difficulty flowing through these partly obstructed vessels to supply the needed oxygen. PolyHb being a solution can perfuse partly obstructed vessels better than red blood cells to supply the needed oxygen. However, when ischemic tissue is perfused with an oxygen-rich solution, reactive oxygen species are generated, resulting in ischemia-reperfusion injury. Unlike PolyHb, PolyHb-SOD-CAT has the dual function of an oxygen carrier with the ability to remove oxygen radicals; like PolyHb, it is a solution and can perfuse partly obstructed vessels. (From Chang 2007, with permission.)

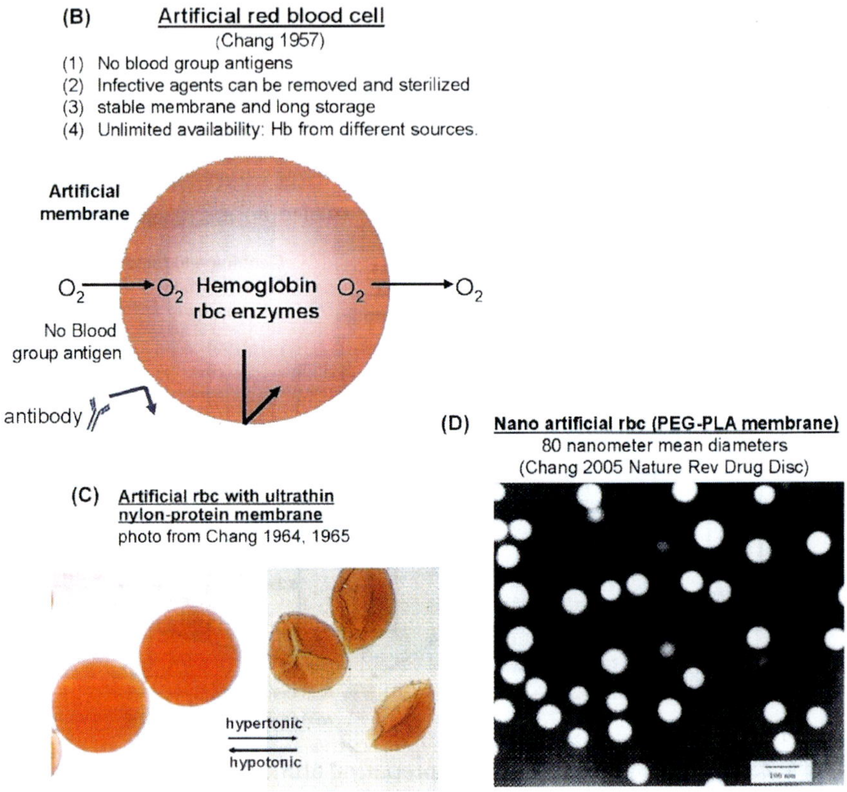

Fig. 6. Complete artificial red blood cells. (B) Schematic representation of artificial red blood cells. (C) Original micron dimension artificial red blood cell and (D) third generation nanodimension artificial red blood cell with 80 nm mean diameter. (From Chang 2007, with permission.)

Chapter 6

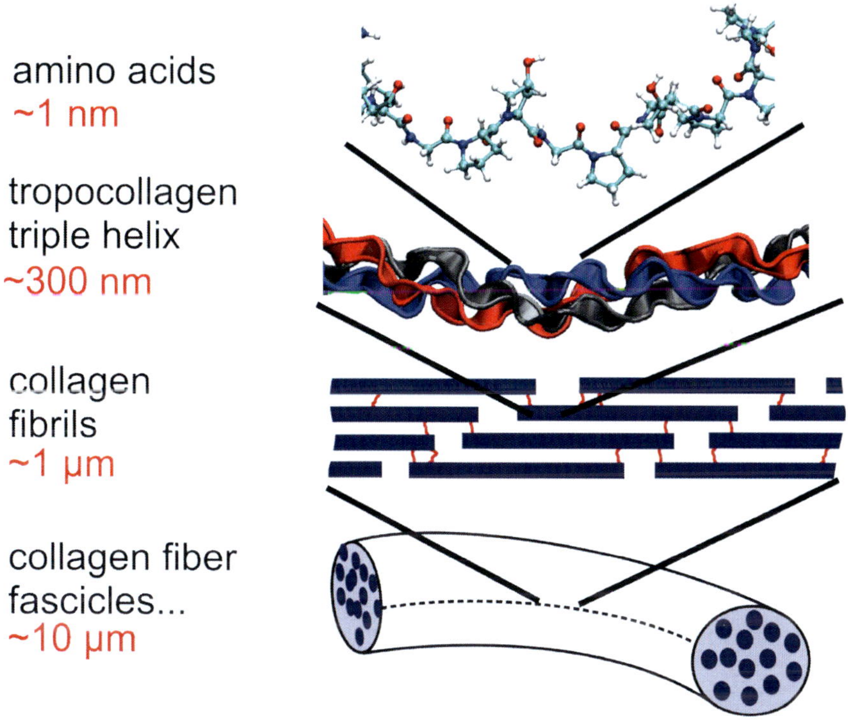

Fig. 2. Schematic representation of a selection of structural features of collagen. Collagen possesses a hierarchical structure with features that range from the characteristic amino acid sequence at the nanoscale, through tropocollagen molecules and fibrils at the sub-micrometre level to fibres measuring tens of micrometres. Published with permission from Elsevier Journals (Buehler 2008).

Chapter 7

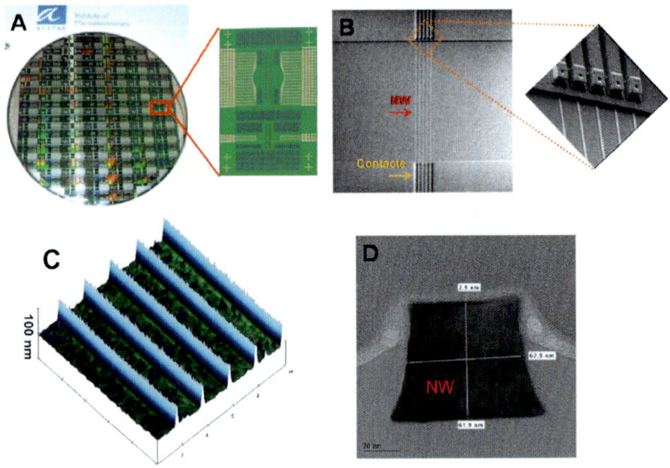

Fig. 2. Silicon nanowire (SiNW) biosensor array chip. (A) Optical images of an 8-inch wafer having 72 SiNW sensor chips (left) and zoom of a single SiNW chip (right) used in the experiment. (B) Scanning electron microscope image of a nanowire cluster composed of five individual nanowires, which are individually addressable. (C) Atomic force microscope image of nanowire arrays. (D) Cross-sectional transmission electron microscope image of a nanowire sensor. (Fig. A, B, D are unpublished materials, Fig. C is reprinted with permission from Gao et al. 2007. Copyright 2009 American Chemical Society.)

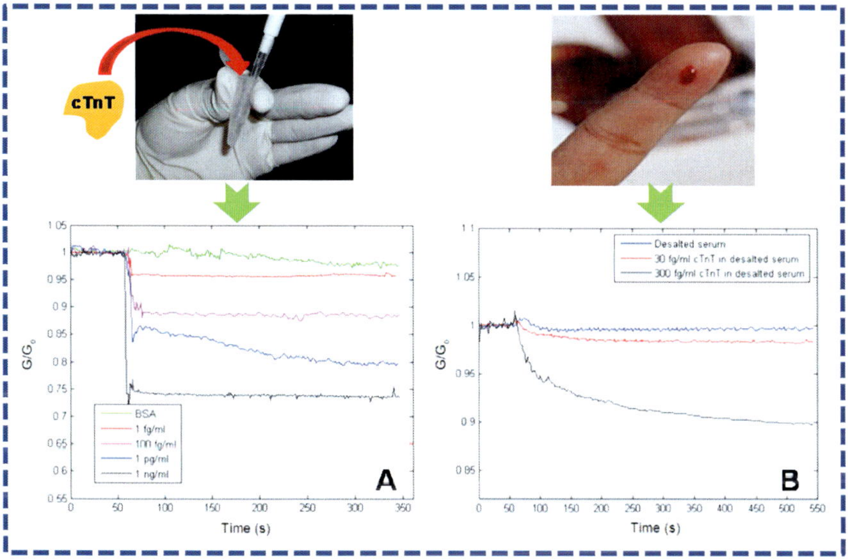

Fig. 4. Real-time detection of troponin T (cTnT) using silicon nanowire (SiNW) biosensor. (A) Real-time conductance response from a SiNW sensor functionalized with anti-cTnT. The response reduces with decreasing cTnT concentration. Addition of 1 μg/ml serum albumin solution leads to negligible signal, pointing to little non-specific binding. (B) Conductance of SiNW sensors functionalized with anti-cTnT and used in experiments to detect cTnT in desalted human serum. Serum solution without cTnT registered negligible change. (Reprinted with permission from Chua et al. 2009. Copyright 2009 American Chemical Society.)

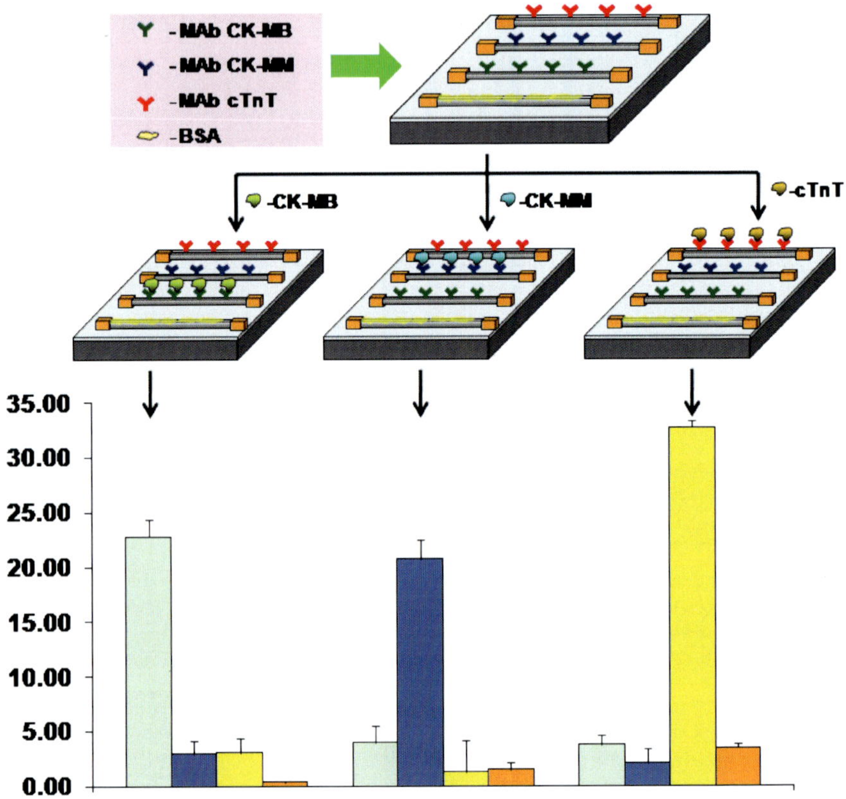

Fig. 5. Specificity of the silicon nanowire (SiNW) biosensor in a multiplexing format. 1 ng/ml of individual biomarker (troponin-T, creatine kinase-MM and creatine kinase-MB) is separately incubated with the SiNW sensor spotted with the three antibodies and serum albumin solution. (Reprinted with permission from Zhang et al. 2009b. IEEE Copyright 2009.)

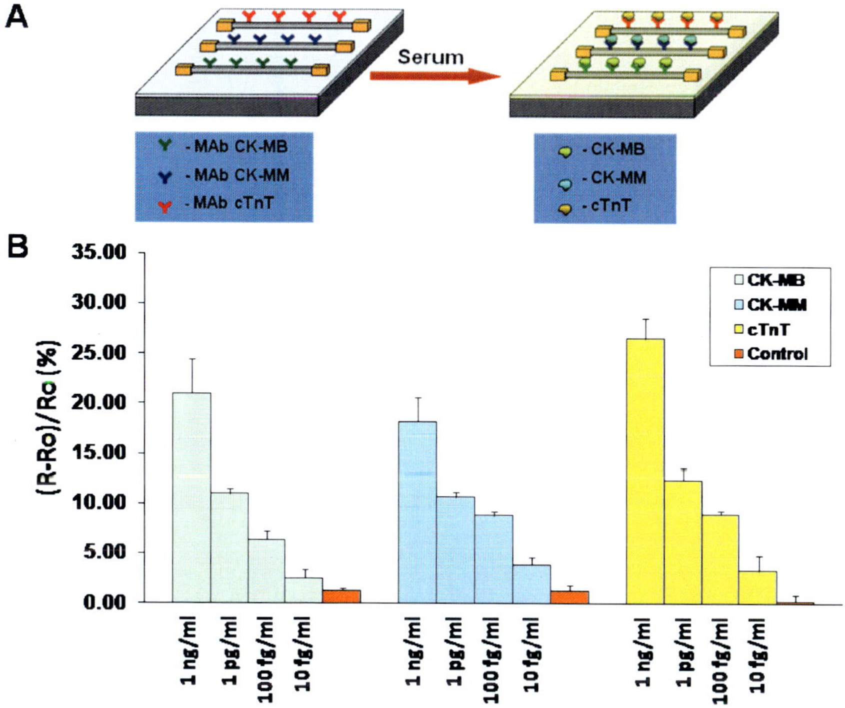

Fig. 6. Sensitivity of the silicon nanowire sensor in a multiplexing format. Response in resistance change versus various concentrations of the three biomarkers (troponin-T, creatine kinase-MM and creatine kinase-MB) spiked in untreated serum. No biomarker is spiked in serum with control. (Reprinted with permission from Zhang et al. 2009b. IEEE Copyright 2009.)

Chapter 8

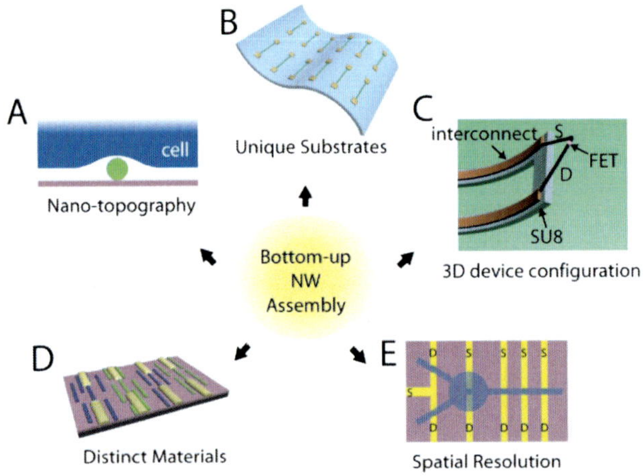

Fig. 2. Unique advantages of bottom-up NW assembly. (A) nanotopographic morphology, (B) ability to assemble devices on flexible, transparent substrates, (C) fabrication of unconventional 3D device configuration, (D) assembly of distinct NW materials on the same chip, and (E) high spatial resolution of NW devices.

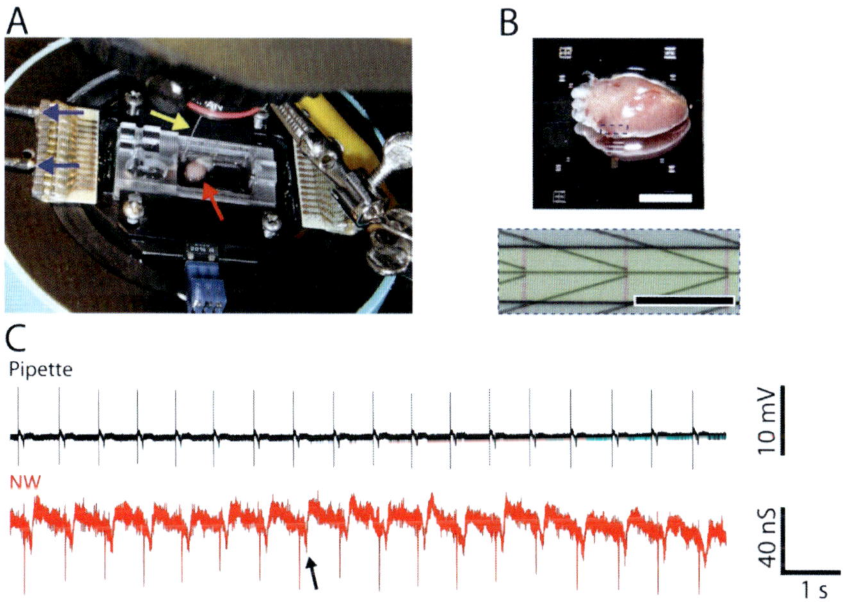

Fig. 3. NW/ Heart interfaces. (A) Photograph of experimental setup showing heart on NWFET chip in temperature regulated cell. Arrows show position of heart (red), reference electrode (yellow) and interconnect wires (blue). (B) (top) Image of heart on surface of a planar chip; scale bar is 4 mm. (bottom) Zoom of dotted region in upper image showing three pairs of NWFETs; scale bar is 150 μm. (C) Simultaneous recordings from a glass pipette (black trace) and a NW device (red trace). Adapted from Timko et al. (2009) (American Chemical Society, copyright 2009).

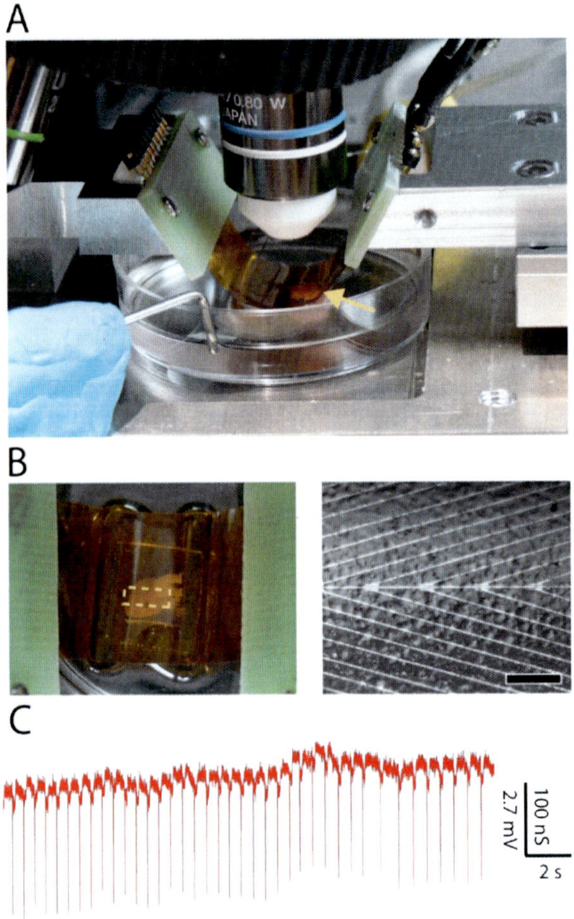

Fig. 4. NWFET array recording in a bent configuration. (A) Photograph of heart (yellow arrow) underneath the concave face of the NFET substrate. (B) (left) Photograph of same system showing registration between heart and markers on the substrate. (right) Optical image taken showing the heart surface versus position of individual NW devices located along the central horizontal axis. Scale bar is 150 μm. (C) Recorded conductance data from a NWFET in this concave configuration. Adapted from Timko et al. (2009) (American Chemical Society, copyright 2009).

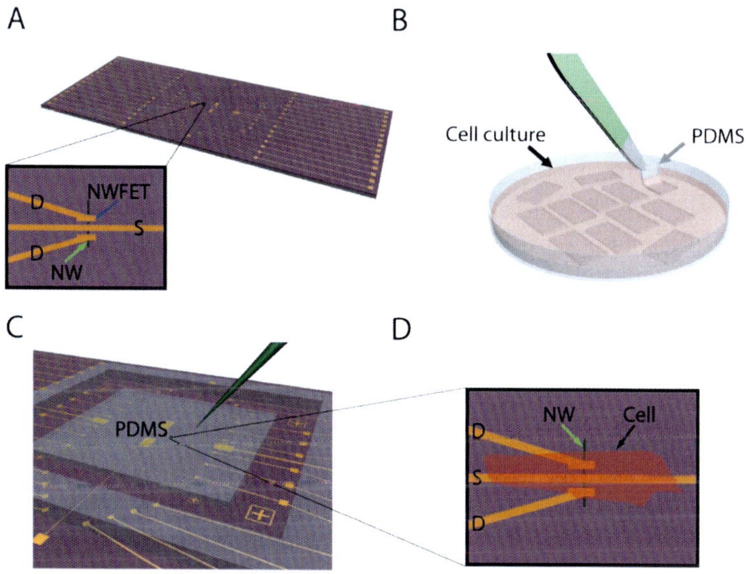

Fig. 5. Controlling nanowire/cardiomycytes interfaces. (A) NWFET chip, where NW devices are located at the central region of chip. The linear gold features correspond to contact and interconnect metal. Zoom shows source (S) and two drain (D) electrodes connected to a NW (blue arrow) defining two NWFETs. (B) Cardiomyocytes cultured on thin flexible PDMS. (C) PDMS substrate with cultured cells oriented over the NWFET chip. The green structure is the probe used to manipulate the PDMS. (D) Schematic of a cardiomyocyte (black arrow) oriented over a NWFET (green arrow). Adapted from Cohen-Karni et al. (2009) (National Academy of Sciences, copyright 2009).

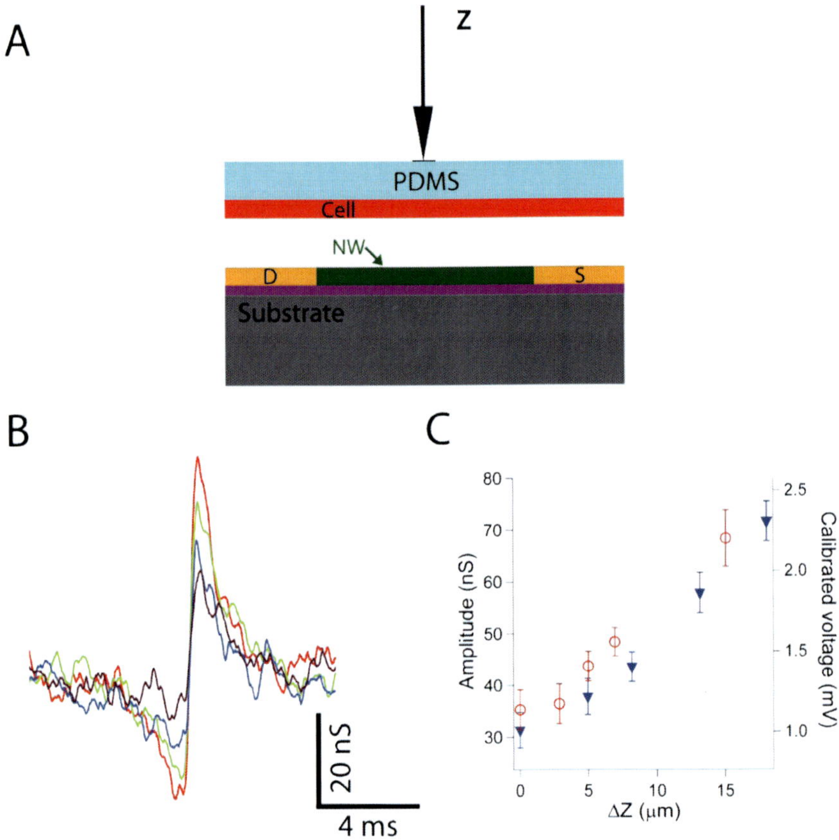

Fig. 6. Effect of applied force on recorded signals. (A) Schematic illustrating displacement (Z) of the cell substrate with respect to a NWFET. (B) Comparison of single peaks recorded with ΔZ values of 0 (purple), 8.2 (blue), 13.1 (green), and 18.0 μm (red). (C) Summary of the recorded conductance signals and calibrated voltages vs. ΔZ, where the open red circles (filled blue triangles) were recorded for increasing (decreasing) ΔZ. Adapted from Cohen-Karni et al. (2009) (Proceedings of the National Academy of Sciences of the United States of America, copyright 2009).

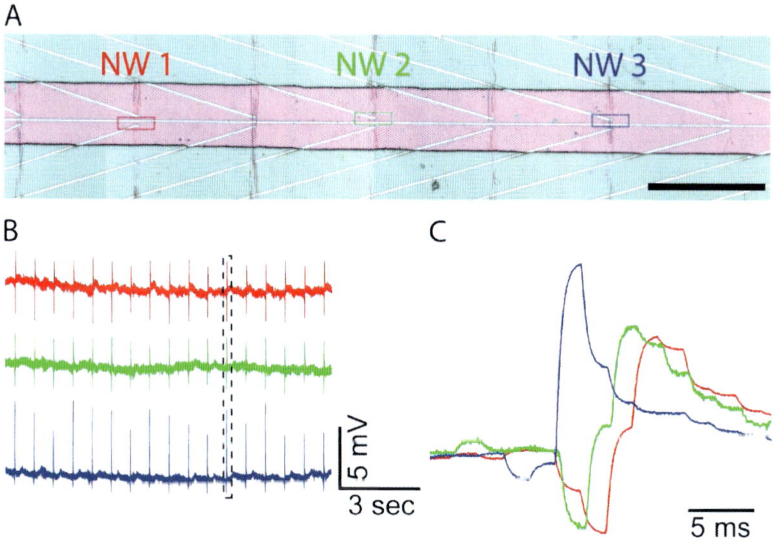

Fig. 7. Multiplexed NWFET recording. (A) Optical micrograph showing 3 NWFET devices (NW1, NW2, NW3) in a linear array; scale bar is 150 μm. (B) Conductance vs. time signals recorded simultaneously from NW1, NW2, and NW3. (C) Comparison of temporally correlated peaks highlighted by the black dashed box in B. Adapted from Cohen-Karni et al. (2009) (Proceedings of the National Academy of Sciences of the United States of America, copyright 2009).

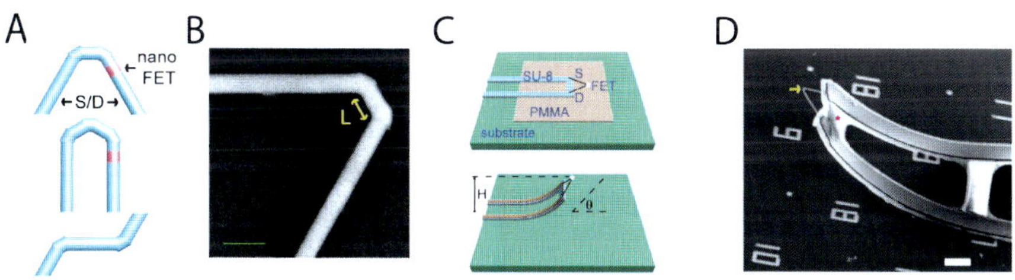

Fig. 8. Synthesis of kinked silicon nanowire probes. (A) Schematics of 60° (top) and 0° (middle) multikinked nanowires, and *'cis'* (top) and *'trans'* (bottom) configurations. The blue and magenta regions designate source/drain (S/D) and nanoFET, respectively. (B) SEM image of a double kinked nanowire with *'cis'* configuration. (C) Schematics of device fabrication. H and θ are tip height and orientation, respectively, and the white dot designates the nanoFET. (D) SEM image of a 3D nanoprobe. The yellow arrow and magenta star mark the nanoFET and SU-8, respectively. Scale bar is 5 μm. Adapted from Tian et al. (2010) (American Association for the Advancement of Science, copyright 2010).

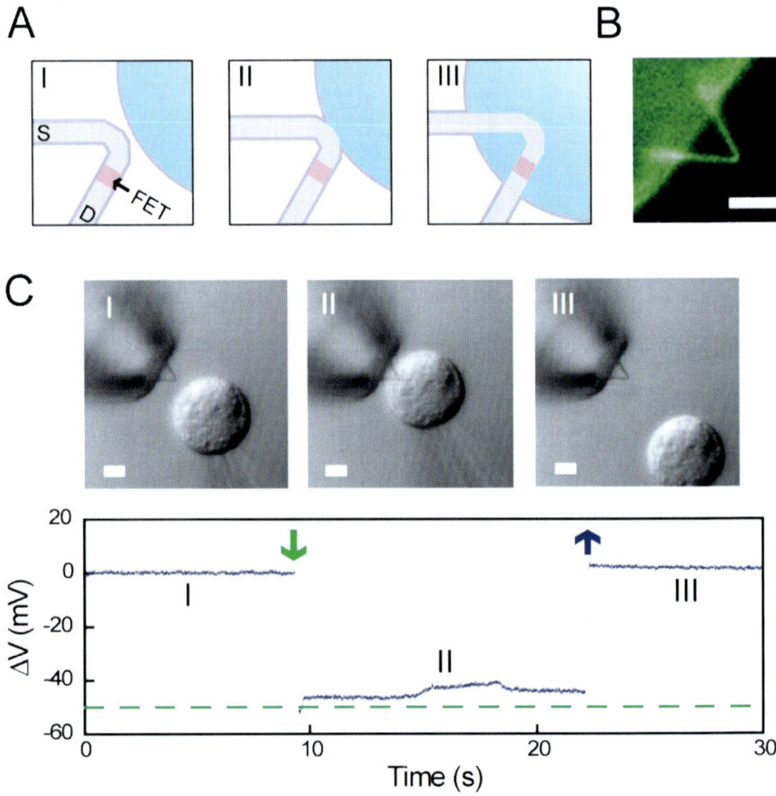

Fig. 9. Surface modification and cellular entry. (A) Schematics of nanowire probe entrance into a cell. Purple, light purple, magenta and blue colors denote the phospholipid bilayers, doped nanowire arms, active sensor segment and cytosol, respectively. (B) Fluorescence image of a lipids-coated nanowire probe. (C) Differential interference contrast (DIC) microscopy images (upper panels) and electrical recording (lower panel) of an HL-1 cell and nanowire probe as the cell approaches (I), contacts and internalizes (II), and is retracted from (III) the nanoprobe. A micropipette was used to manipulate and voltage-clamp (broken green line) the cell. Scale bars: 5 μm. Adapted from Tian et al. (2010) (American Association for the Advancement of Science, copyright 2010).

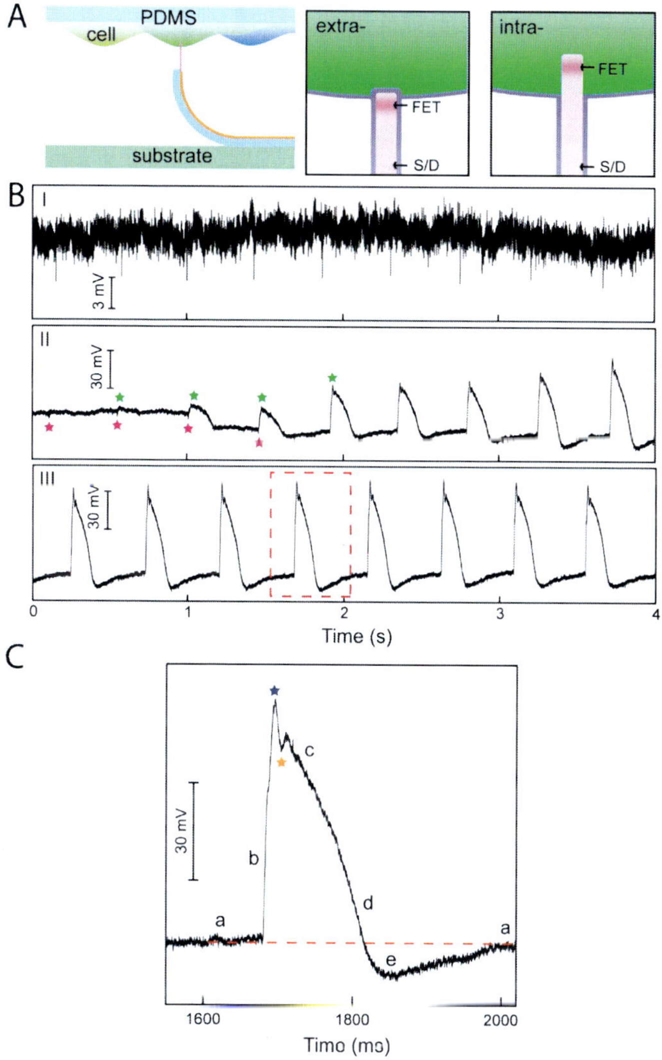

Fig. 10. Electrical recording from beating cardiomyocytes. (A) Schematics of cellular recording from cardiomyocyte monolayer (left panel), and extracellular and intracellular nanowire/cell interfaces. The cell membrane and nanowire lipids coatings are marked as purple lines. (B) Recording from beating cardiomyocytes: (I) extracellular recording; (II) transition from extracellular to intracellular recording; and (III) steady-state intracellular recording. Green and magenta stars mark peak positions of intracellular and extracellular signal components, respectively. (C) Zoom-in signal from the corresponding red-dashed region in (B). Blue and orange stars designate features associated with inward sodium and outward potassium currents, respectively. a-e are defined in the text. Adapted from Tian et al. (2010) (American Association for the Advancement of Science, copyright 2010).

Chapter 9

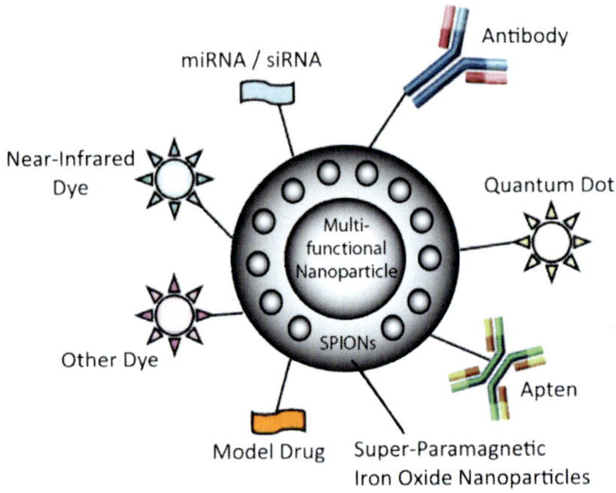

Fig. 5. Schematic representation of a multifunctional nanoparticle having different properties due to its numerous derivatization with antibodies, fluorophores, contrast agents and other small biologically relevant molecules. (Unpublished material of the author.)

Chapter 10

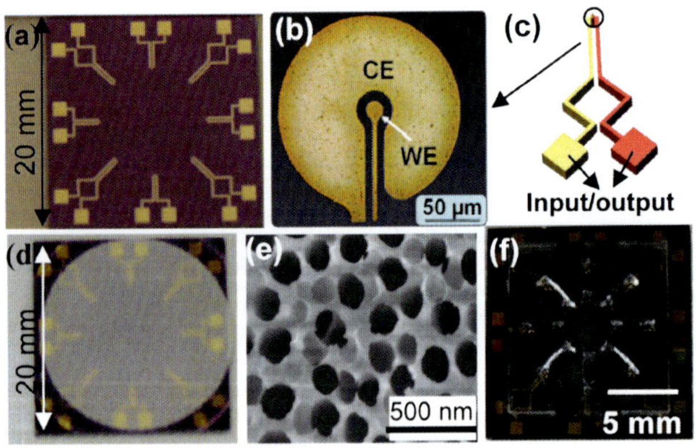

Fig. 2. Optical micrograph representing the manufacturing of a nanomonitor platform. (a) Nanomonitor (NM) device. (b) Sensing site—working electrodes (WE) and counter electrodes (CE). (c) Input/output pads connected to sensing site through interconnects. (d) Nanoporous membrane overlying on NM. (e) SEM of membrane. (f) Microfluidic device encapsulating the NM.

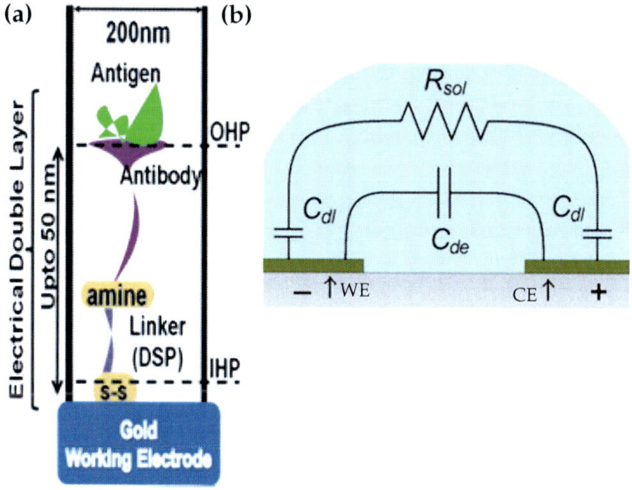

Fig. 3. Representation of biomolecule binding within the nanowells of the nanomonitor. (a) Schematic showing the antibody-antigen binding inside a nanowell. (b) Equivalent circuit of a sensing site showing a resistive capacitive circuit, where the capacitor correlates to the double layer capacitance.

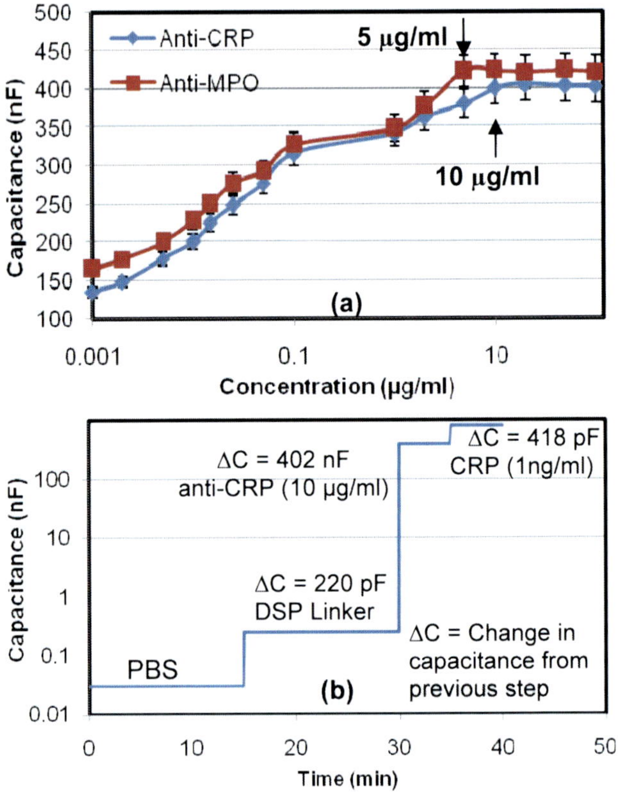

Fig. 4. Antibody saturation cure and representation of a time series for the dose response. (a) Antibody saturation curves for Anti-C-reactive protein and Anti-Myeloperoxidase. (b) "Real- time" detection showing stepwise change in capacitance with time.

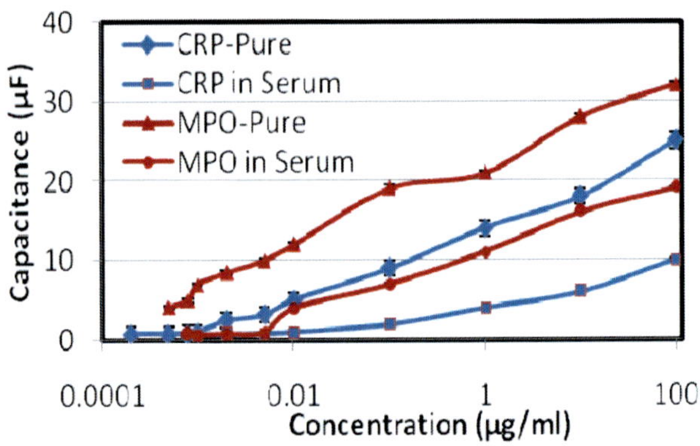

Fig. 5. Dose response curves for the inflammatory proteins. Dose response for C-reactive protein and myeloperoxidase detection from pure and serum samples.

Table 2. Digital micrograph of the 48 well gold Nanomonitor-ELISA plate experiment.

	1	2	3	4
A	Au, HRP-Ab, TMB **(No DSP, No anti-CRP)**		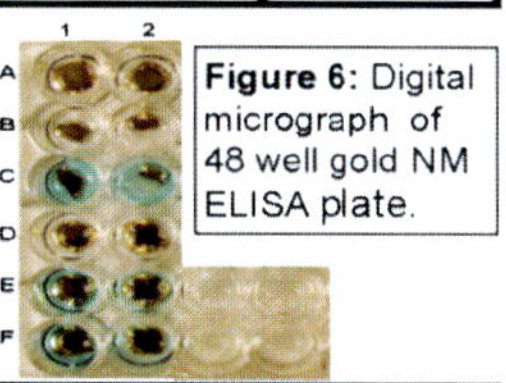	
B	Au, DSP, HRP-Ab, TMB **(No anti-CRP)**			
C	Au, anti-CRP (50 μg/ml), HRP-Ab, TMB **(No DSP)**			
D	Au, DSP, anti-CRP (50 μg/ml), TMB **(No HRP-Ab)**			
E	Au, DSP, **anti-CRP (10 μg/ml)**, HRP-Ab, TMB		Blank	
F	Au, DSP, **anti-CRP (50 μg/ml)**, HRP-Ab, TMB		Blank	

Details of the solutions added for the 12 samples in the validation experiment. The important steps in each cell are in bold type. Purple cells represent the controls, red cells represent the anti-c-reactive protein tests, and white cells are blank.

Chapter 11

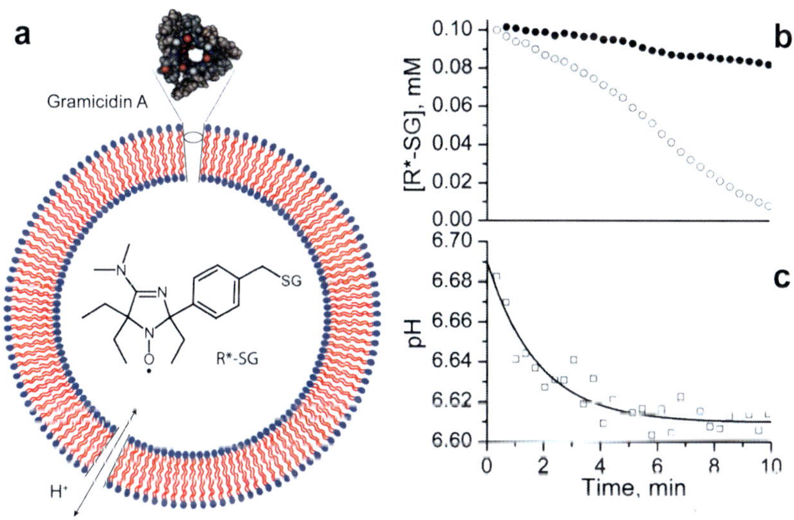

Fig. 9. NanoSPINs and their application for EPR monitoring of ischemia-induced acidosis in rat heart homogenate. (a) Schematic design of pH-sensitive nanoSPIN: the phospholipid membrane of the liposome with incorporated gramicidin A provides selective permeability to H^+, while protecting encapsulated nitroxide, R*-SG, from biological reductants. (b) Kinetics of the reduction of free (○) and encapsulated (•) R*-SG probe in the rat heart homogenate measured from its EPR signal intensity decay. (c) Ischemia-induced pH decrease calculated from the EPR spectra of the encapsulated R*-SG. (From Woldman et al. 2009, reproduced by permission of The Royal Society of Chemistry.)

Chapter 12

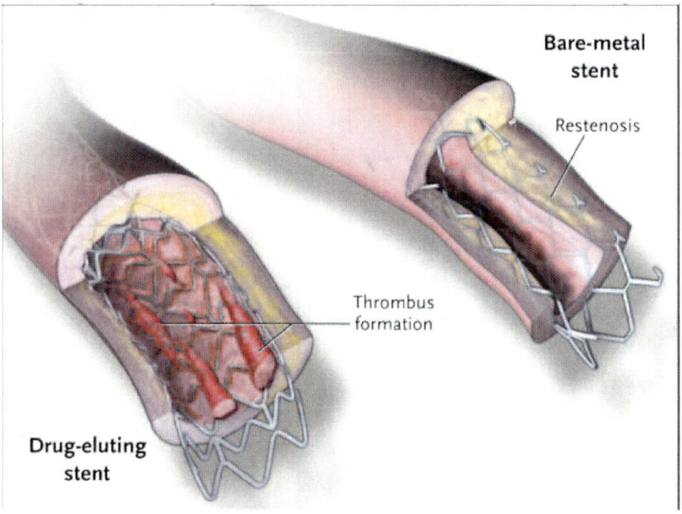

Fig. 1. Contrasting mechanisms of obstruction of bare metal stents and drug-eluting stents. Bare metal stents may be narrowed or obstructed by ingrowth of tissue. With drug-eluting stents, this process is inhibited, but since the struts remain uncovered, they may be prone to thrombosis after antiplatelet therapy is discontinued. Reproduced with permission (Shuchman 2006).

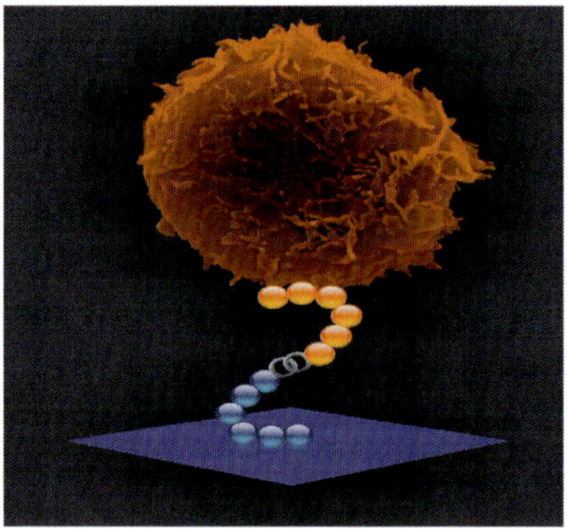

Fig. 4. Schematic diagram of an IFBM at the interface between the biologic and material realms. Used with permission (Meyers et al. 2009).

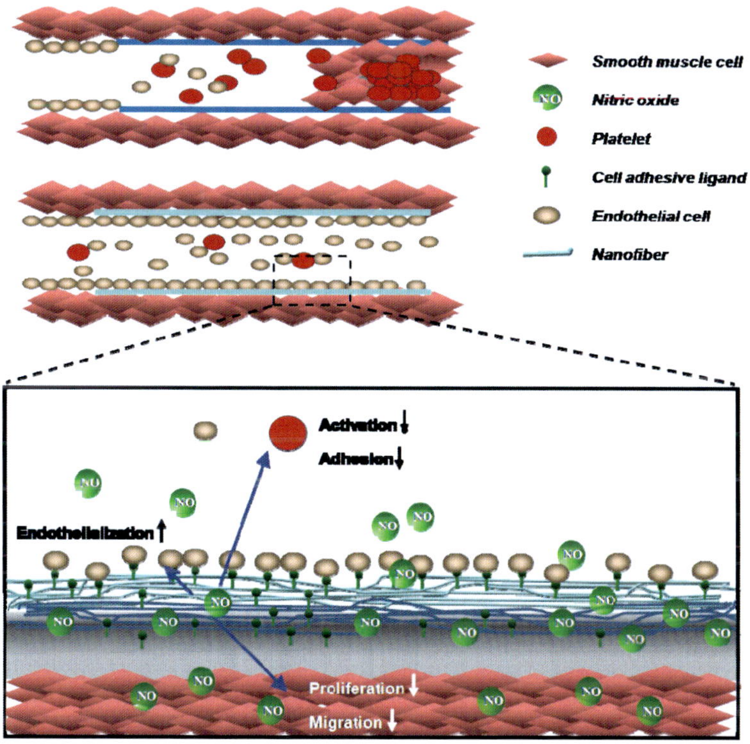

Fig. 5. Schematic comparison of current stents and endothelial ECM-mimicking PA coated stents. (a) Schematic comparison of current stent/vascular grafts (blue) and endothelial ECM-mimicking nanofibrous matrix (green). (1) Endothelial disruption exposes vessel wall, leading to restenosis, thrombosis and incomplete endothelialization. (2) Native endothelial ECM-mimic nanofibrous matrix will prevent restenosis and thrombosis while promoting endothelialization. (b) Components of the native endothelial ECM-mimic nanofibrous matrix. NO released from the nanofibrous matrix prevents platelet activation, adhesion, smooth muscle cell proliferation and migration and promotes endothelialization. Used with permission (Kushwaha et al. 2010).

Chapter 15

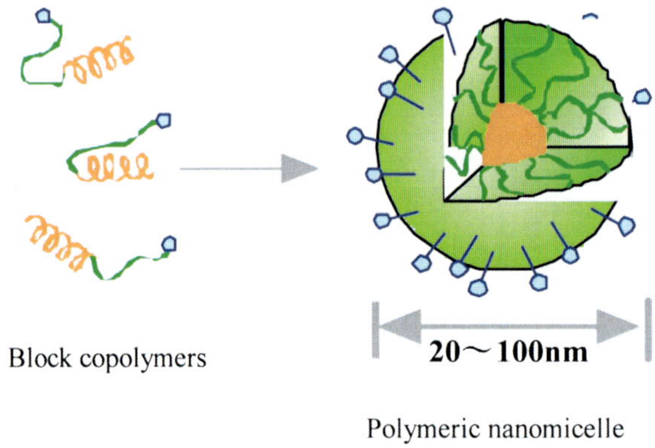

Fig. 1. The schema of the structure of polymeric nanomicelle. Polymeric nanomicelle consists of DNA and block copolymer showing a core shell structure.

Chapter 17

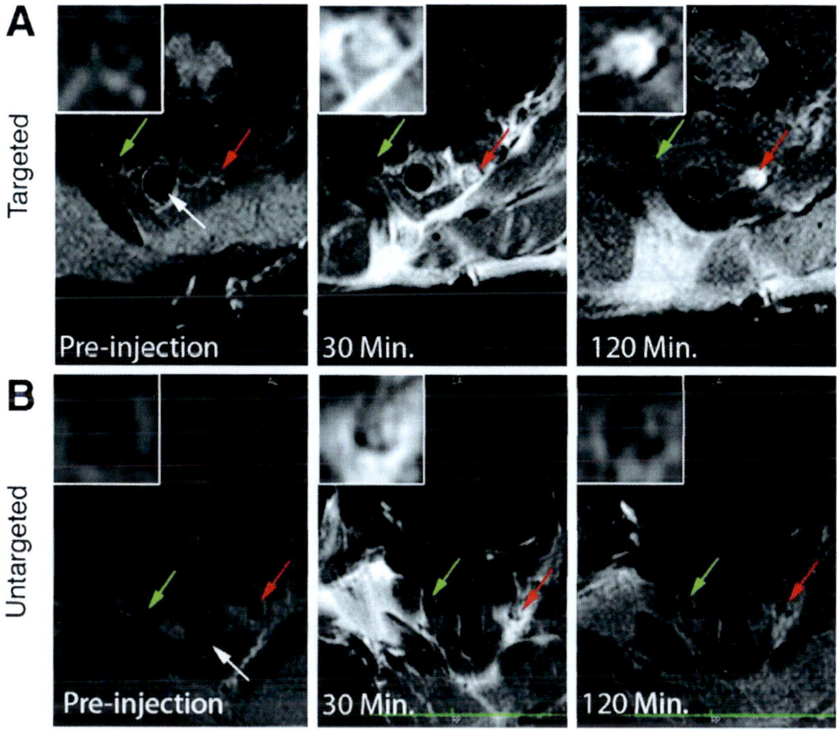

Fig. 3. *In vivo* MRI images of AA-induced arterial thrombosis. The trachea (white), the non-injured left carotid (green) and the injured right carotid (red) are indicated with arrows. Inserts on the upper left corner of each image are the magnified image of the injured carotid. (A) 30 min after the injection of P975, T1 signal enhancement was observed and the signal intensity was persistent throughout 120 min. (B) Following the Gd-DOTA (control) injection, signal intensity increased for the first 30 min, yet it rapidly decreased in 120 min due to the clearance of the control agent from the circulation. Reproduced from Klink et al. (2010), with permission.

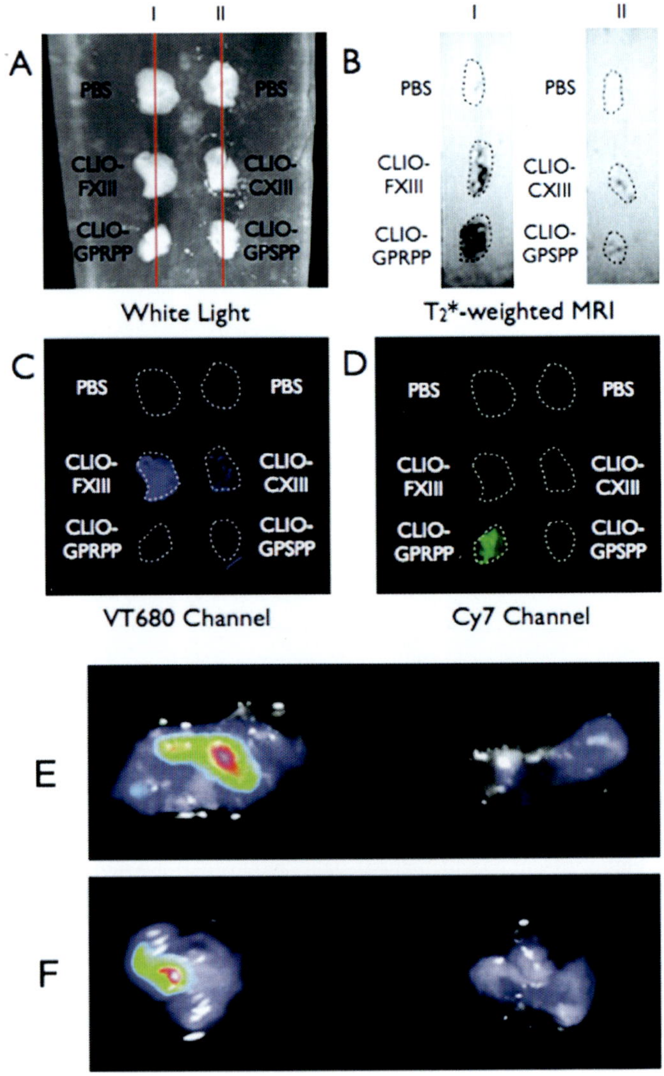

Fig. 4. *In vitro* and *in vivo* imaging of thrombi. (A) White light image of FFP clots. Lines indicated the sagittal view of the clots acquired in MRI. (B) T_2-weighted MRI image of CLIO-FXIII and CLIO-GPRPP showed T_2 signal decrease as a result of strong binding of agents to the clots. (C) FRI image of the CLIO-FXIII incubated clots in VT680 channel. (D) FRI image of the CLIO-GPRPP incubated clots in Cy7 channel. (E) *Ex vivo* FRI image of fibrin-targeted nanoagents in Cy7 channel indicated strong binding of the agent to the clots. (F) Increased FRI signal in VT680 channel confirmed selective binding of FXIII targeted nanoagents to the clots. Reproduced from McCarthy et al. (2009), with permission.

Chapter 18

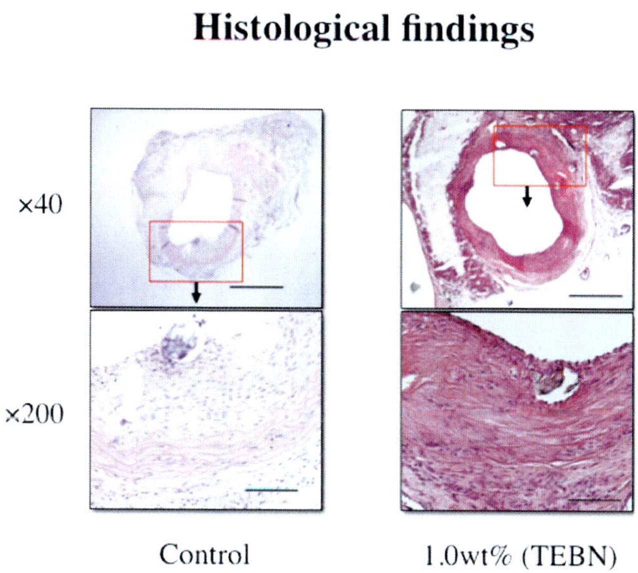

Fig. 5. Histological findings in the anastomotic site 2 wk after operation (arterial model). H&E staining of control group magnification ×40 (above) ×200 (below) and TEBN containing 1.0 wt% tacrolimus magnification ×40 (above) ×200 (below). Neo-intimal hyperplasia is significantly decreased in TEBN containing 1.0 wt% tacrolimus. Scale bar = 100 μm (above), scale bar = 500 μm (below) (Mutsuga et al. 2009a).

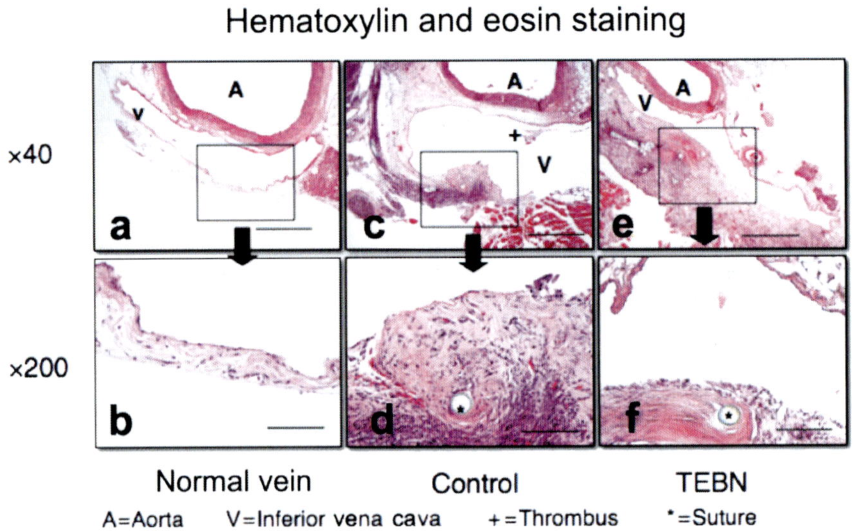

Fig. 6. Histological findings in the anastomotic site 1 wk after operation (venous model). HE staining of normal group (a) ×40 (b) ×200, control group (c) ×40 (d) ×200 and TEBN group (e) ×40 (f) ×200. Neo-intimal hyperplasia occurred around the surgical suture (d) and clearly decreased in TEBN group (f). Scale bar in (a), (c) and (e) 500 μm. In (b), (d) and (f): 100 μm (Mutsuga et al. 2009b).

Chapter 19

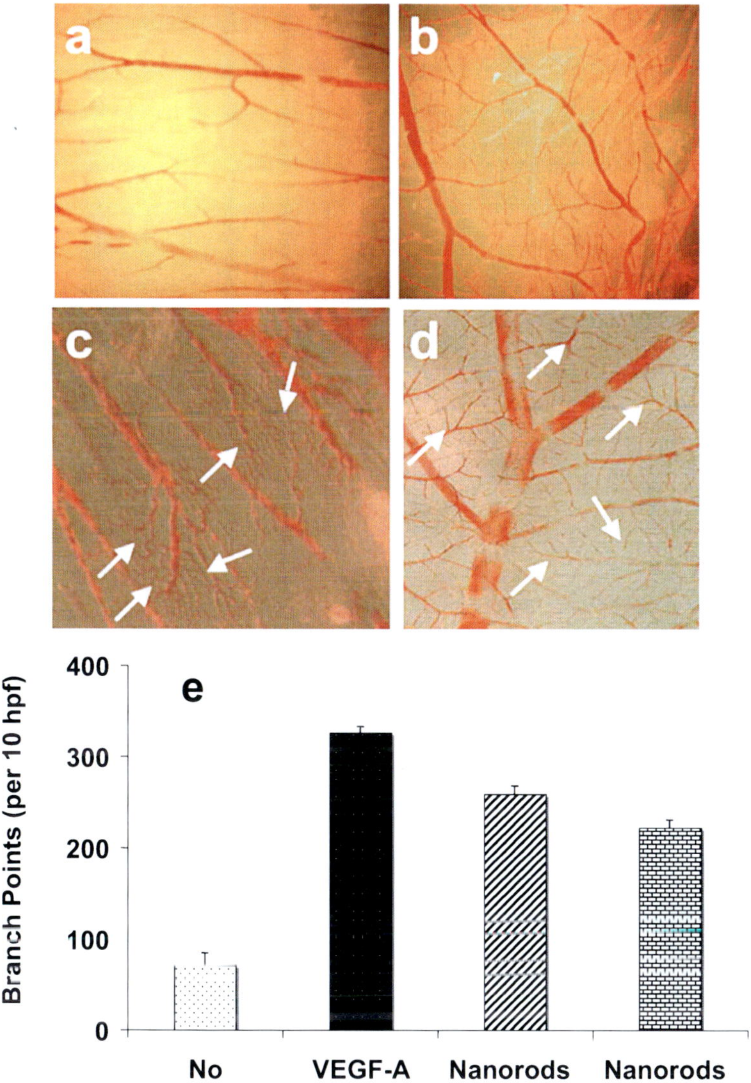

Fig. 3. Inorganic nanorods induced *in vivo* angiogenesis observed by CAM assay. CAMs treated with (a) TE (tris-EDTA) buffer, (b) VEGF (50 ng), (c, d) 1 μg and 10 μg of nanorods in TE buffer. (e) Angiogenesis was quantified by counting branch points arising from tertiary vessels from a minimum of 10 specimens from three separate experiments. Reprinted with permission from Patra et al. (2008b). Copyright ©2008, Wiley-VCH Verlag GmbH & Co. KGaA.

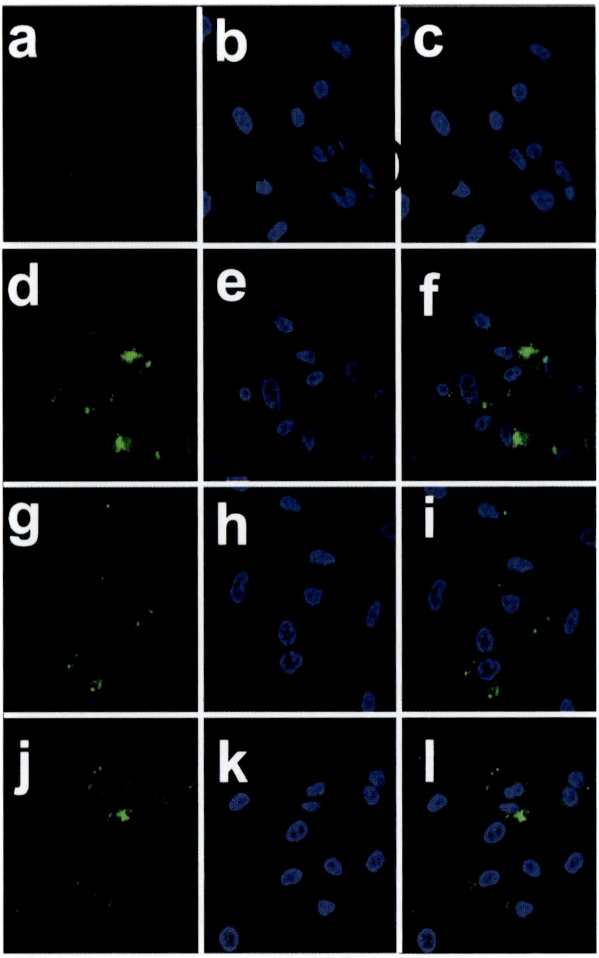

Fig. 5. Generation of ROS in HUVECs in presence of $Eu^{III}(OH)_3$ nanorods. ROS are produced during the endothelial cell proliferation in presence of $Eu^{III}(OH)_3$ nanorods at different concentrations. (a-c) control untreated HUVECs, (d-f) HUVECs treated with 100 μM/ml of tert-butyl hydro peroxide (TBHP) as positive ROS inducer. (g-l) HUVECs treated with 20 μg/ml (g-i) and 50 μg/ml (j-l) of $Eu^{III}(OH)_3$ nanorods. Reprinted with permission from Patra et al. (2008b). Copyright ©2008, Wiley-VCH Verlag GmbH & Co. KGaA.

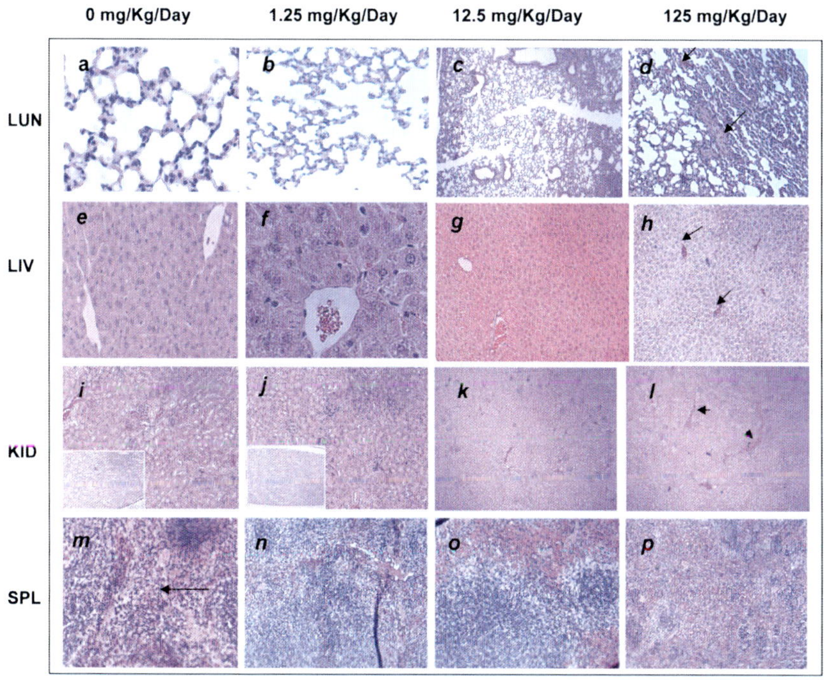

Fig. 6. Histological specimens of mice tissues (lung, liver, kidney and spleen) collected from mice sacrificed on day 60 stained with hematoxylin and eosin (H&E). Long-term survival of mice was also observed even at highest does of $Eu^{III}(OH)_3$ nanorods (125 $mgkg^{-1}day^{-1}$). (a) Control animal lung section showing normal alveolar geometry and normal-looking alveolar septum. (b) Normal alveolar geometry and normal-looking alveolar septum with dose of 125 $mgkg^{-1}day^{-1}$. (c) Normal lung parenchymal blood vessels with dose of 125 $mgkg^{-1}day^{-1}$. (d) Control animal liver section showing normal hepatic architecture. (e) Normal portal tract, sinusoids and hepatocytes with dose of 125 $mgkg^{-1}day^{-1}$. (f) Normal central vein with dose of 125 $mgkg^{-1}day^{-1}$. (g) Kidney sections from control animals show normal renal cortex with normal-looking glomerular tufts and tubules. (h) Normal-looking glomerular tufts with dose of 125 $mgkg^{-1}day^{-1}$. (i) Normal-looking renal tubules and blood vessels. (j) Splenic sections from control animals show normal splenic lymphoid follicles. (k) Normal splenic lymphoid follicles after a dose of 1.25 $mgkg^{-1}day^{-1}$. (l) Normal splenic sinuses are still seen after a dose of 1.25 $mgkg^{-1}day^{-1}$. The histological pictures were taken at following magnifications: a at > x10, b and c at > x100, d at > x40, e and f at > x100, g at > x40, h and i at > x100, j at > x10, k at > x 40, i at > x100. Reprinted with permission from Patra et al. (2009). Copyright ©2009, Elsevier Limited.

Chapter 20

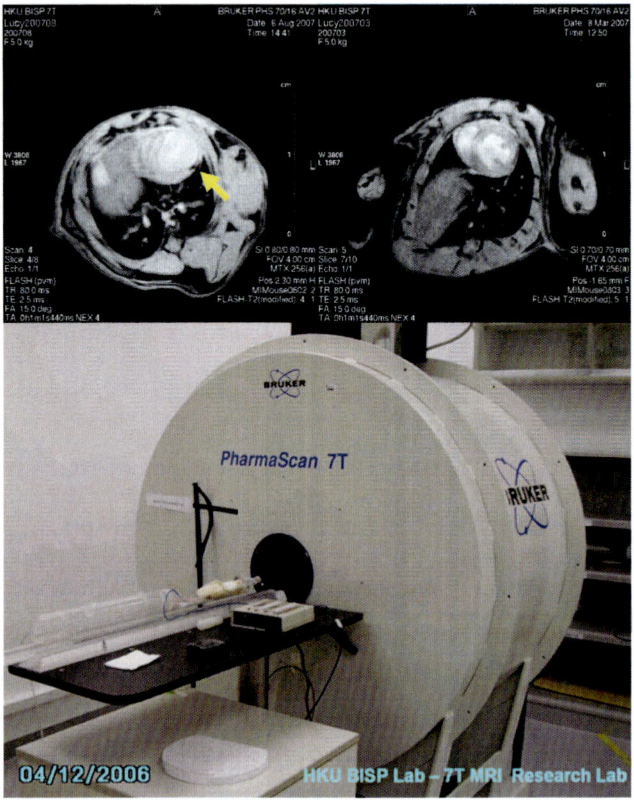

Fig. 1. (a) MION-PLL labeled mouse embryonic stem cells were observed as hypointensities at the injection site (yellow arrow) in myocardial infarcted mouse with T_2^*-weighted MR image, while (b) no apparent hypointensities were observed in myocardial infarcted mouse without cell transplantation using (c) 7 Tesla MRI scanner.

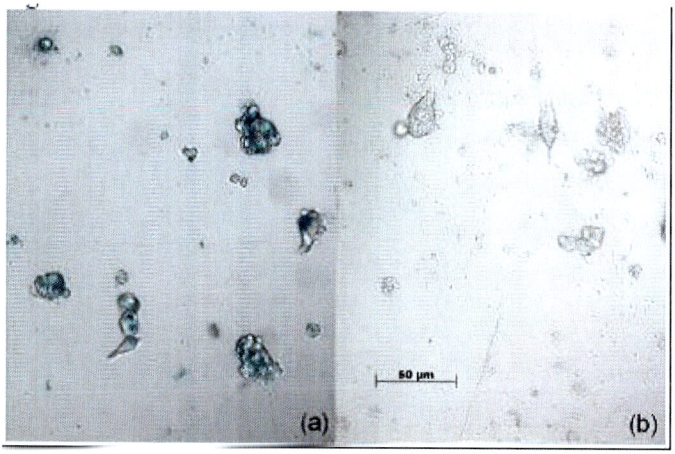

Fig. 3. Micrographs show (a) clumps of mouse ESCs stained with Prussian blue to demonstrate the uptake of MION particles. MION are visible as blue iron deposits; (b) mouse ESCs without labeling of MION-PLL showing no apparent MION uptake.

Chapter 21

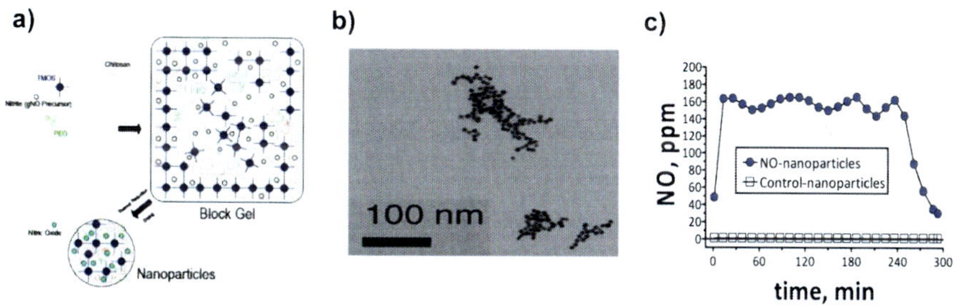

Fig. 1. NO-releasing nanoparticle synthesis and morphology. (a) Schematic of NO-releasing nanoparticle synthesis. The various components are prepared and combined to form a block gel consisting of a silica hydrogel matrix encompassing the other ingredients. The block gel is dried and the nitrite precursor converted to NO via thermal reduction, resulting in the formation of the nanoparticles. (b) NO-releasing nanoparticles. The scale bars represent 100 nm, the bottom one representing the lower left and right panes and the upper one the upper left image. (c) NO gas levels were measured using a chemiluminescent NO analyzer at 7.4 pH, 1 mg of NO-np. Control nanoparticles are prepared identically, but without nitrite.

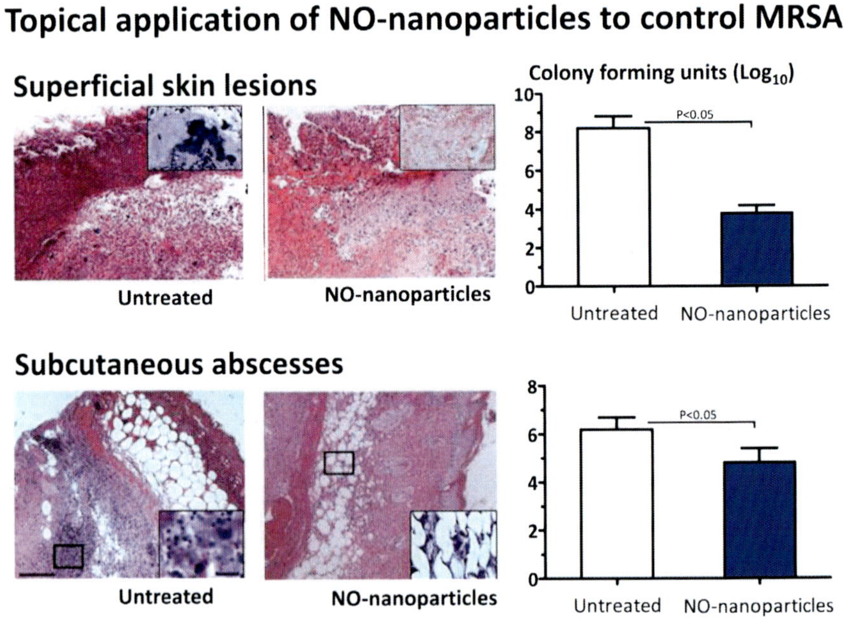

Fig. 2. Topical application of NO-releasing nanoparticles (NO-np) to treat methicillin-resistant Staphylococcus aureus (MRSA), superficial and subcutaneous abscesses in mice, showed that the NO-np exerted antimicrobial activity against MRSA abscesses. Topical or intradermal NO-np treatment of abscesses reduced the area and bacterial load while improving skin architecture. Notably, inflammatory cytokines involved in immunomodulation and wound healing revealed that NO-np led to a reduction in angiogenesis, preventing bacterial dissemination. Data previously published by the authors (Han et al. 2009a, b).

Microvascular changes to infusion of NO-nanoparticles

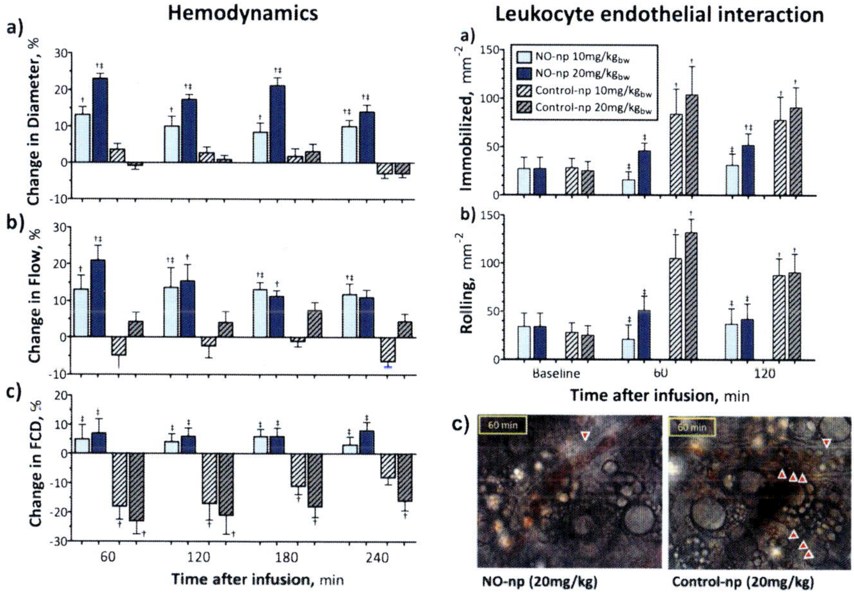

Fig. 4. Left panels, microvascular hemodynamic changes after infusion of NO-releasing nanoparticles (NO-np) and control nanoparticles (Control-np, without NO). Circulating NO-np induced vasodilatation and increased microvascular perfusion during their circulation lifetime of several hours. (a) Changes in arteriolar diameter after infusion of 10 and 20 mg/kg of NO-np (no pattern) and Control-np (pattern upward diagonal). (b) Changes in arteriolar blood flow after infusion of NO-np and Control-np. (c) Changes in functional capillary density (FCD) after infusion of NO-np and Control-np. †P < 0.05 to baseline; ‡P < 0.05 to Control-np at same concentration. Right panels, rolling and immobilized leukocytes after infusion of NO-np. (a) Immobilized leukocytes after infusion of 10 and 20 mg/kg of NO-np (no pattern) and Control-np (pattern upward diagonal). (b) Rolling leukocytes after infusion of NO-np and Control-np. Rolling leukocytes were quantified in the same animals and locations as Immobilized. †P < 0.05 to baseline; ‡P < 0.05 to Control-np at same concentration. (c) Immobilized and rolling leukocytes after infusion of 20 mg/kg of NO-np and Control-np. Red arrows point to immobilized leukocytes on the images, other leukocytes in the images are rolling or flowing. Investigations were performed using the hamster chamber window model, widely used for microvascular studies in the unanesthetized state, and the complete surgical technique is described in detail elsewhere (Endrich et al. 1980). Data previously published by the authors (Cabrales et al. 2010b).

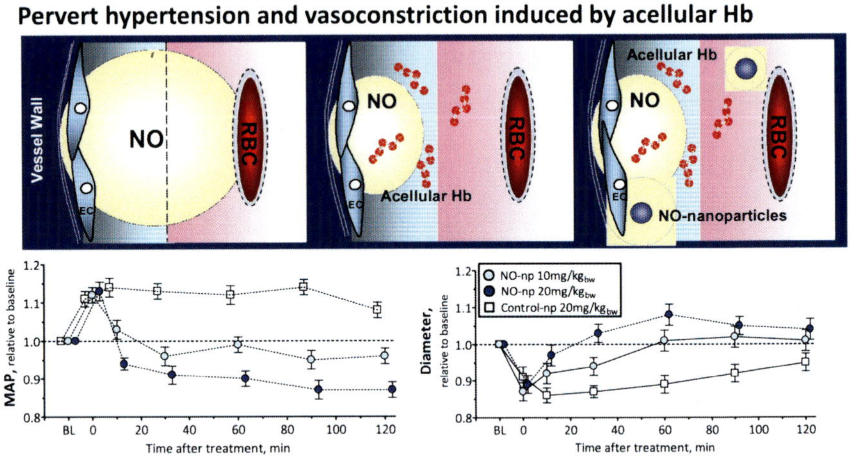

Fig. 5. Acellular Hb induces vasoconstriction due to NO scavenging, limiting the NO available for vascular smooth muscle. The central objective of this study was to restore NO bioavailability using NO-releasing nanoparticles (NO-np), to reverse the vasoconstriction and hypertension induced by polymerized bovine Hb (PBH) NO scavenging. PBH (13 g/dl) was infused in a volume equal to 10% of the animal blood volume. NO-np were provided at 10 or 20 mg by kg body weight. The study was performed using the hamster window chamber model to concurrently access systemic and microvascular hemodynamics (Endrich et al. 1980). Infusion of PBH increased blood pressure and induced vasoconstriction. Treatment with 10 and 20 mg/kg of NO-np reduced the blood pressure and vasoconstriction induced by PBH. Moreover, the higher dose of NO-np decreased blood pressure and induced vasodilation compared to baseline, respectively. Treatment with NO-np to decrease PBH-induced vasoconstriction increased metHb. NO-np has potential as a therapeutic agent to replenish NO in situations where NO production is impaired, insufficient or consumed, thereby preventing vascular complications. Data previously published by the author (Cabrales et al. 2010a).